For John and Delores White.

First published in 2010 by Jason White, 460 West Canfield Street #311
Detroit, Michigan United States 48201

All drawings and photographs by Jason White.

ISBN-13: 978-0-9846220-0-9

Printed in the United States

Old School Viscom

20 Renderings in 20 Steps

by Jason White

Table of Contents

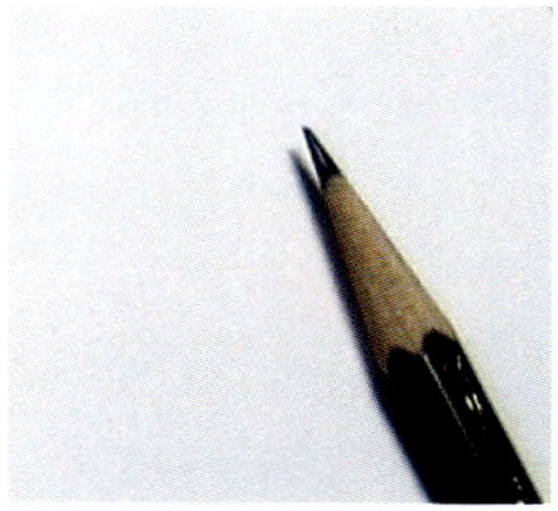

Preface

The objective of this book is to teach first and second year automotive design students some of the basic rendering skills which all design professionals should know.

As an automotive designer and an instructor at the College for Creative Studies in Detroit, I've watched the industry migrate from analog to digital. In the midst of this important transition, one key element is often overlooked: The fundamentals of visual communication, also known as Viscom.

Knowing a variety of Viscom skills provides a designer with multiple options for communication. While today's design students are talented and hardworking, their drawing styles are increasingly less diverse. This is an alarming trend because part of what makes a designer valuable is their individuality. Without it, distinction in the field becomes an even greater challenge. The remedy lies in preserving rendering techniques that promote individuality and experimentation-and by extension, encourage creativity.

The tutorials created for this book present a variety of "old school" techniques. But the beauty of these techniques is that they can be new school too. They are easily transferred to digital media. Most are rooted in logical planning and observation of details. All reinforce a basic understanding of how to use light, shadow, mid-tone, and reflection to describe surfaces. It is my hope that you will try all of these techniques, step by step, and see which work best for you. I think you will find that they still have a place in today's digital design world.

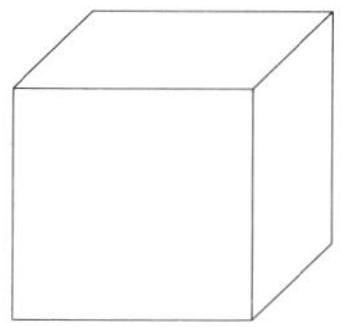

A Few Words About Perspective

If you've never drawn a car in perspective, it can feel like a daunting challenge. But don't worry. It's just like any other skill; the more you practice, the better you'll get. And with the volume of drawings you'll be doing as a car design student, you'll have plenty of time to hone your perspective skills. At the high school level, some of you may have learned about perspective drawing with two vanishing points. The basic principles of drawing a car in perspective are similar to this. However, since you'll be drawing cars in multiple views and in various sizes on marker paper, vellum or Canson, setting up vanishing points each time is not really all that practical. Typically, they would end up off the page (see below).

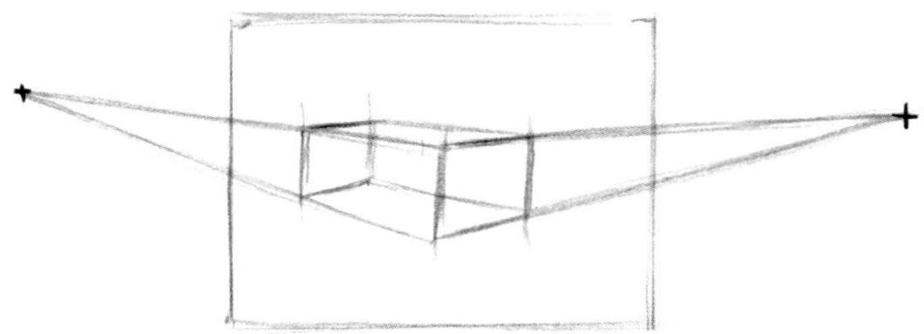

You should strive to develop a technique that allows you to draw in perspective quickly and loosely so you won't have to rely on vanishing points at all.

Before we get into the basics, let's take a look at a perspective view that is common throughout the industry, but should not be the only view in your repertoire: the dog's eye view (see next page).

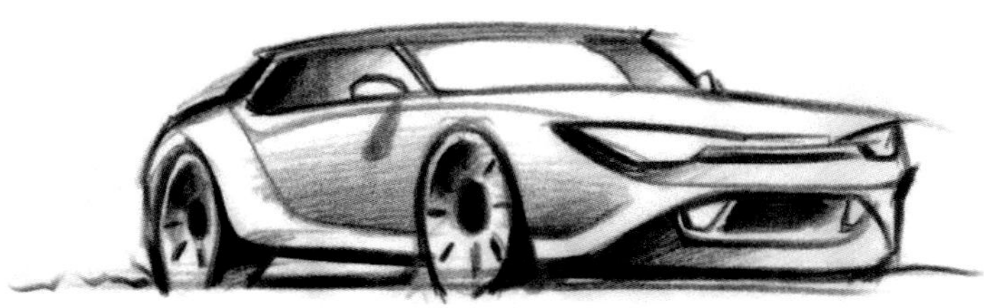

This is called a dog's eye view because the point of view we are looking from is almost at ground level, like a dog would see. This view is typically appropriate for vehicles like sports cars and pick-up trucks because it gives us a dramatic, planted impression of the vehicle. You'll find it very tempting to draw cars in this perspective all the time, and you should resist this temptation for a couple of reasons: 1) it can become a crutch because it's so easy to do, and would therefore keep you from learning more advanced techniques, and 2) the flatness of this perspective offers the viewer a very limited impression of what the vehicle looks like. Think of the dog's eye view as a jelly doughnut. Indulging in a couple of these once in a while won't kill you, but don't make a habit of it.

FIRST - DRAW A BOX

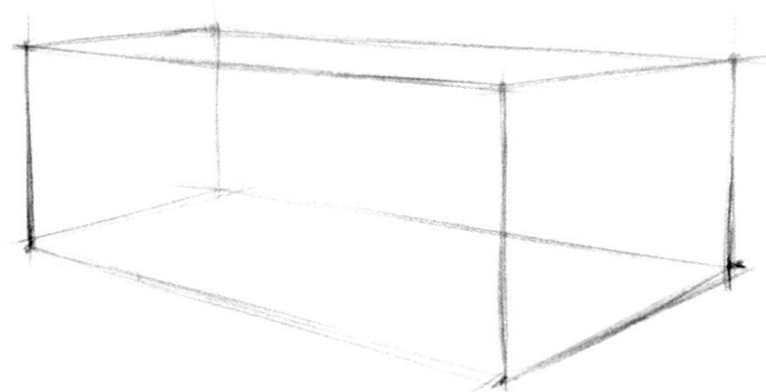

The best way to set up a perspective sketch is to draw a box like the one above. This box gives us a typical three-quarter view of a car, meaning that three quarters of the view show us the bodyside and one quarter shows us the front or rear of the car. Draw the box lightly with a Verithin pencil; this way the box won't overpower the car itself. For beginners, the vertical lines in the box should be parallel with the sides of the page; the horizontal lines should converge slightly towards the sides of the page. Getting the convergence of the horizontal lines right is probably the hardest part of drawing a car in perspective. If they converge too little, the drawing will look very clinical, like an engineering drawing. If they converge too much, it will appear as if we are viewing the car through a fish-eye lens. Try to strike a balance between these two extremes. If it helps, look at photographs of cars and try drawing boxes around them.

NEXT - DRAW THE WHEELS

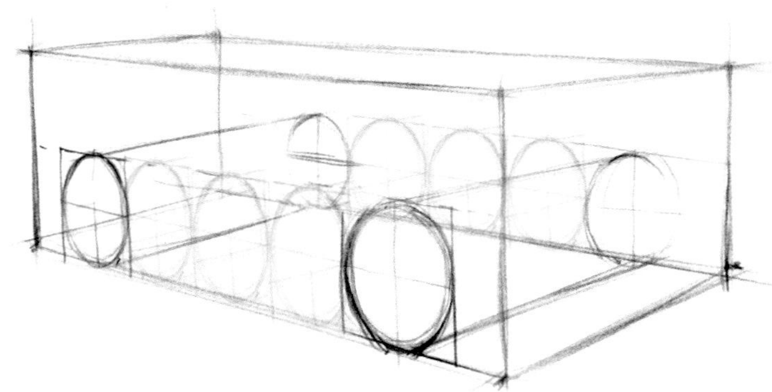

Think of the wheels as stone cylinders like those on the Flintstone's cars. In most cases, they should be roughly three wheels apart in side view. Before you draw the wheels, strike a line about midway through both sides of the box so the wheels will be the correct height. Draw the wheel ellipses in a circular motion. Keep the motion of your arm in your shoulder and elbow, not your wrist. It is vital that the ellipses be square with the side of the box. If they aren't, it will look like the wheels are broken or tilted. To prevent this, you may want to draw squares to put the ellipses in. Be sure to draw through the box to the other side.

NEXT - DRAW THE CAR WITHIN THE BOX

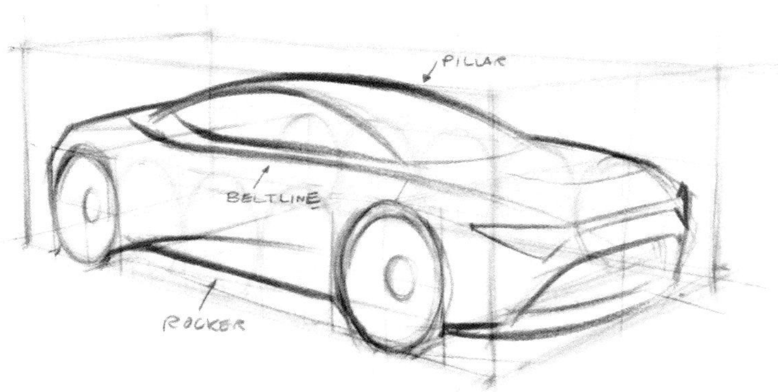

At this point, you can begin drawing the basic lines of the car; things like the pillars, the belt line and the rocker line. You can then draw details like the lamps and vents. It's not necessary to draw through the car like you did with the box, but use the box and your cylinder lines to guide you when drawing the car. For example, let the wheel on the left front guide you as to where the fender is on the far side of the car. Draw lightly at first so you can make corrections if you need to. If you want your car to have dished wheels, draw the hub to the inside of the wheel ellipse to give the impression that the wheel has a dished surface.

ABOUT FORESHORTENING

Foreshortening is an important part of perspective drawing. With the exception of a Hummer or the Land Rover Defender, most cars do not have flat front and rear ends. They all have varying degrees of wrap to them. In the last view, you can see that the right-hand headlamp is quite visible, but the left-hand headlamp is foreshortened from our point of view. This is to indicate that the front end has some curvature to it. Without the foreshortening, the front end would look dead flat, which is usually not what we want. Foreshortening takes a lot of practice to get right. Again, referencing photographs of cars may be helpful.

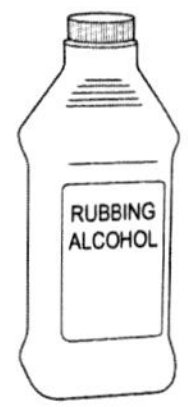

A Few Words About Materials and Tools

In any creative discipline, the choice of materials and tools can make a world of difference. Depending on where you live, some of the items mentioned in this book may be right around the corner; others may require mail order. But they all share one thing in common: they've been mainstays in the car design business for generations. The reason is simple. Under the right circumstances, they produce superior results. Here is a brief introduction to some of the more prominent items featured in this book.

Materials

Canson

This brand of paper comes in a variety of vivid colors. Each page features a rough texture on one side and a smooth texture on the other. For the work in this book, everything is done on the smooth side.

Marker Paper

For a layered marker technique, a high-quality paper is essential. In the author's opinion, *Copic* brand is by far the best. If your local art supply store doesn't carry it, you can always order it through the mail.

Craft Paper

This material is surprisingly inexpensive considering how versatile it is. Chrome details and period subjects are very much suited to its use.

Vellum

Translucent paper that allows you to work both sides for added depth. For CCS students, only one brand will do: *George Vincent*. In the event you can't find any, the *Clearprint* brand should suffice.

Tools

Verithin pencils

These pencils have a distinctive hexagonal section and come in a variety of colors. They are good for laying in loose construction lines and applying fine details. Mistakes clean up easily with an eraser.

Prismacolor pencils

Far softer than the Verithins, these pencils produce bold, expressive lines that grab the viewer's attention. When used in conjunction with Verithins, your drawings will have far more variation in line weight. Due to their waxy nature, Prismacolor pencils do not erase easily.

Fine black pens

Some details are so fine that even Verithin pencils won't do. In this case, a fine black pen such as *Uniball* or *Copic* will do the trick. For bolder lines, try a felt-tipped pen from a brand like *Staedtler.*

Prismacolor markers

These markers come in colors matching the pencils of the same name. While they aren't the juiciest markers around, they do offer consistent performance at an affordable price. Be warned, however: the color on the label may be slightly different from the color you see on the page. Always test your markers before you buy.

Copic markers

These lozenge-shaped markers are famous for their color accuracy. When they dry out, they can be replenished with small ink wells; these are typically displayed next to the markers in the stores.

Nupastels

These long, rectangular pastels are for soft, subtle gradations. They can be applied directly to the paper like a pencil, or they can be scraped into a fine powder with an X-acto knife and applied with a cotton pad. A pack of twelve colors should be sufficient for your work.

Gouache

This is a type of opaque watercolor which can be used for highlights and lamp details. With a little water and a steady hand, it can add the perfect finishing touch. *Windsor & Newton* is a trusted brand.

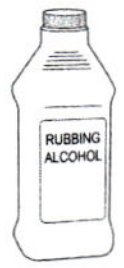

Rubbing Alcohol

Typically found in your neighborhood pharmacy, this liquid can be combined with powdered pastels to create dramatic backgrounds. You can get similar effects with solvents like *Bestine,* but rubbing alcohol is more affordable and far safer to use.

Additional Tools

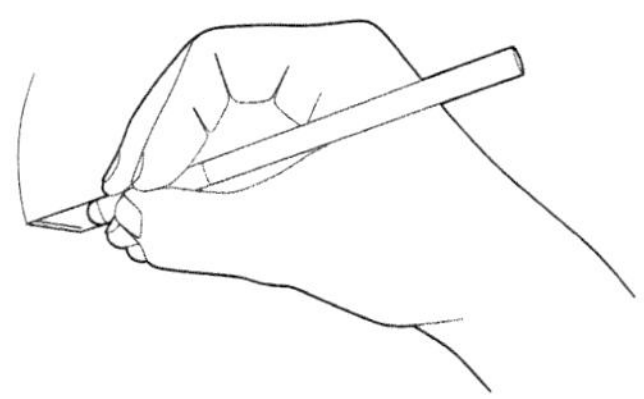

X-acto knife

This brand of knife provides the precision that a pair of scissors can't match. But with this precision comes added responsibility: The triangular blades are extremely sharp and should be handled with the utmost care. Number 11 blades are an ideal size.

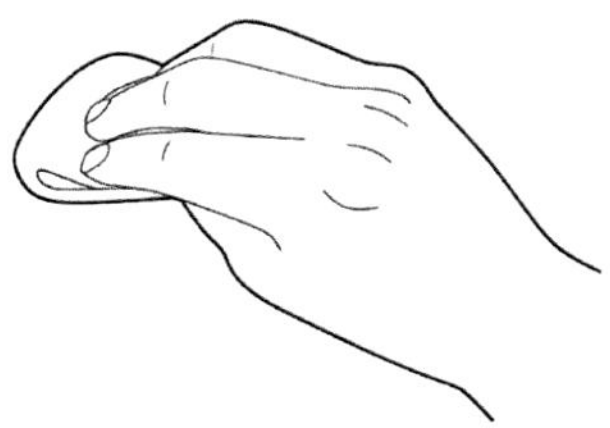

Webril pads

These cotton pads come in rectangular bags or large rolls. They provide a soft, refined texture for applying powdered pastel to paper. The more you use a Webril pad, the softer it gets. So if you're careful to only use certain colors with specific pads, you shouldn't have to throw them away.

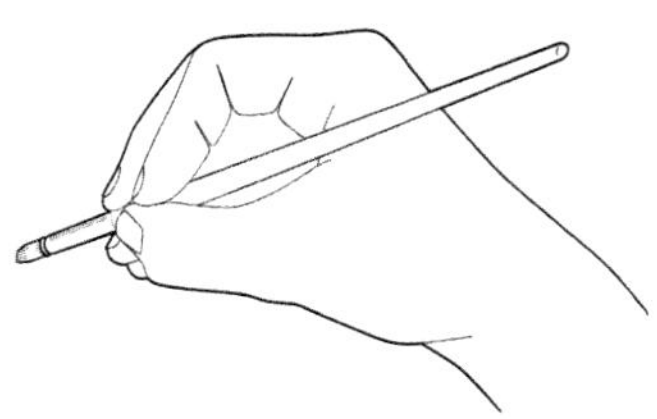

Paintbrush

For the application of highlights, be sure to select a small, flat-edged watercolor brush with soft bristles.

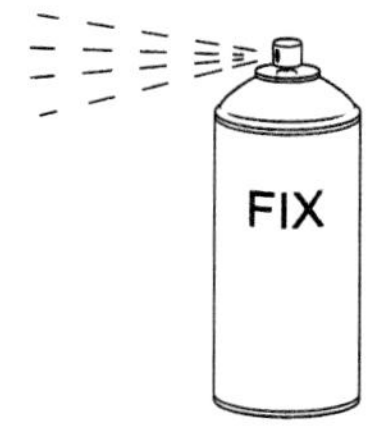

Fixative

Sometimes paper can get overloaded with pastels, leaving no tooth to accept additional material. Fixative adds a new layer of tooth to the paper, while at the same time preventing the pastels from smudging. Fixative can be highly toxic, so read the warning labels!

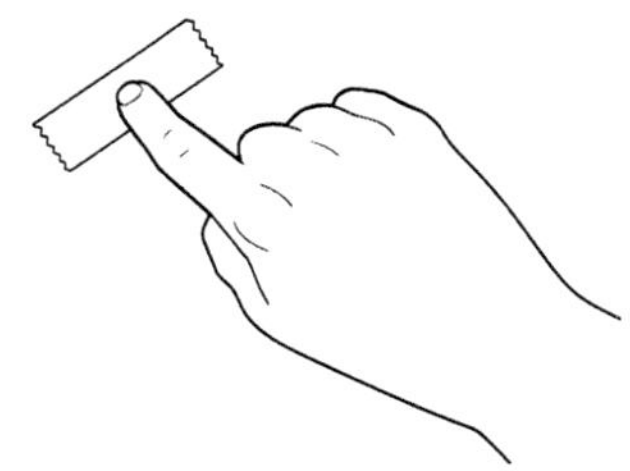

Tape

For spliced Canson renderings, having a role of tape on hand is essential. Ordinary masking tape or wide painter's tape should be sufficient in most cases. If your paper is thinner than Canson, clear cellophane tape may be more appropriate.

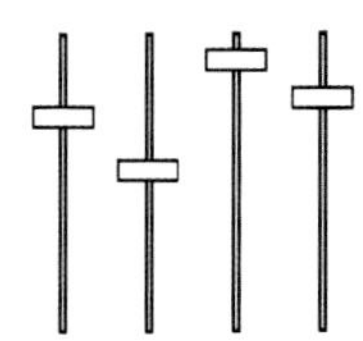

Photoshop

Once your renderings are finished, you may wish to scan them in and clean them up a bit using Adobe Photoshop or a similar program. You can also adjust the color saturation to your liking.

The Tutorials

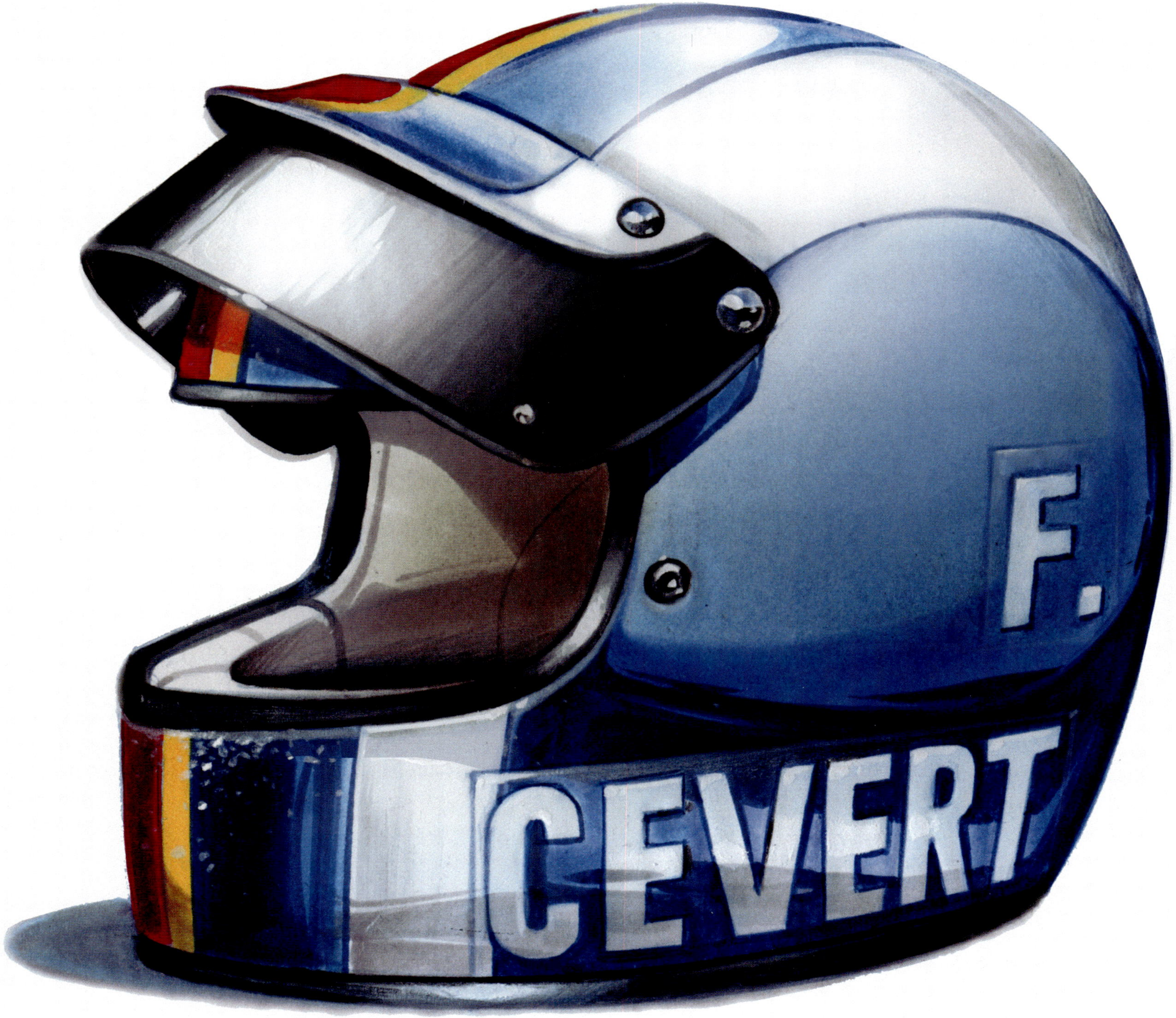
F.
CEVERT

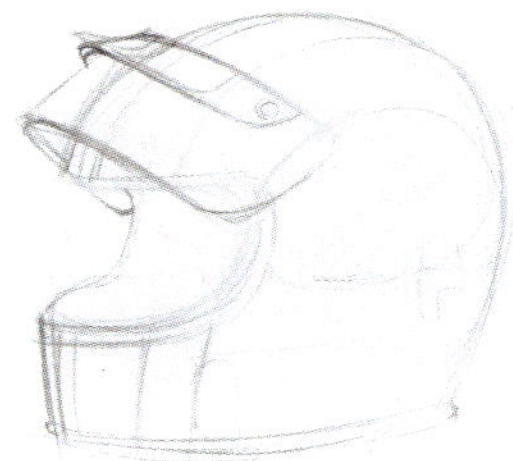

1. Basic Shapes

For those who have never attempted industrial design renderings before, describing complex shapes can be a daunting task. But you have to learn to walk before you can run. A helmet is a good, simple rendering exercise because it comprises two basic shapes: A cylinder and a sphere. Once you've practiced rendering these shapes, moving on to more complex ones may not seem so daunting.

The Tools

Prismacolors:
White.

Gouache:
White, Black.

Verithins:
Indigo Blue,
Black.

Nupastels:
Indigo Blue, Sienna Brown.

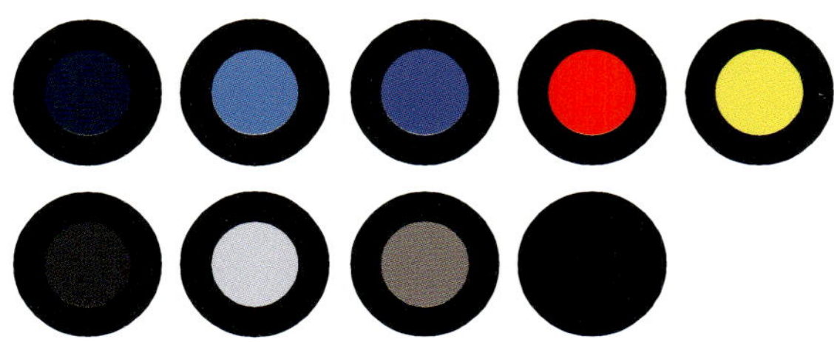

Prismacolor Markers: Indigo Blue, Cerulean Blue, Blue Slate, Crimson Red, Canary Yellow, Cool Grey 70%, Cool Grey 20%, Warm Grey 40%, Black.

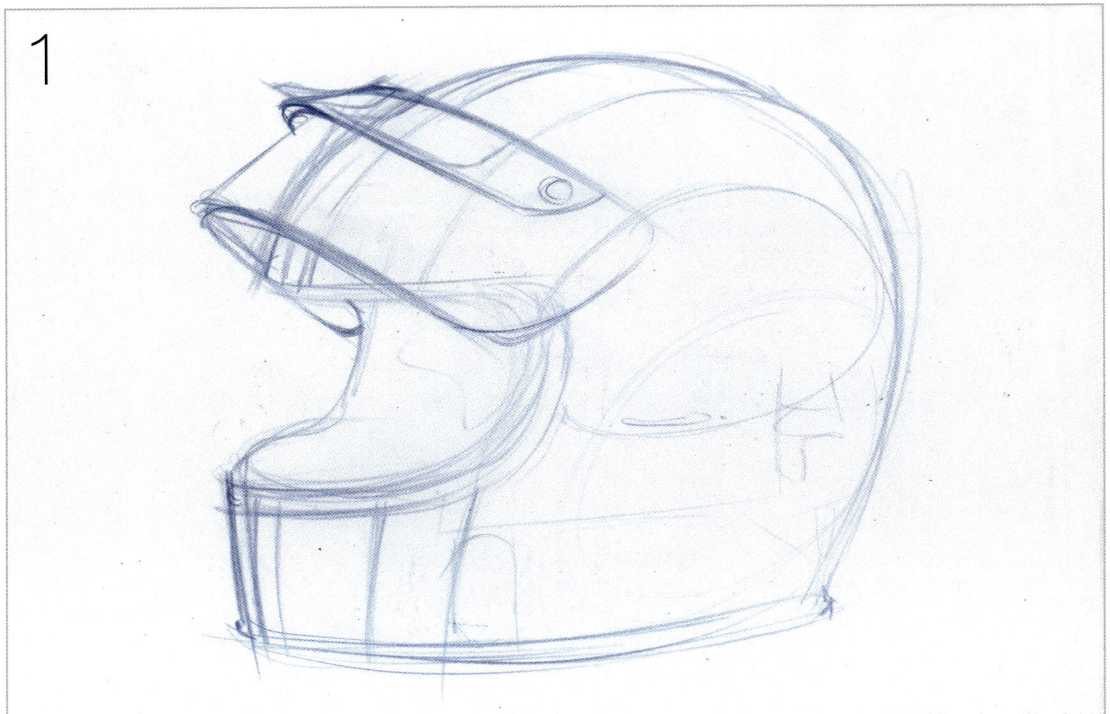

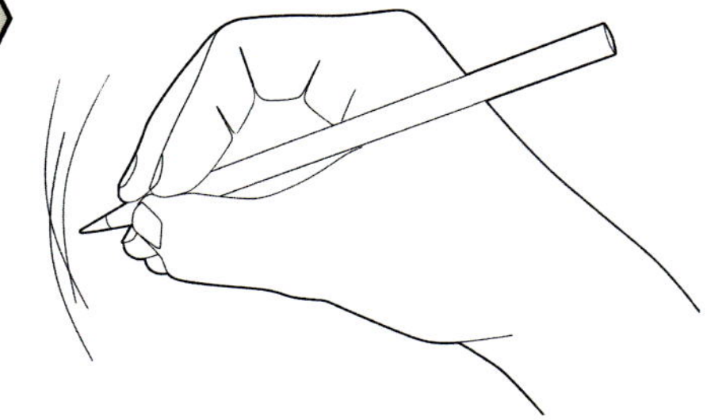

Start by loosely sketching out the helmet on a clean sheet of marker paper – preferably Copic brand. At this point, you should not be concerned with the linework or the details; just try to get the proportions and the angle of the helmet right. Although the visor is tinted and won't be transparent from our side, be sure to draw through it so you can get a complete sense of the overall shape.

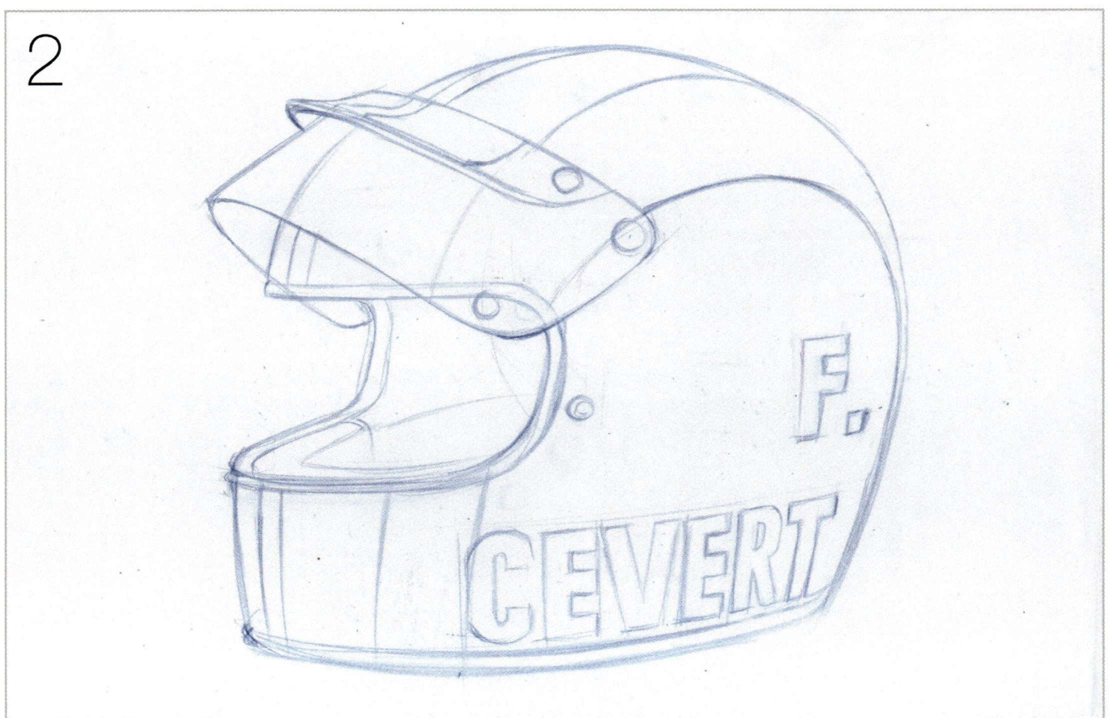

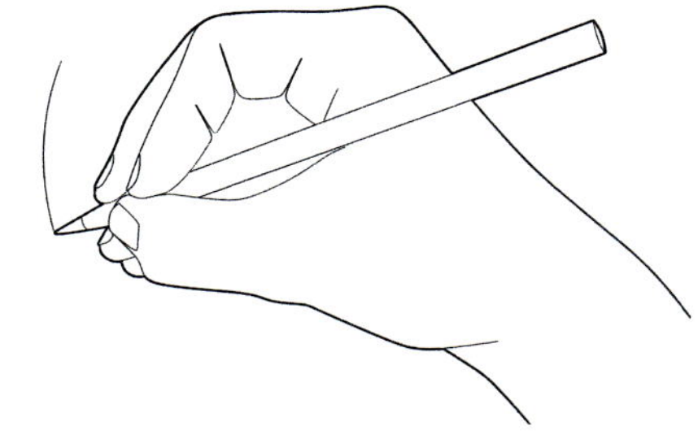

Now take another sheet of paper and do a tighter overlay of your rough drawing. This time, pay close attention to the details, especially the block lettering. Notice how the letters get progressively more angled as they wrap towards the back of the helmet.

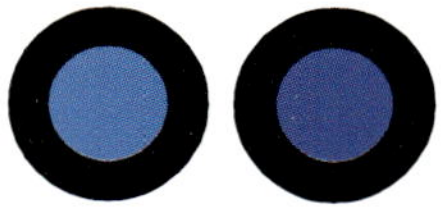

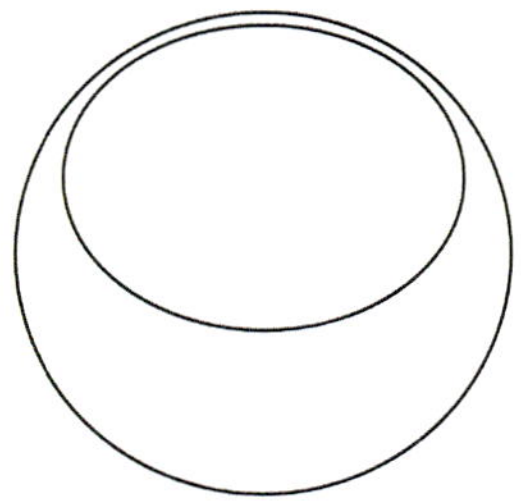

Now, using the blue markers, lay in the helmet's main reflection. A glossy sphere will tend to have a horizon that wraps up and around the shape; the figure above illustrates this. To create a sense of warm to cool in the reflection, concentrate the Cerulean towards the front and Blue Slate towards the back.

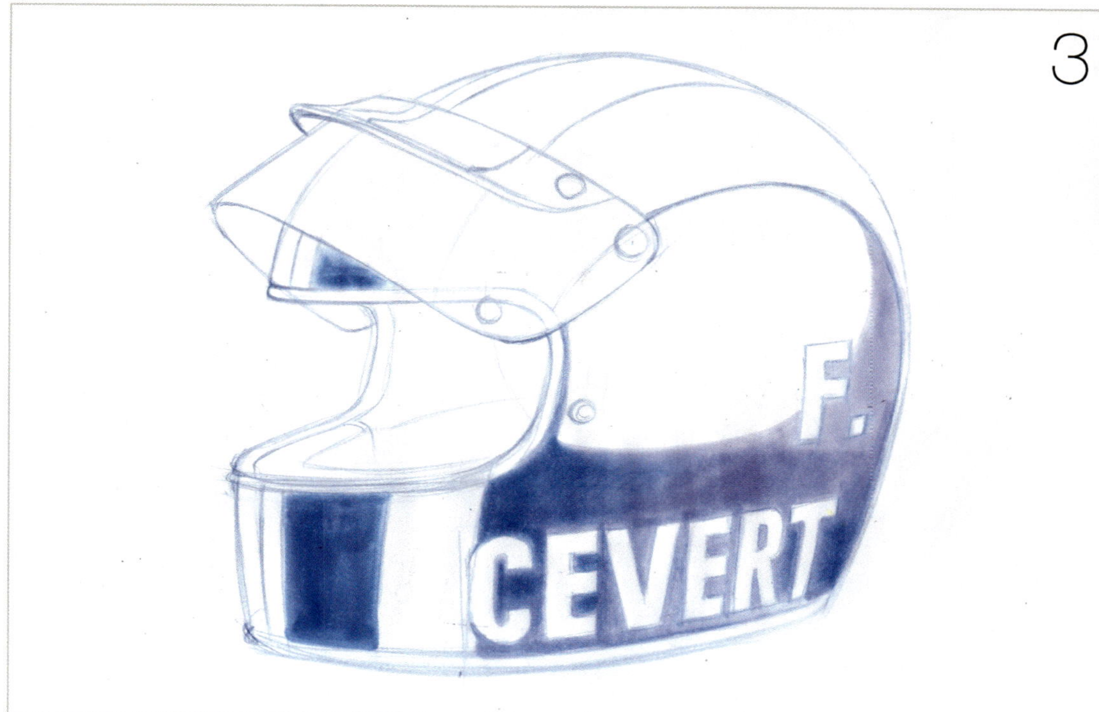

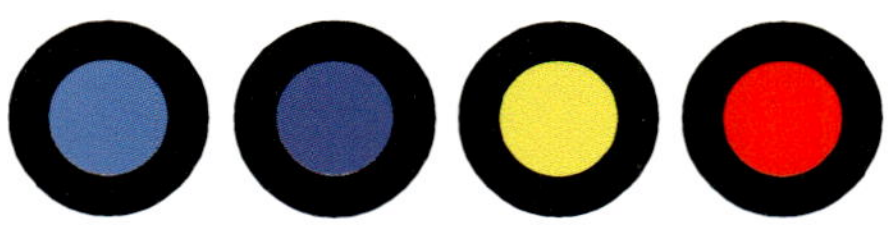

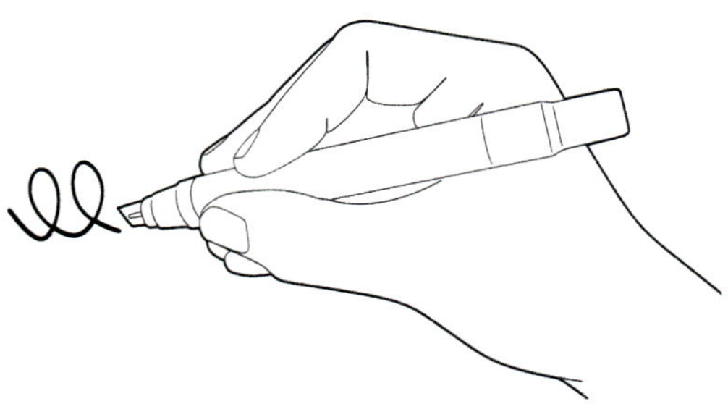

Using another blue marker that is almost dried out, blend between the two blues across the main reflection. Making circular motions will help minimize any streaking. This is also a good time to lay in the red and yellow stripes, but don't let the yellow pick up any of the blue.

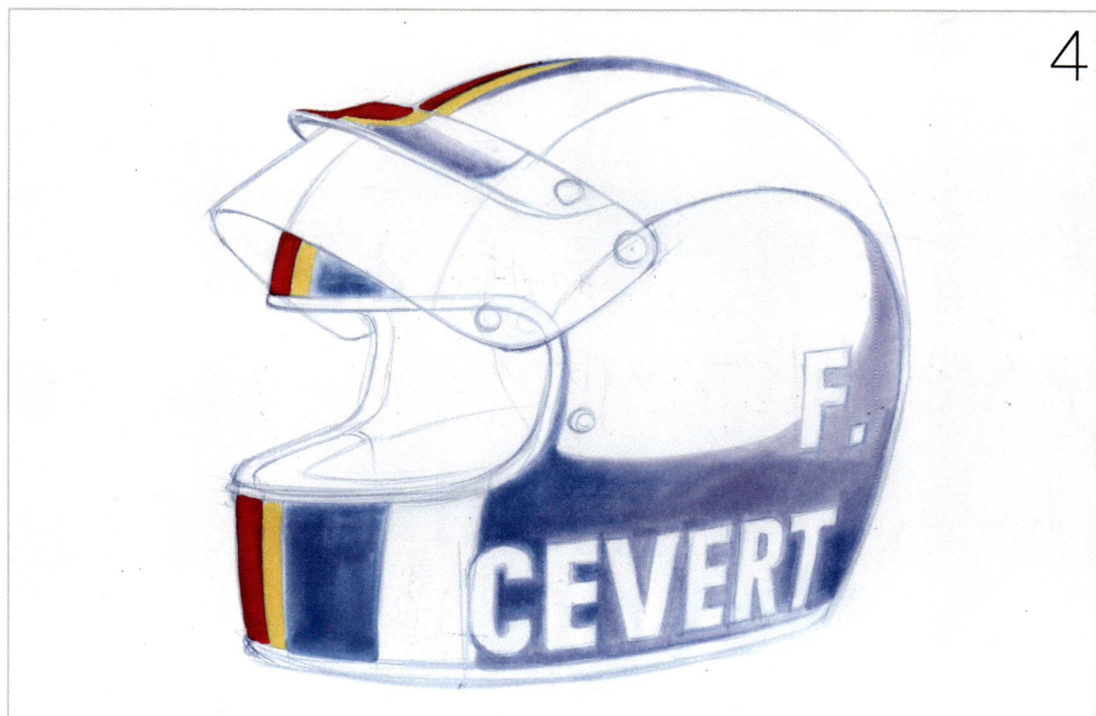

5

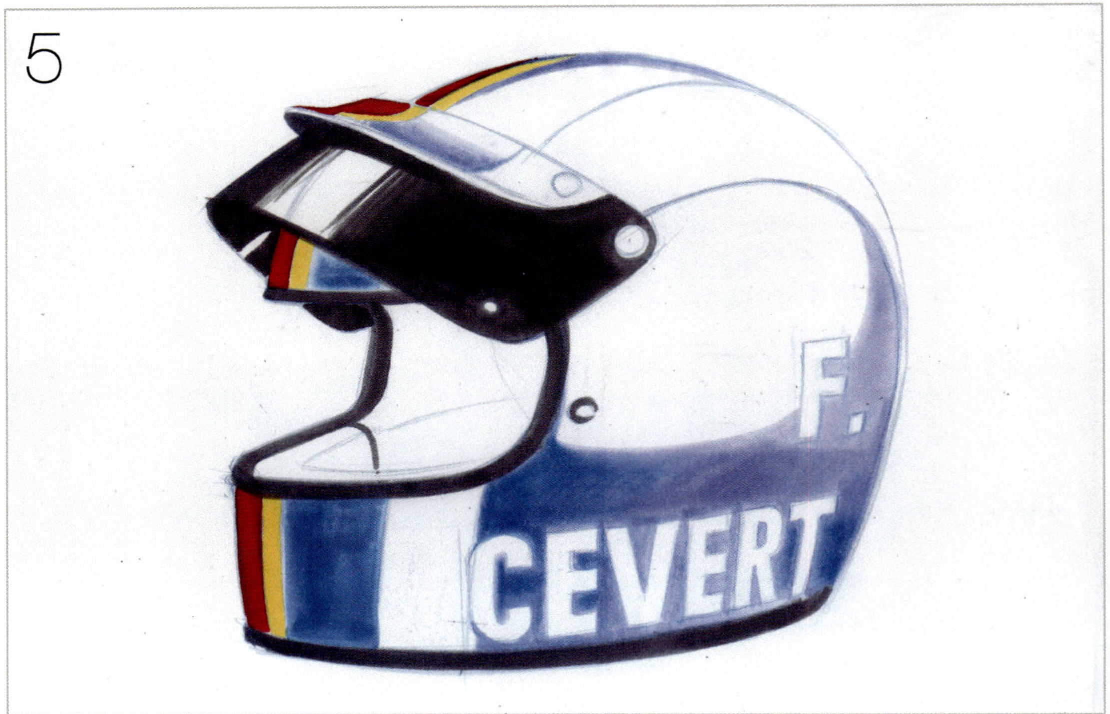

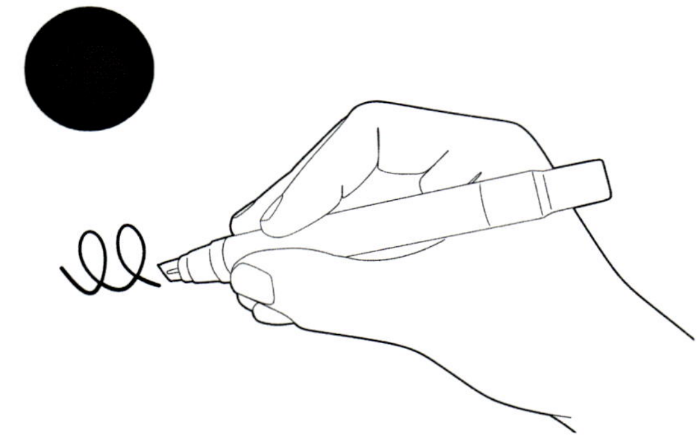

Begin laying in the visor and the rubber seals with Cool Grey 70%. Be sure to leave a column of clean paper for the main highlight. In the dark parts of the visor, use circular motions to get full coverage and minimize streaks.

6

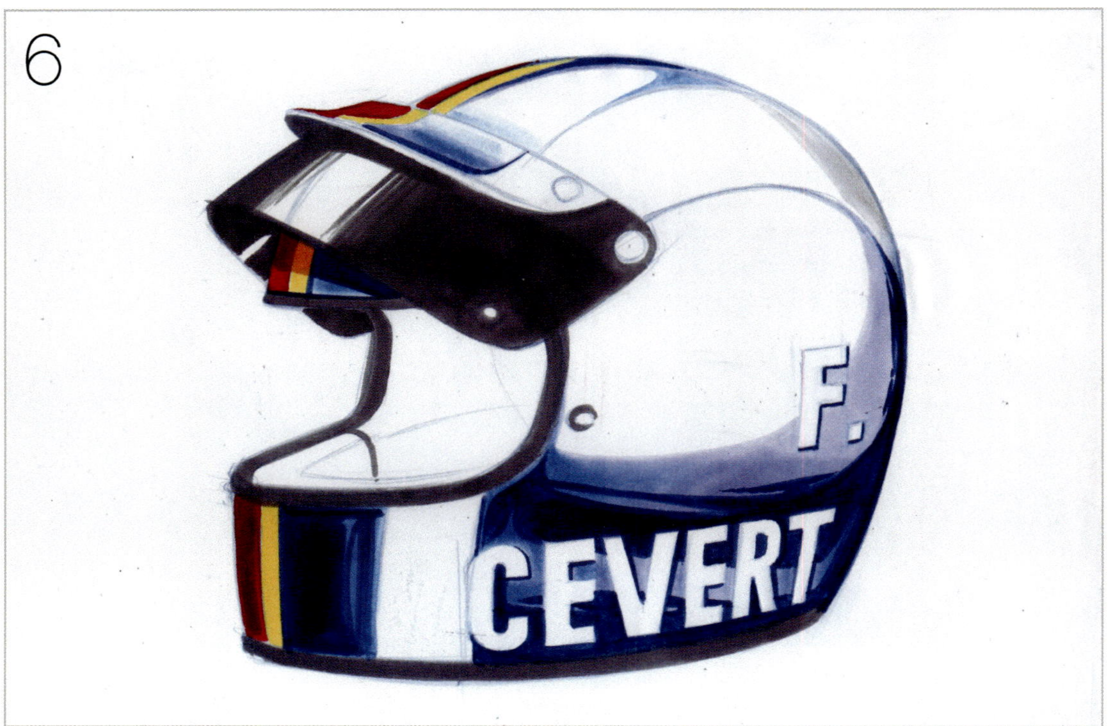

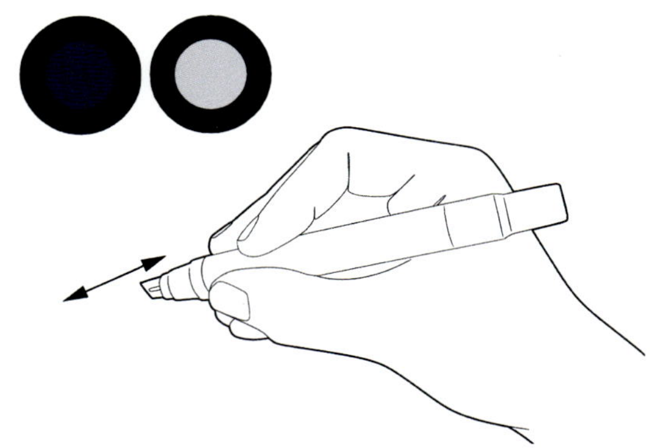

Since a simple reflection on a simple shape tends to look a bit dull, use the Indigo Blue marker to add a little bit more complexity. The stretched out forms may be trees or signs in the foreground, but be sure to keep them loose and abstract. Also, use Cool Grey 20% to carry the horizon from the blue graphic into the white of the bare helmet.

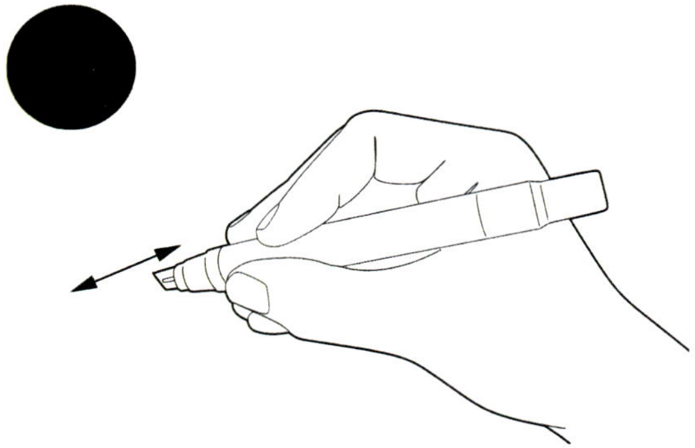

Now, use the black marker to add more depth and contrast to the visor. You can also darken areas of the rubber seals that fall under the shadow of the visor. Make sure any marker linework runs in one direction, since the visor is essentially a cropped cylinder.

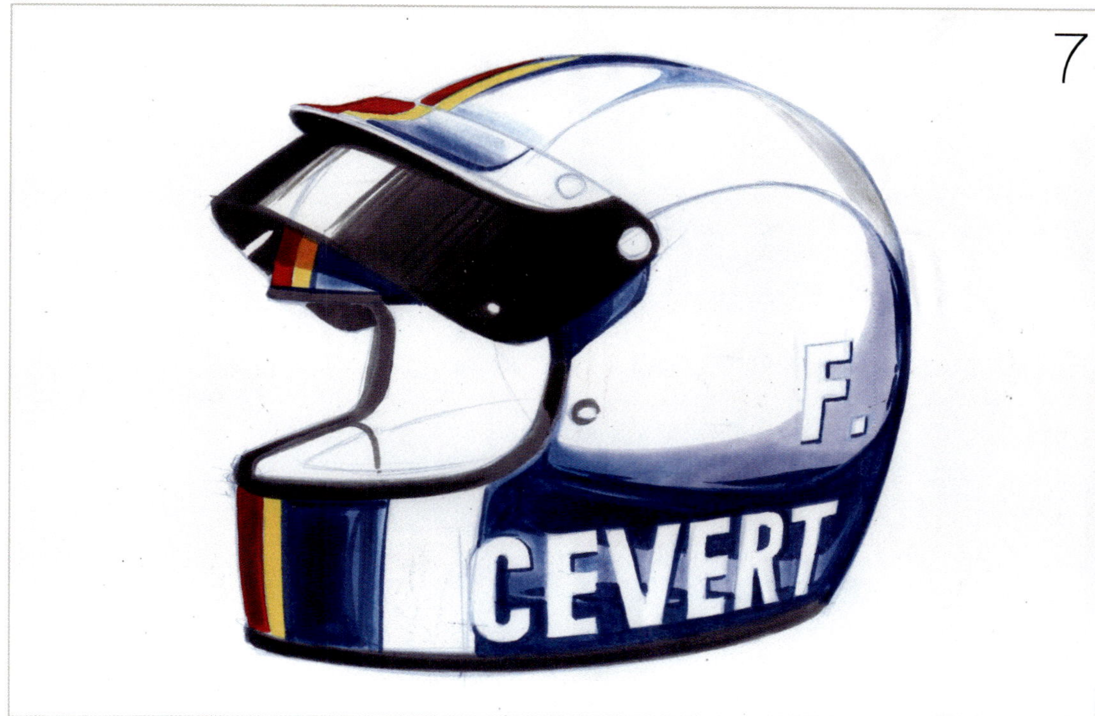

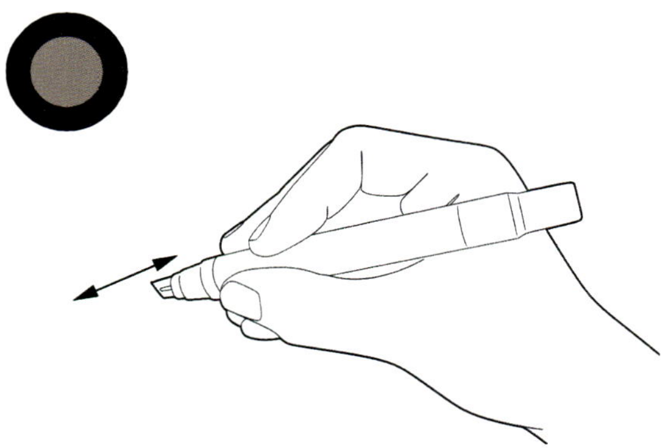

Take the warm grey marker and drop a shadow along the inside of the helmet. The warmth of the soft, padded inside surface will help distinguish it from the cool, glossy exterior.

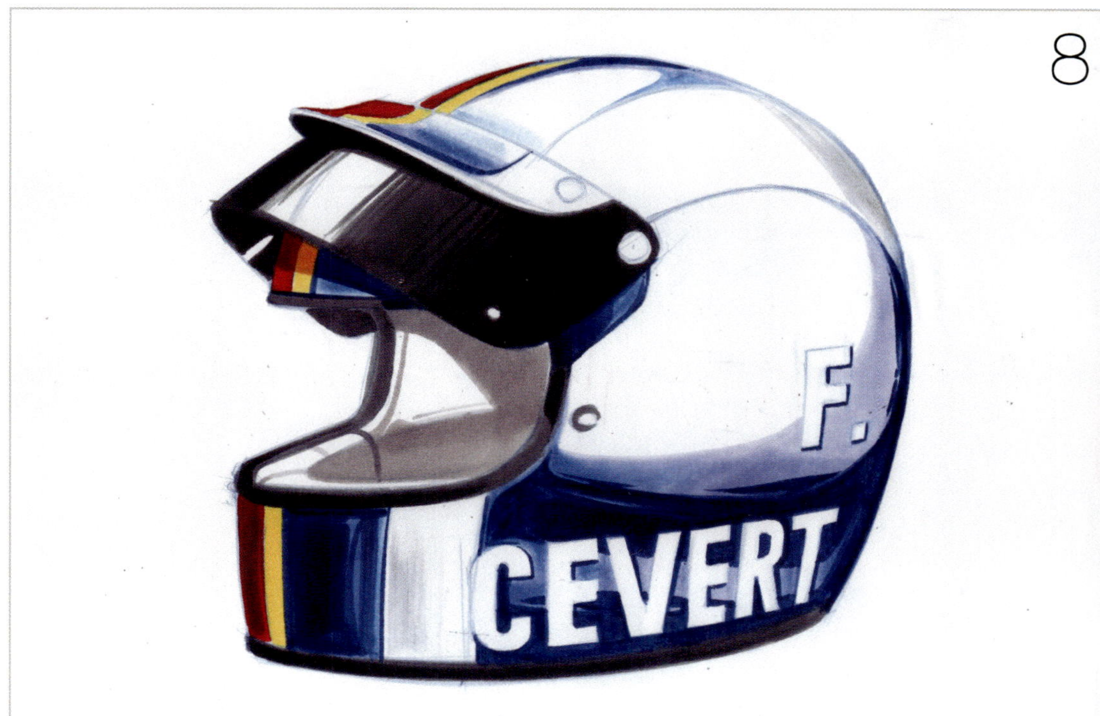

9

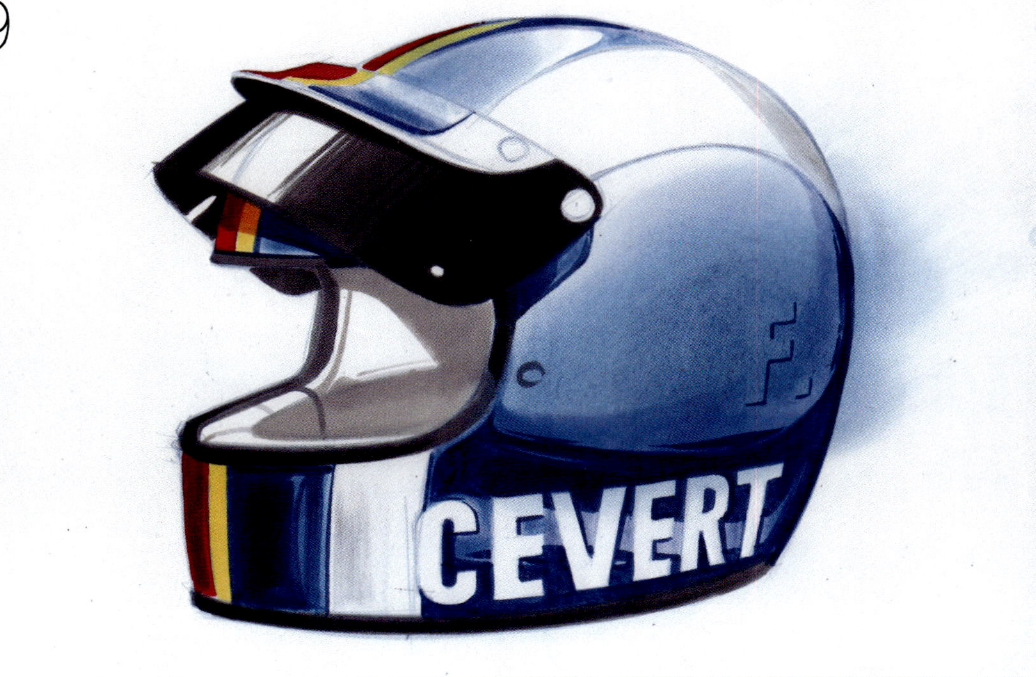

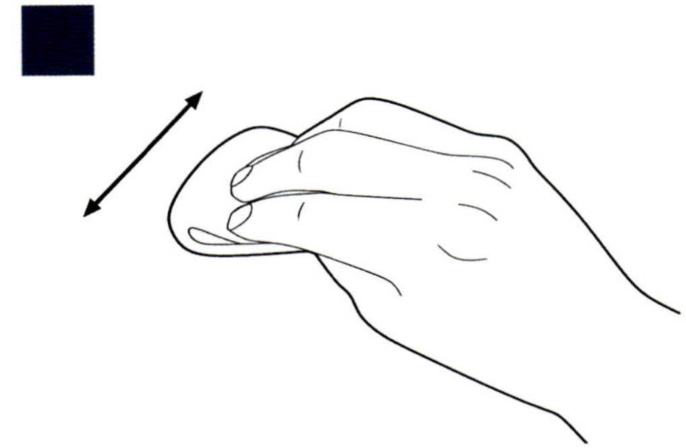

Begin laying in the sky tone with pastel powder and a Webril cotton pad, but be sure to keep the white part of the helmet clean. Since the light is coming from above, this will be the brightest part of the helmet surface.

10

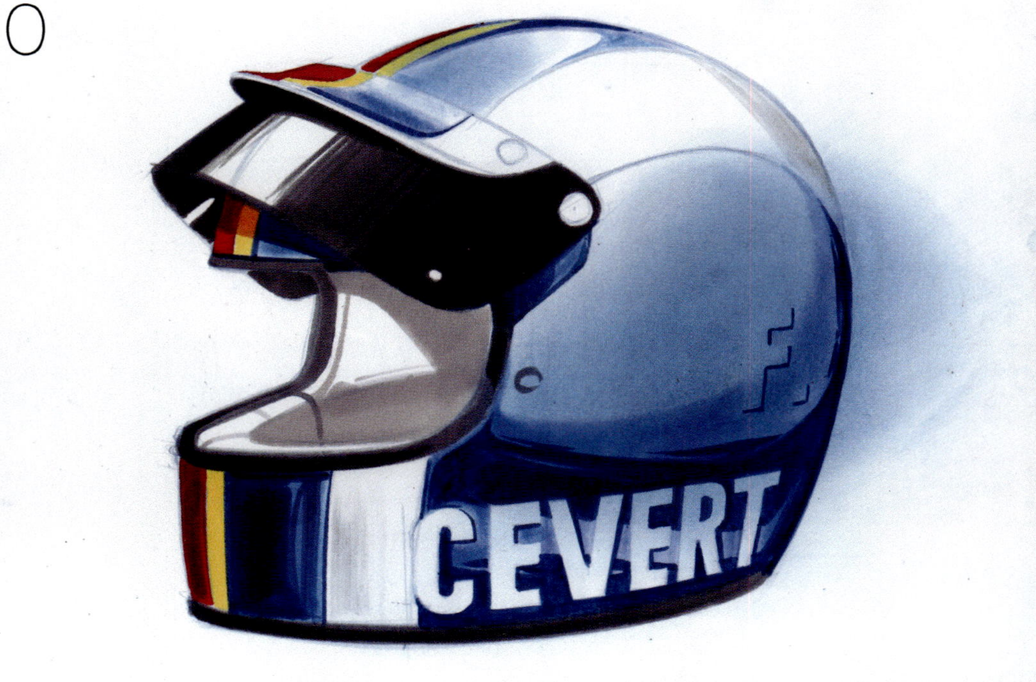

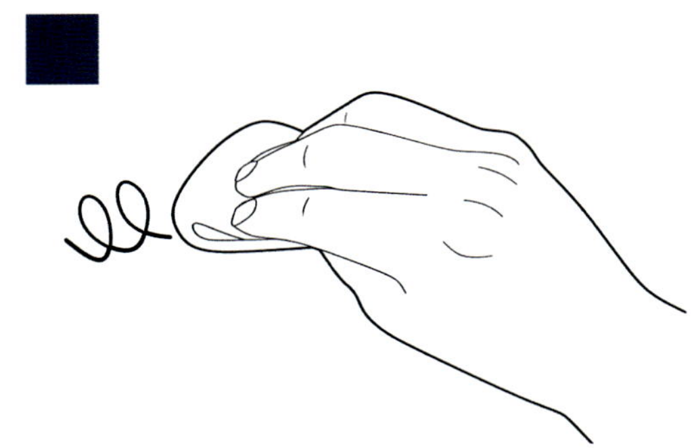

Using circular motions, smooth out imperfections with the Webril pad. You may try cutting the pastels with baby powder to expedite the process.

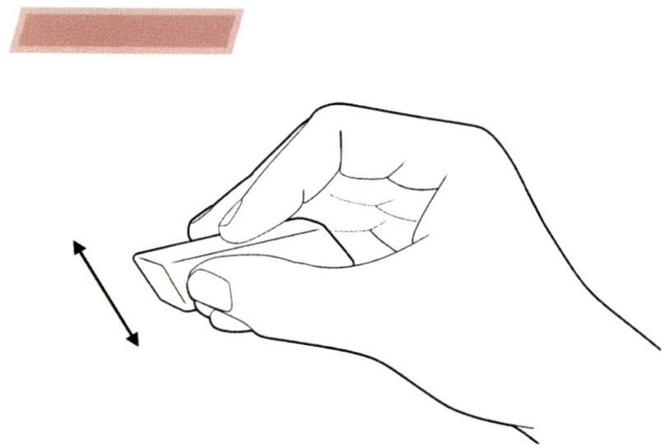

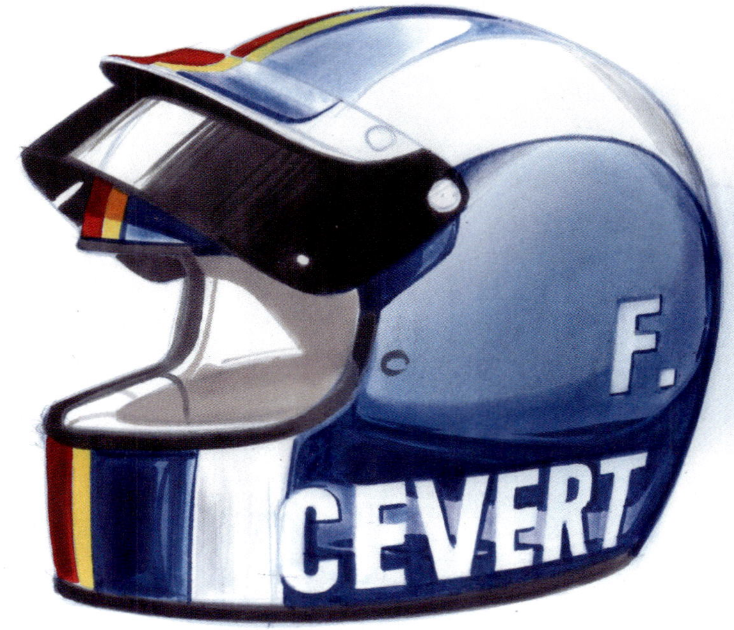

11

Using a clean eraser, remove the excess pastel. Take a moment to clean up any ragged-looking lines as well.

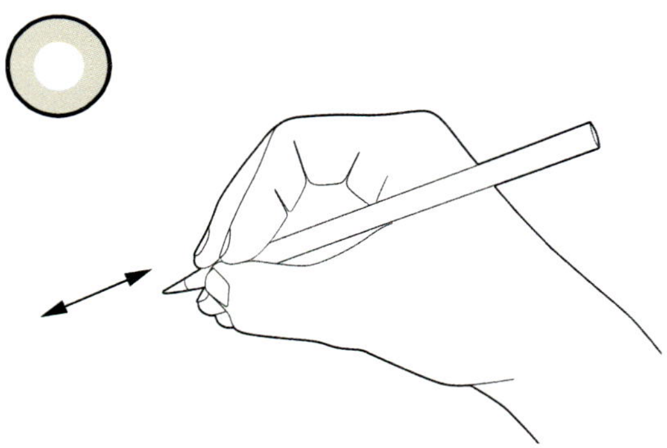

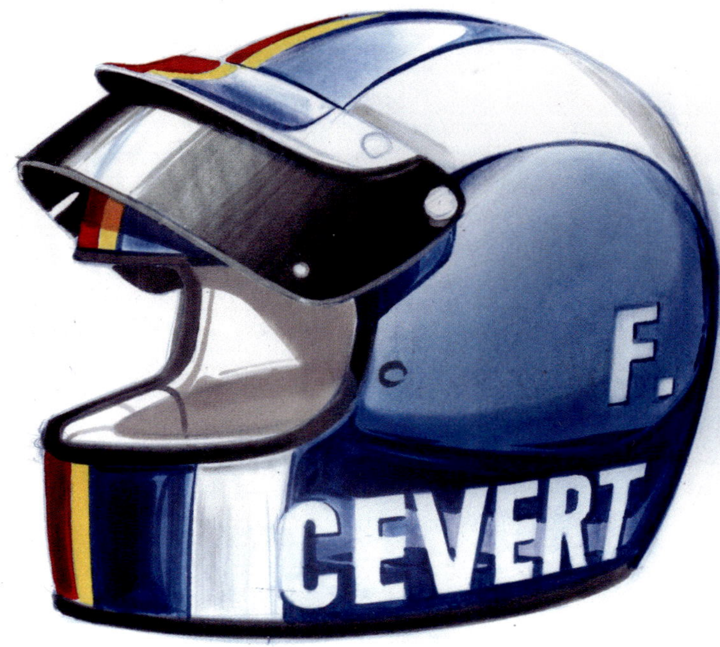

12

With deliberate, straight strokes, add white Prismacolor pencil near the visor highlight. This will further describe the curvature of the surface.

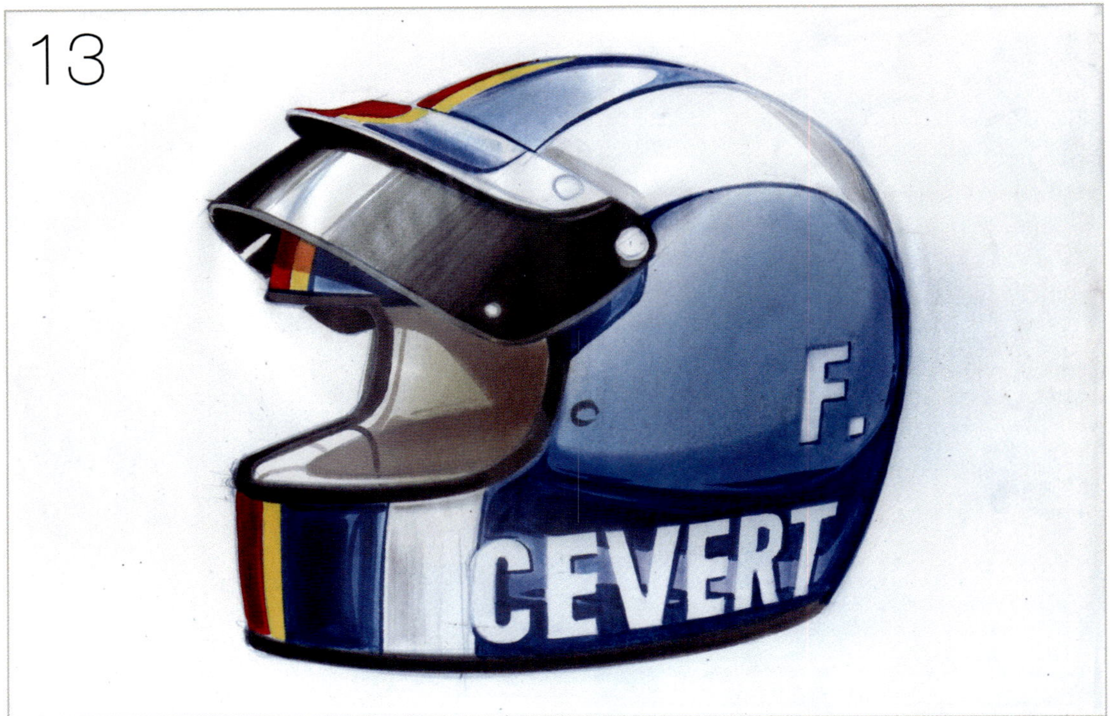

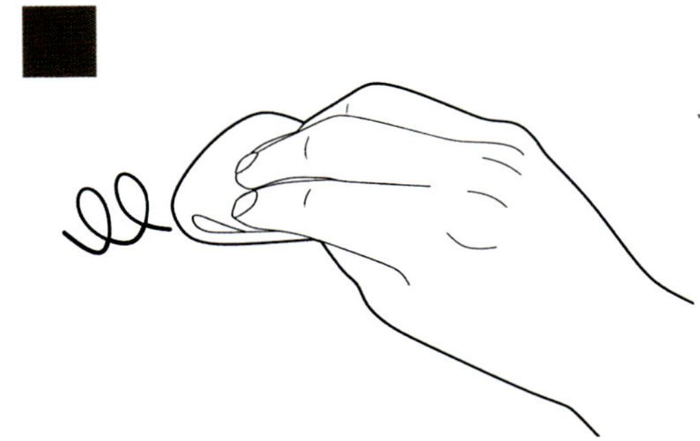

To soften the inside surface of the helmet, apply a small amount of brown pastel. The warmth of the brown pastel will further distinguish the inside from the outside.

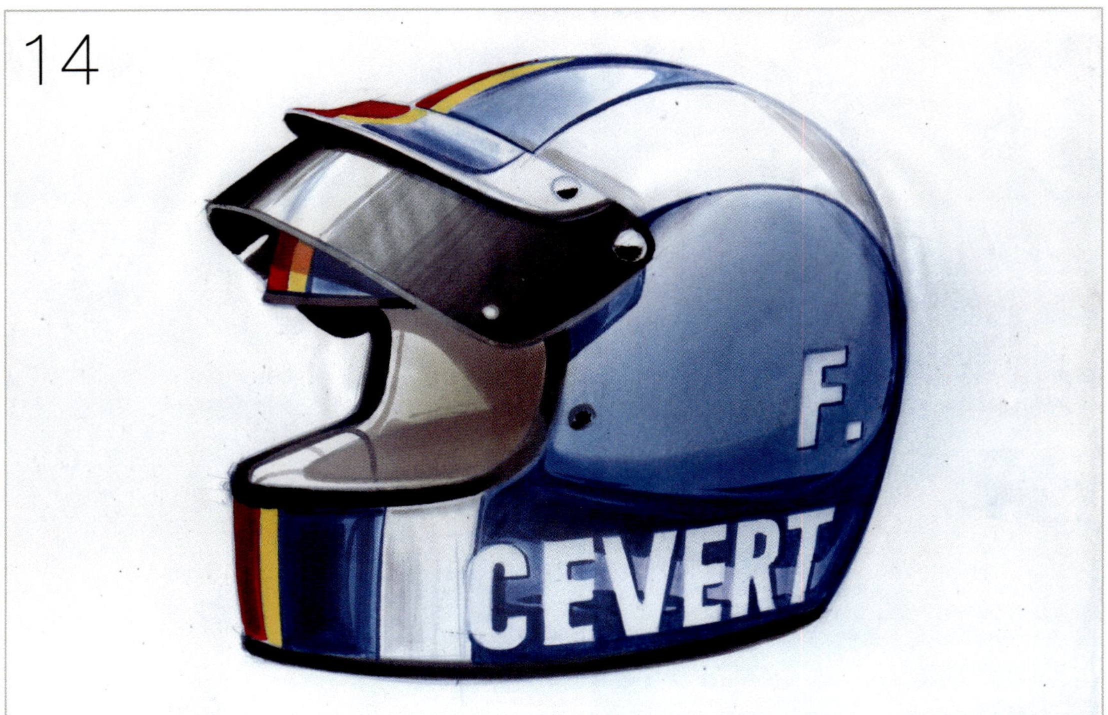

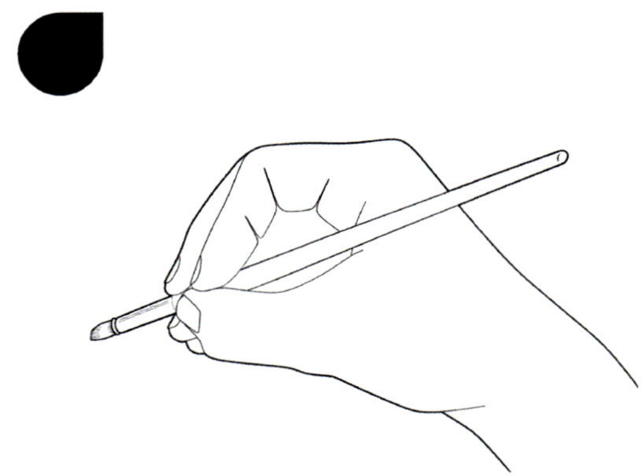

Bump up the contrast by adding some black gouache. The edges of the visor and the parts of the seal in shadow are good candidates for this.

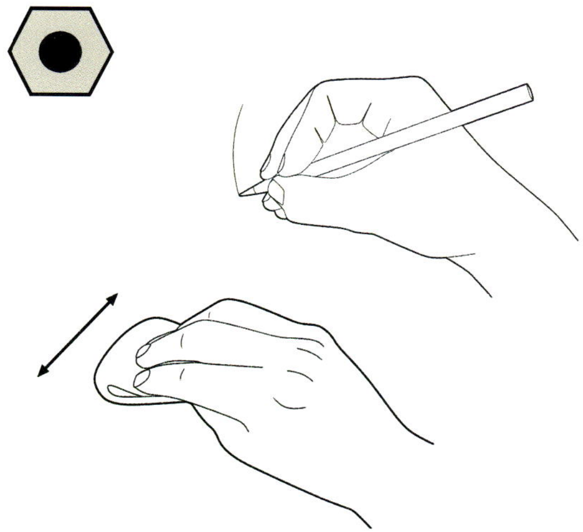

Now it's time to sharpen the drawing with some linework. Begin by adding more definition to the inside edge of the helmet. You can use a Webril pad to soften the pencil a bit.

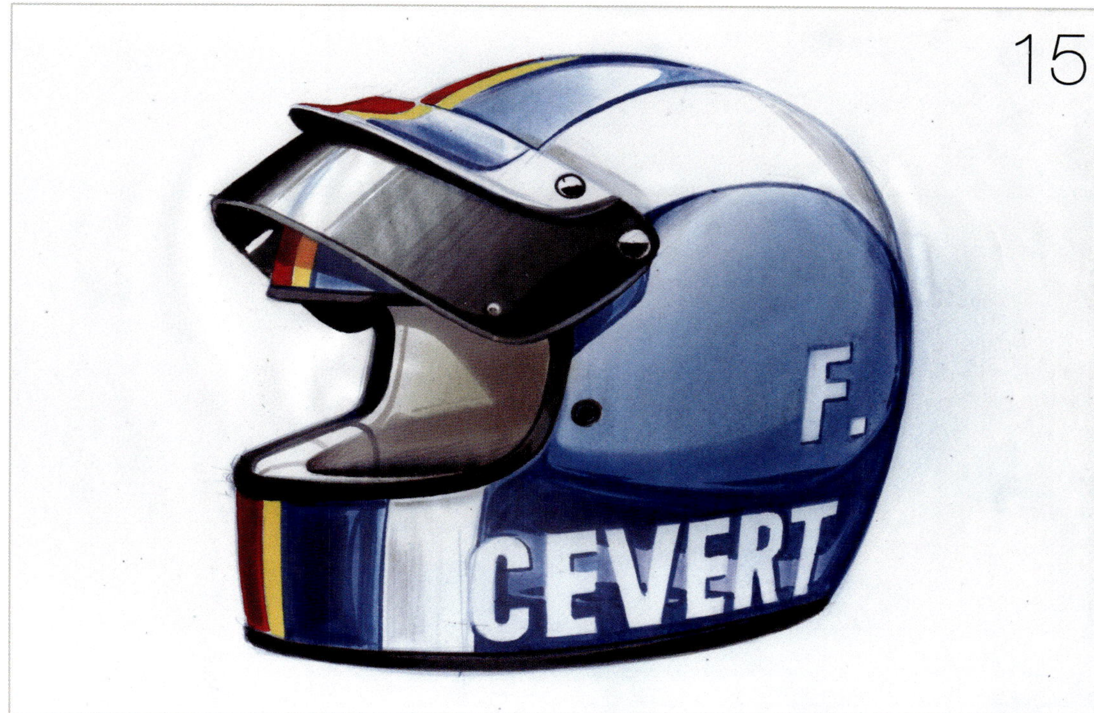

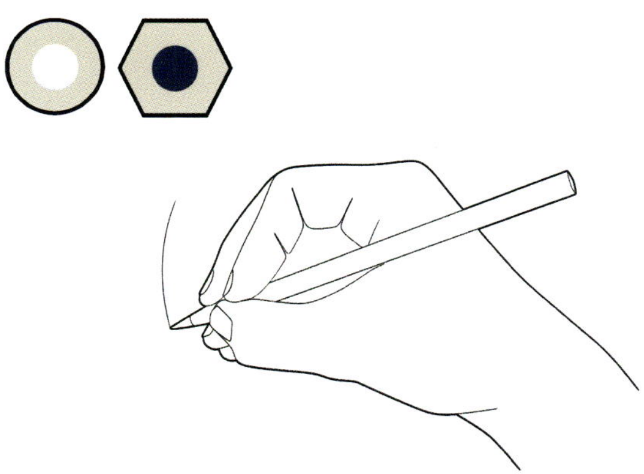

The lettering on the helmet is actually a rectangular plastic decal, so use white pencil to give the illusion of the clear portion. Use a bit of blue pencil to define the edge by the "C"

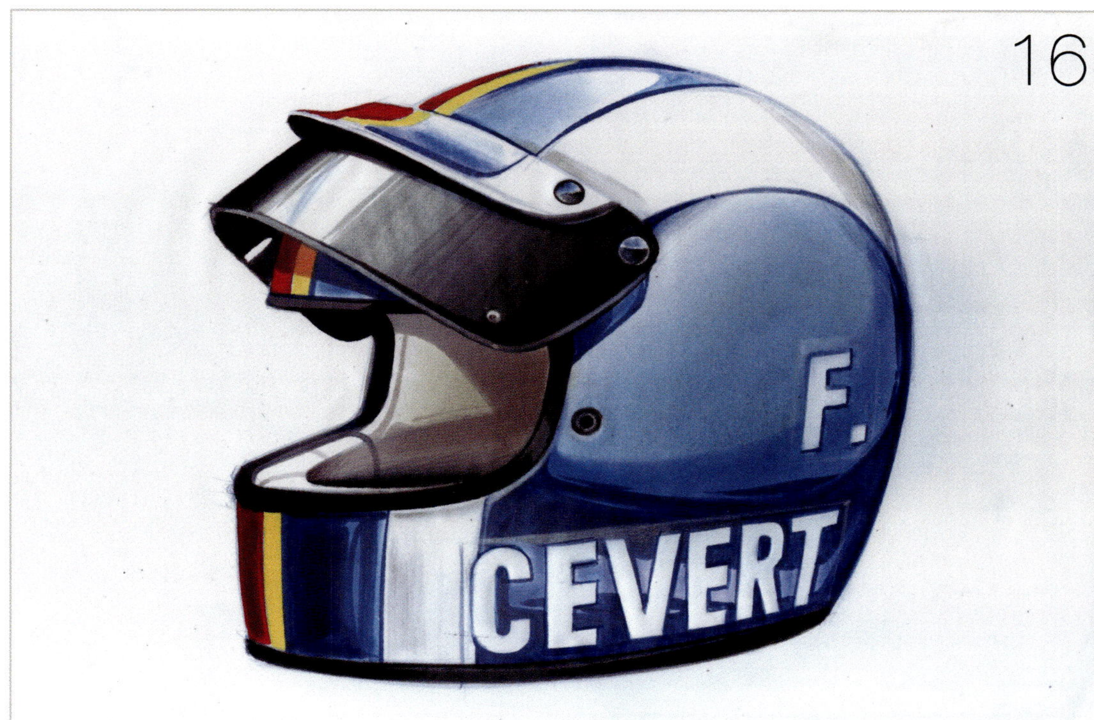

17

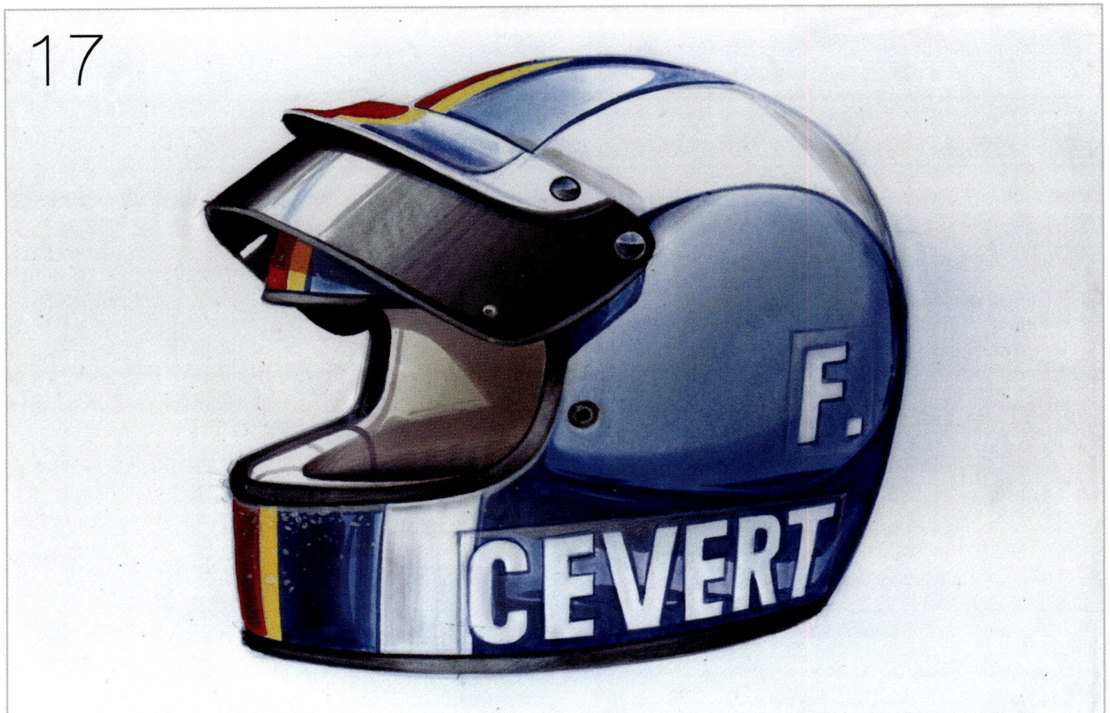

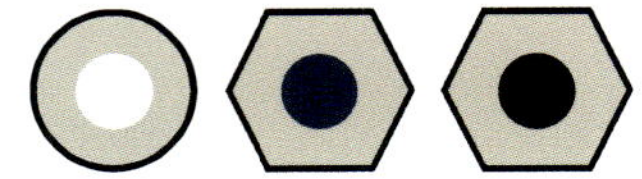

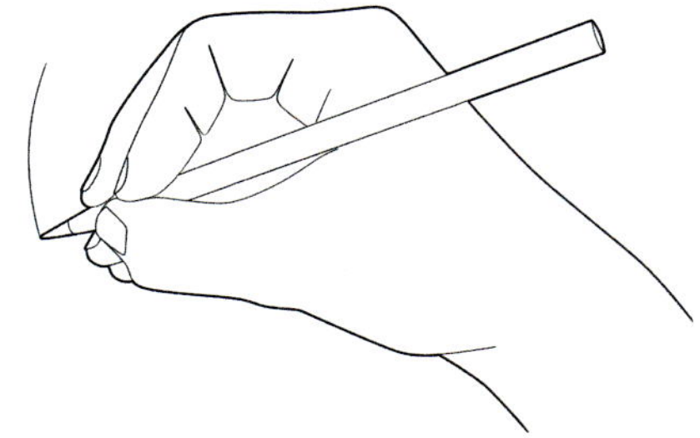

Continue defining edges around the piece: The lettering, the seals, the snaps. Use the white pencil to highlight the edge of the helmet by the seal; stab at the surface below the seal to simulate wear and stone pecking.

18

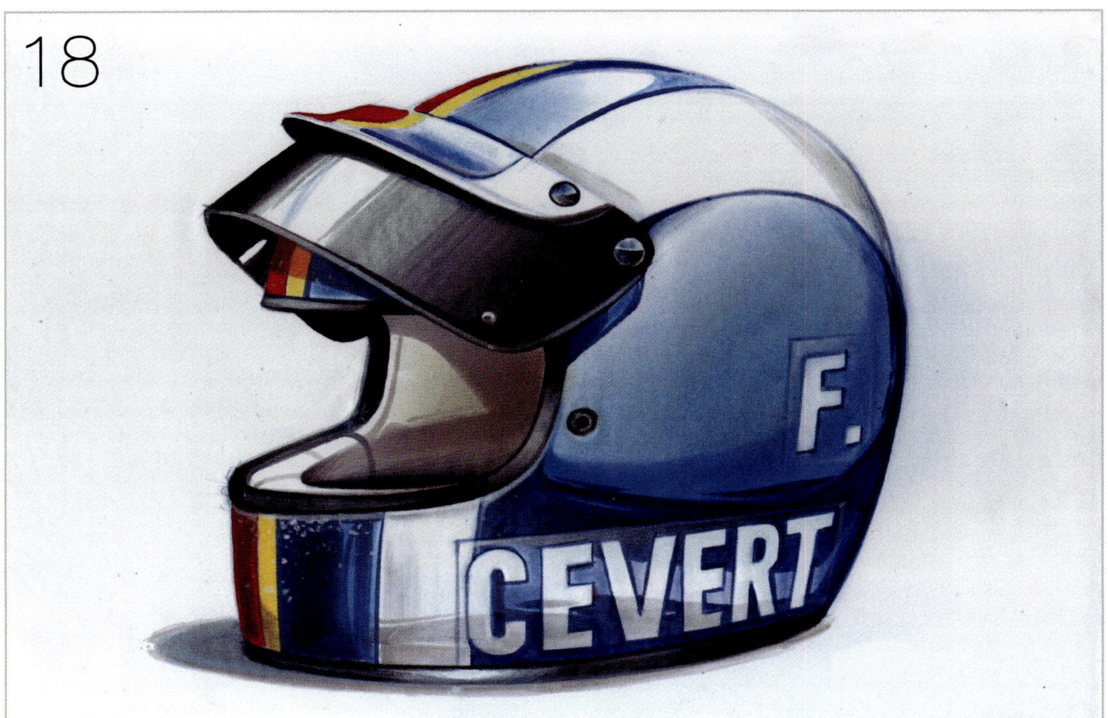

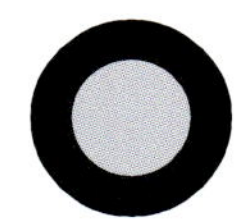

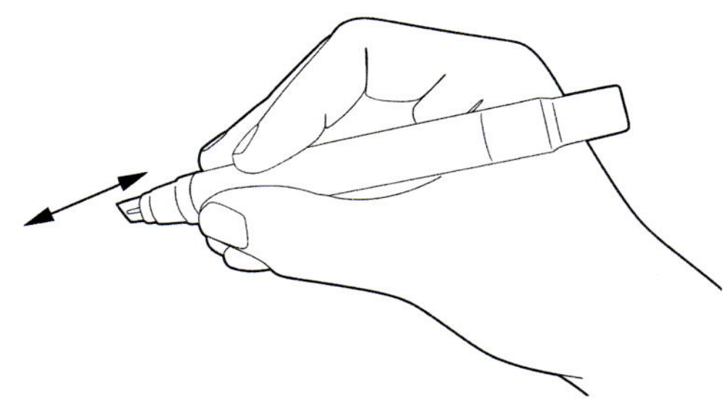

Take a moment to make sure that the reflections are travelling across the helmet in spite of the color changes. For example: Apply Cool Grey 20% on the lettering and inside the white stripe. Also, take this opportunity to drop a simple shadow to the left.

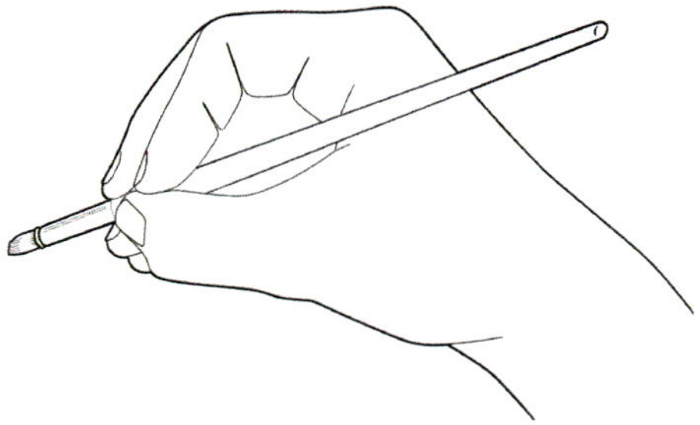

Finally, apply sparing amounts of white gouache for the highlights. Make sure the hot spots on the visor are moving in the same direction as the pencil and marker strokes.

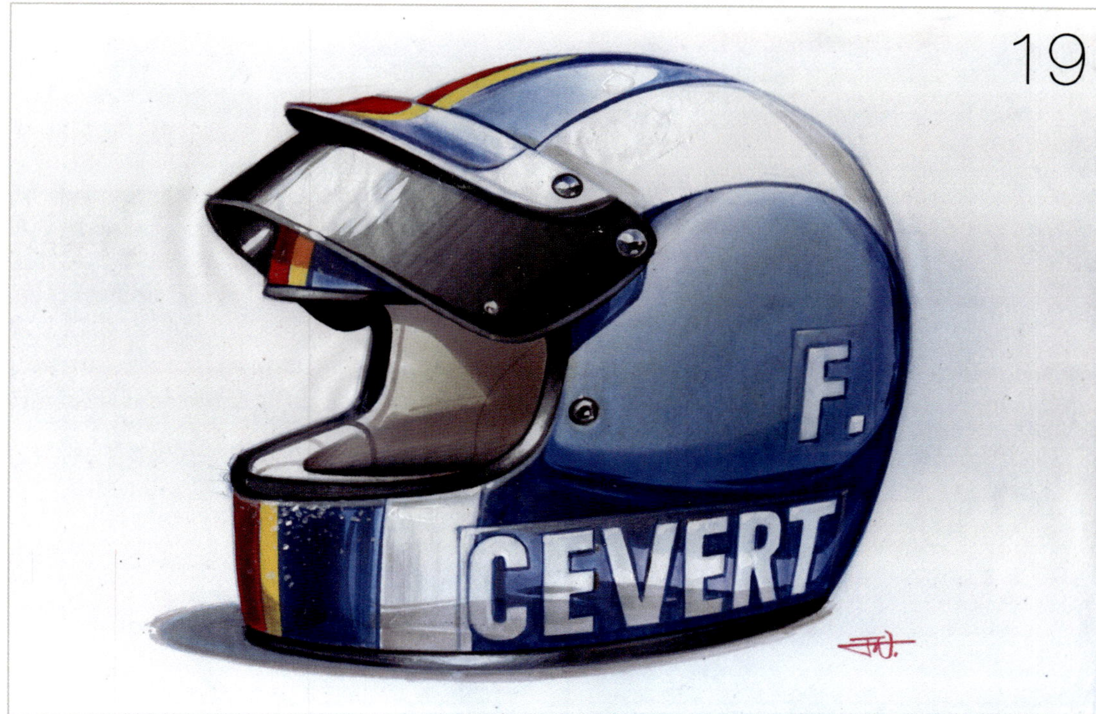

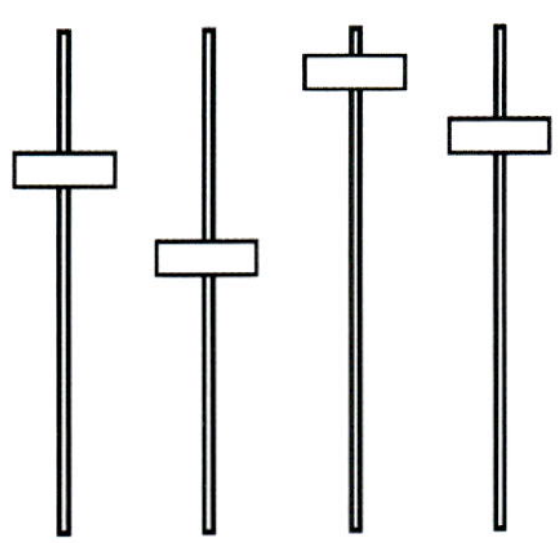

Scan in the image and clean up the background in Photoshop. You can also turn up the color saturation to give the finished piece some extra punch.

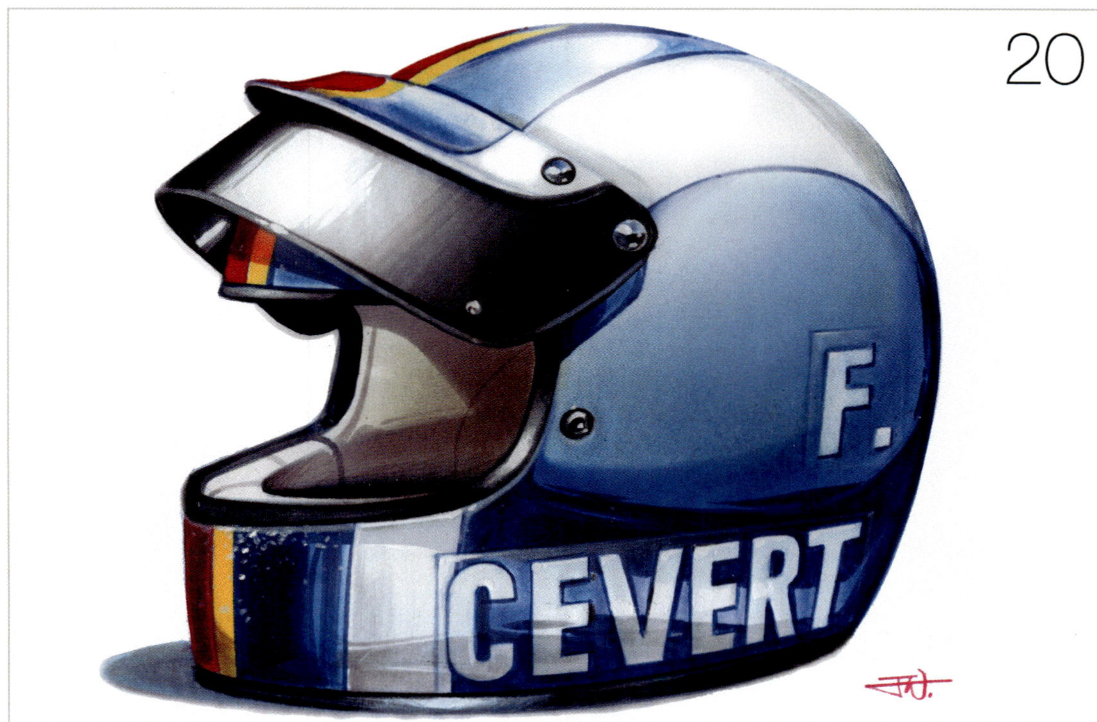

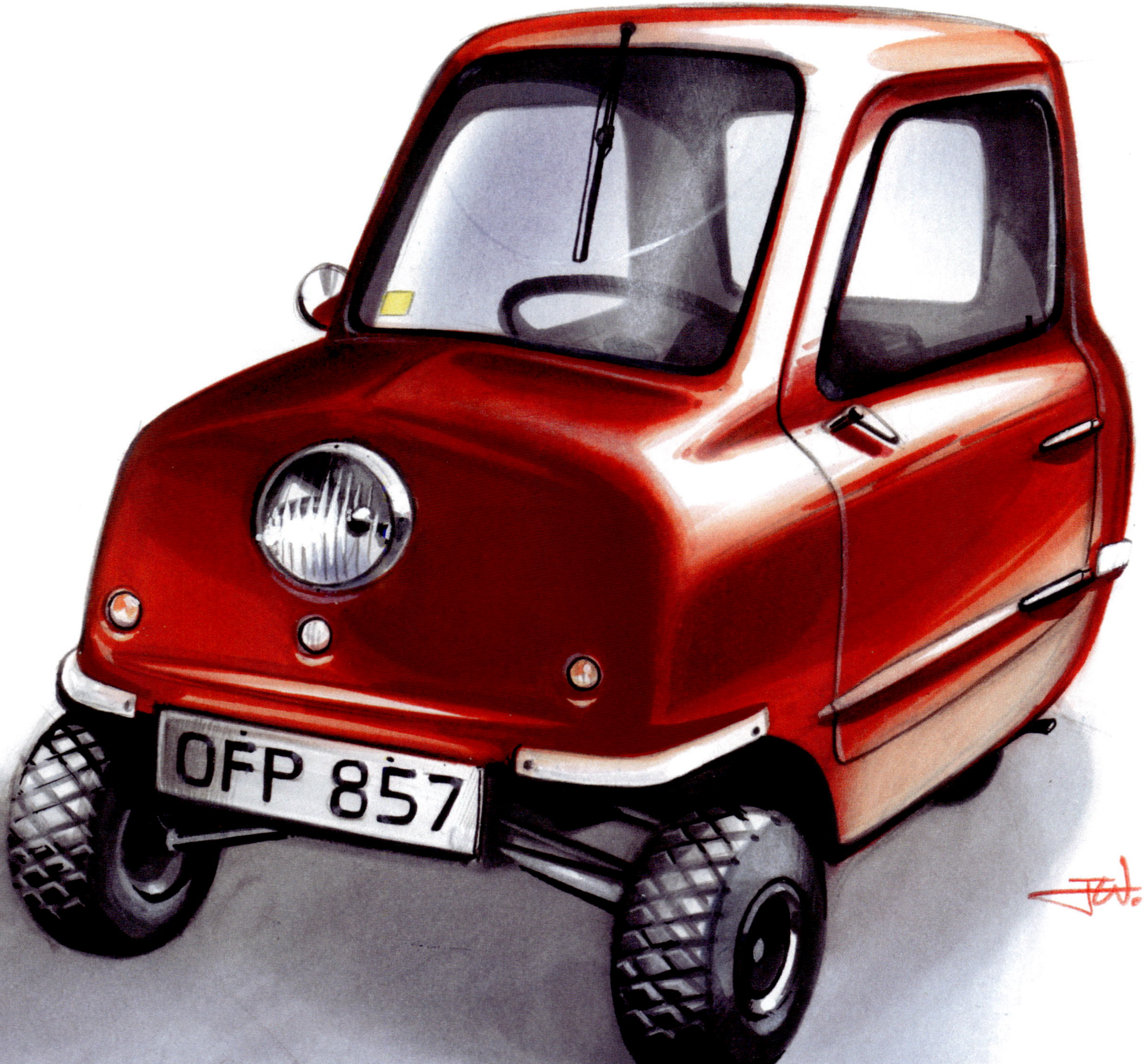
OFP 857

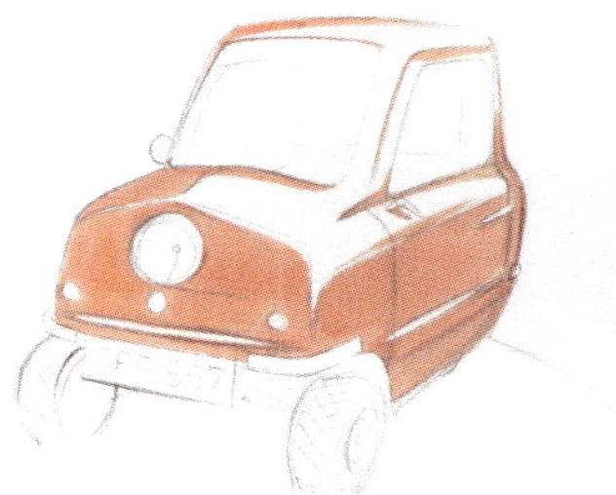

2. Layering

Now let's move on from glossy spheres and cylinders to glossy cubes. The subject of this rendering is a slightly re-imagined version of the Peel P50, a British micro-car from the 1960s. Just because the subject has basic, simple shapes doesn't mean that the rendering needs to be clinical or boring. In this tutorial, you'll learn how to methodically layer your markers and pencils. You'll find this technique can infuse any subject with richness and depth.

The Tools

Prismacolors: White.

Gouache: White, Black.

Verithins: Black.

Nupastels: Indigo Blue.

Prismacolor Markers: Salmon Pink, Crimson Red, Mahogany Red, Canary Yellow, Cool Grey 60%, Cool Grey 20%, Cool Grey 90%, Black.

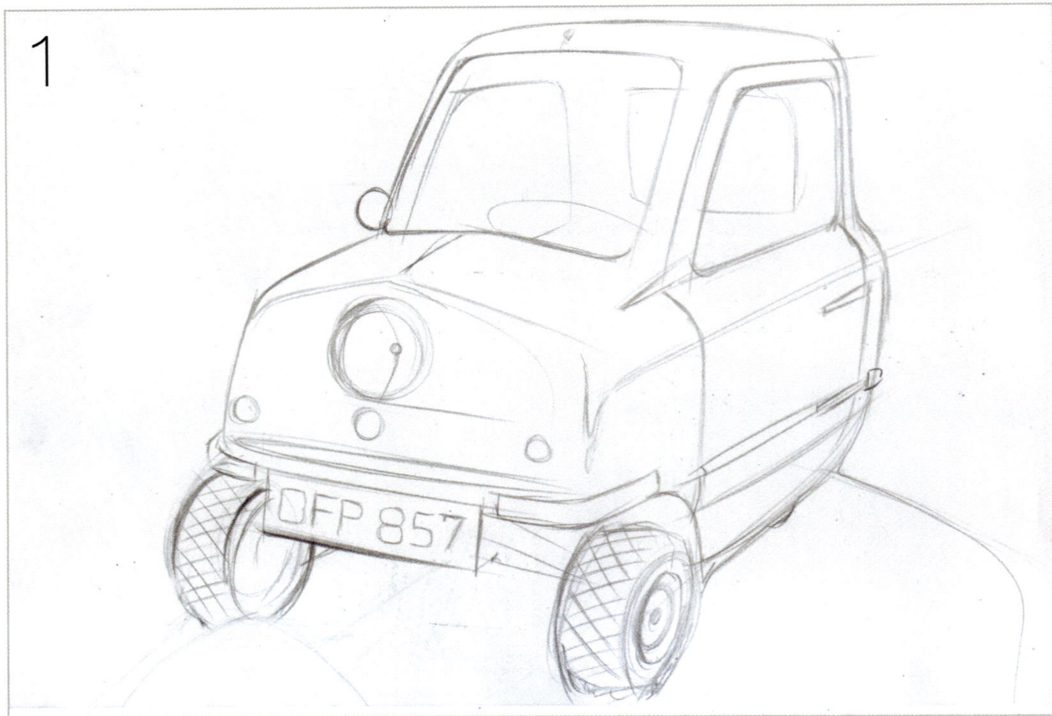

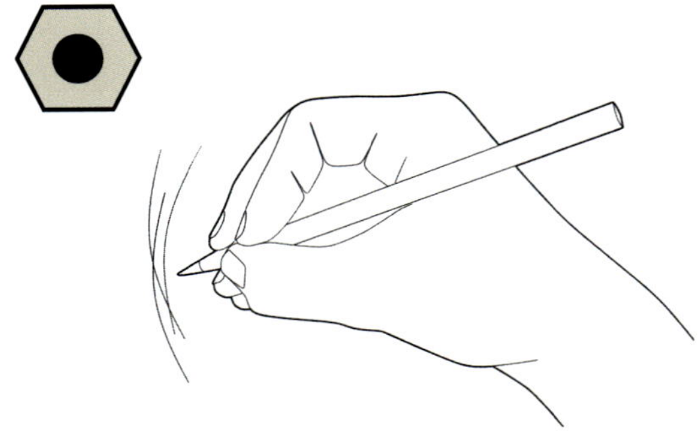

Use a black Verithin to loosely outline the car on marker paper. Take the time to include some details, but only in a cursory manner.

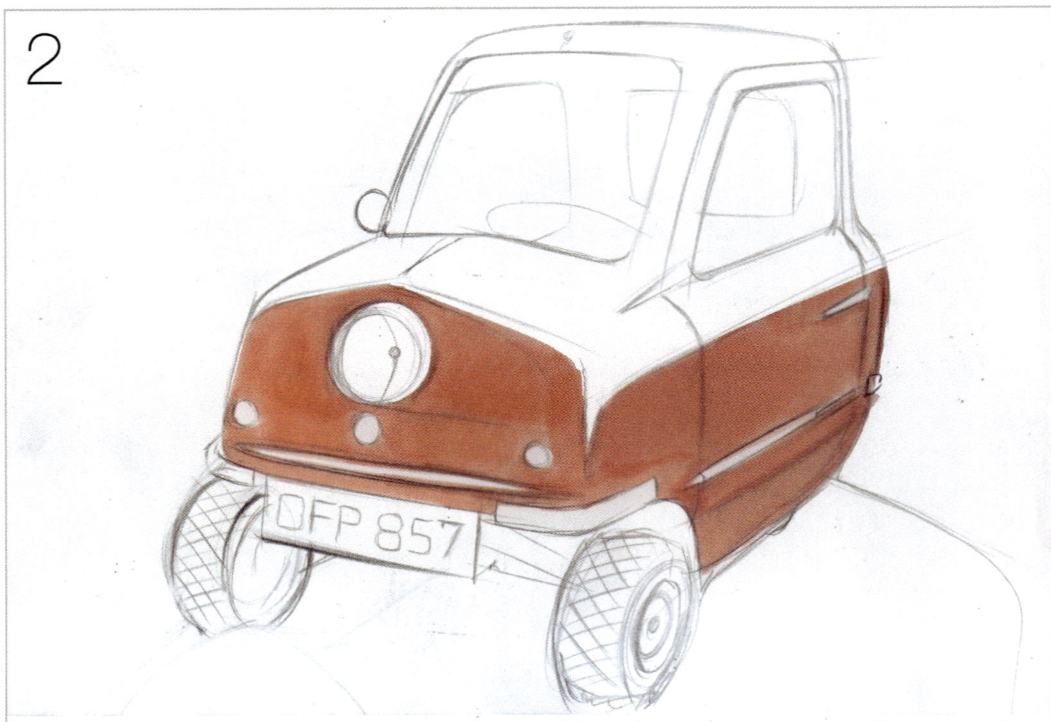

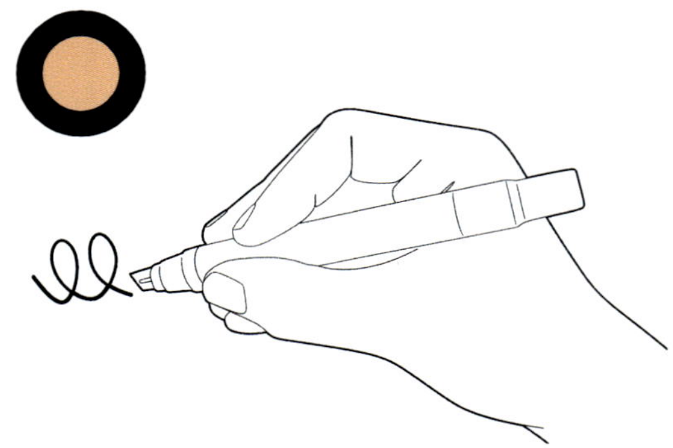

Now you can begin laying in the main reflection with a Salmon Pink marker. Although the car will ultimately be bright red, it's critical to start with a lighter shade to get a rich, layered effect.

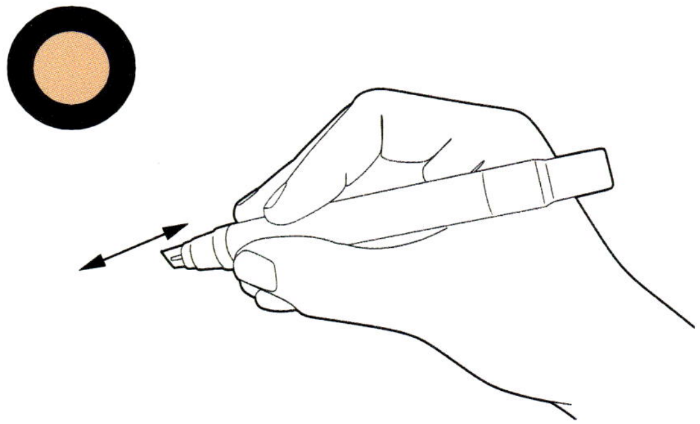

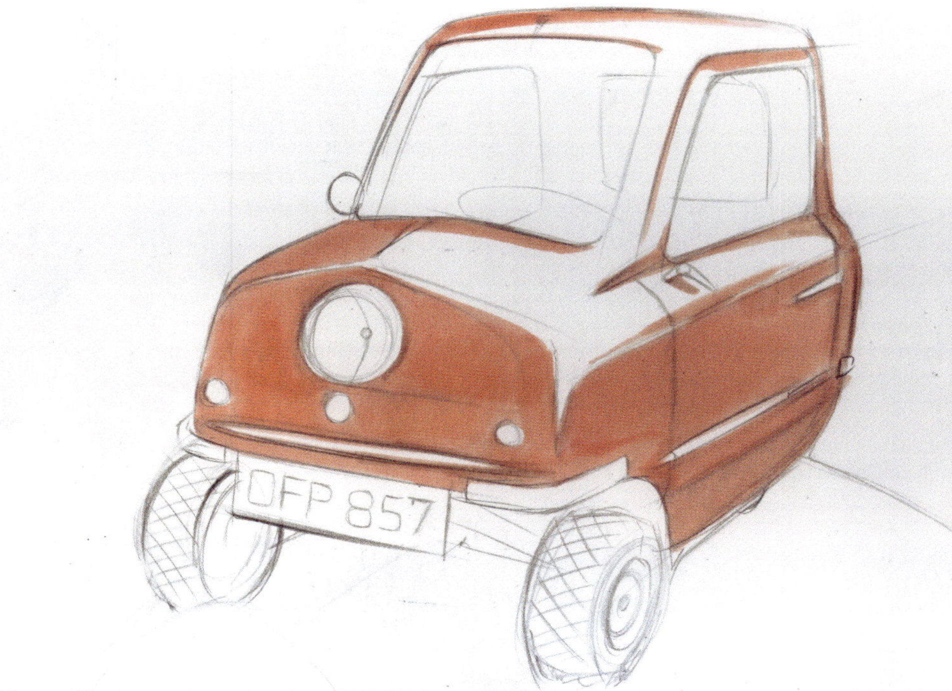

3

Continuing with the Salmon, lay in some secondary and detail reflections: A tree in the foreground, the reflection on the far side of the hood and some streaks along the shut lines to indicate underflush conditions.

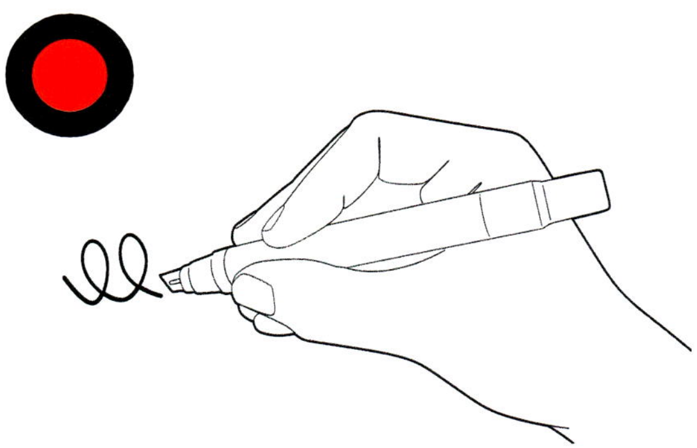

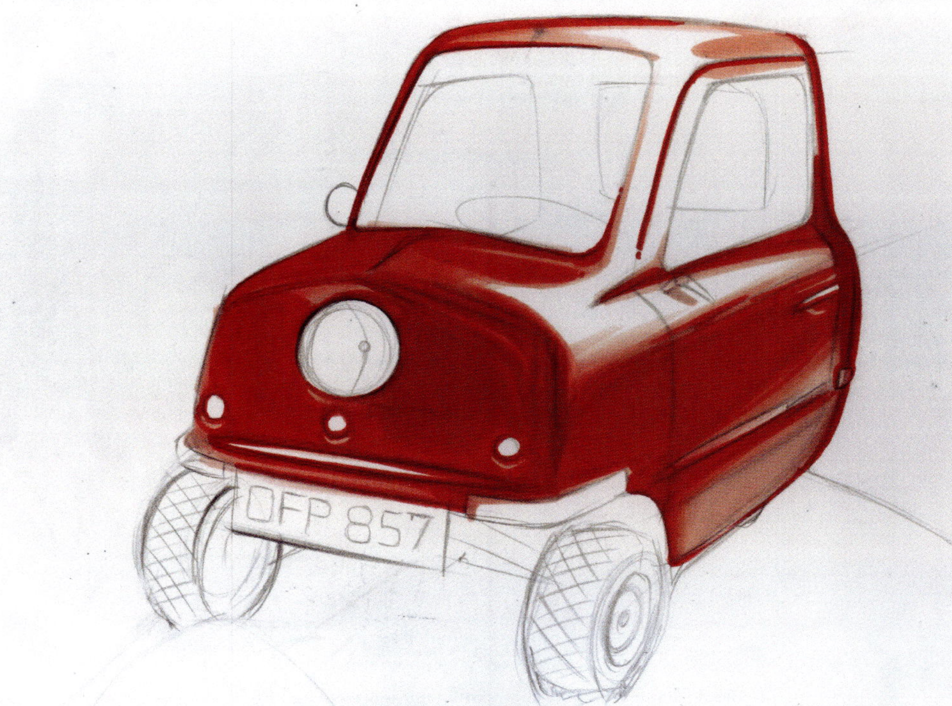

4

Now you can layer the Salmon marker with Crimson Red. The trick here is to leave a bit of Salmon peeking through along the edges and within the reflections. Leave a large Salmon space on the left rear corner; this will represent light reflecting up from the white floor.

5

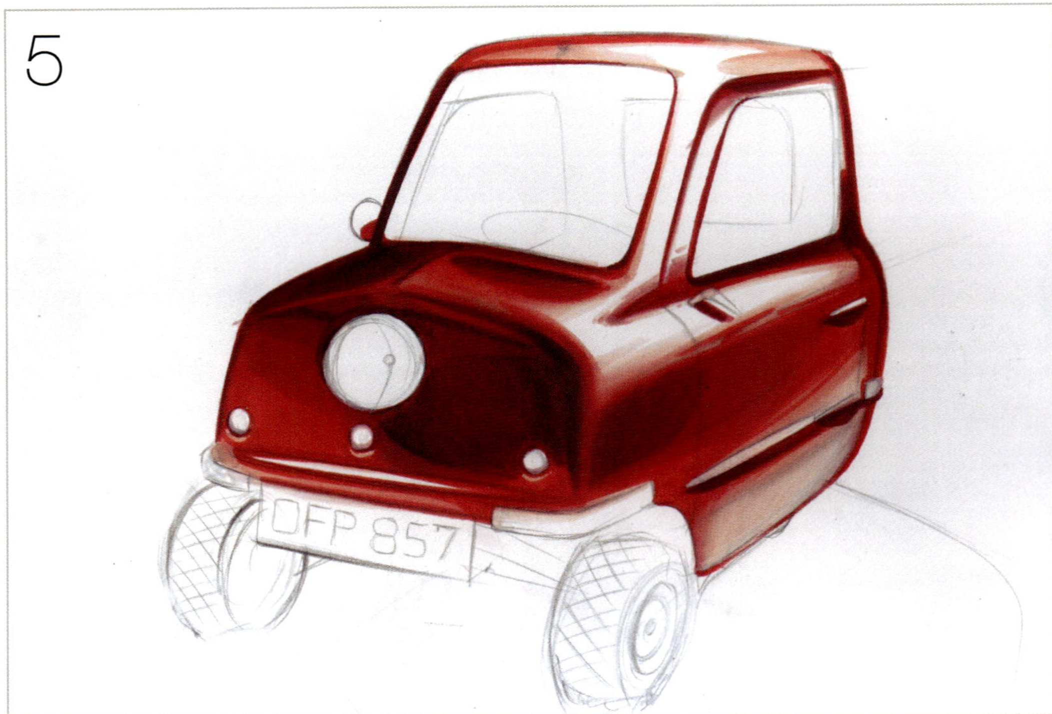

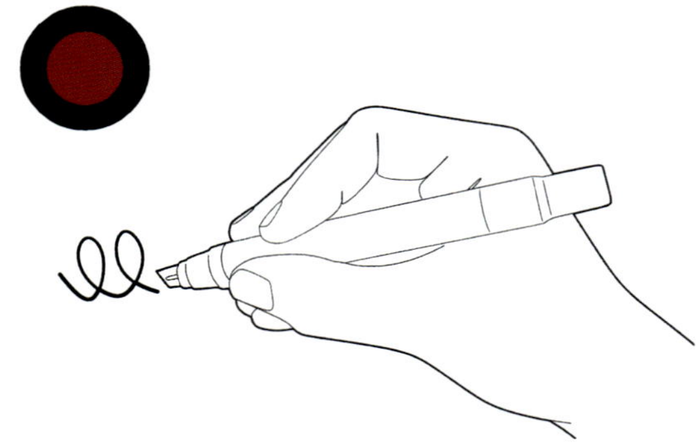

Now add a third layer of marker, this time in Mahogany Red. As with the Crimson Red, you should leave spaces for the lighter shades to peek through. This three-layered technique produces reflections far more sophisticated than those created with just one marker.

6

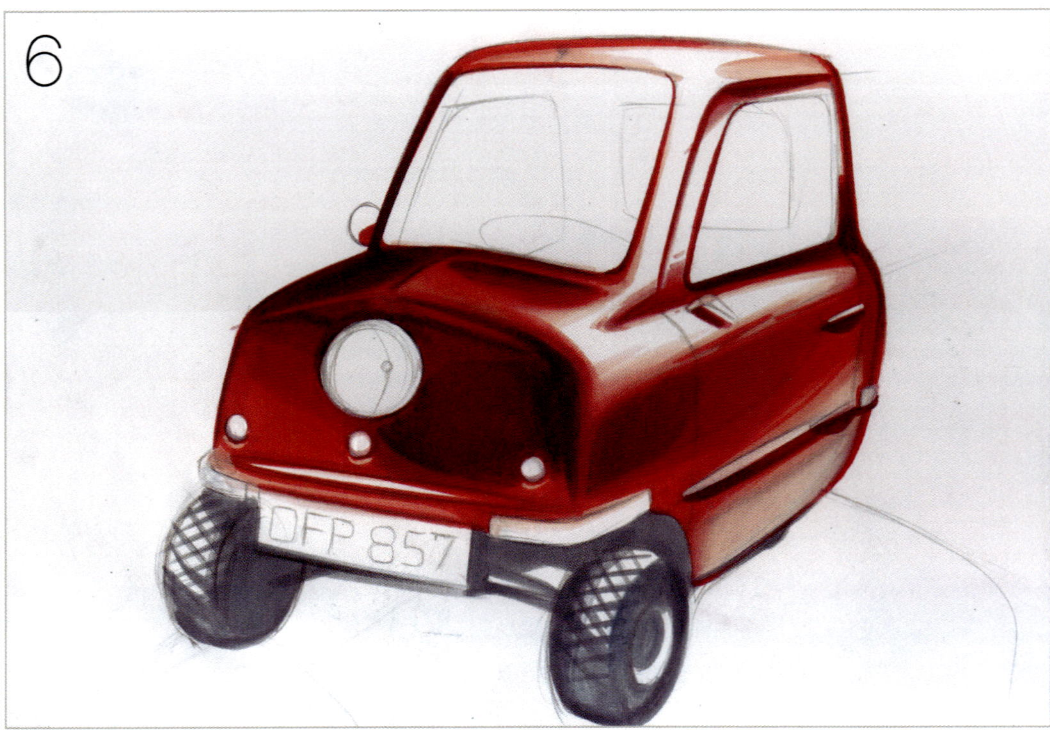

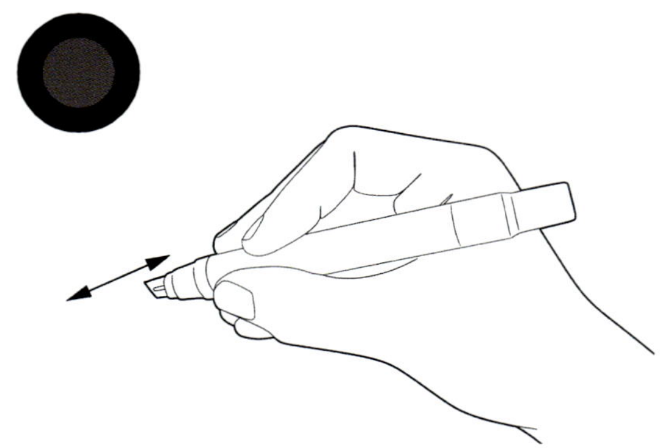

Use Cool Grey 60% to block in the tires and the undercarriage. Be sure to leave some paper showing through for the lit portion of the tread; this will make detailing the tires much easier.

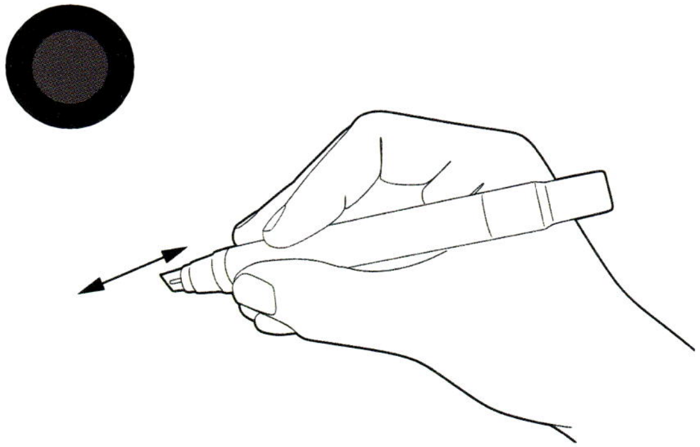

Continue with Cool Grey 60% and block in the interior of the car: The A pillars, the B pillars and a portion of the steering wheel. The cool temperature of the tires and the interior will give this rendering some subtle color contrast.

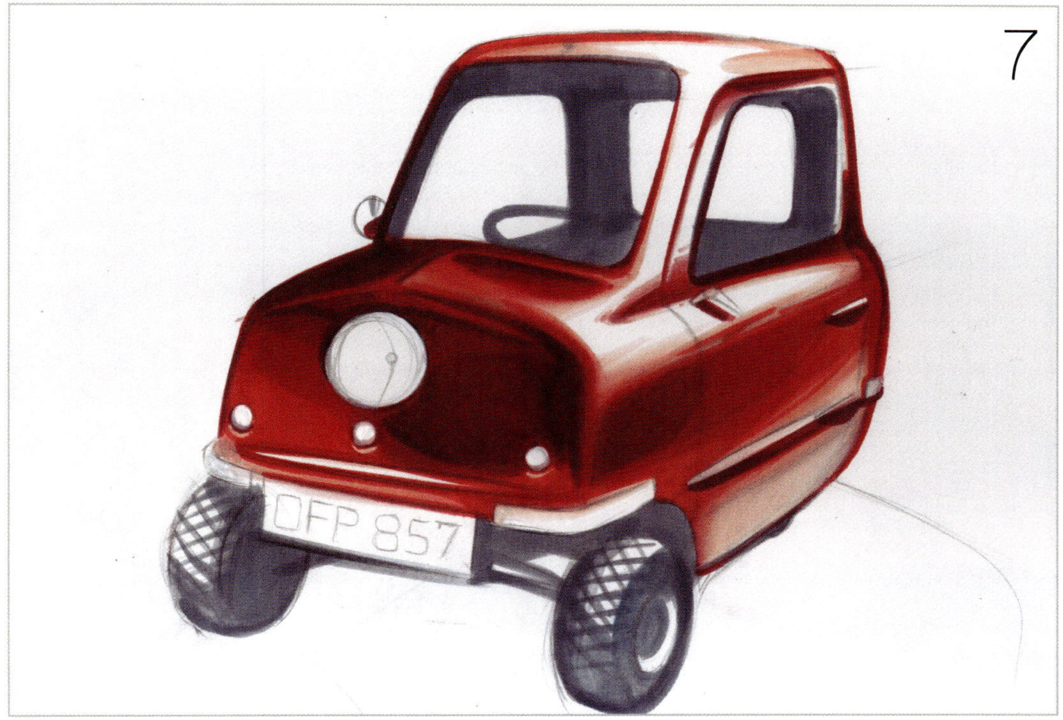

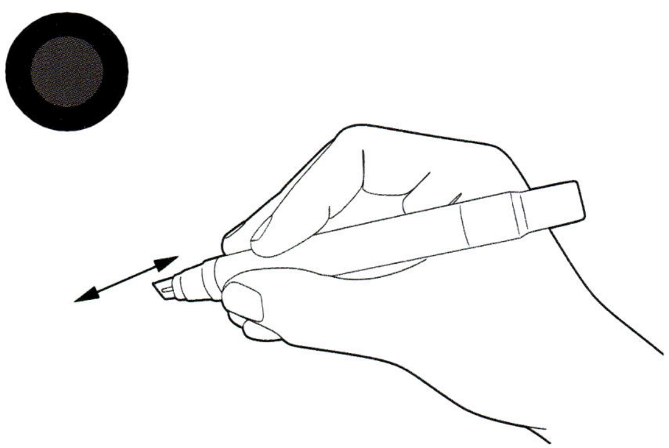

The classic round headlamp can be indicated using a simple, tried-and-true technique. The main surface is like a spoon; it flips reflections upside down. So make the top dark and leave the bottom light. The bulb shield is a half sphere, so render it the opposite way.

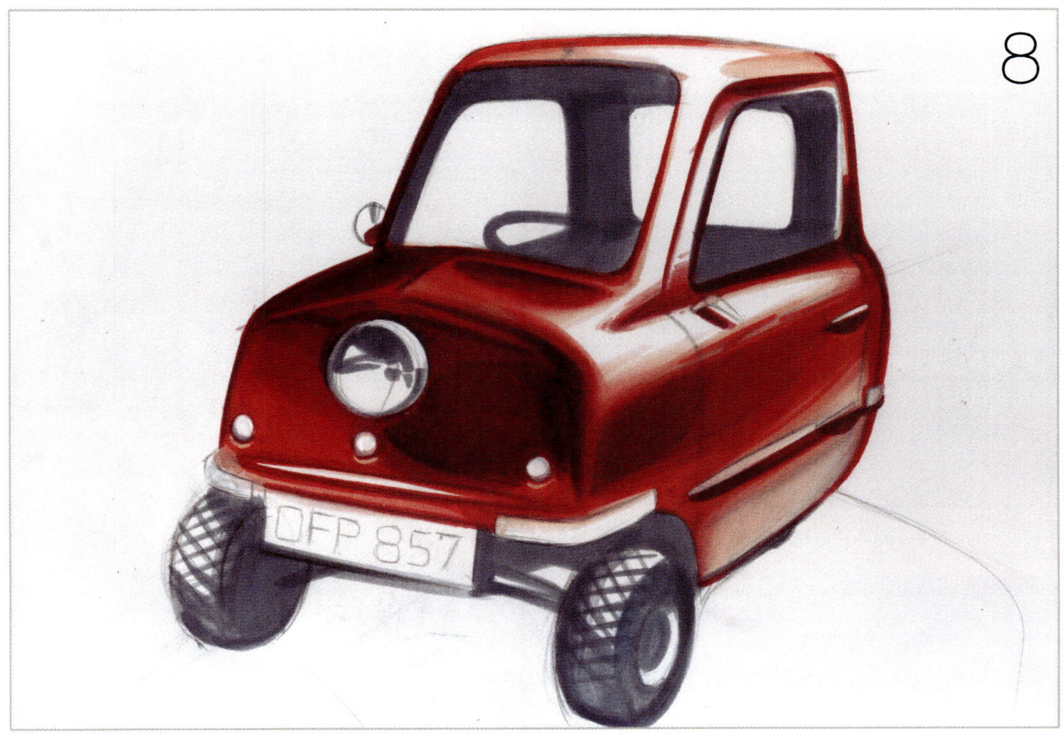

9

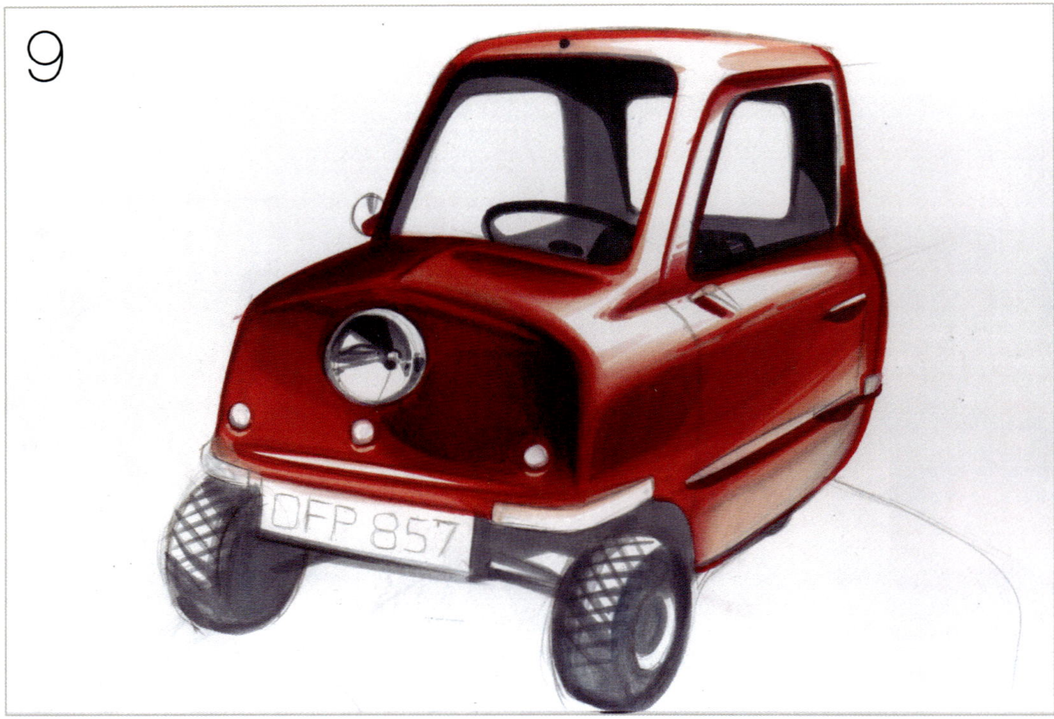

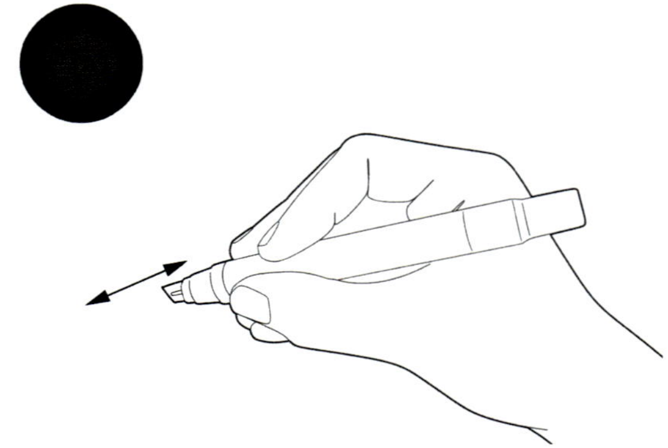

Use Cool Grey 90% to add more detail: A shadow emphasizing the light coming through the windows and small detail reflections in the headlamp.

10

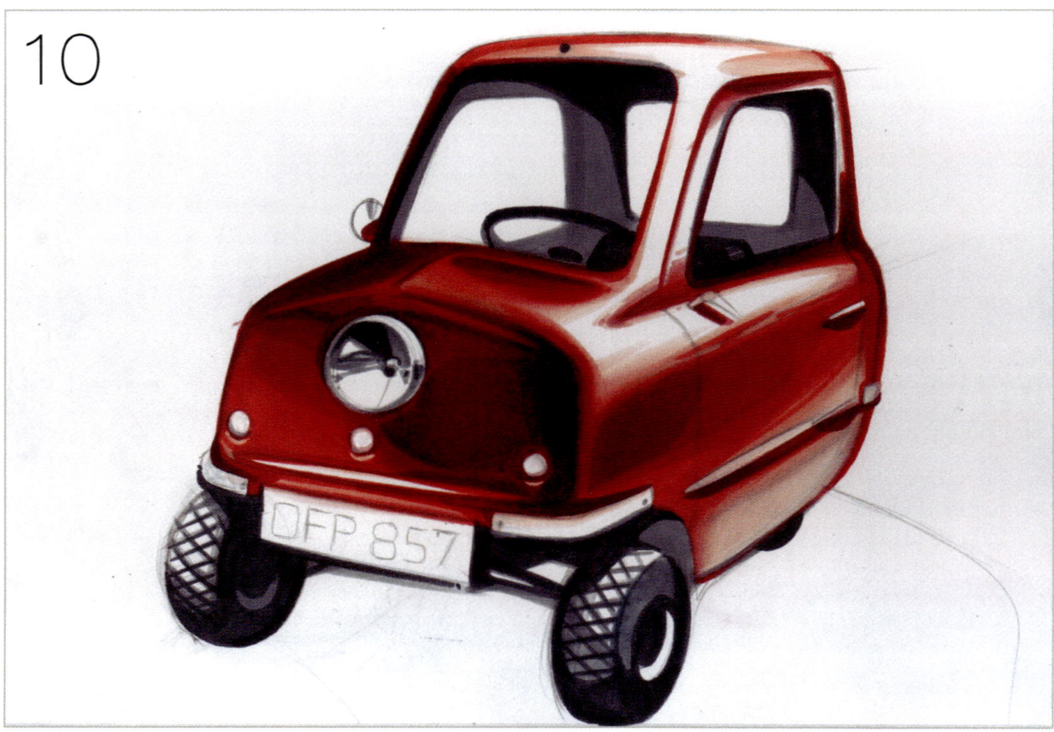

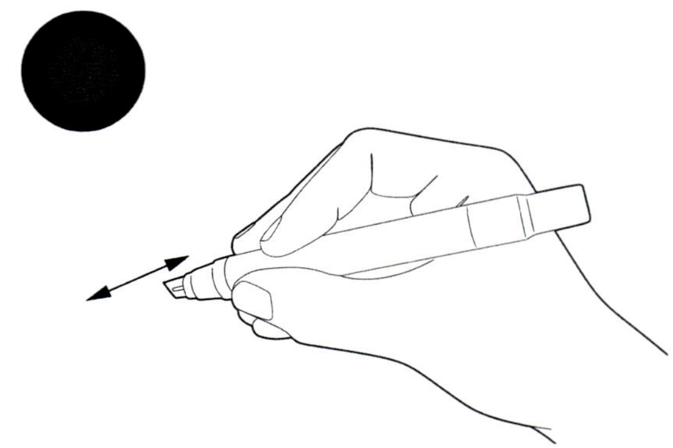

Continuing with Cool Grey 90%, lay in the darkest part of the tires and the undercarriage. The same three-layered principles apply here; leave some paper and light grey showing for the full effect.

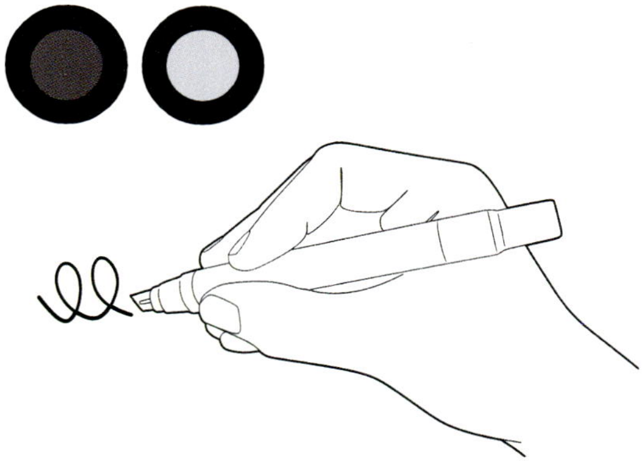

Using circular motions, lay in the darkest part of the floor shadow with Cool Grey 60% and the lightest part with the 20%. Then, while the marker is still wet, use the 20% as a blender to soften and smooth out the shadow.

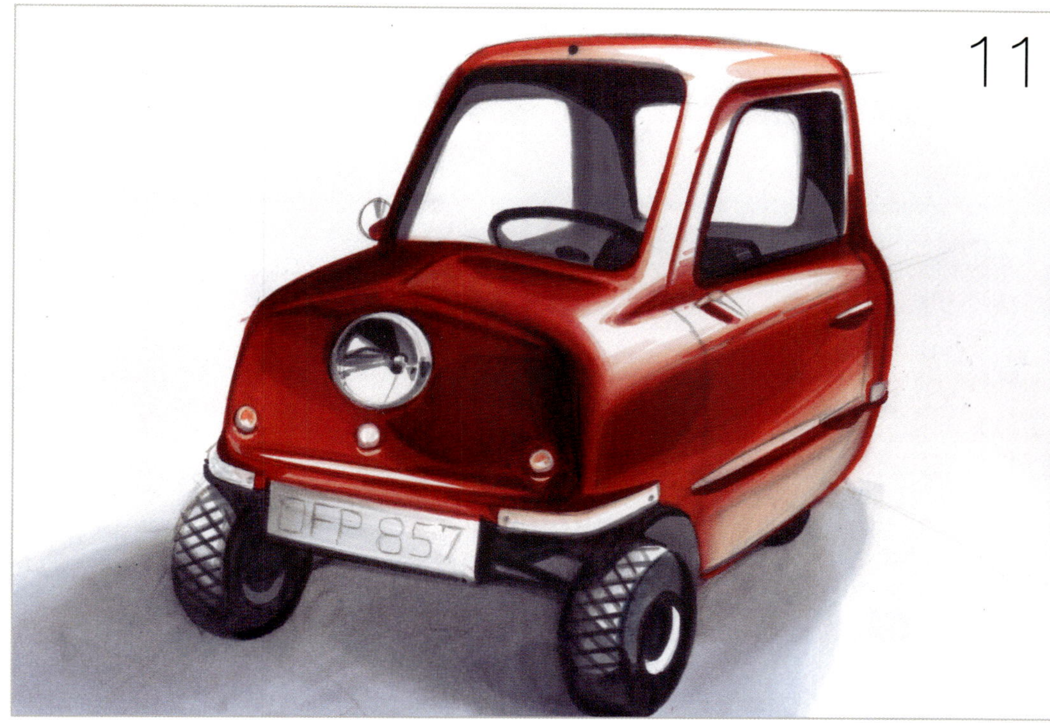

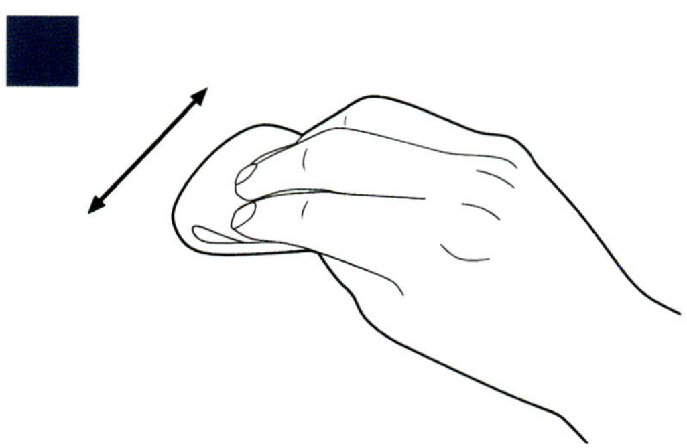

Since the marker portion of the rendering is so prominent, the sky tone in the window should be more subdued.

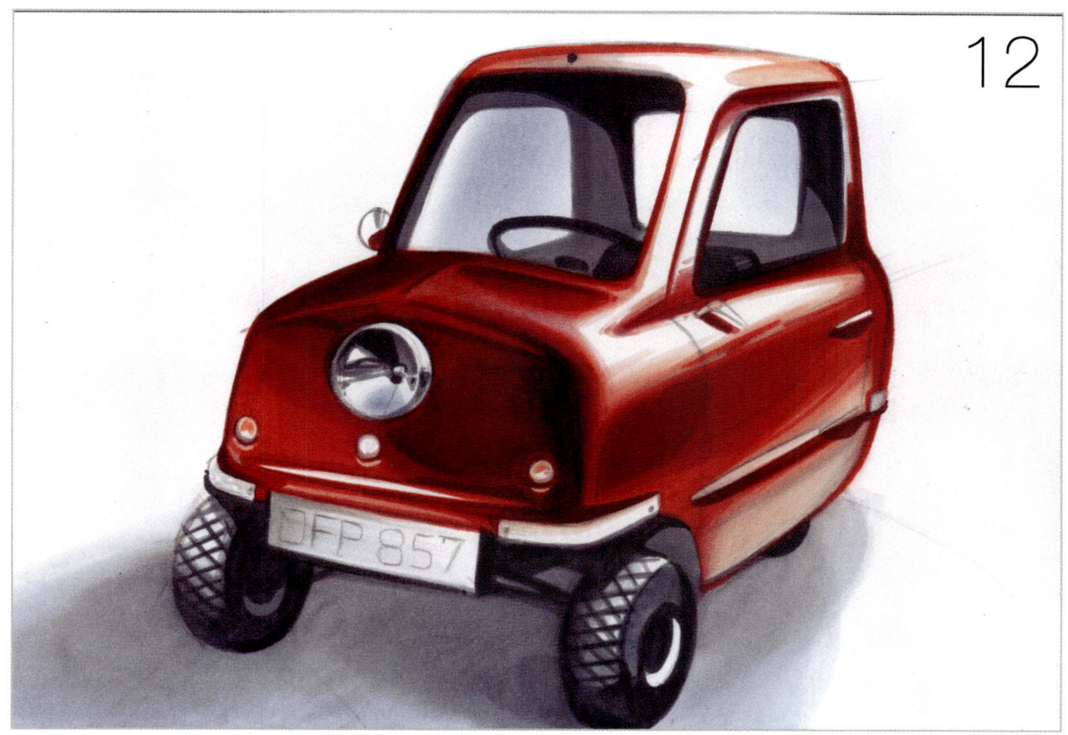

13

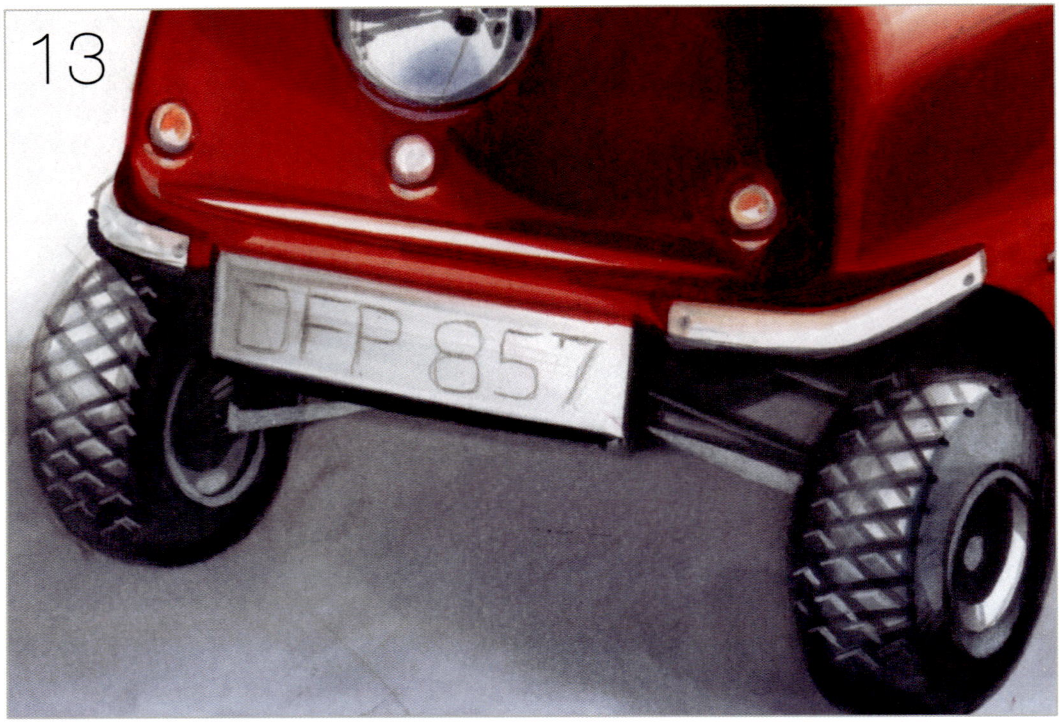

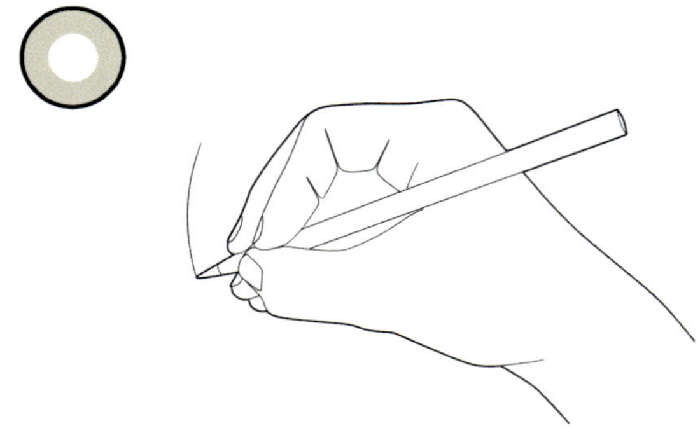

Apply white pencil on the edges of the tire tread. You can also scratch in a bit of white pastel and blend it out with your finger to give the tread some shine.

14

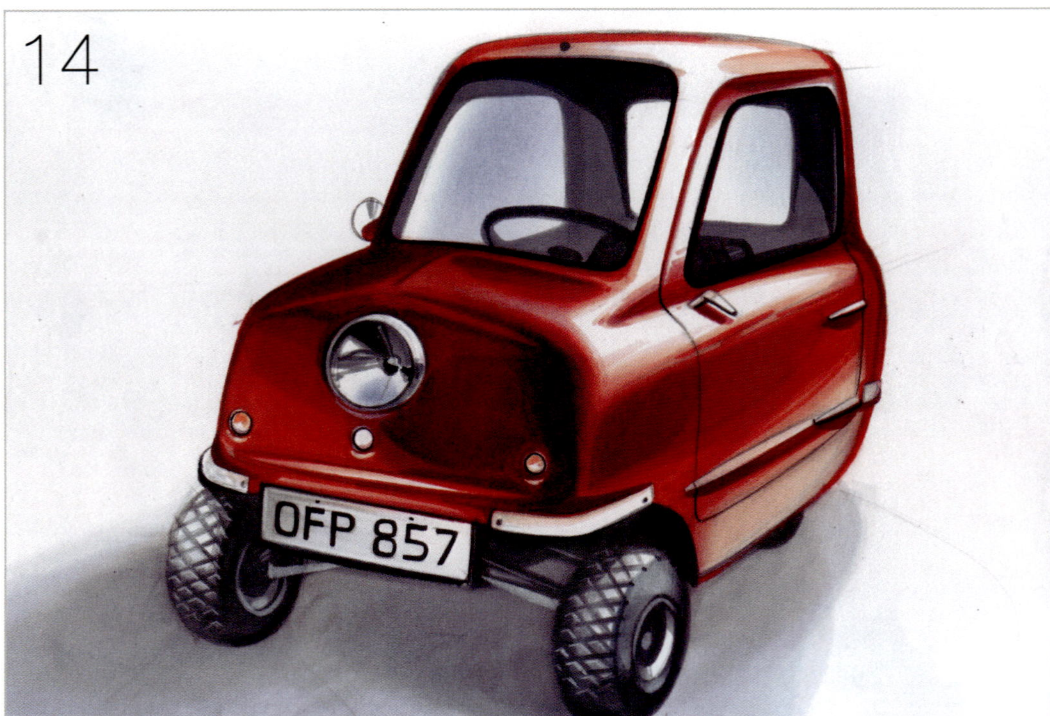

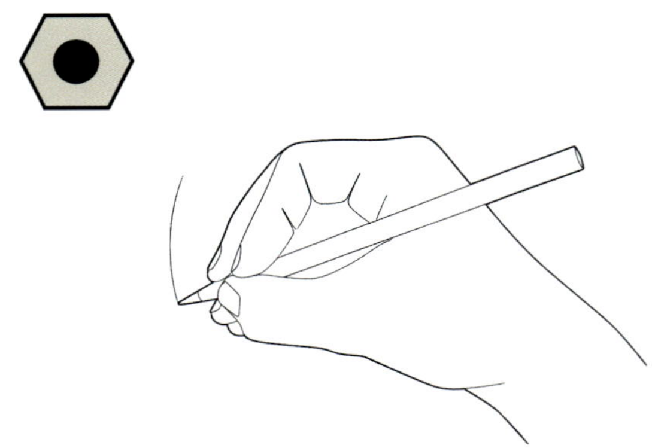

With a steady hand, begin laying in the final linework: The license plate characters, the window seals, the door shuts and the headlamp gaps.

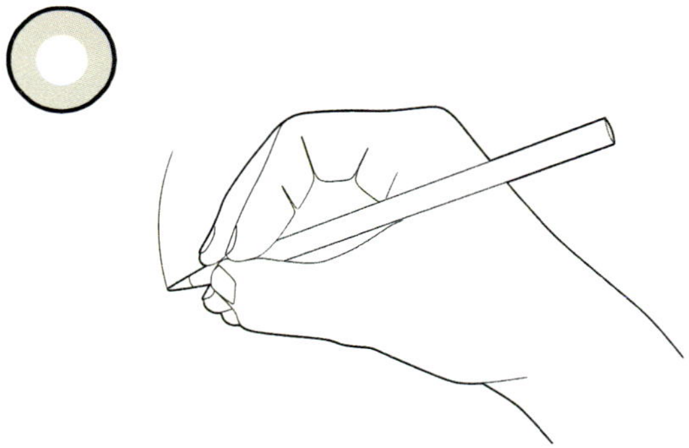

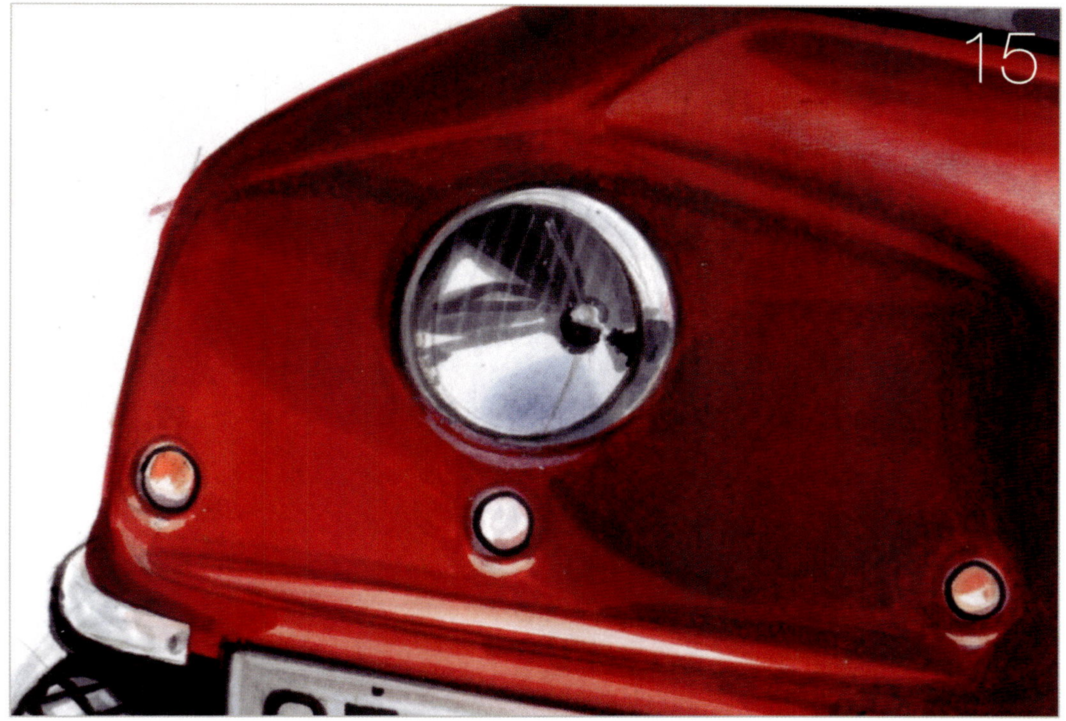

Add some subtle white lines to the headlamp lens to create an illusion of transparency. Be sure to give the lines some crown, or the lens will appear flat.

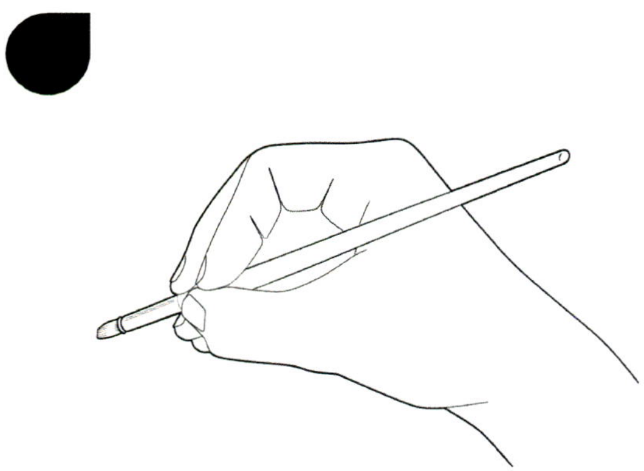

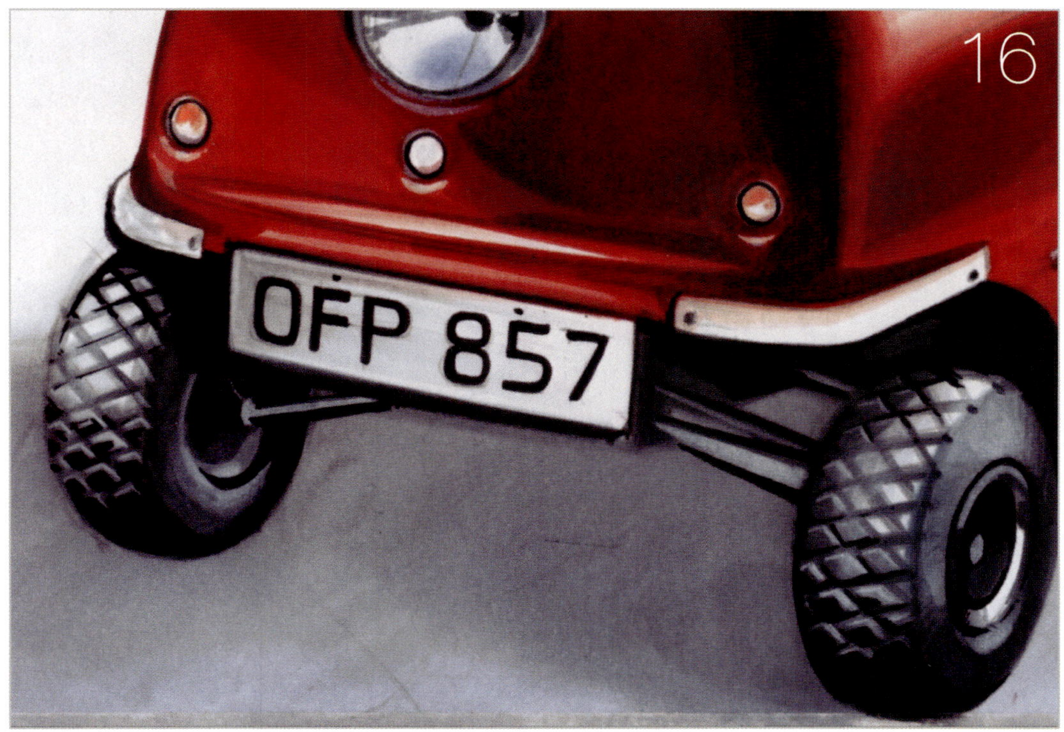

With a small brush, apply a bit of black gouache in the darkest areas of the tires and wheel wells. This will yield far more contrast than a black marker.

17

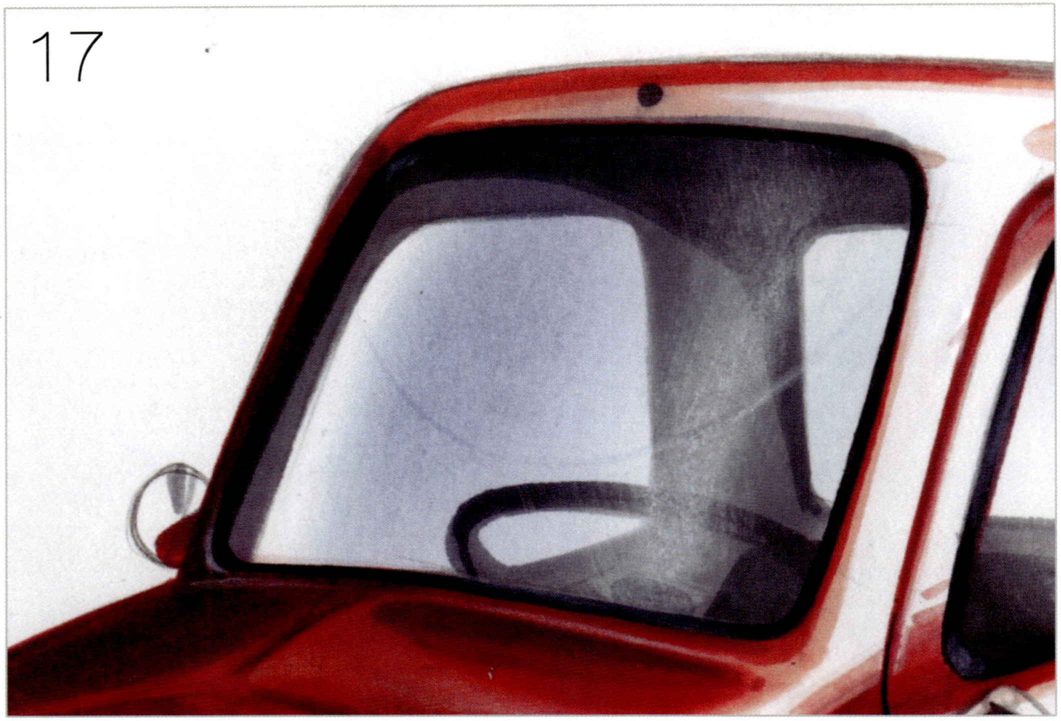

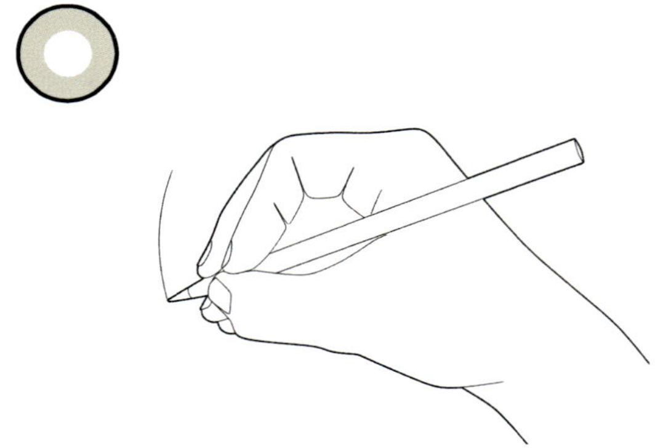

Since this is an old car, you can use the rough texture of the white pencil to give the windshield a weathered appearance. Strike a sharp, curved line to give the indication of a wiper scratch.

18

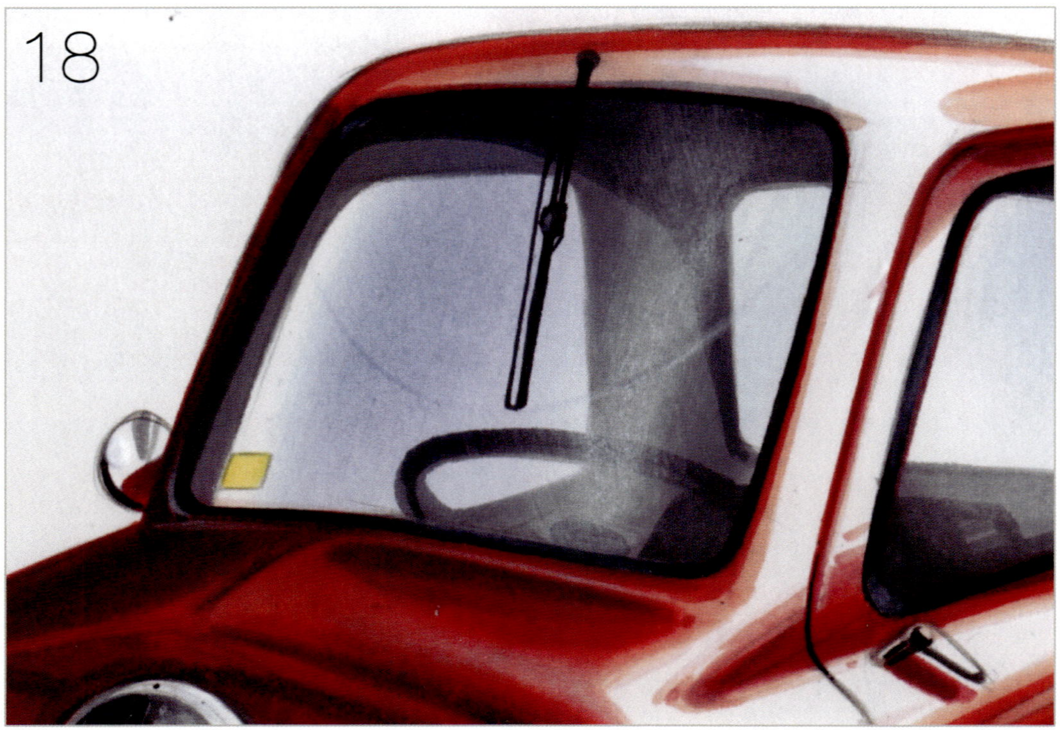

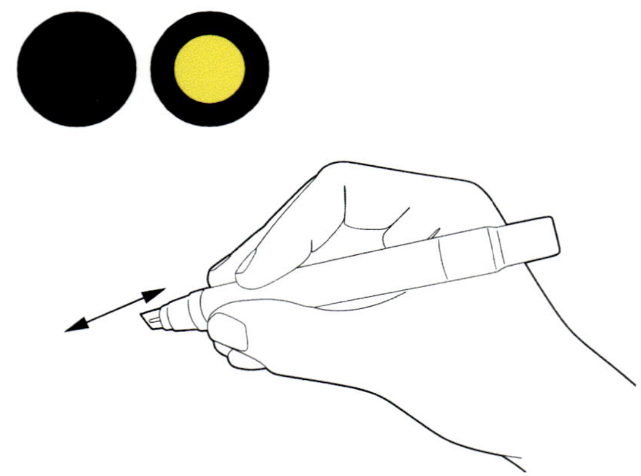

Use the fine tip of a black marker or a felt pen to draw the wiper. You may wish to add weight to the window seals and a few reflections on the chrome door handle. Use a bit of yellow marker to indicate a parking sticker.

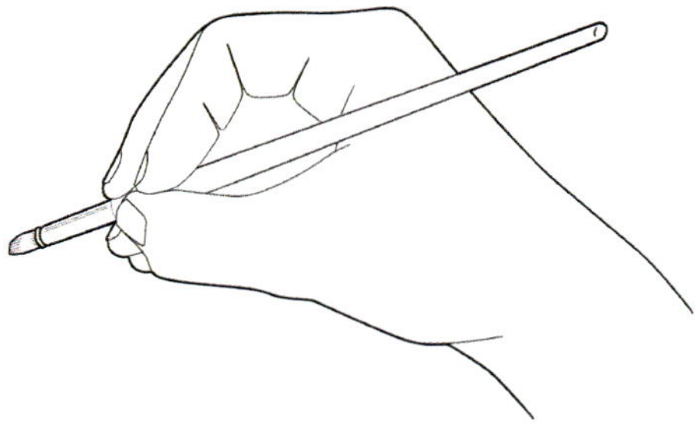

Apply small amounts of white gouache for the highlights. Strategically placing speared lines on the headlamp lens can add a nice crowning touch.

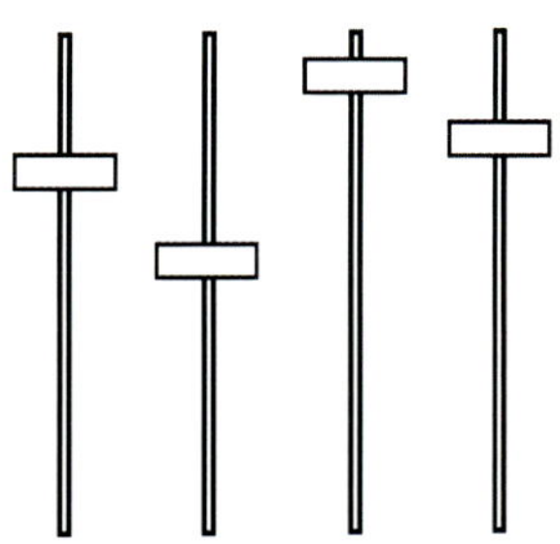

Scan in the image and fix any minor flubs in Photoshop. You can also shift the color balance a bit if you want the paint to appear more red-orange.

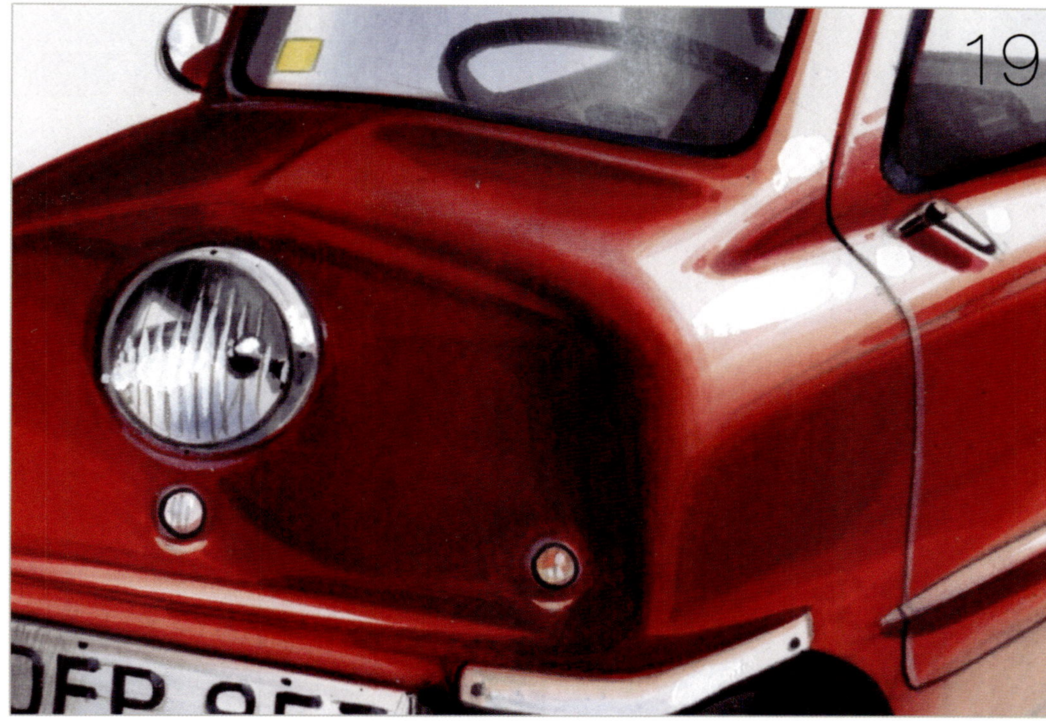

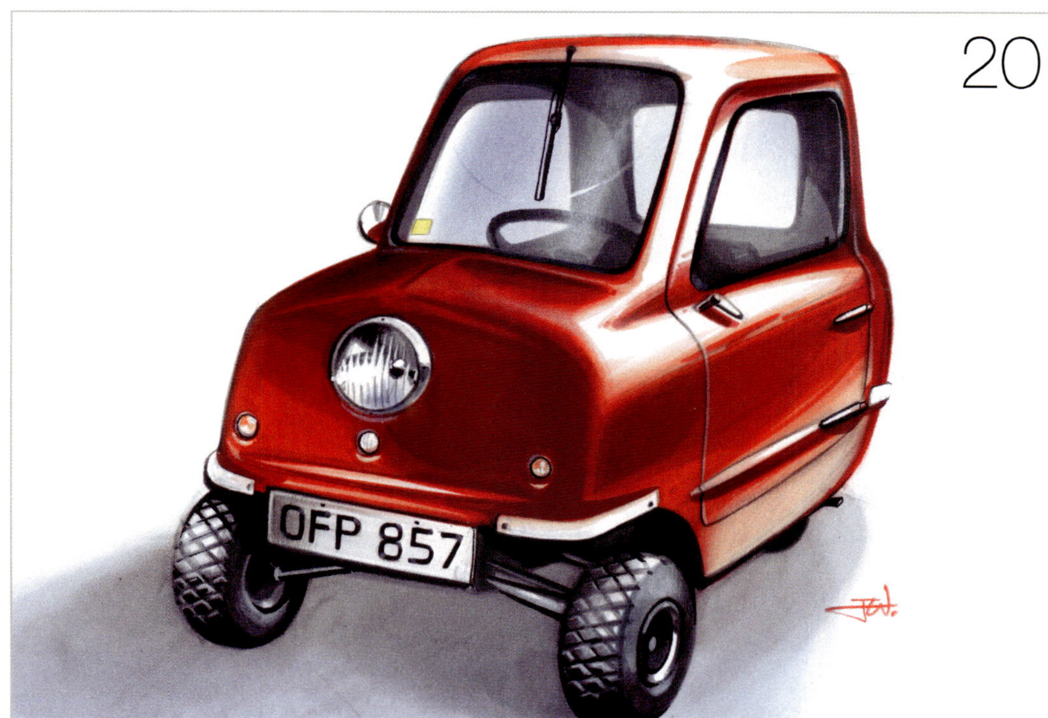

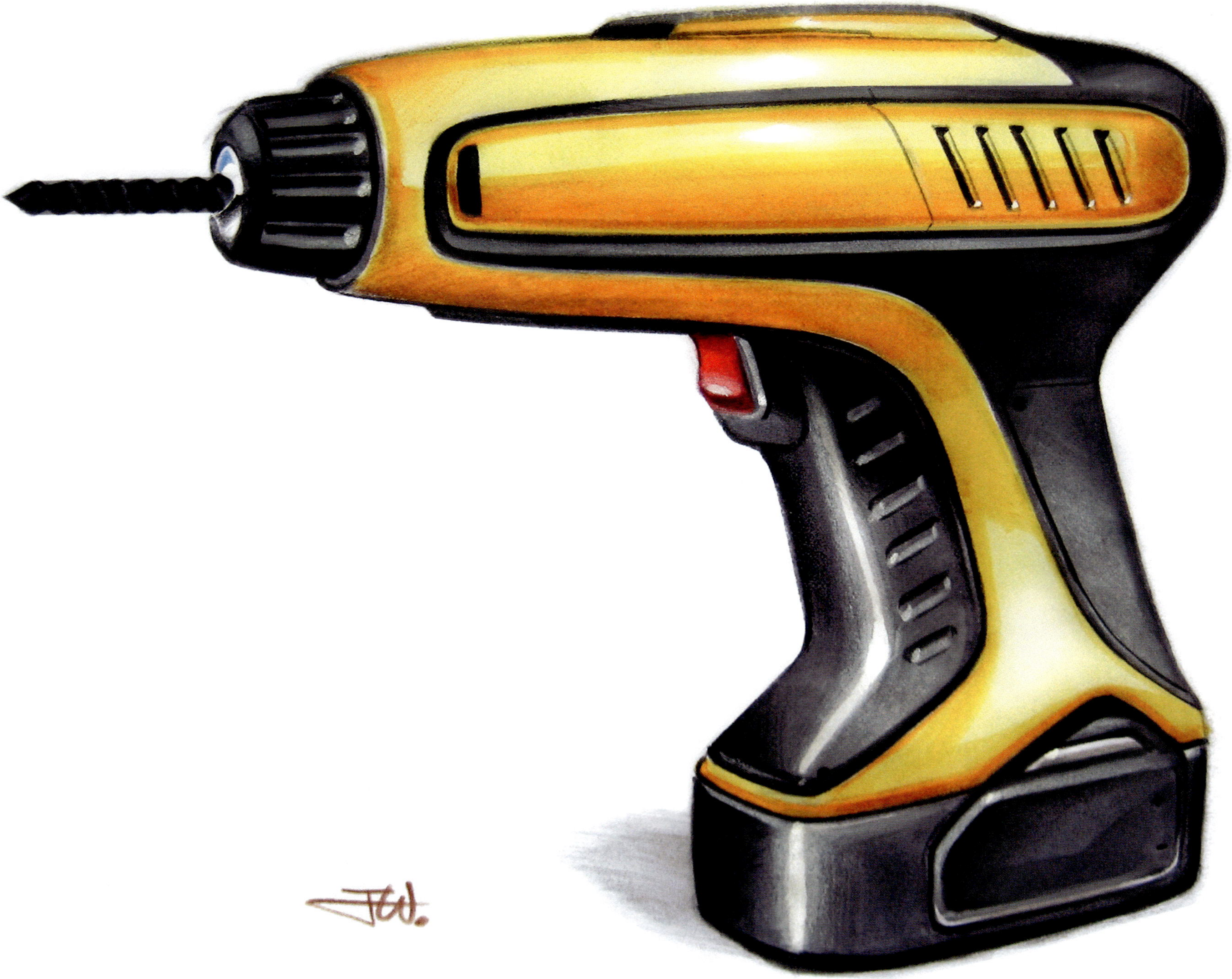

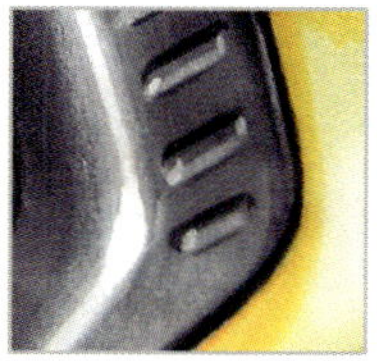

3. Plastics

Accurately depicting different kinds of plastic requires paying close attention to two things: texture and gloss level. In this tutorial, you'll learn how to render shiny yellow plastic and semi-matte black plastic. Once again, layering will play a critical role in the process.

The Tools

Prismacolors: White, Yellow Orange, Black, Deco Pink, Mahogany Red, Cool Grey 20%, Sienna Brown.

Gouache: White.

Verithins: Black.

Prismacolor Markers: Crimson Red, Cerulean Blue.
Copic Markers: Pale Yellow, Cadmium Yellow, Chrome Orange, Cool Grey 70%.

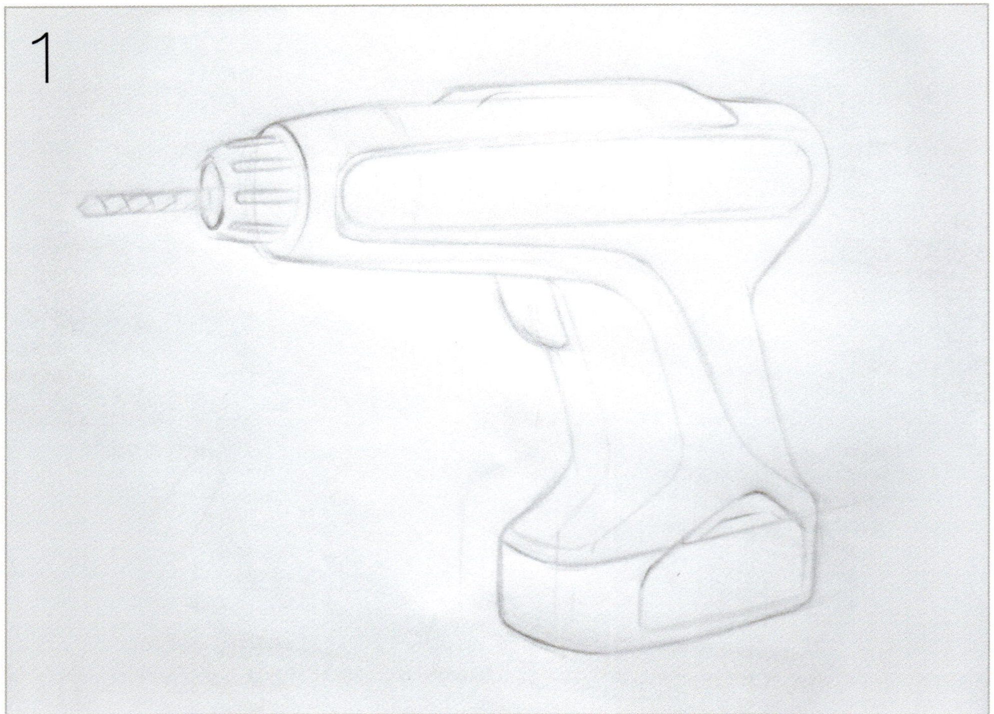

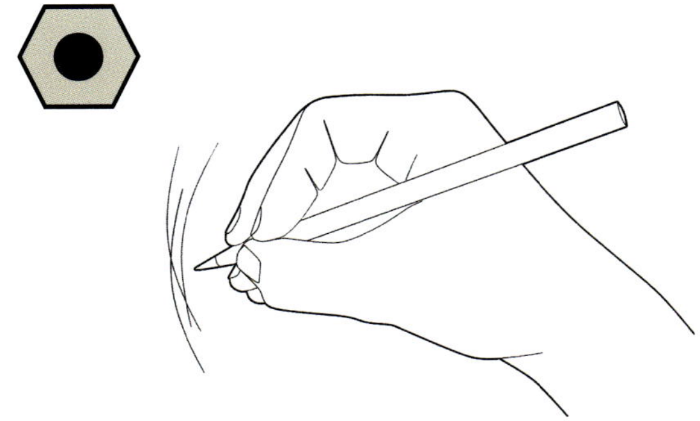

Use a black Verithin to set up the drawing, but try to use as few lines as possible (this will make it easier to apply the Pale Yellow marker in step two).

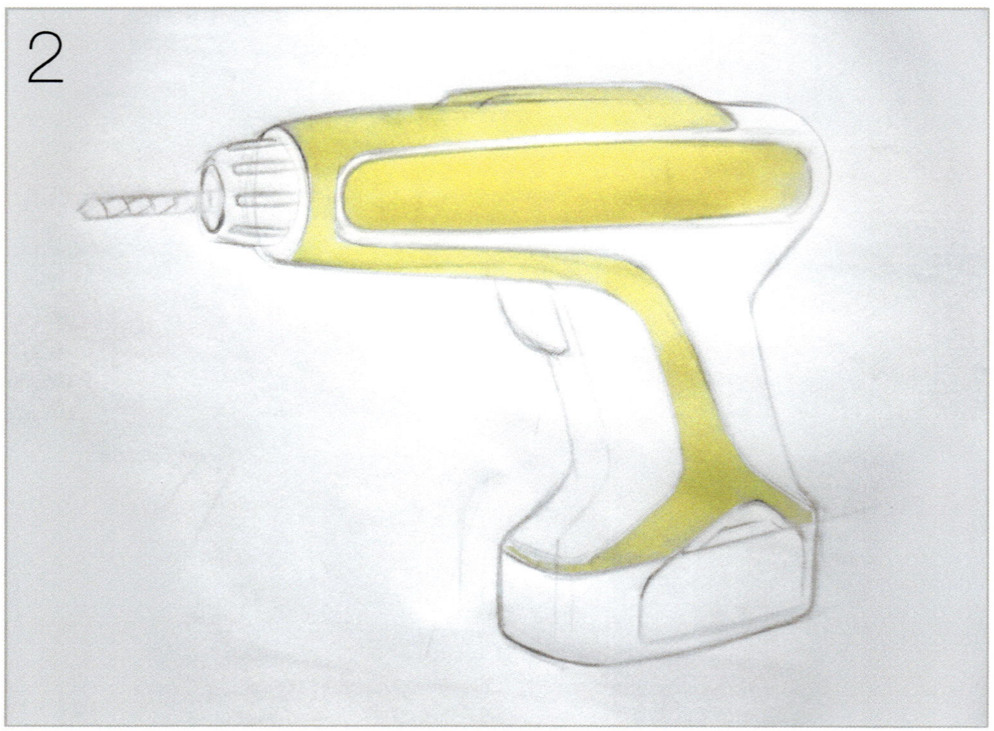

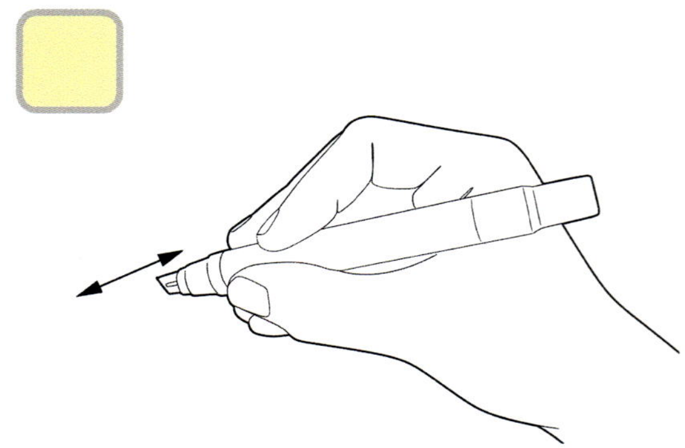

When applying yellow markers for the body of the drill, try not to touch the pencil lines. Yellow markers, more than any other, will pick up pencil marks and turn discolored.

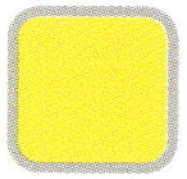

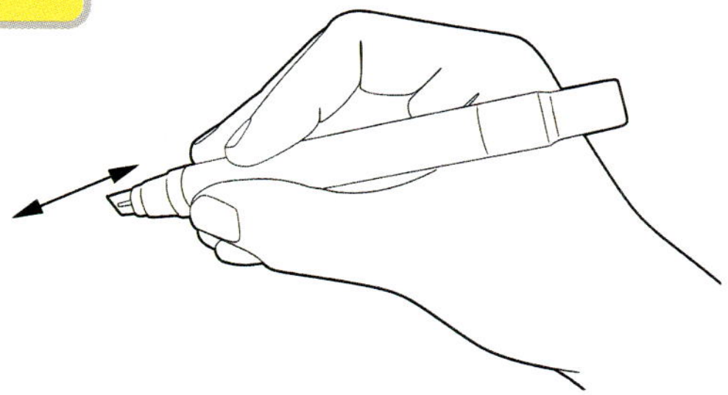

Now, switching over to Cadmium Yellow, block in the reflections. Think of the body of the drill as the bodyside of the car, but with one key difference; the handle portion will pick up light once it swings out from under the body surface.

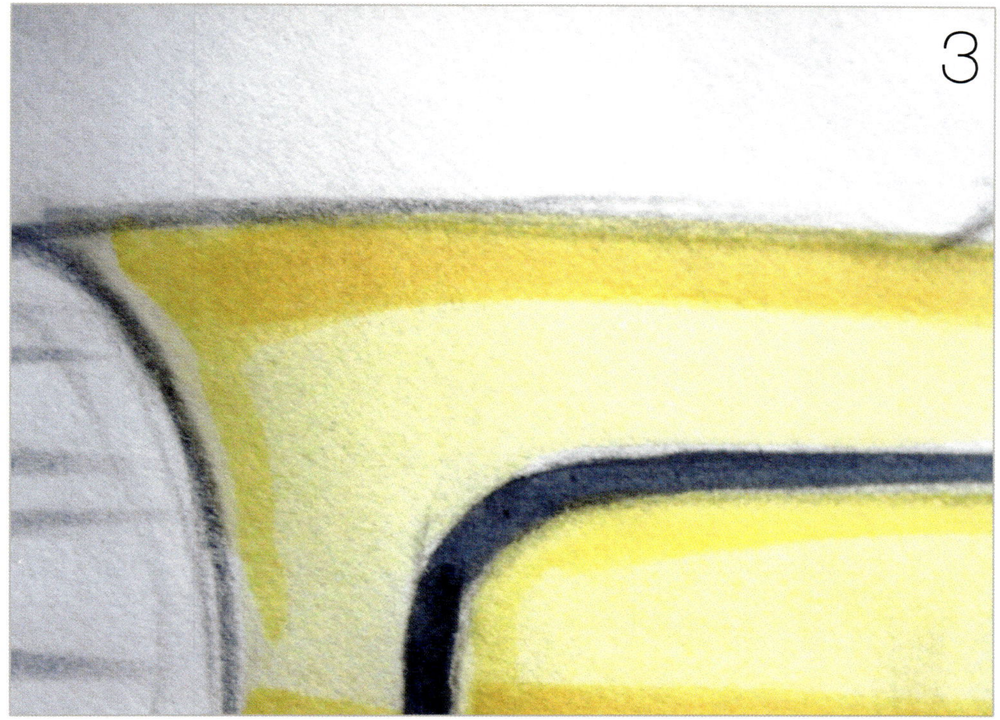

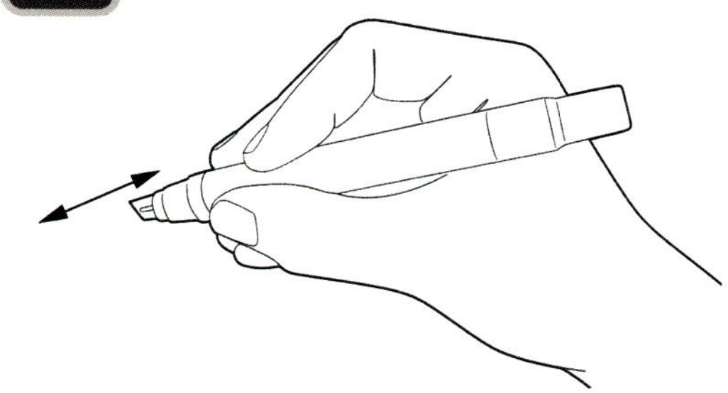

Begin blocking in the matte black portion of the drill with Cool Grey 70%. A steady hand will be necessary because these graphics are both thick and thin. Be sure to leave some paper showing through on the edge of the handle (see step five); this will be a highlight later on.

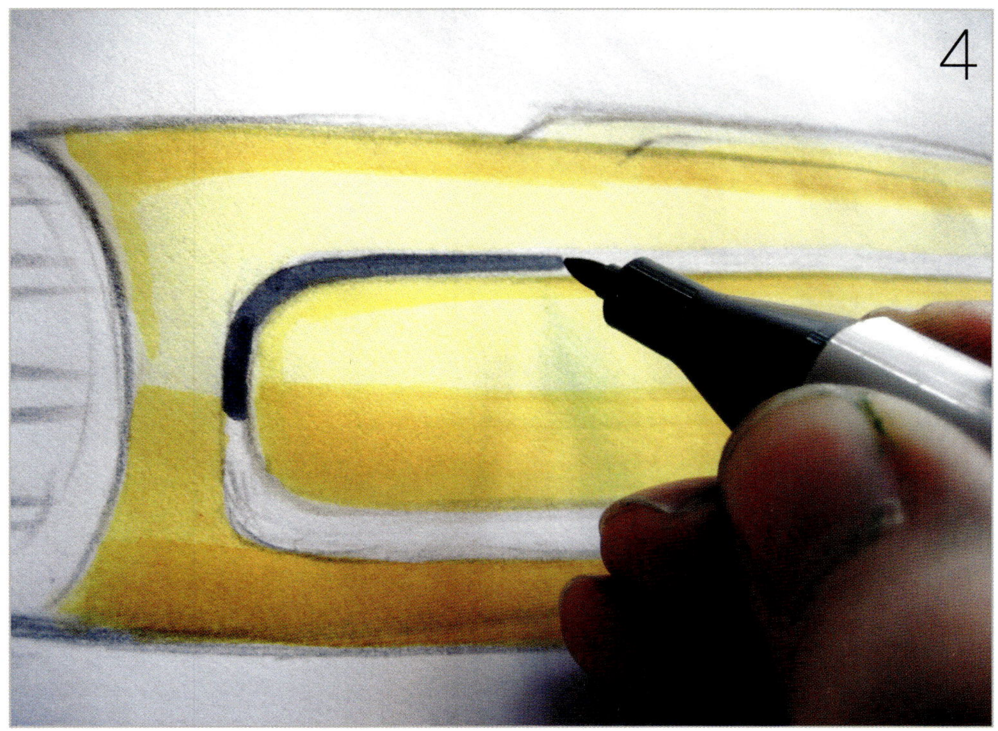

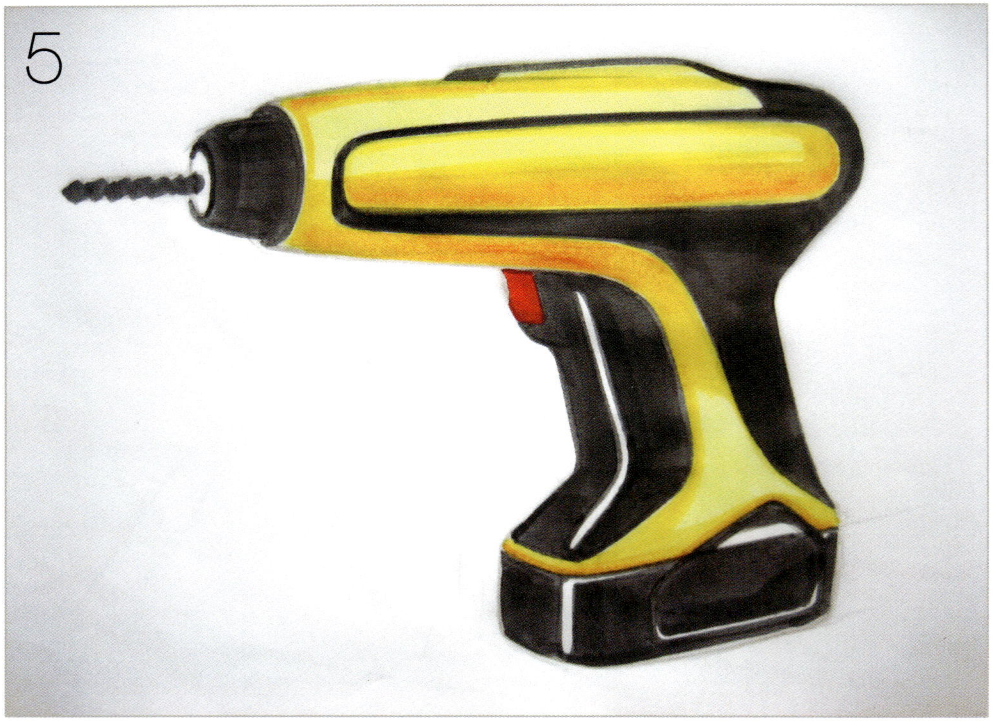

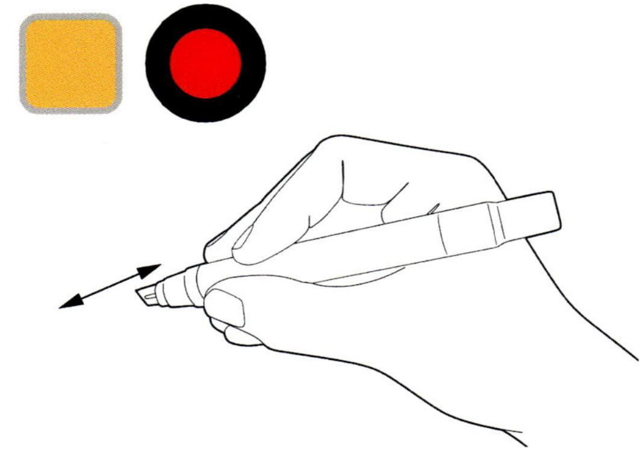

After you've finished blocking in the matte black plastic, punch up the yellow reflections with a few strokes of Chrome Orange. Color the switch red.

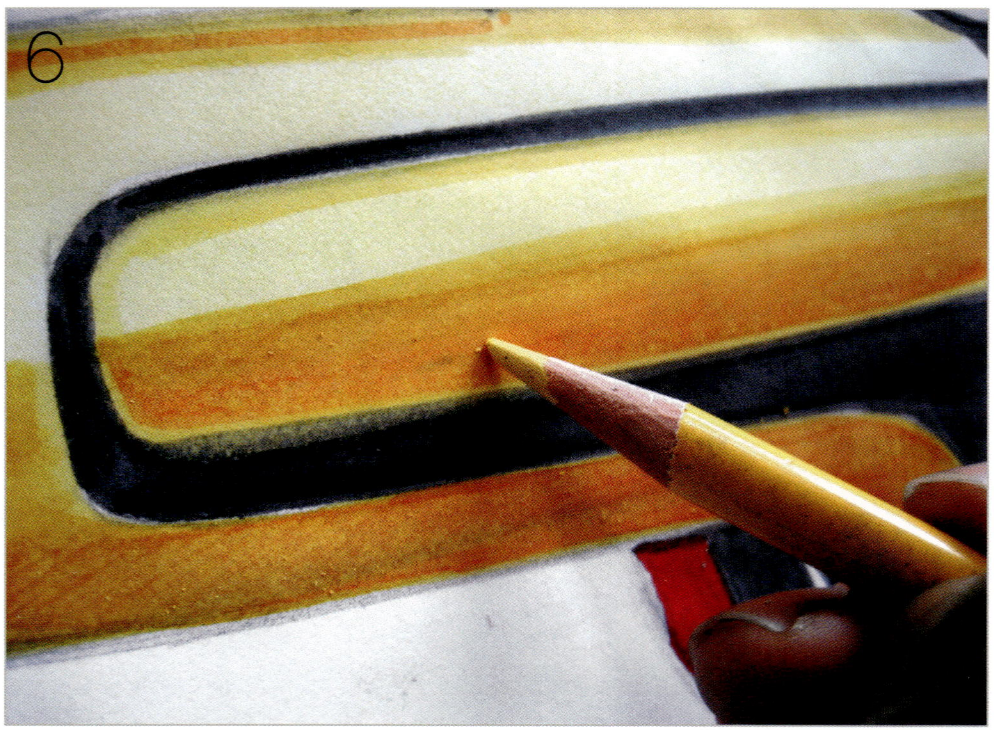

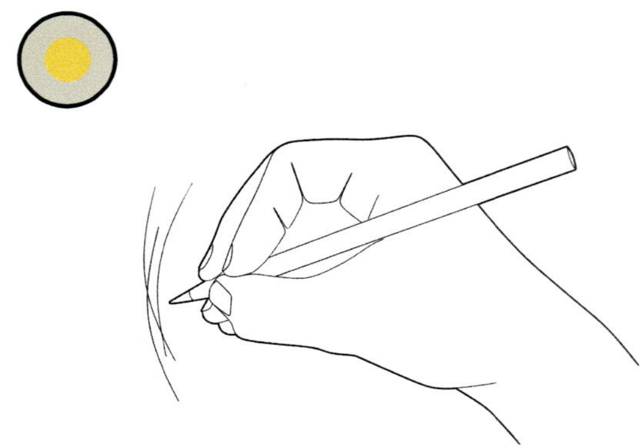

If the orange marker appears too far removed from the yellow of the drill, come back in with a Yellow Orange Prismacolor to balance out the palette.

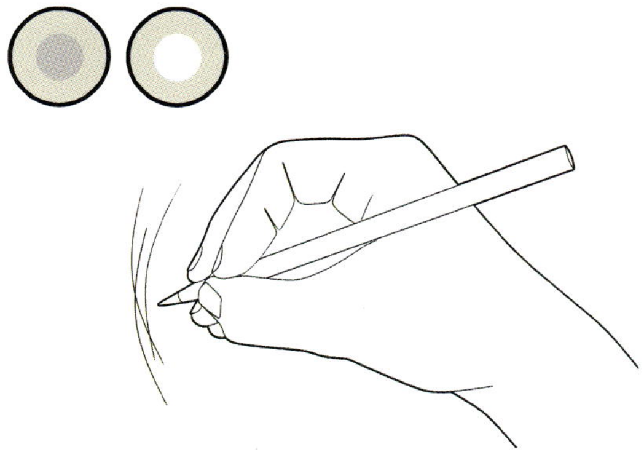

To give the black plastic a semi-matte feel, use Cool Grey 20% around the edges of the highlight. You can supplement this with some white pencil on the very edge of the highlight.

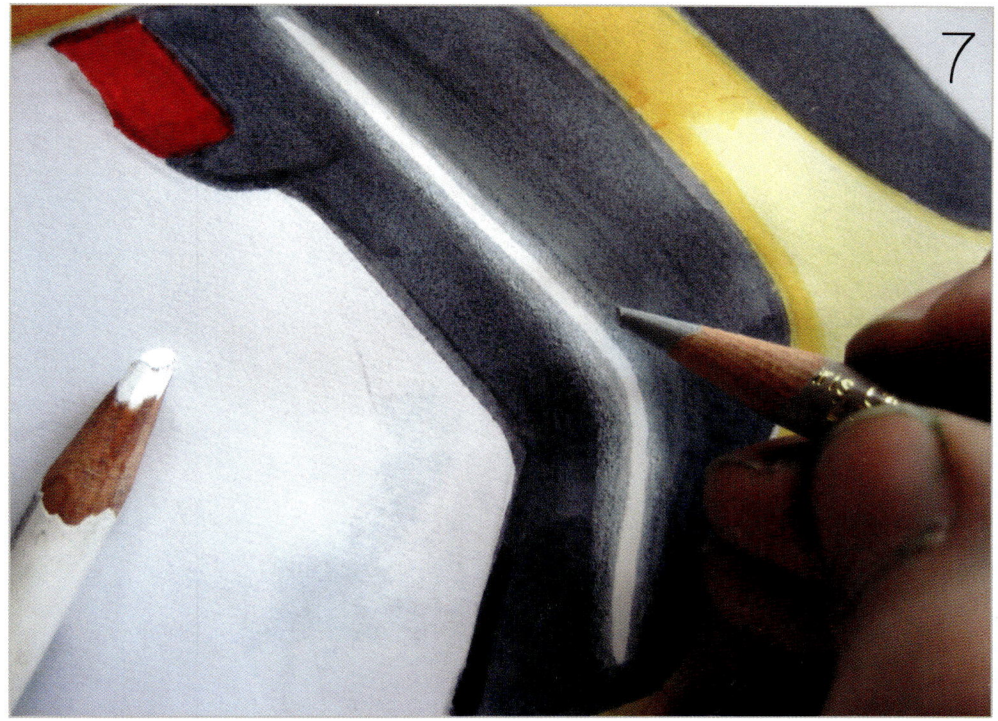

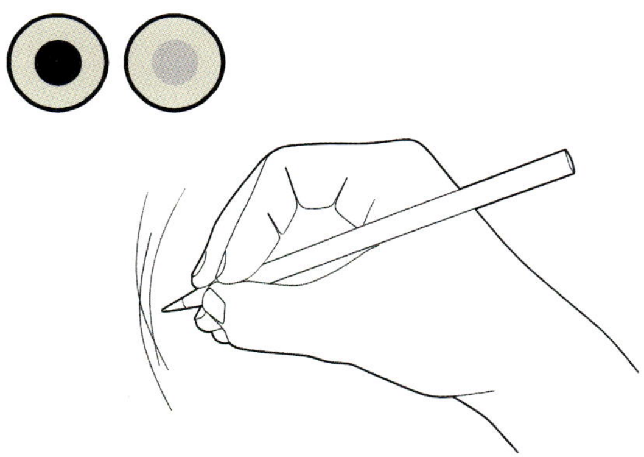

Use black Prismacolor and Cool Grey 20% to define the surface detail of the handle. Resist the urge to use white Prismacolor for the embossed highlights because they will compete with the main highlights for attention.

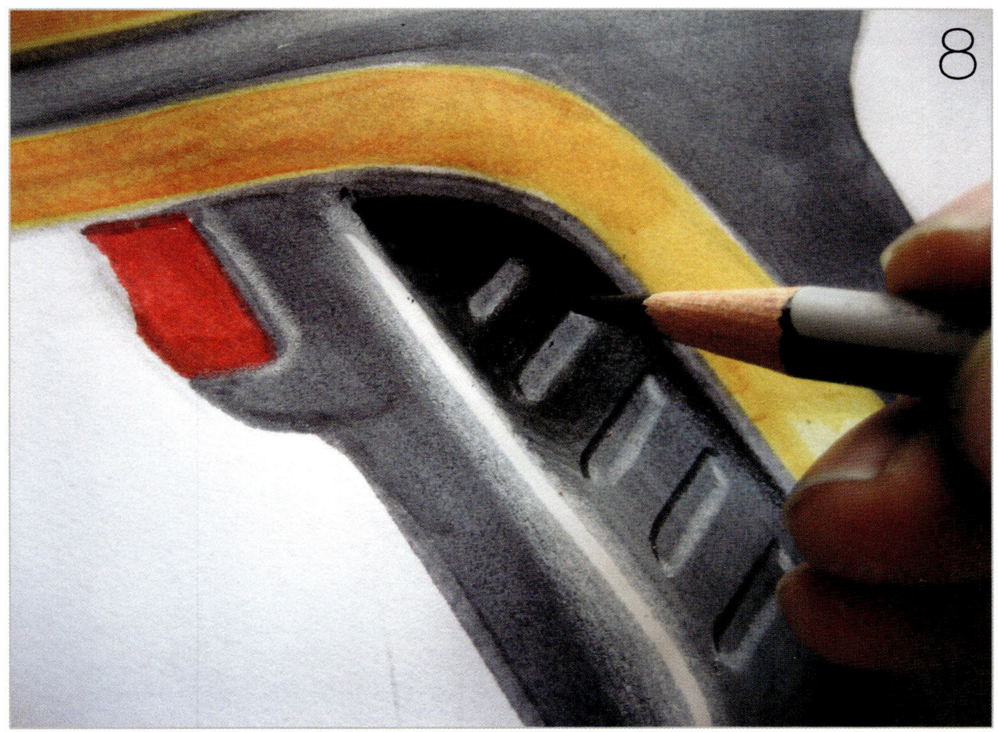

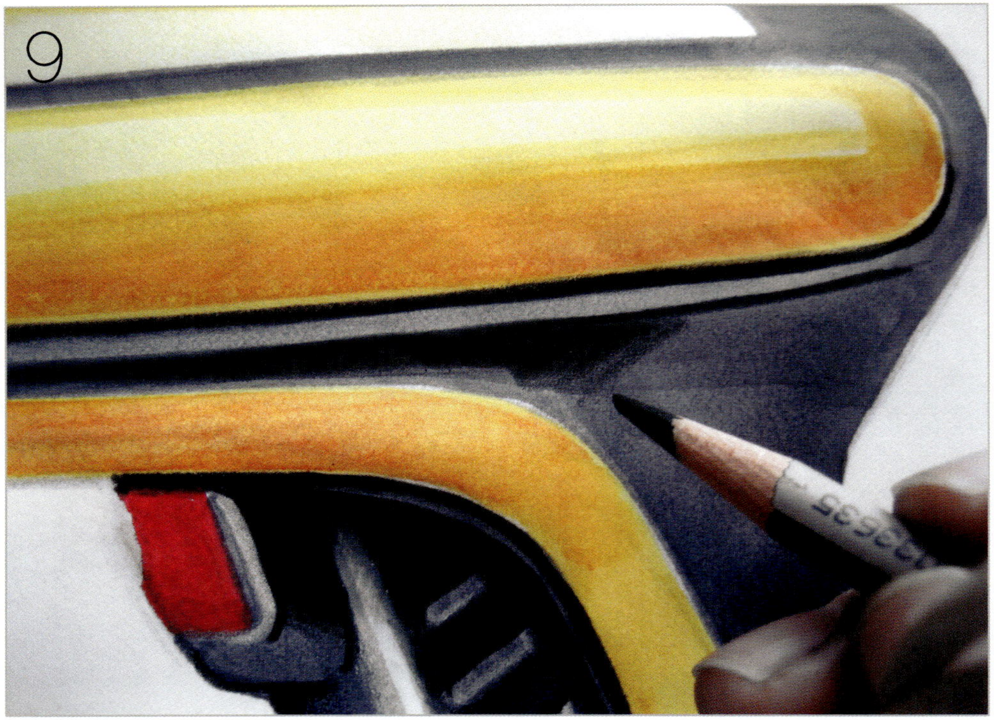

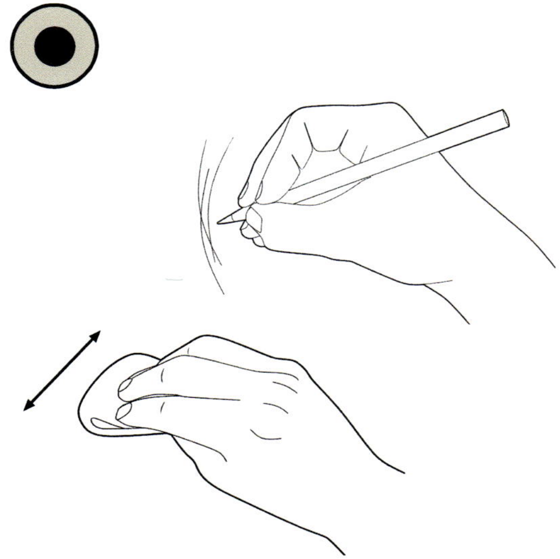

Continue with black Prismacolor along the underside edge of the drill. You can soften the pencil after the fact with a clean Webril pad.

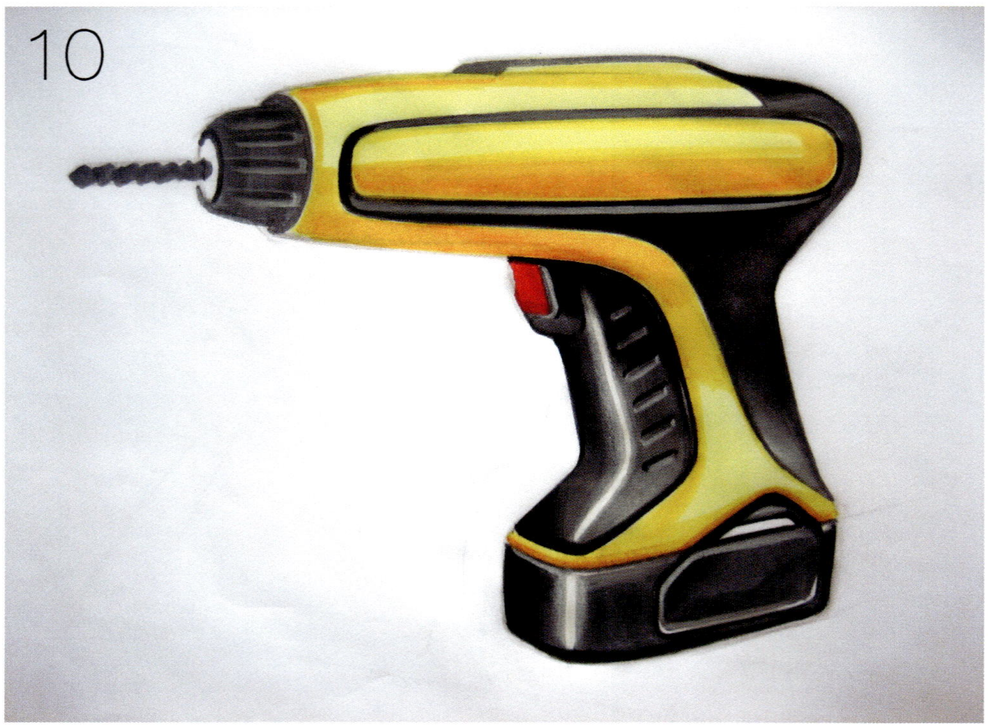

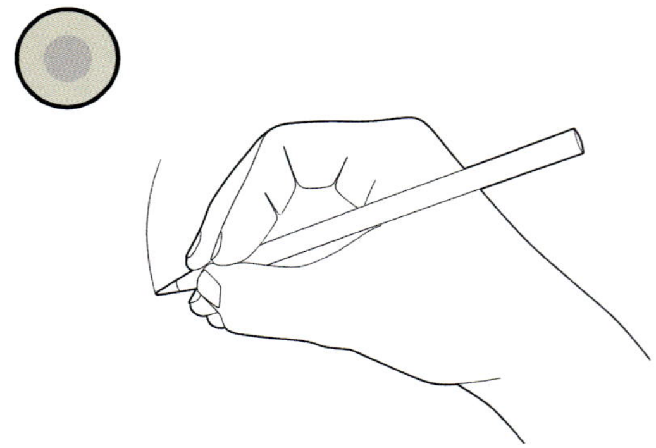

Now go around the rendering with Cool Grey 20% and define the other beveled edges. By sticking with Cool Grey, the main highlight should still be the focal point of the piece.

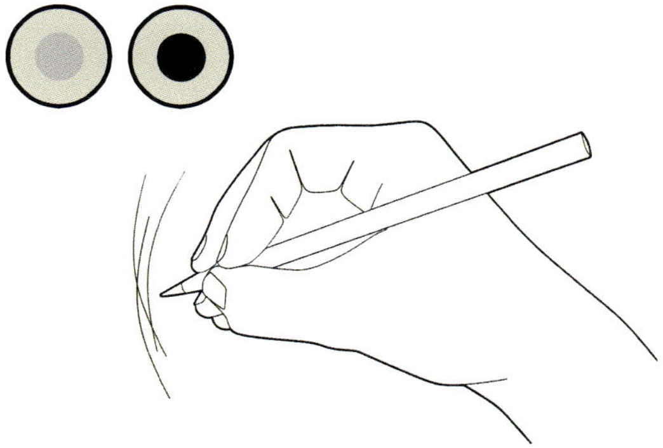

Take a moment to analyze the progression from light to dark across the black plastic. You may need to balance out some areas with either black or Cool Grey 20%.

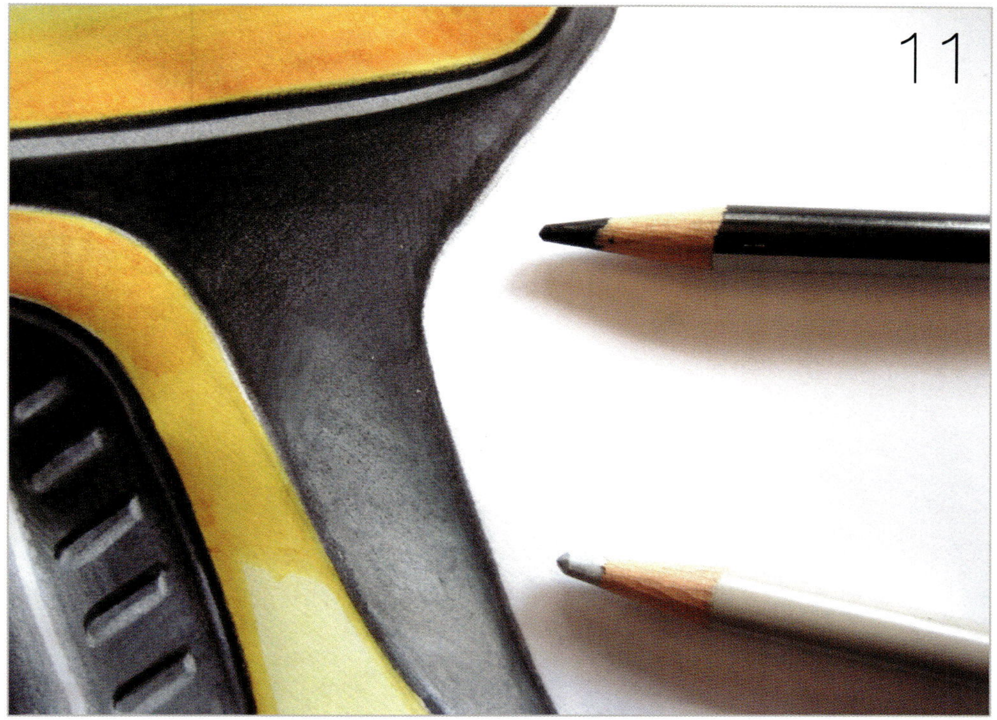

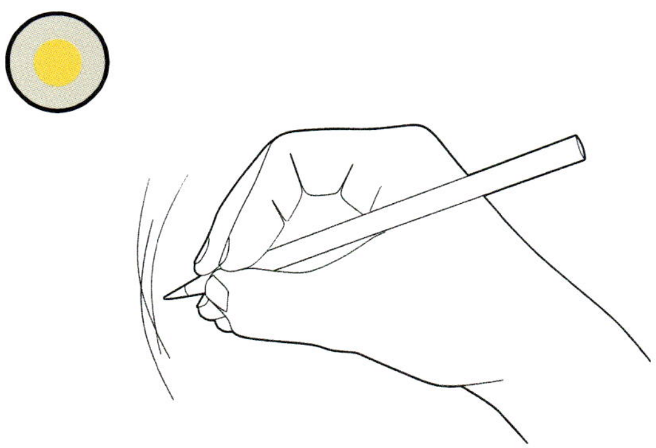

Create some detail reflections on the body portion of the drill with a Yellow Orange Prismacolor pencil.

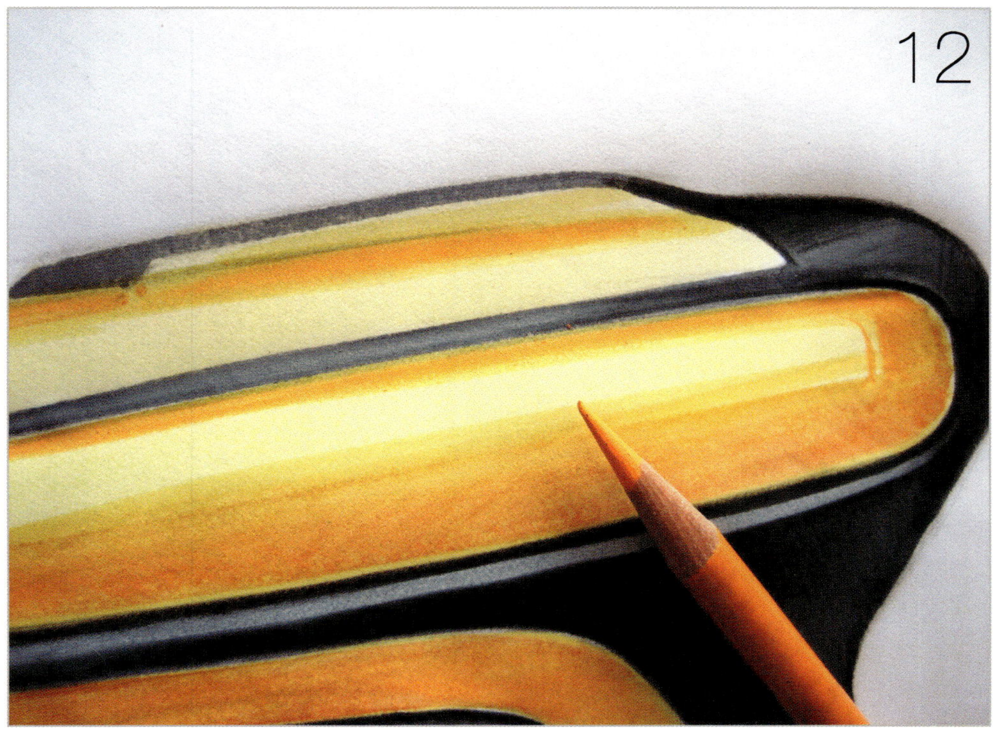

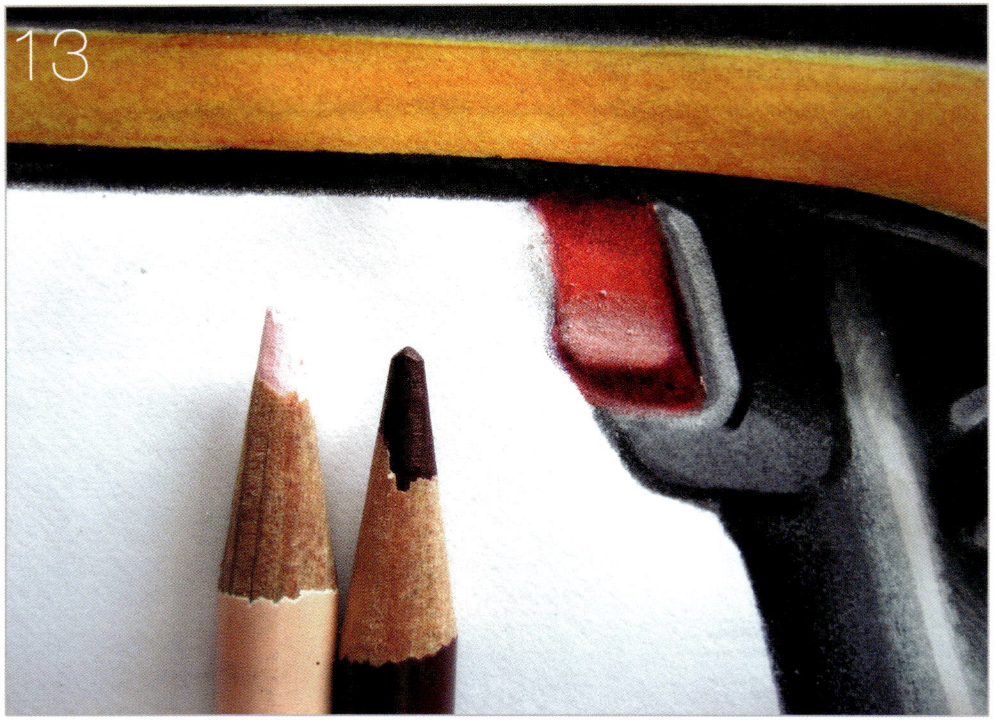
13

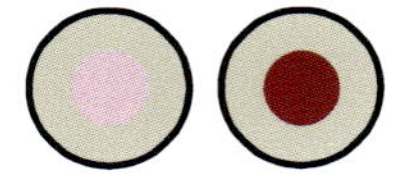

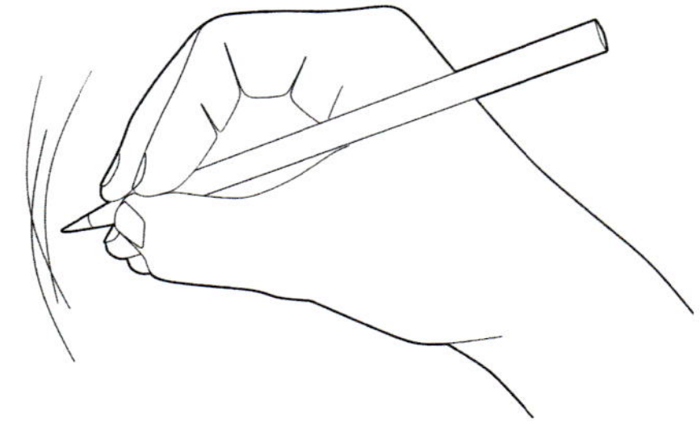

Using Deco Pink and Mahogany Red Prismacolor pencils, give the switch some simple definition. Employing these colors is a far more sophisticated choice than plain black and white.

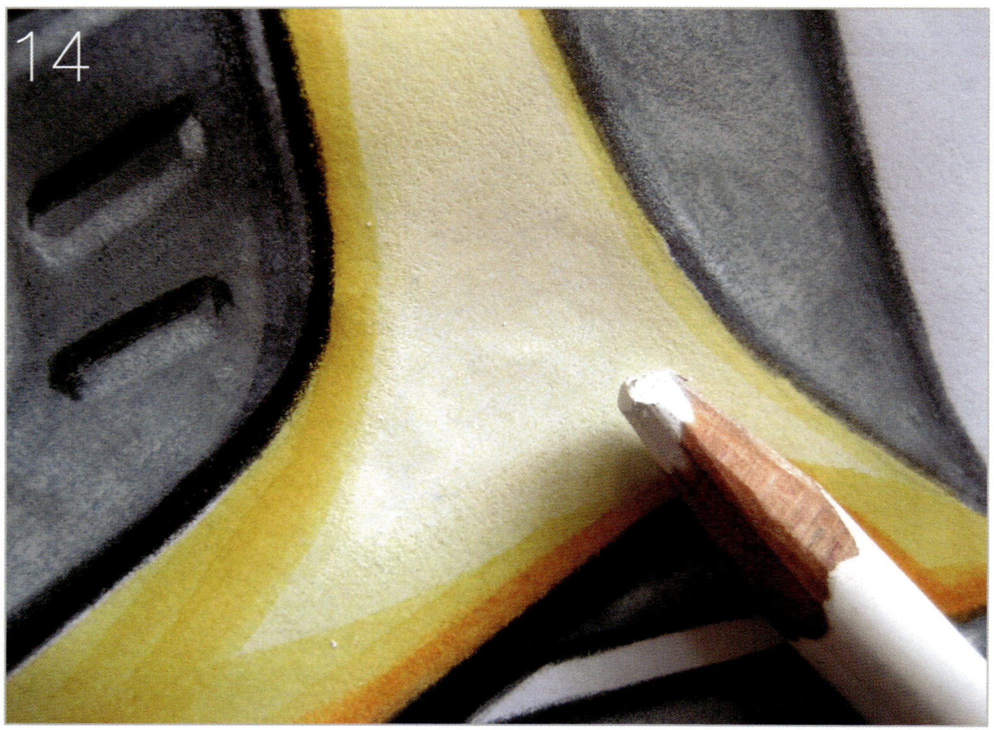
14

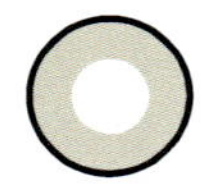

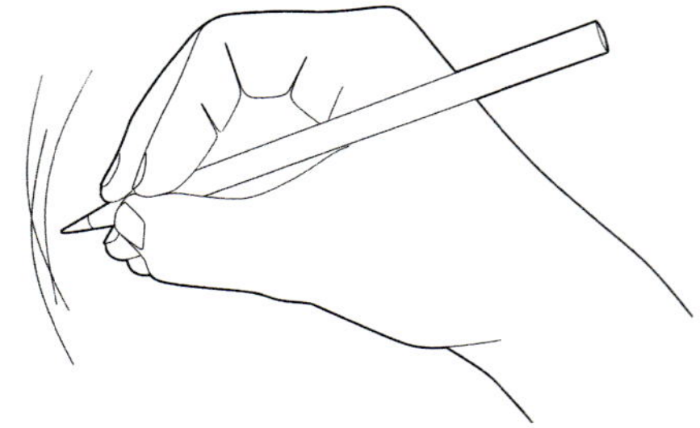

Apply some white Prismacolor pencil in the lightest parts of the drill's yellow body. Avoid pressing too hard with the white pencil, as this might cause the page to buckle.

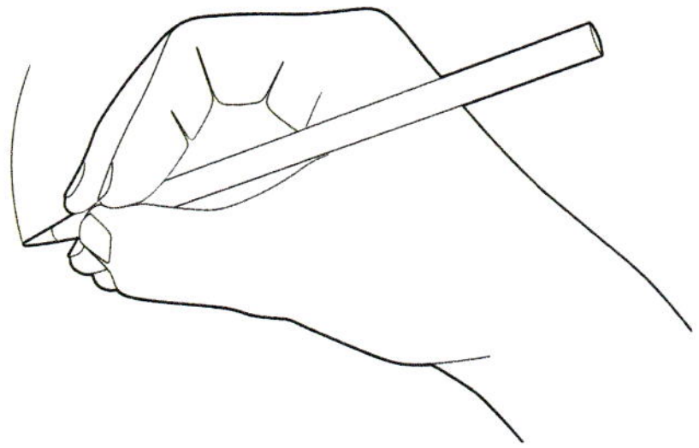

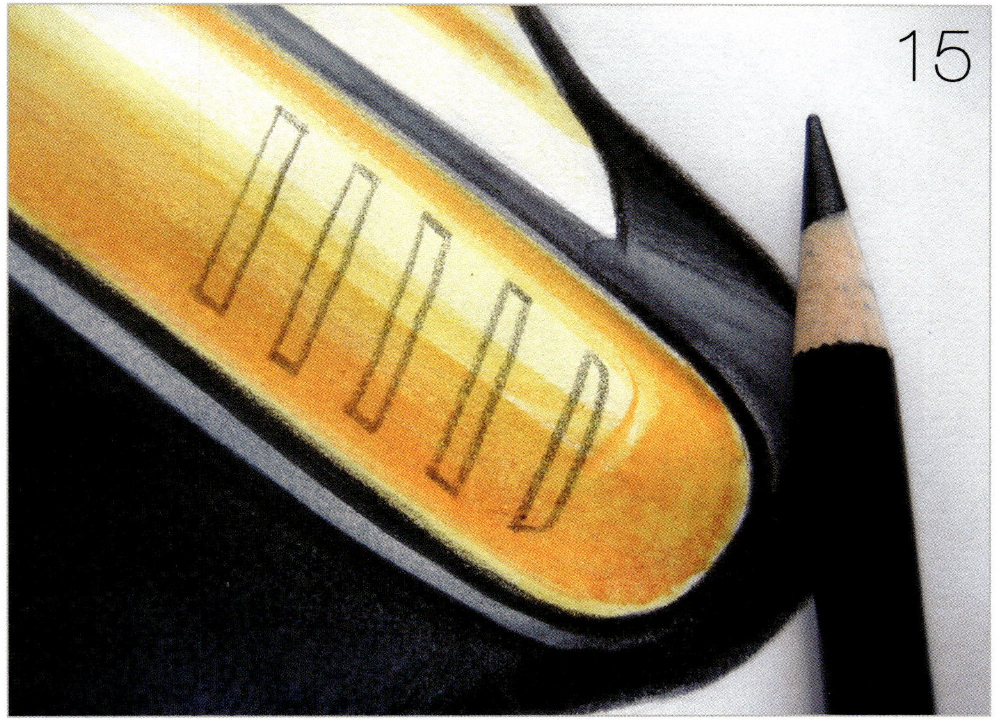

Use black Prismacolor to define the vents on the side of the drill.

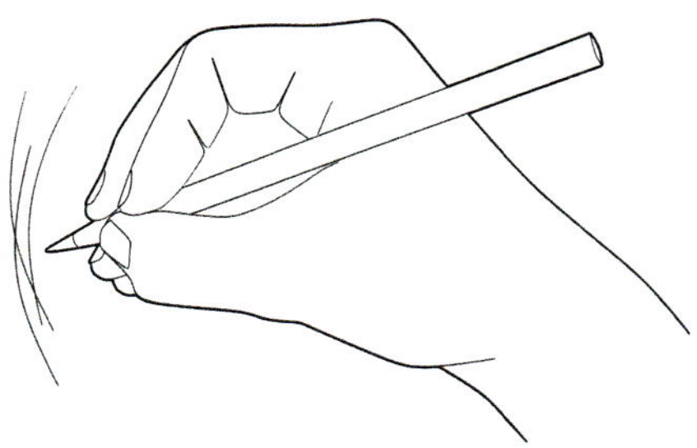

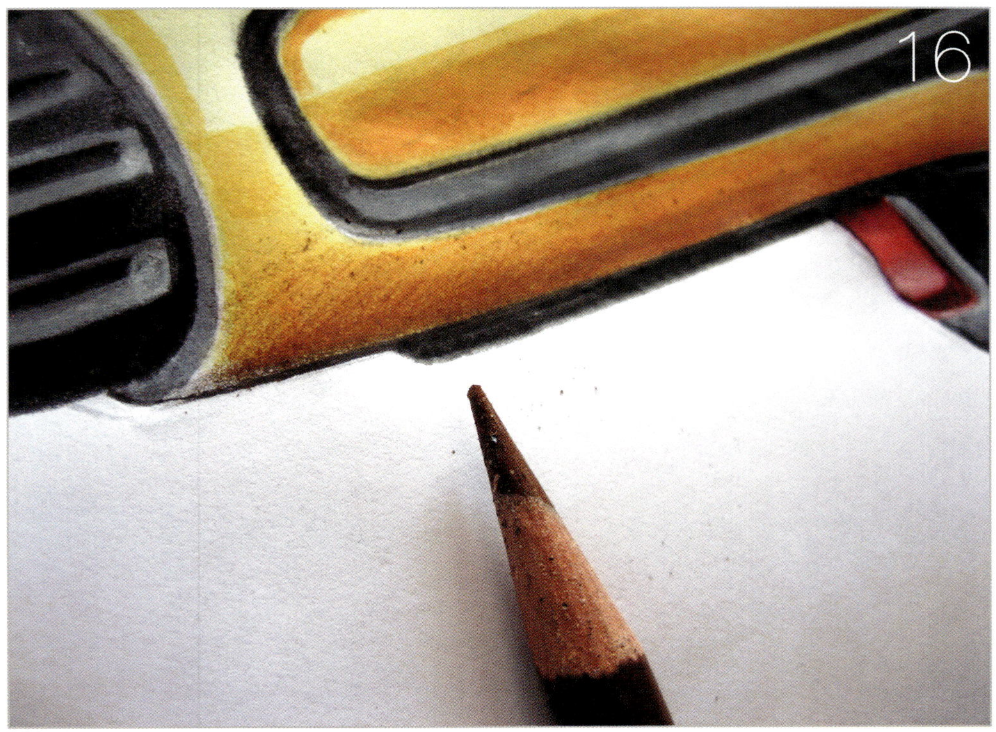

Try applying just a hint of Sienna Brown to the underside of the drill body. This will describe the shape even further.

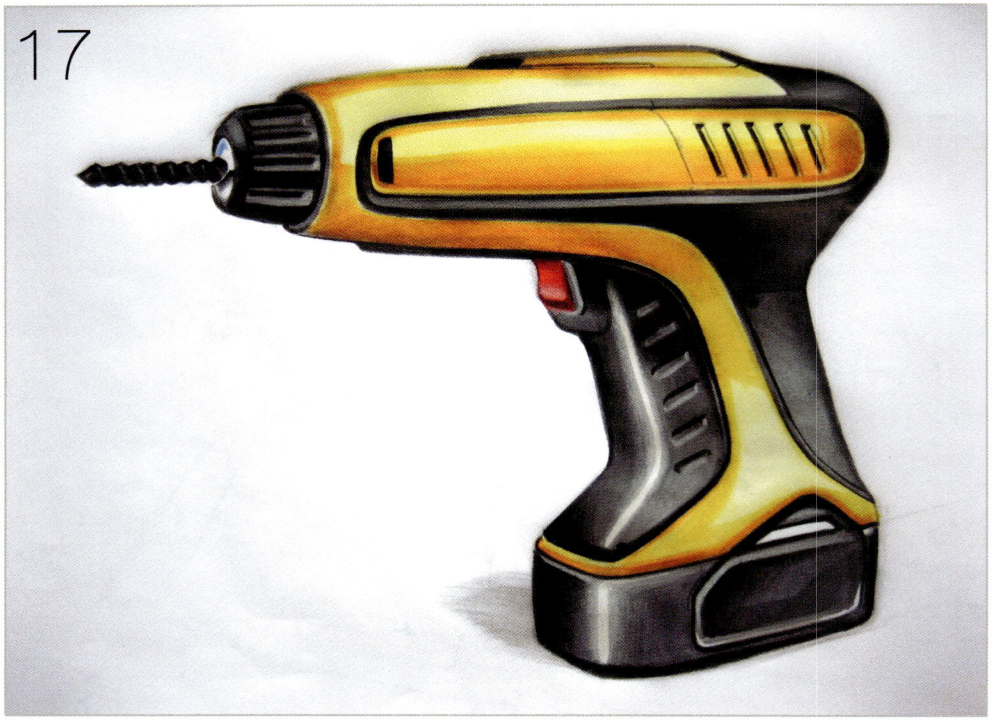

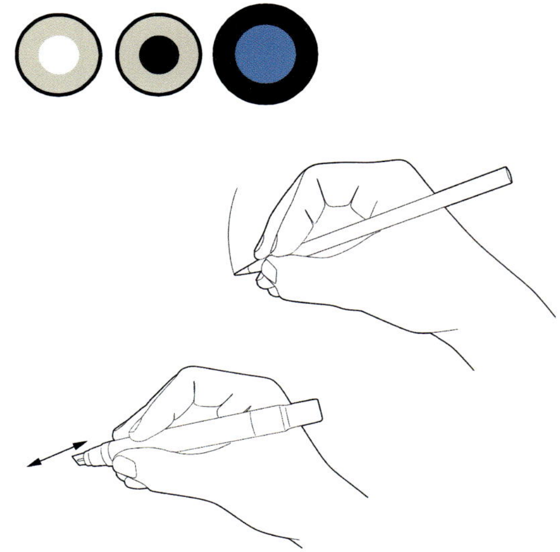

In the late stages, you can take time to add some smaller details -- like blue sky tone in the chrome drill hub and various black bolts around the edges.

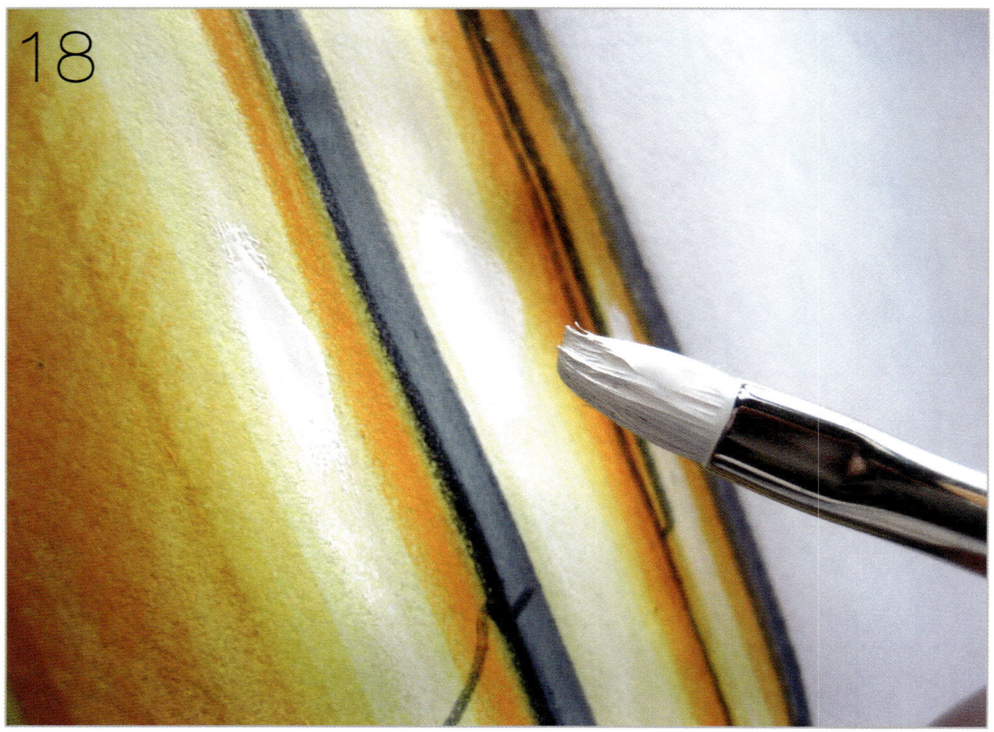

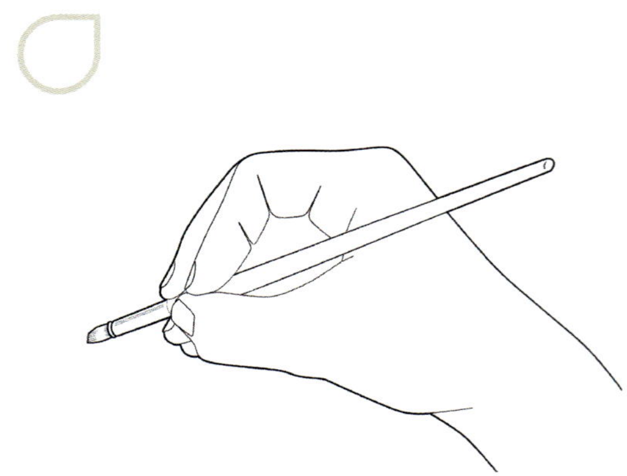

Begin applying white gouache in limited quantities. Try using your fingertip to soften the edges of the highlights.

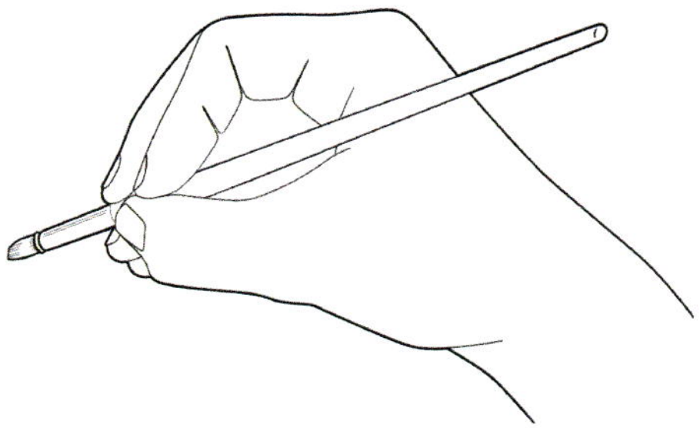

After applying the main highlights, turn your attention to the smaller ones, like those on the lower edges of the vents.

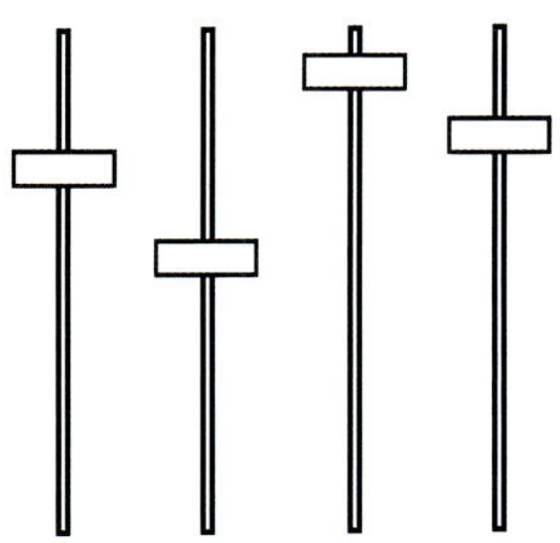

Finally, scan in the rendering and clean up the background in Photoshop. You can also increase the color saturation to make the yellow more vibrant.

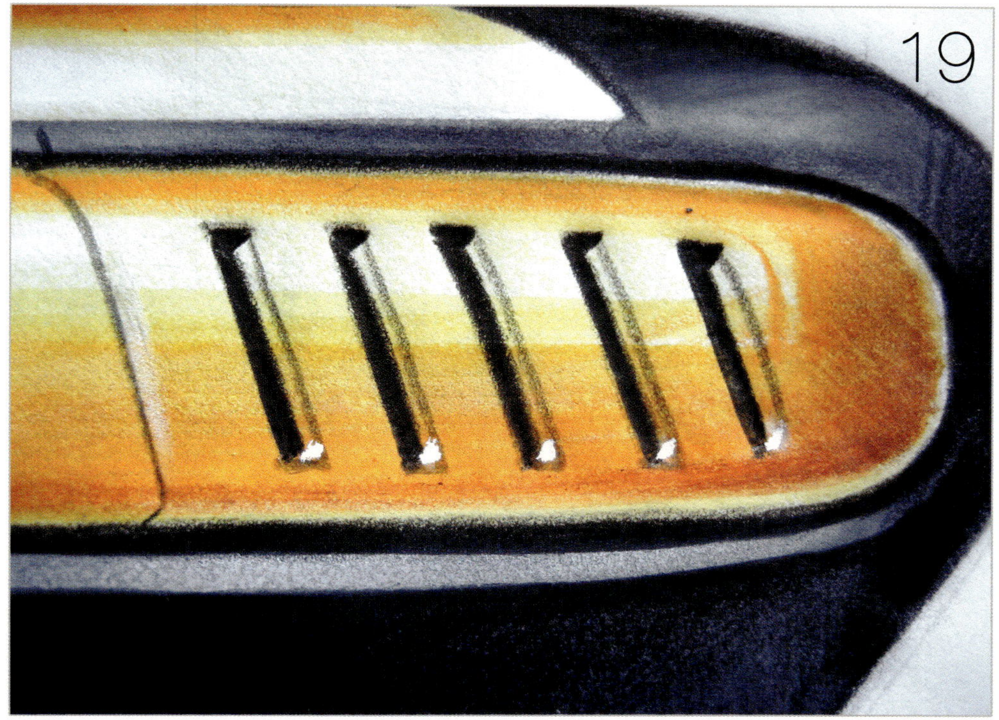

19

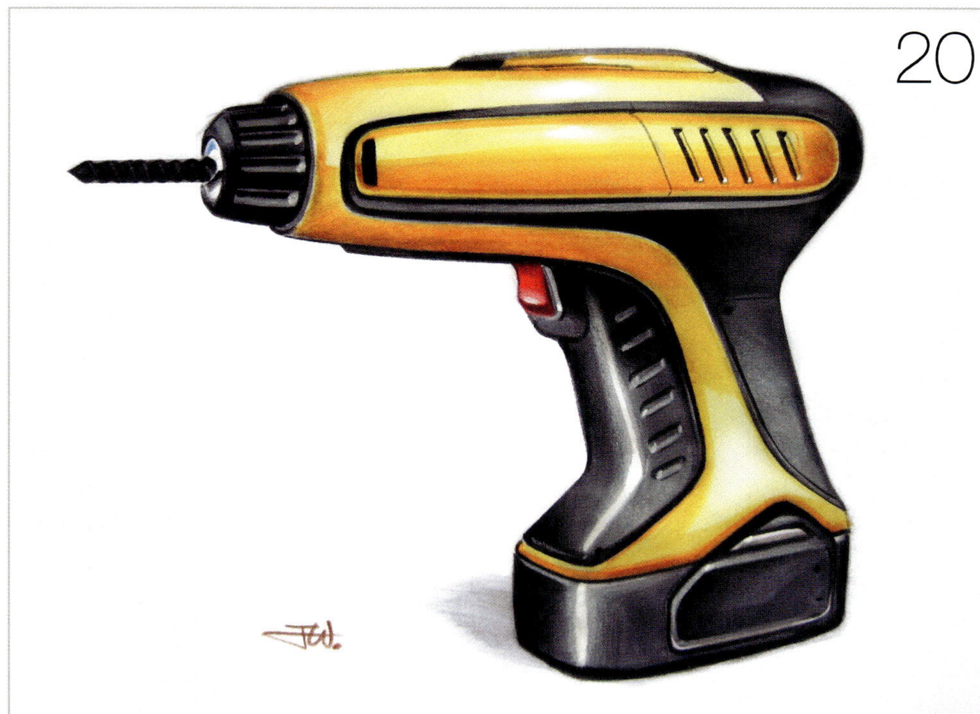

20

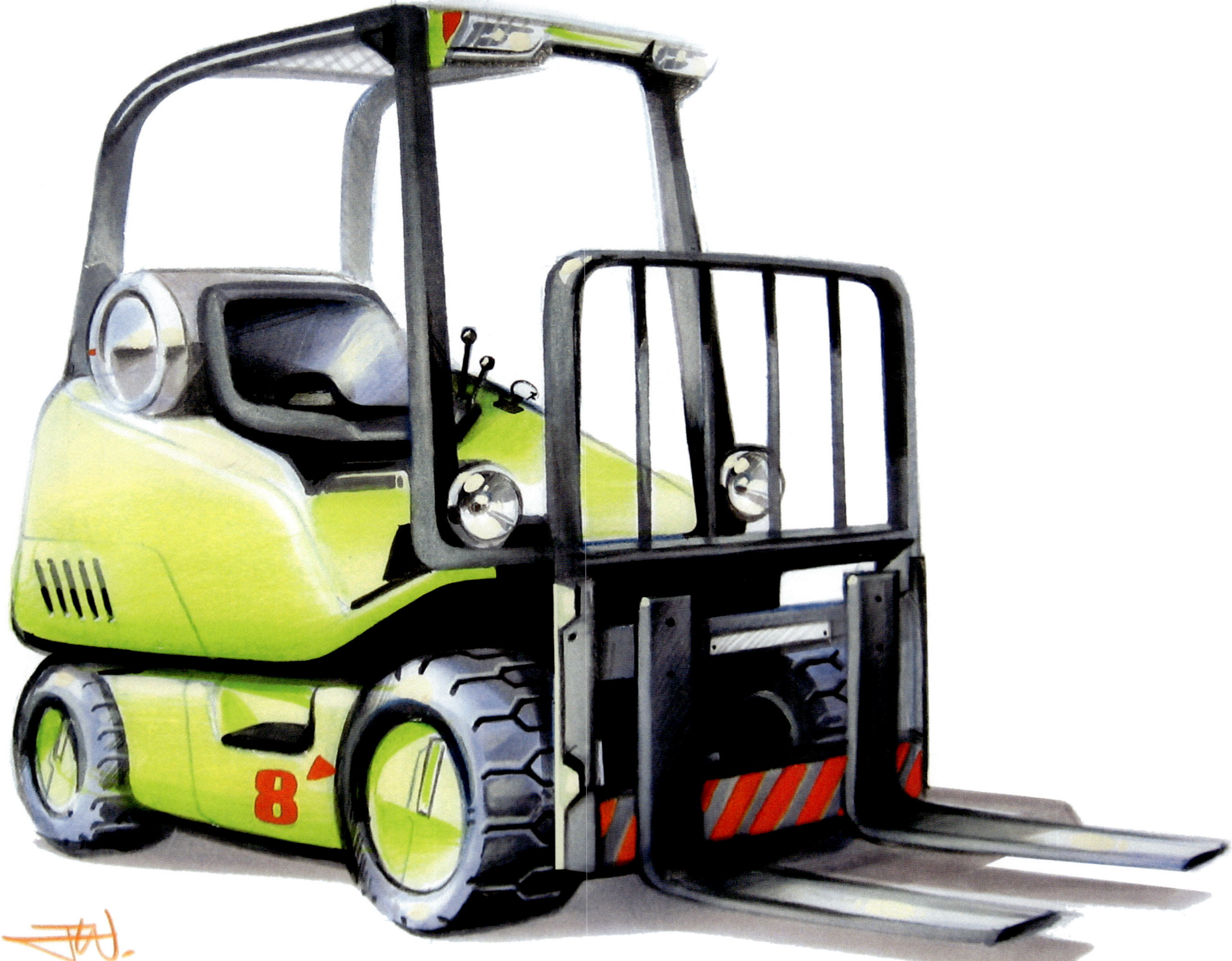

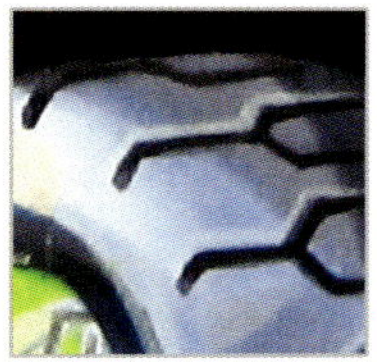

4. Multiple Materials

Industrial equipment can be a challenge to render because of the multiple materials involved. In the case of this forklift, we have the Day-Glo yellow body, rubber tires, forged steel bars, a bright metal propane tank and some flashes of safety red. Distinguishing one from the other is not as simple as it looks.

The Tools

Prismacolors:
White, Yellow Chartreuse, Black.

Verithins:
Black,
Indigo Blue.

Gouache:
White, Black, Red.

Prismacolor Markers: Cool & Warm Grey 40%, Blue Slate.
Copic Markers: Pale Yellow, Cool Grey 70%, Black.

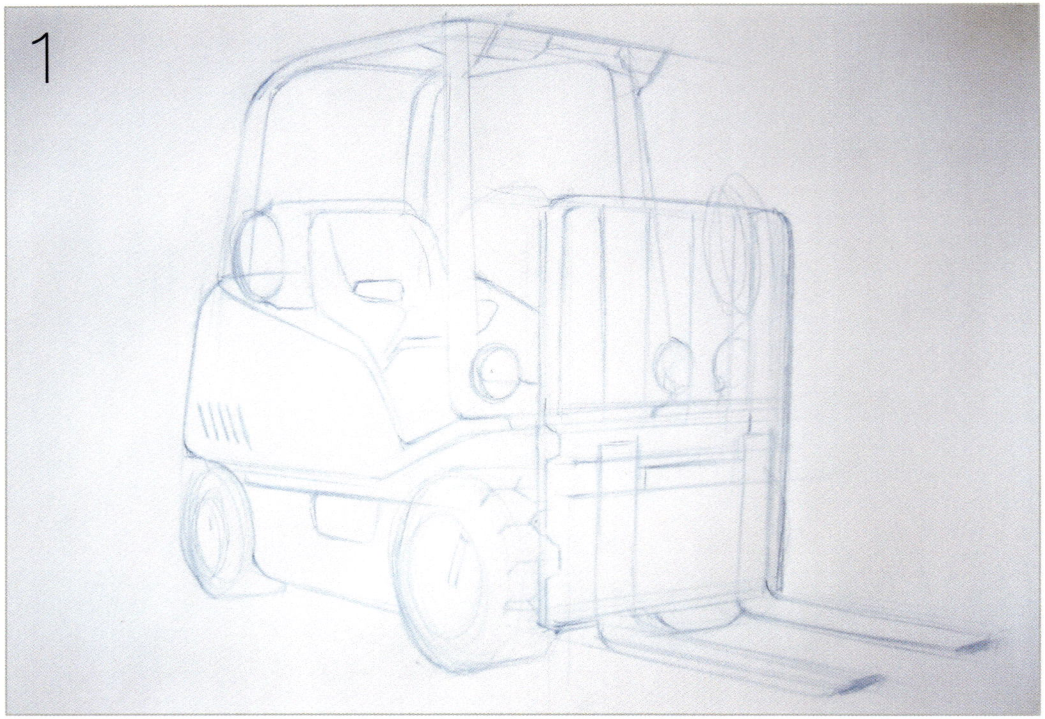

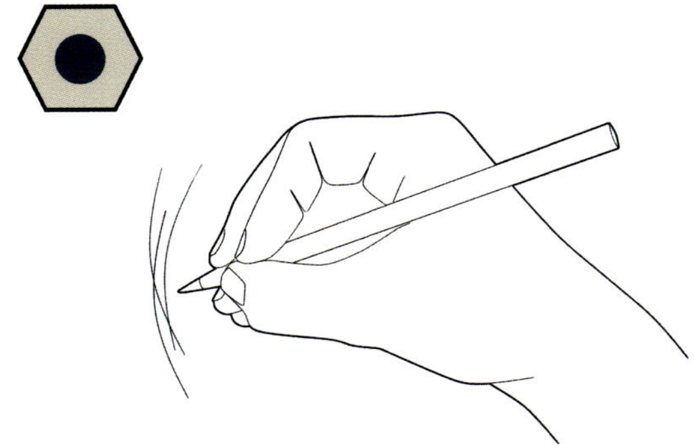

Normally, pairing blue pencils with yellow markers would be risky. But since this is a yellow green forklift, the blue pencil will help bring out the green.

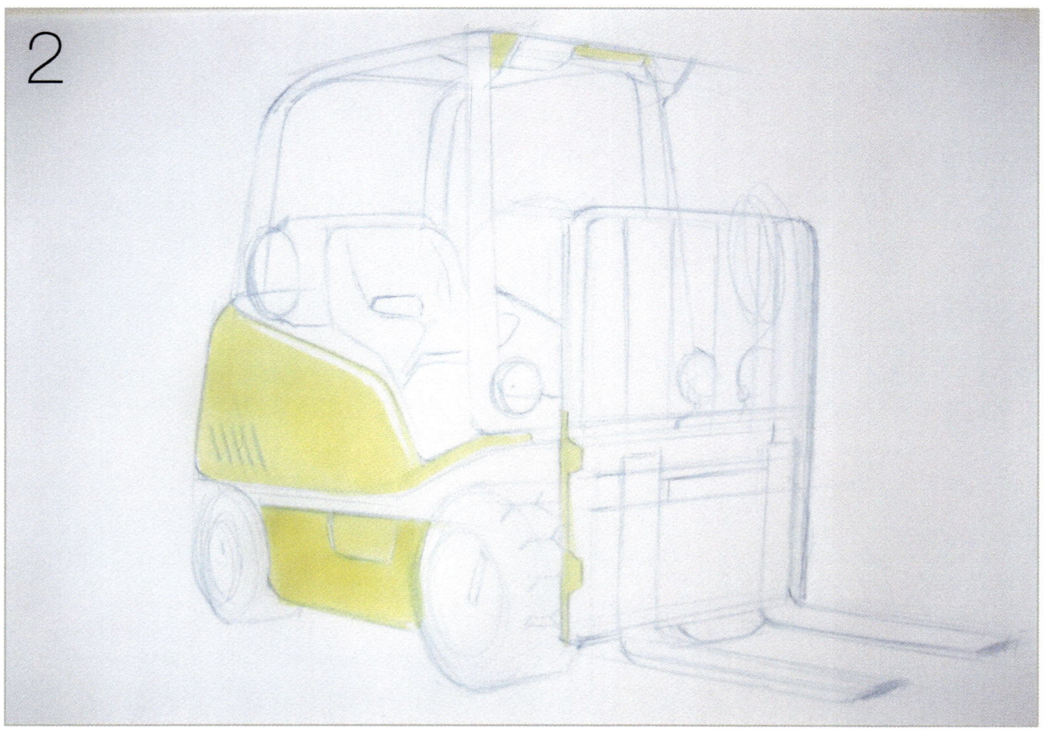

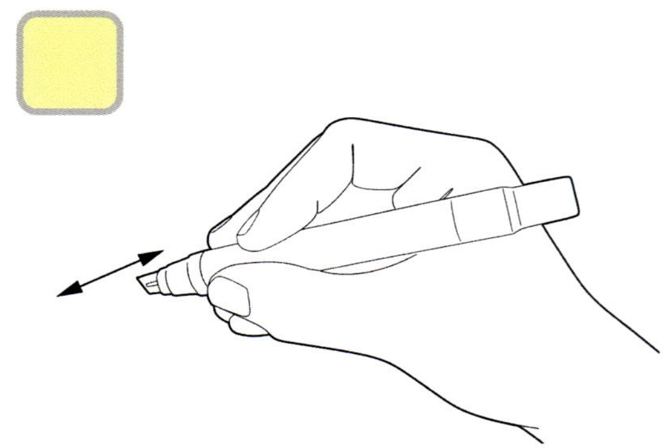

Start by laying in Pale Yellow as a base color. Later on, you can use yellow green pencil to get the tint just right.

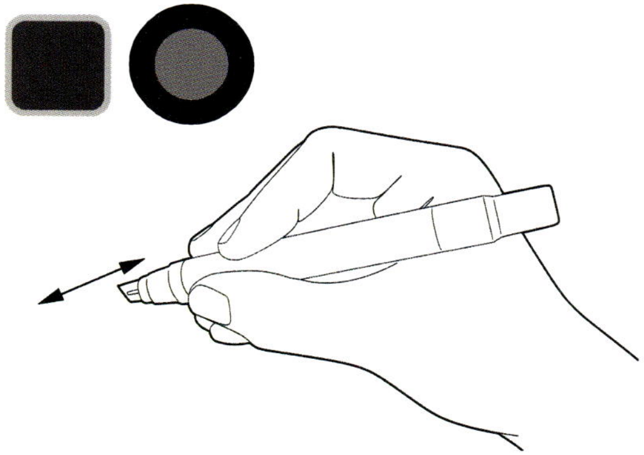

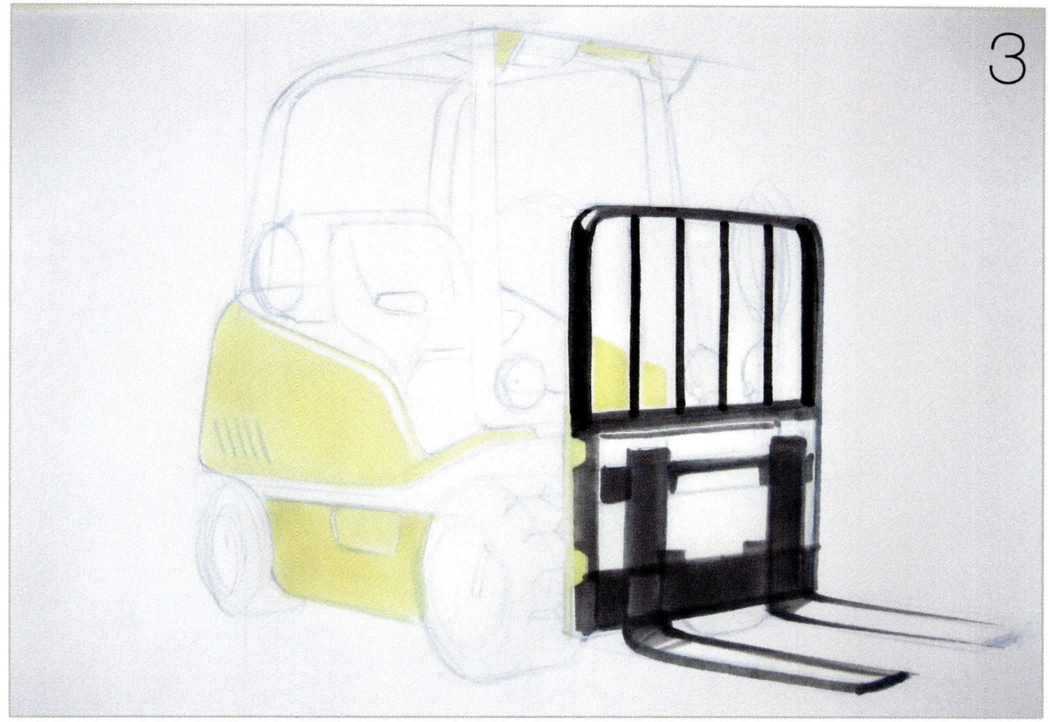

Block in the lift portion with Cool Grey 40% and 70%. As with previous renderings, leave paper showing through on the bits facing the light source so the highlights will be easier to execute later.

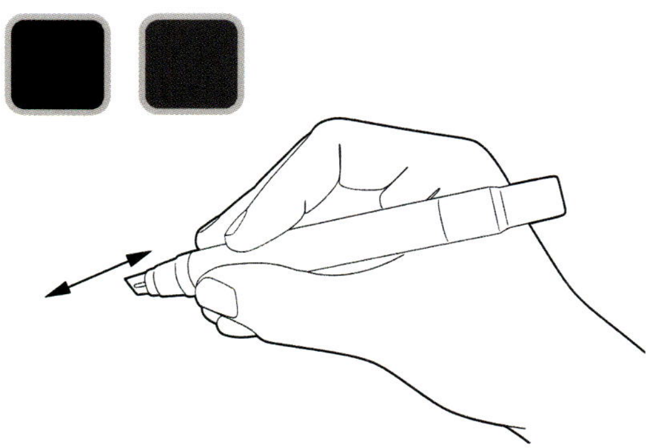

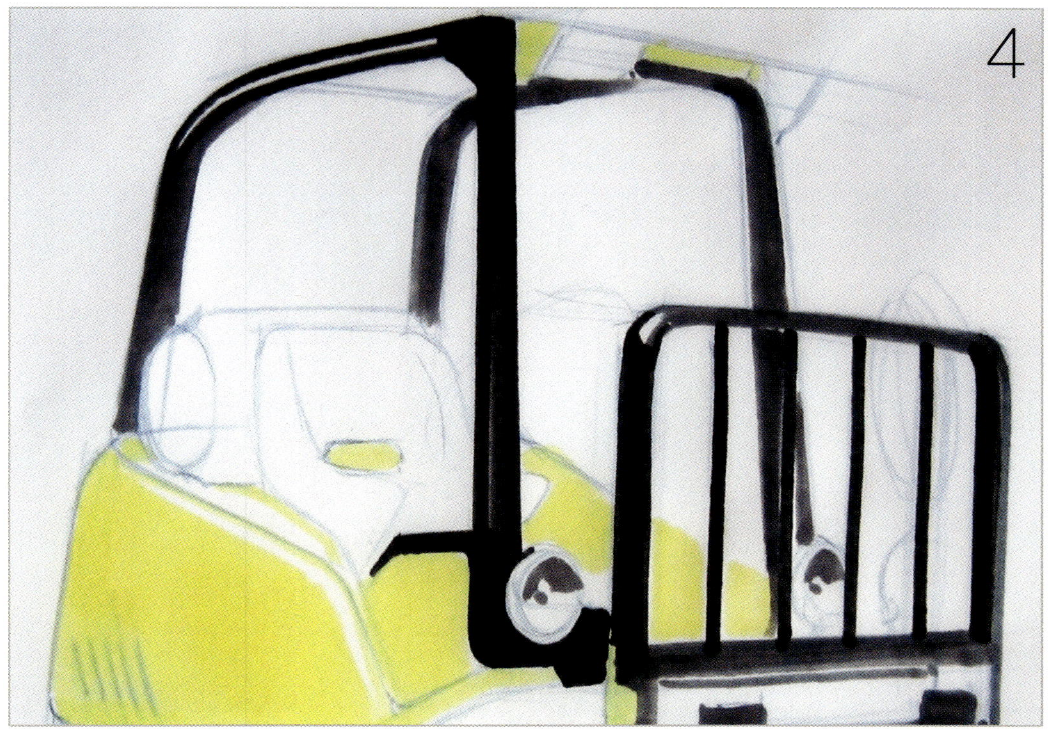

Now render the roll cage slightly darker to help distinguish it from the lift.

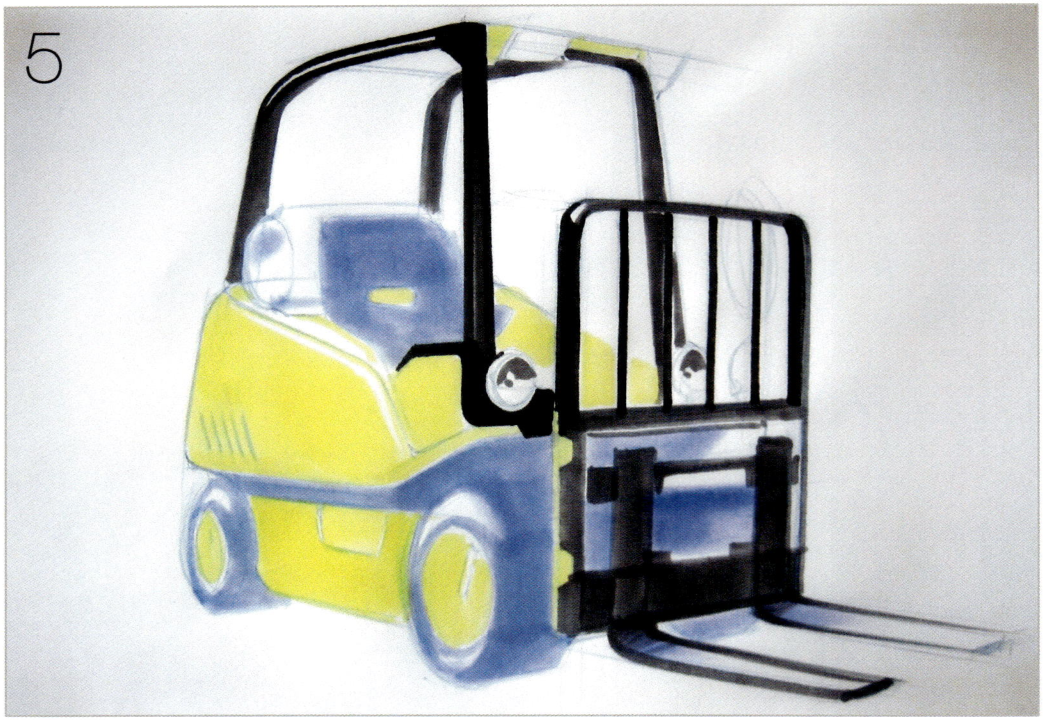

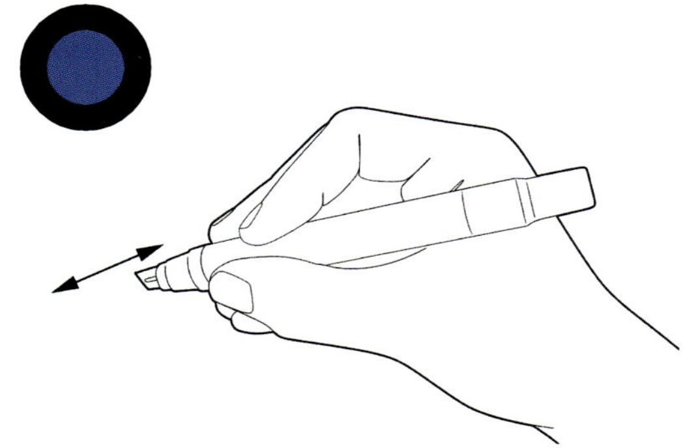

Lay in all the rubber bits with a Blue Slate marker. By using this as a base color, it will be far easier for the viewer to distinguish metal from rubber.

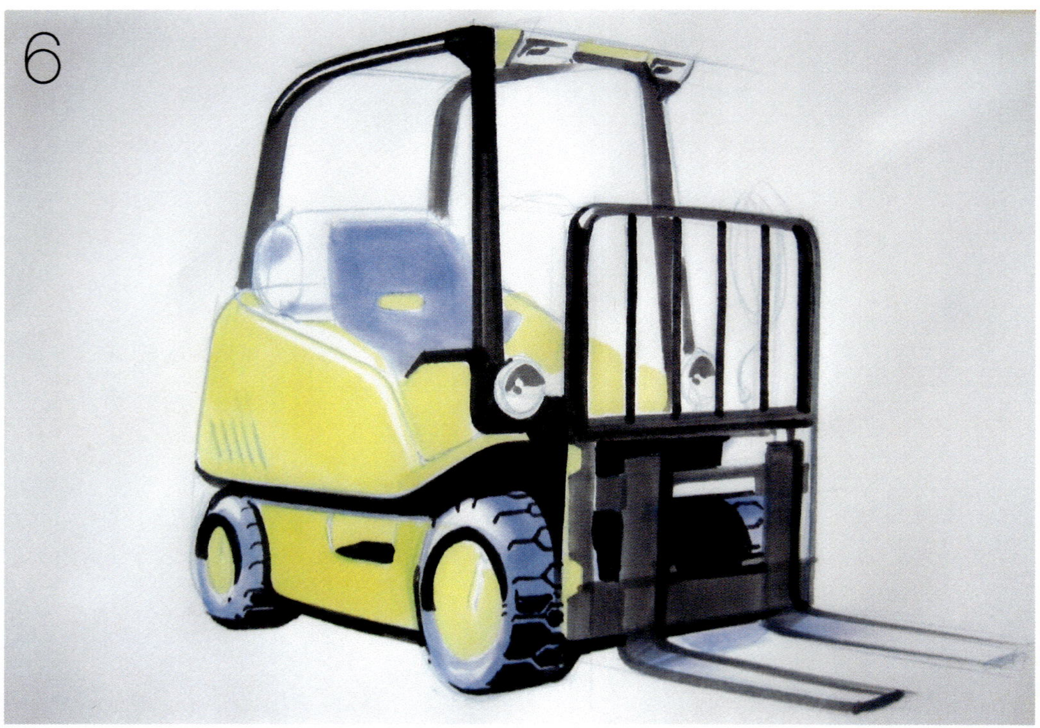

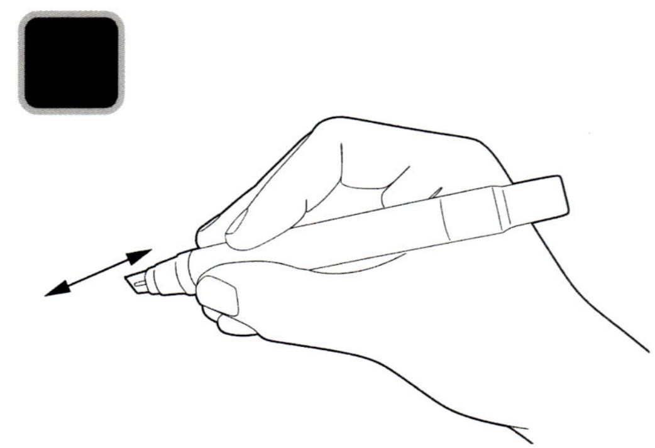

For the moment, treat the tires and the shadows as if the contrast is turned all the way up and block in the darkest darks.

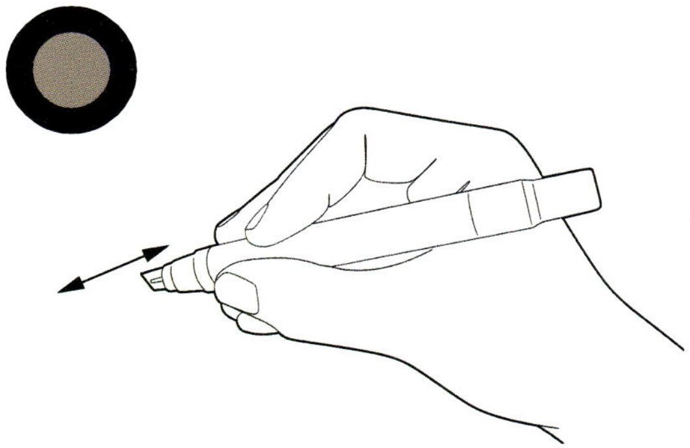

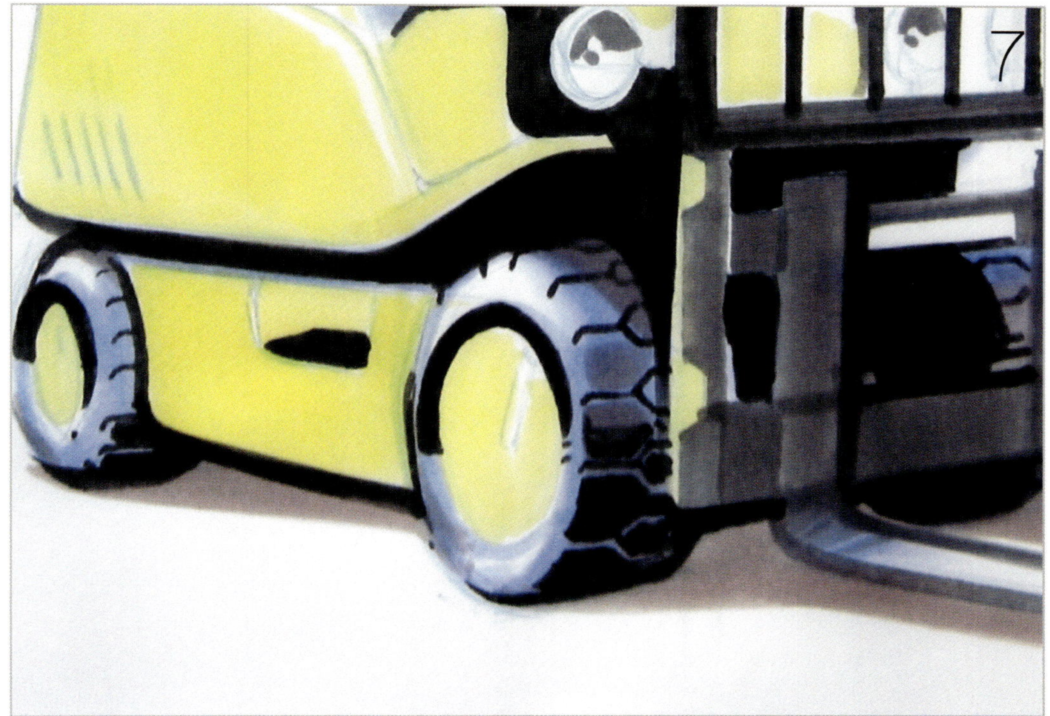

For the floor shadow, use a Warm Grey marker instead of a Cool Grey. Try sneaking a bit into the tire treads as mid-tone; this will add another dimension to the tire color and serve to link the shadow back to the subject.

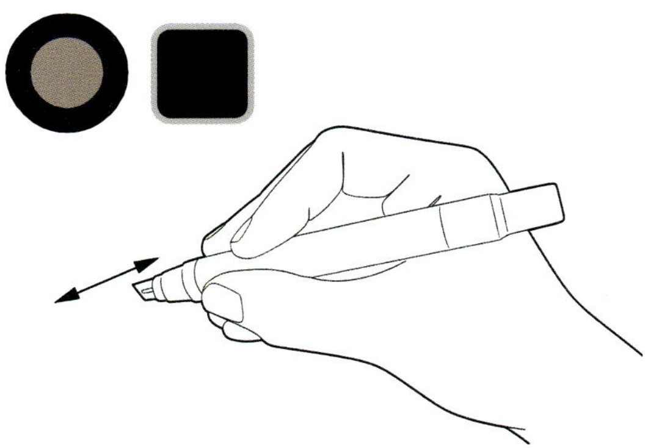

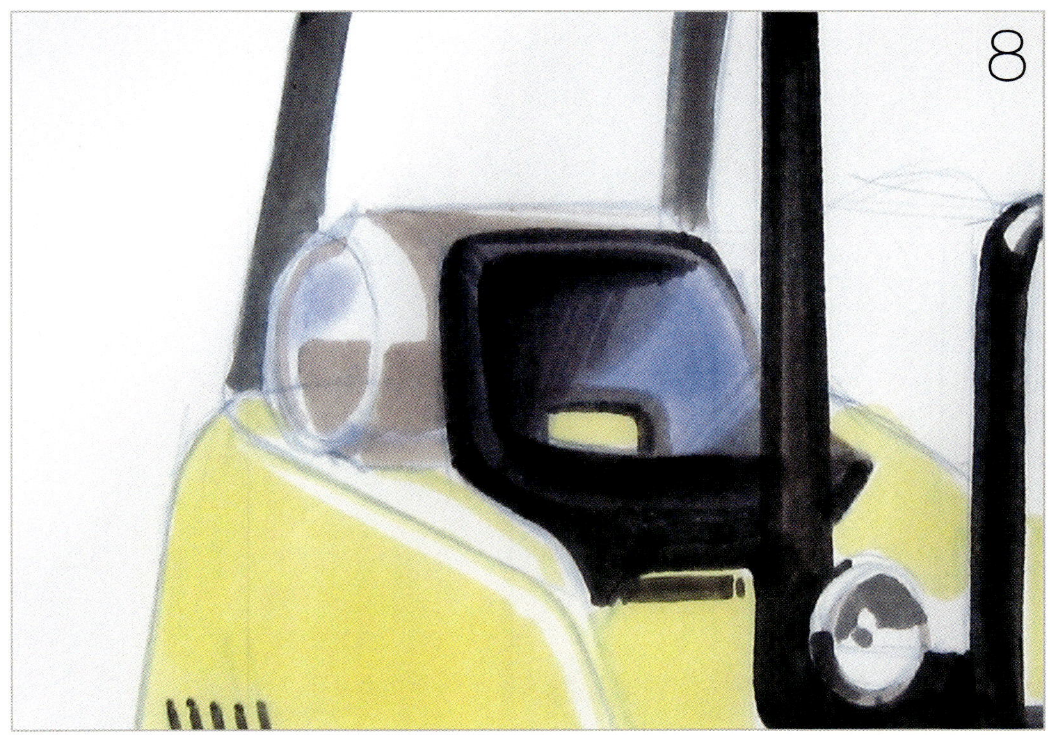

Continuing with the Warm Grey, treat the propane canister as a chrome cylinder and give it a sharp, flat horizon. Use both Warm Grey and black markers to define the darker parts of the seat.

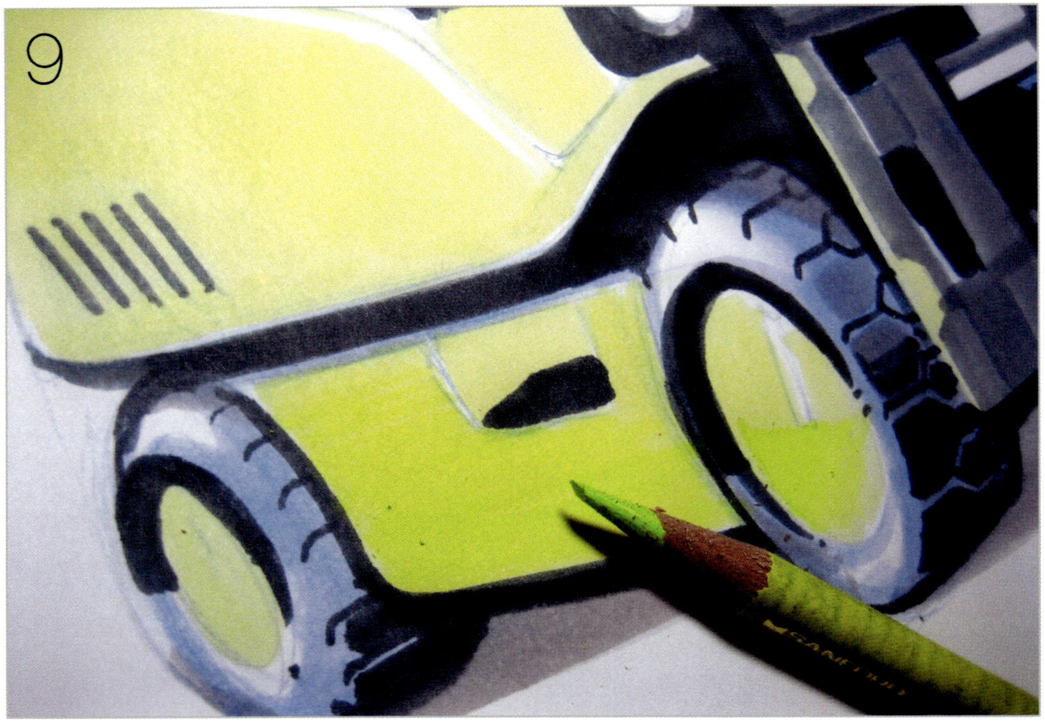

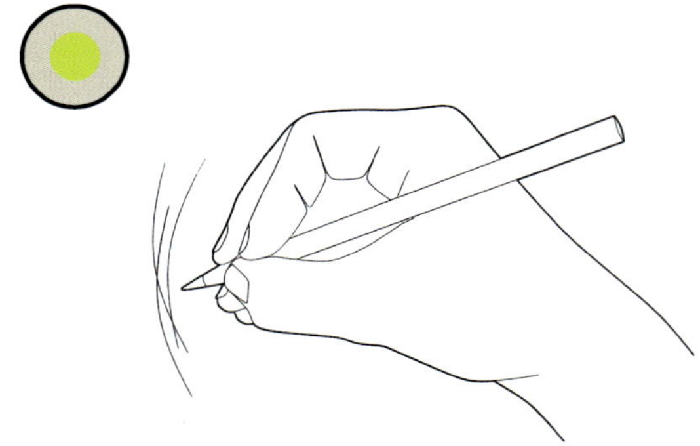

Now, gently apply some Yellow Chartreuse Prismacolor pencil over the Pale Yellow marker.

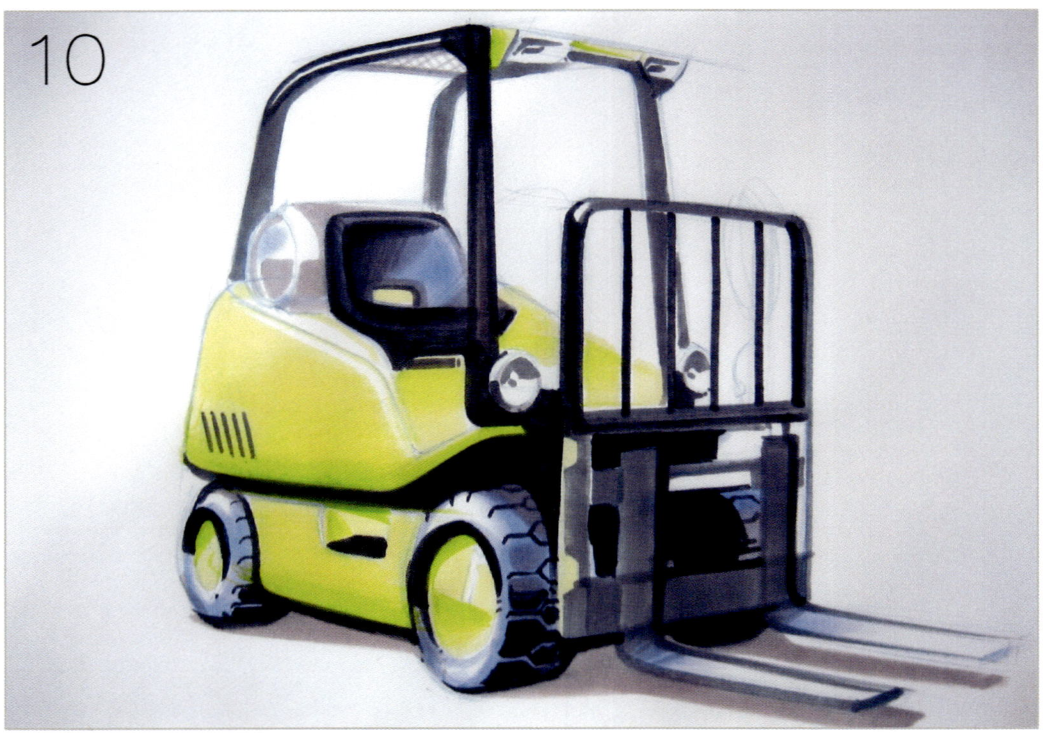

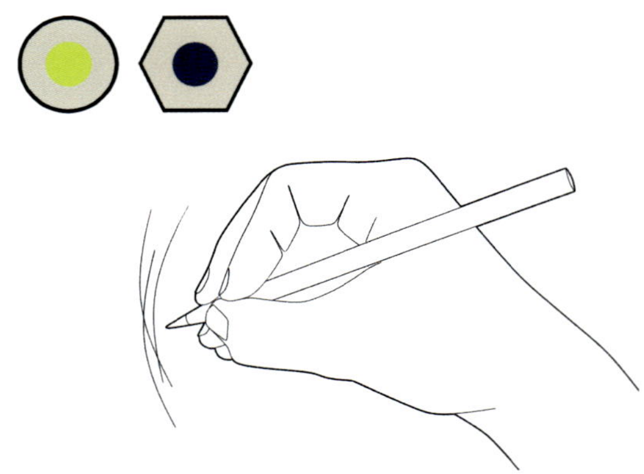

In the darker portions of the body, try blending some of the Indigo Blue Verithin with the Yellow Chartreuse.

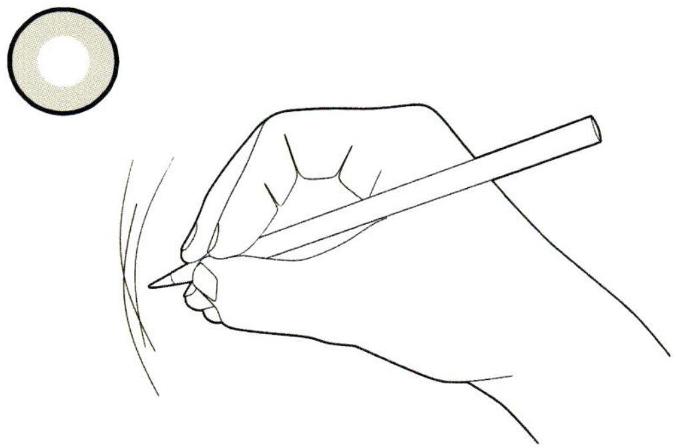

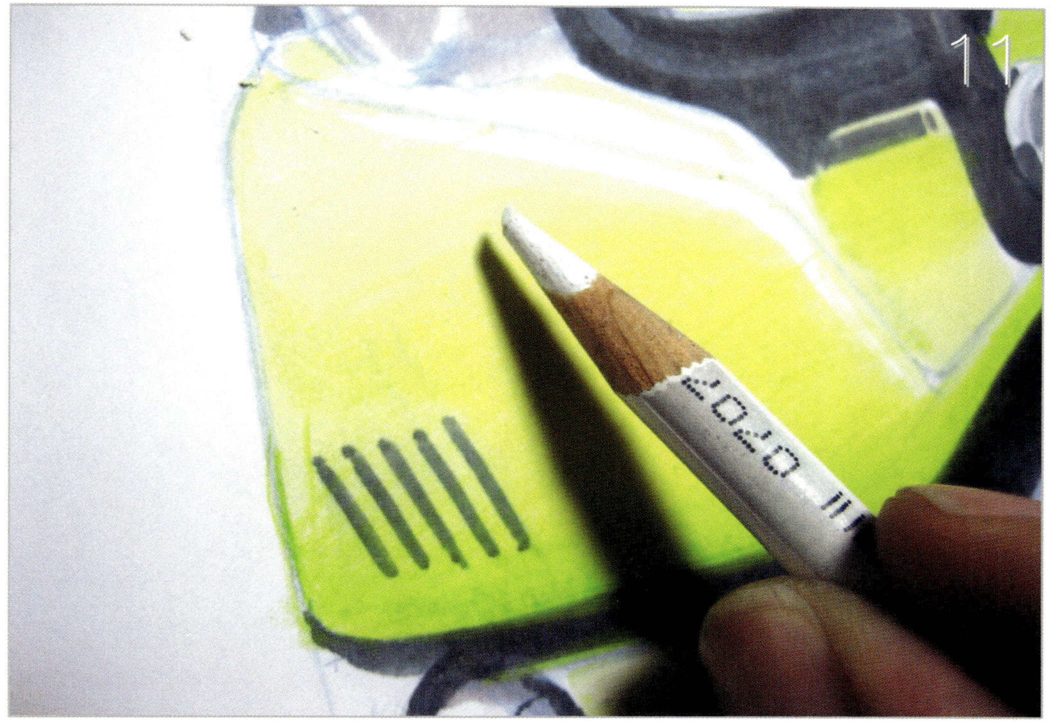

Near the highlighted corners of the body, ghost in some white pencil to give the Day-Glo surface a plastic look.

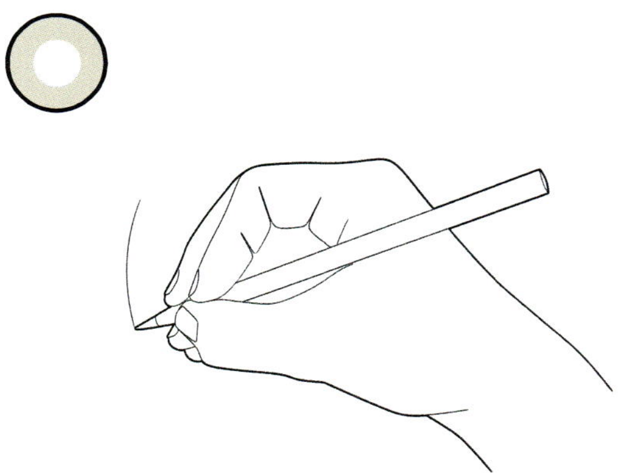

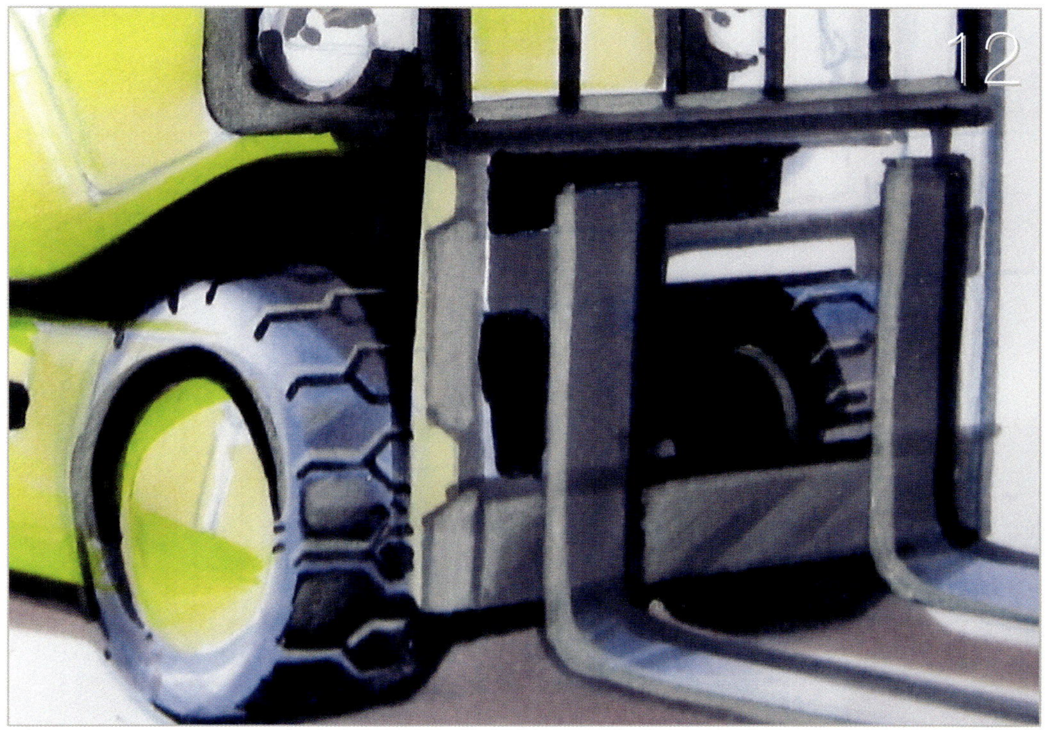

White pencil is also useful for showing the lit edge of the tire treads. Make sure you highlight the right side of the groove (the one facing you). Also, try adding a bit of sheen to the metal bars.

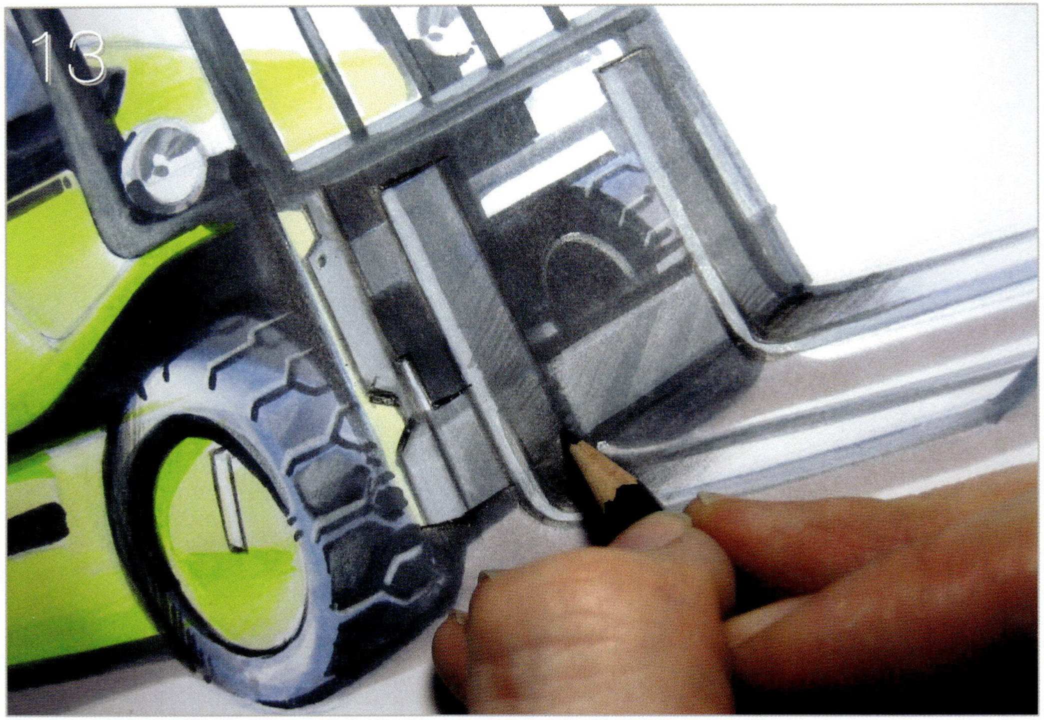

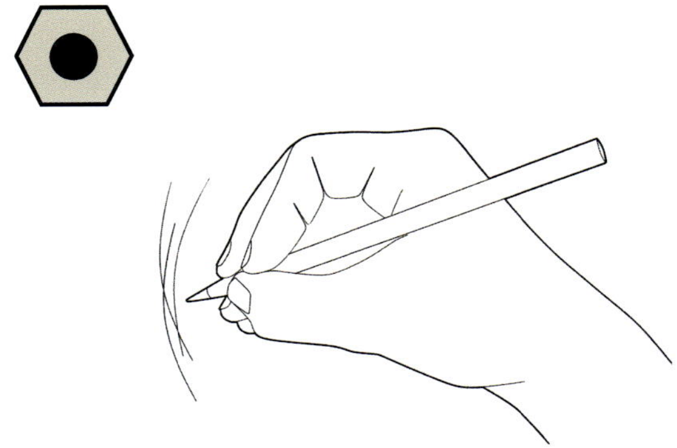

With the main colors in place, take some time to sharpen the linework. Black Verithins are the perfect tool for this job.

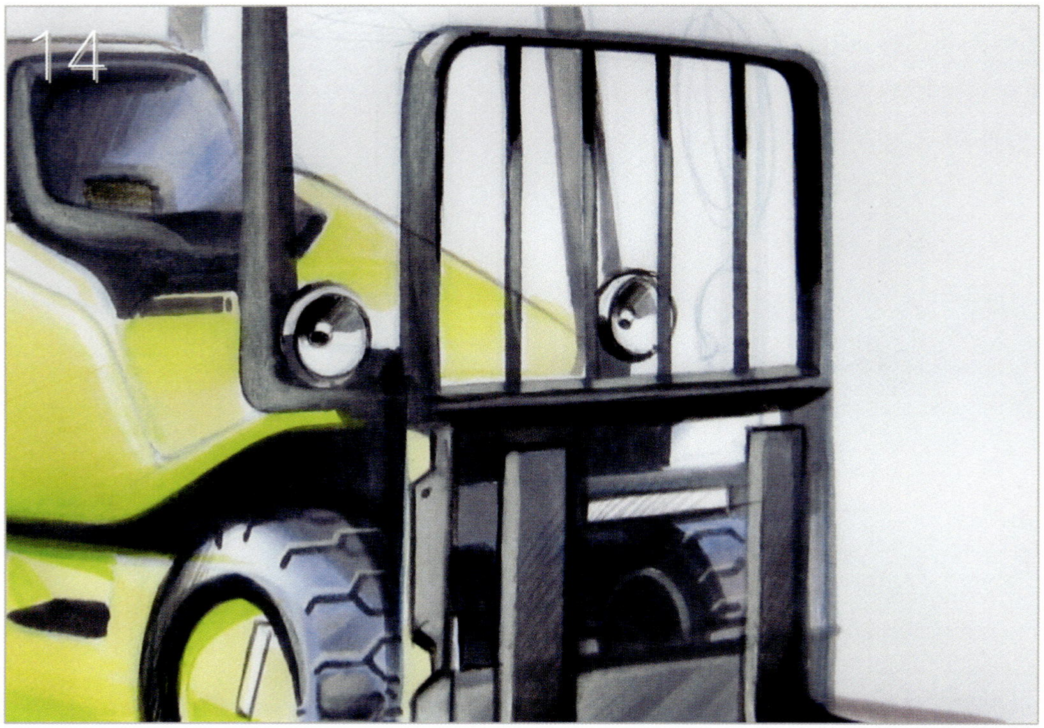

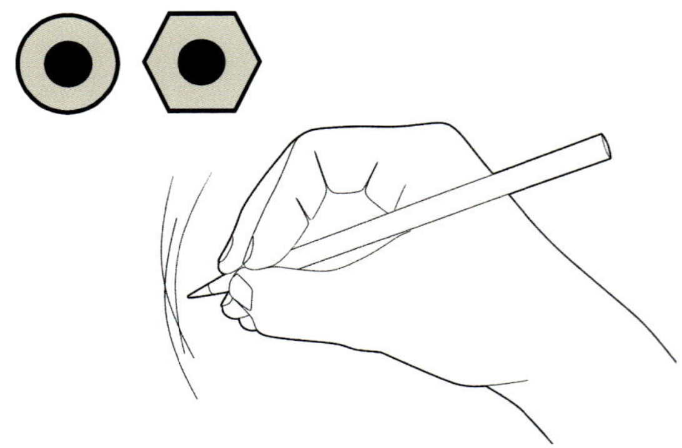

Use black pencils to add detail to the headlamp cans. Also, cast a shadow on the tops of the vertical bars to show they're inset.

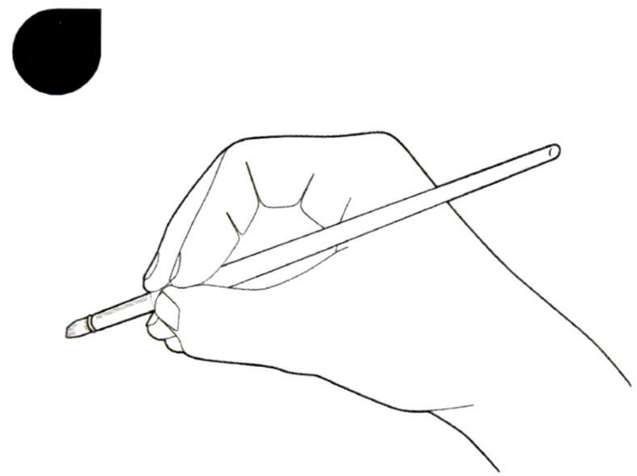

Take the contrast to another level by applying some black gouache to the underside. The lower fringes of the tires can also benefit from these accents.

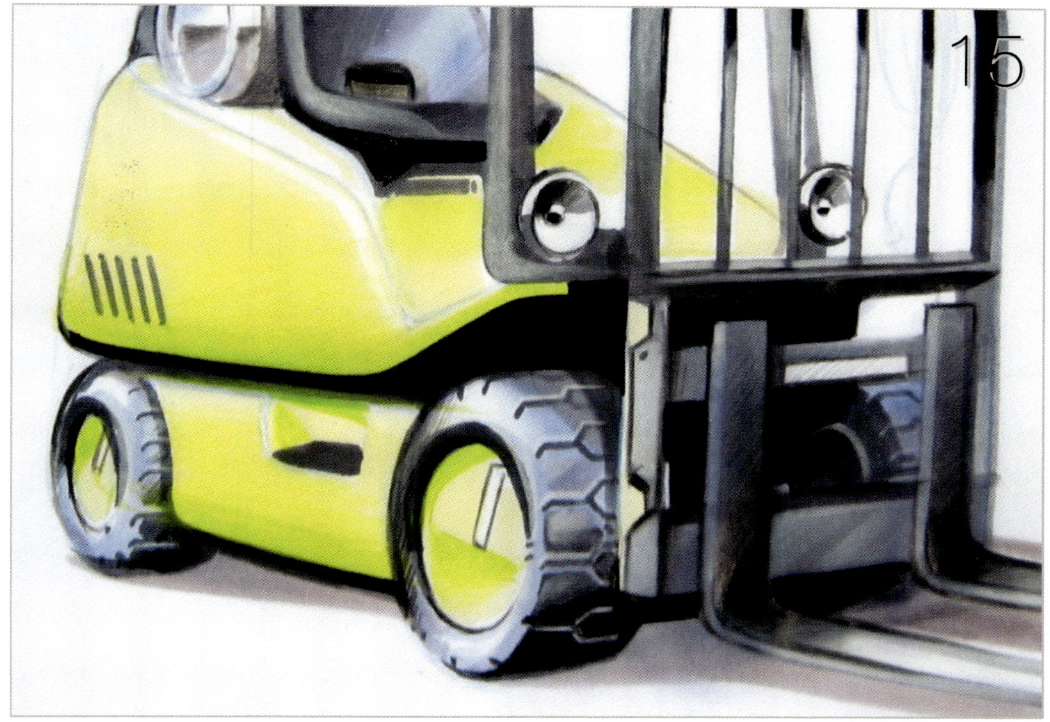

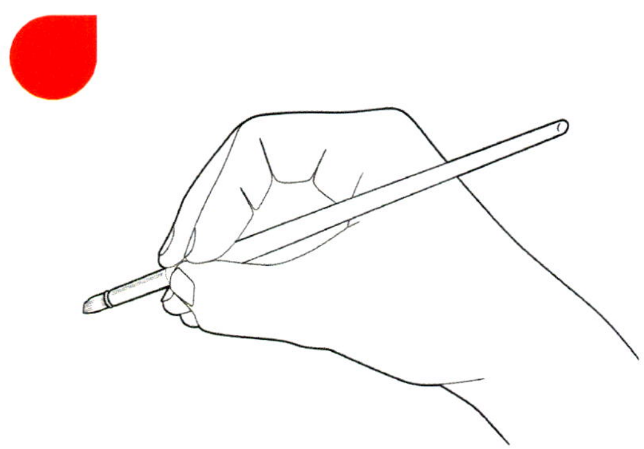

With a steady hand, brush in some diagonal, red caution stripes for a splash of color. Drop in a couple small flashes of red in other places for some visual balance.

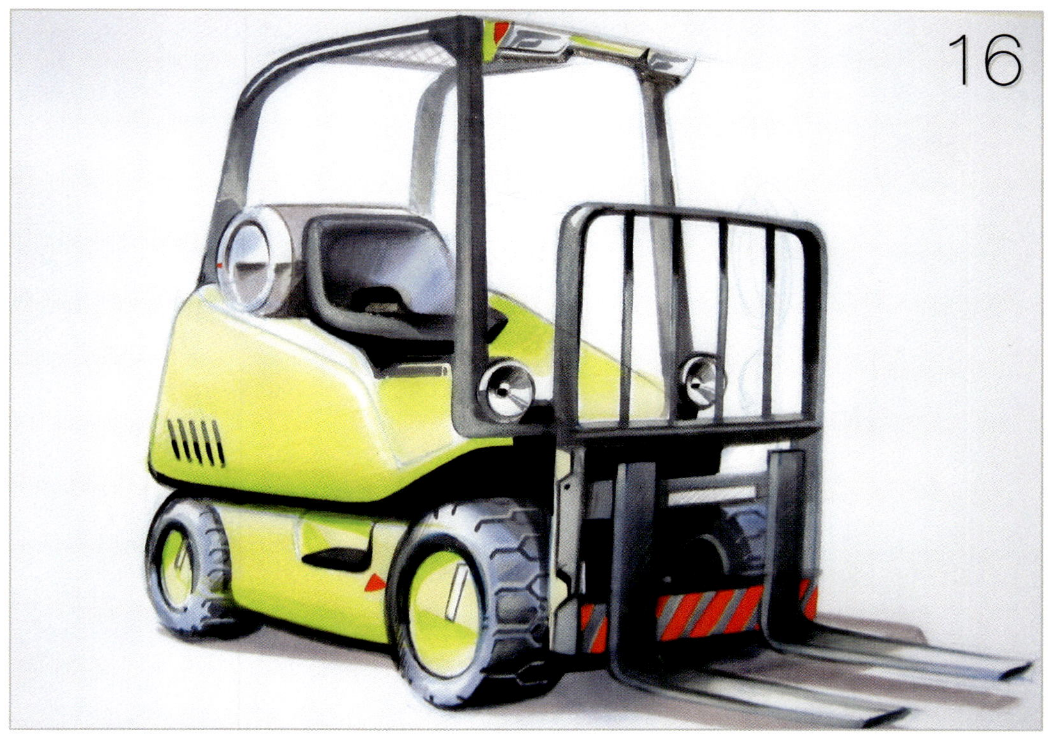

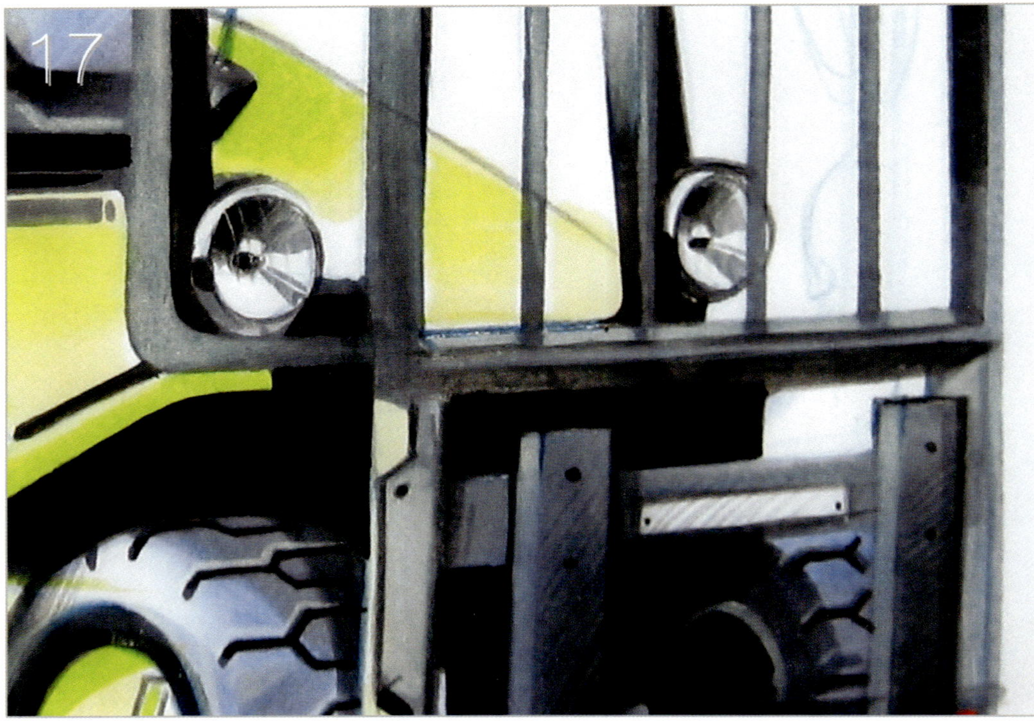

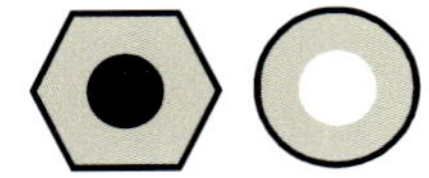

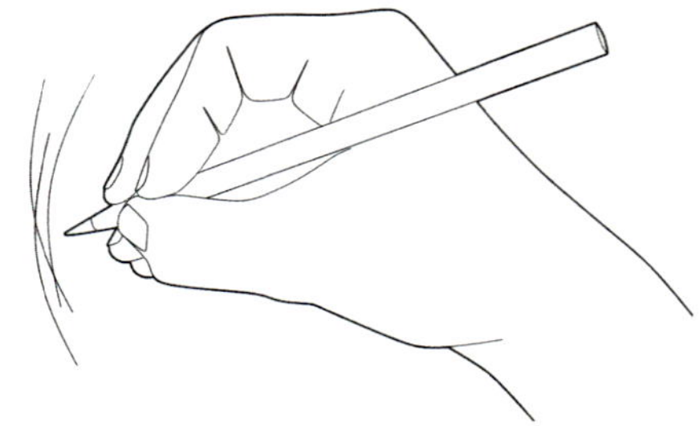

Continue adding details, like the white lines on the headlamp lenses and the black bolts holding the lift arms to the cross members.

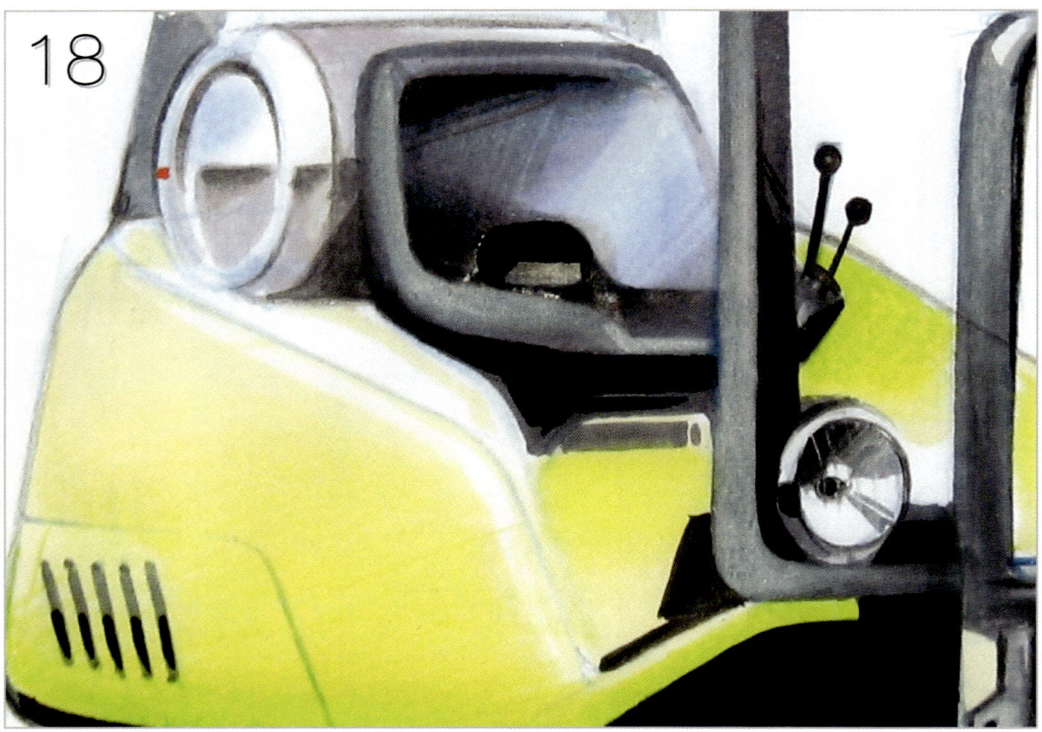

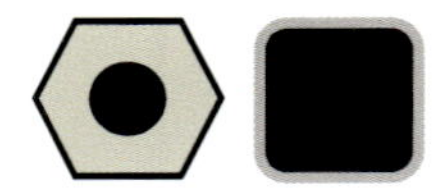

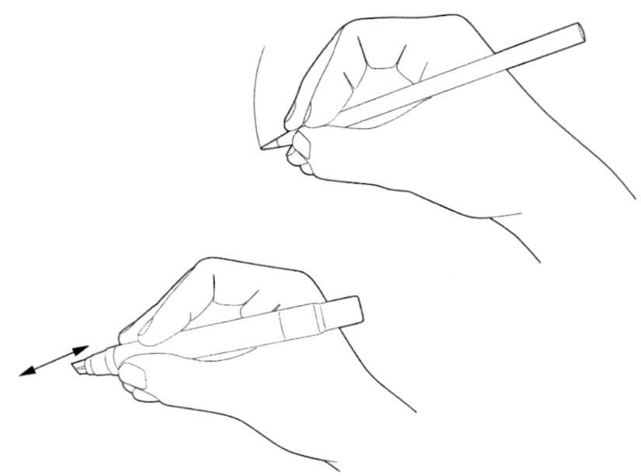

Draw a couple of levers next to the operator's seat; block in a brake pedal on the floor.

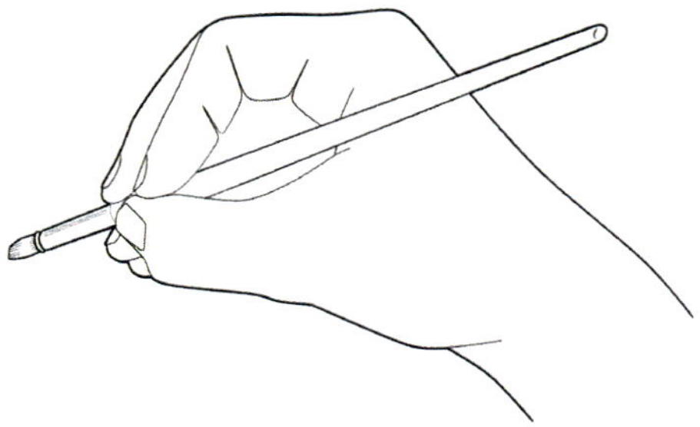

There aren't many shiny surfaces in this rendering, but there are still a few opportunities for gouache highlights.

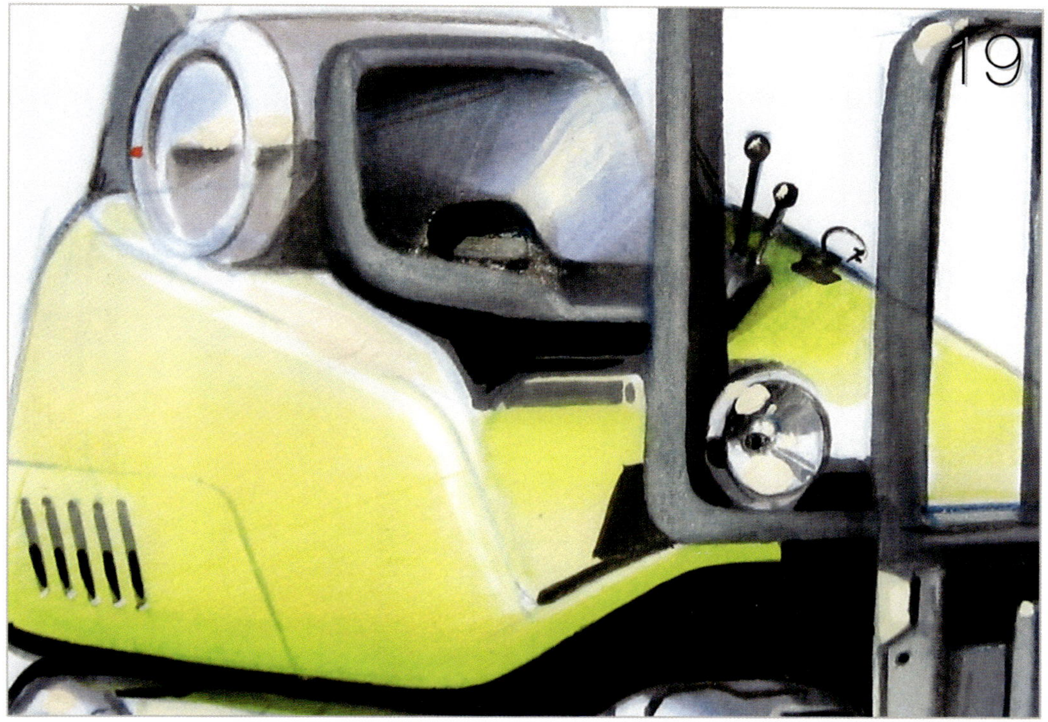

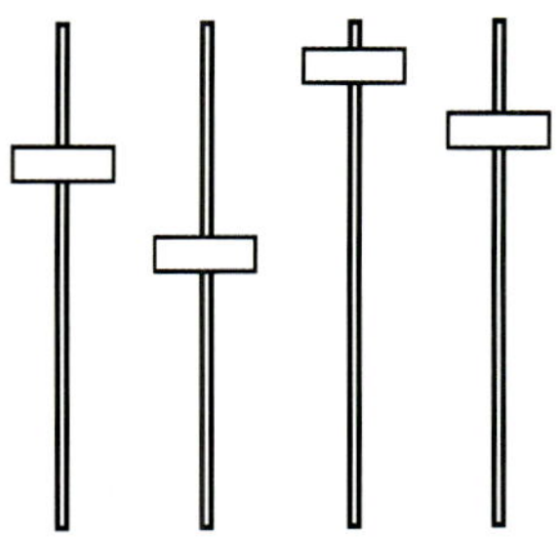

Scan in the rendering and clean up the background in Photoshop. You can also adjust the color levels to make the body more or less green.

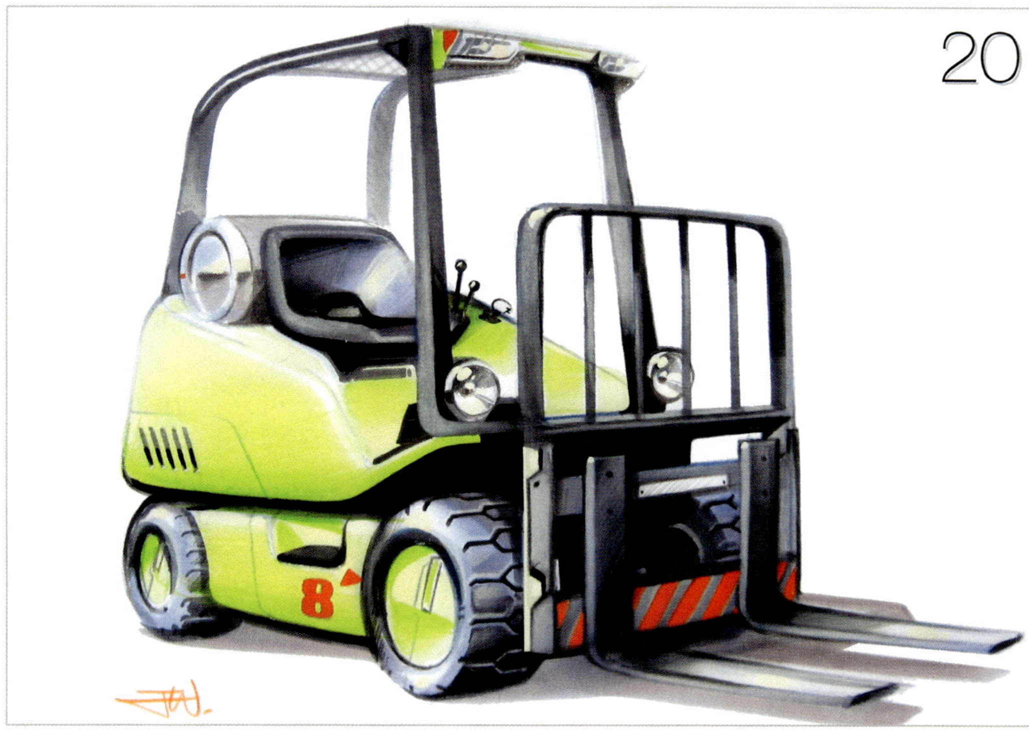

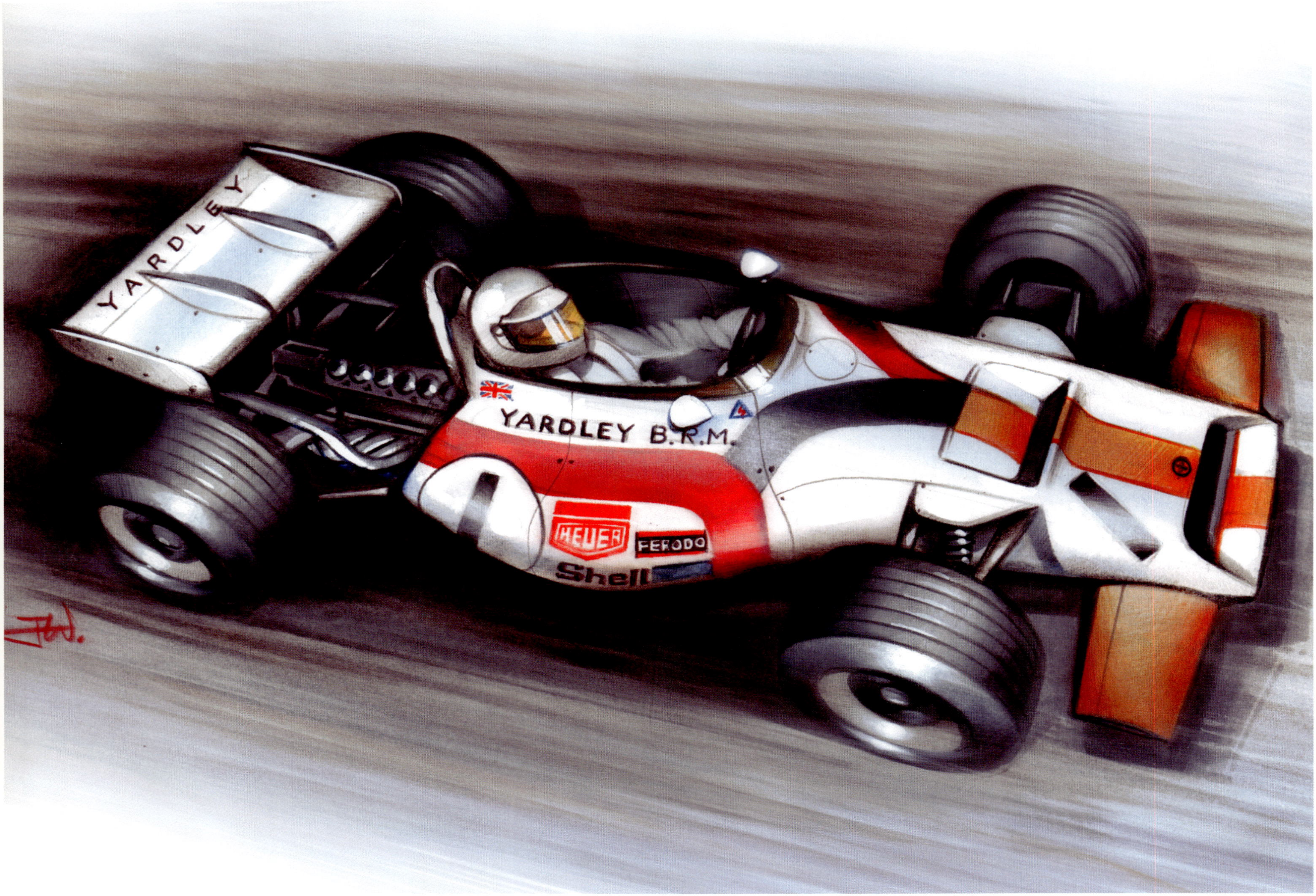
YARDLEY
YARDLEY B.R.M.
HEUER
FERODO
Shell

5. Road Surface Vignettes

In car design, a vignette is a background shape or a wash of color that helps pop the subject off the page. Vignettes are especially useful if the subject of a rendering is light in color. But a simple, abstract vignette is sometimes not appropriate, especially if the subject is in motion. In this tutorial, you'll learn how to create a road surface vignette; one that will help pop the car off the page and convey a sense of motion at the same time.

The Tools

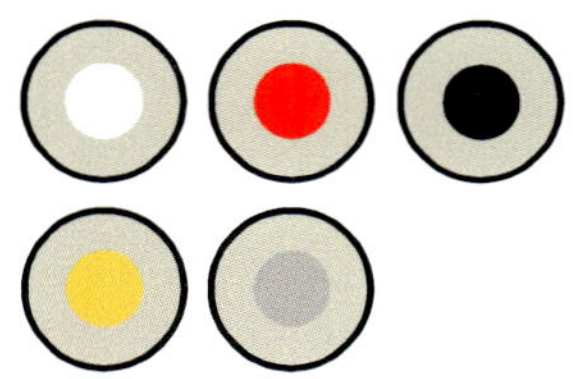

Prismacolors: White, Crimson Red, Black, Yellow Orange, Cool Grey 20%.

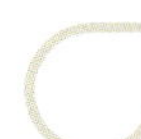

Gouache: White.

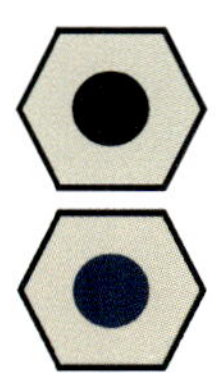

Verithins: Black, Indigo Blue.

Nupastels: Indigo Blue, Sienna Brown, White.

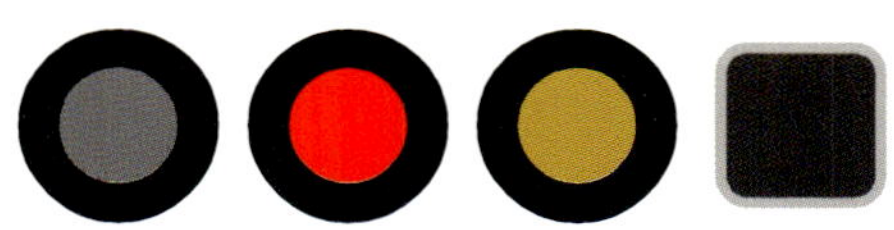

Prismacolor Markers: Cool Grey 40%, Crimson Red, Goldenrod. Copic Markers: Cool Grey 70%.

1

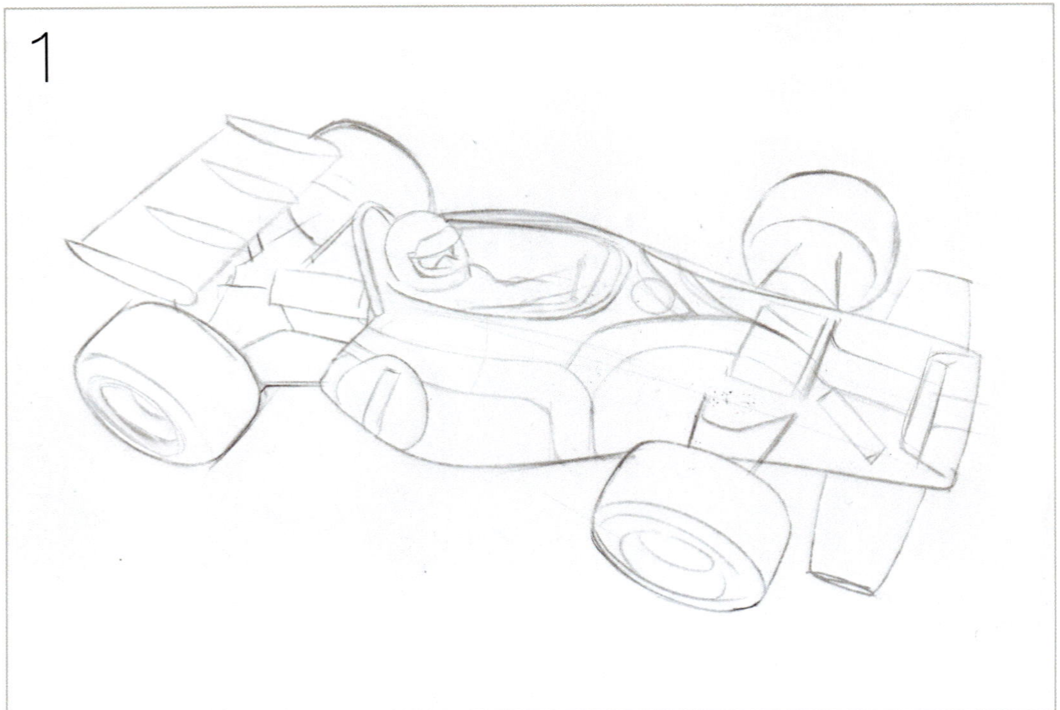

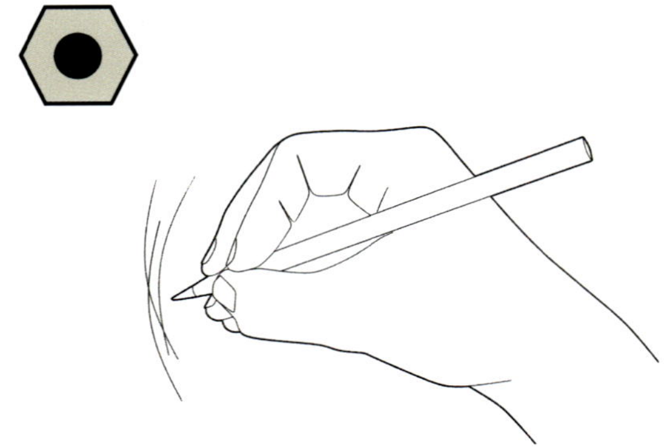

Start by laying out the rendering on marker paper. Pick an angle that (1) has a generous amount of top view, and (2) communicates speed.

2

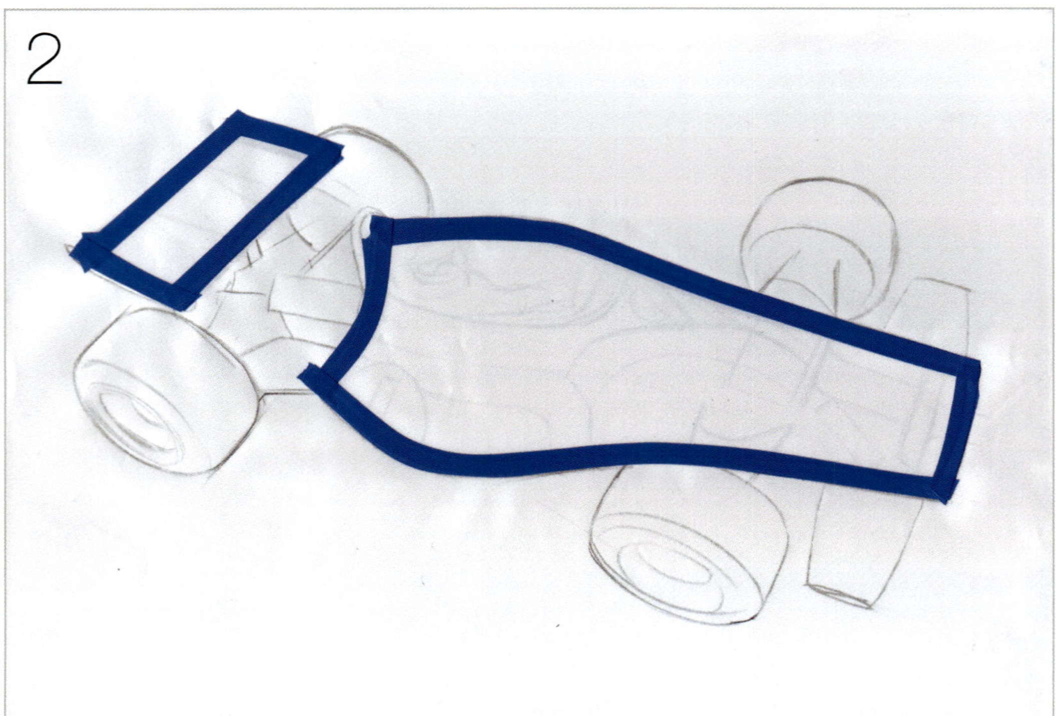

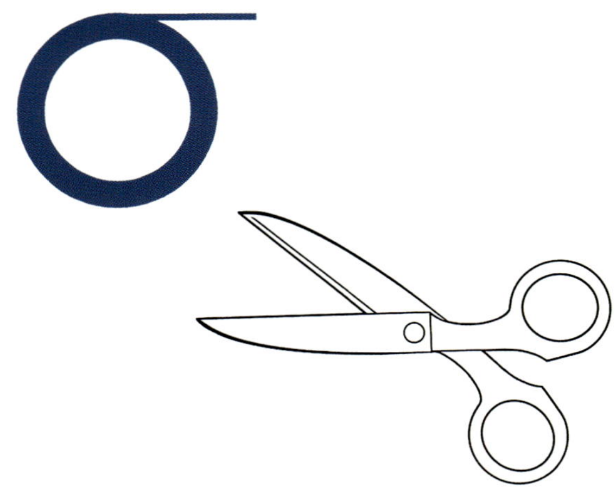

Trace the white sections of the car onto a sheet of drafting mylar. Cut out and secure the pieces of mylar with blue painter's tape.

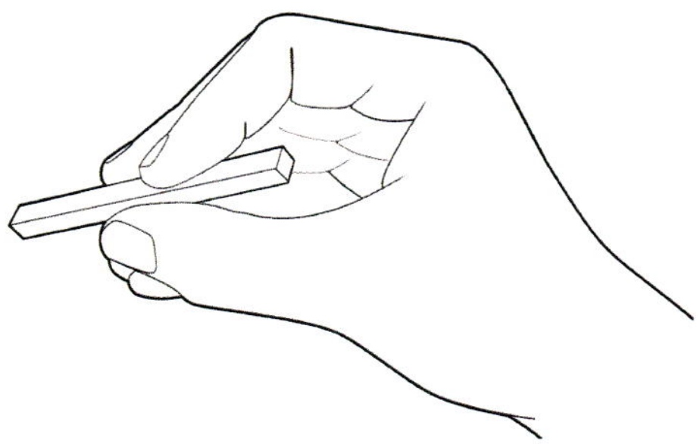

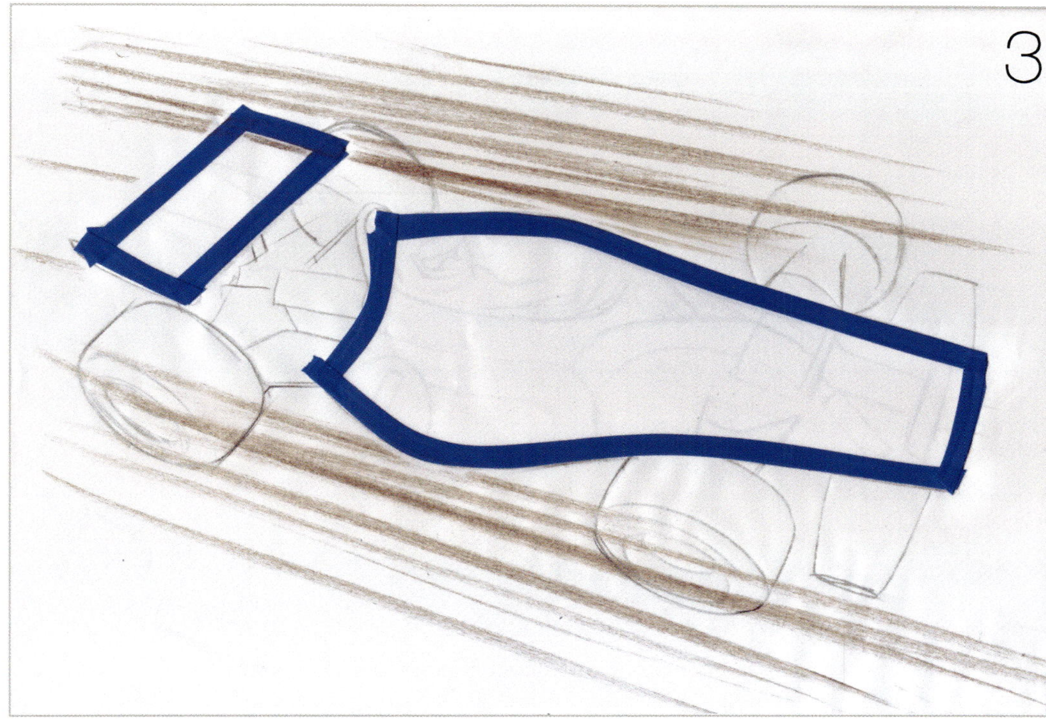

Begin laying in pastel strokes by hand. Be sure to keep these strokes in line with the perspective of the car.

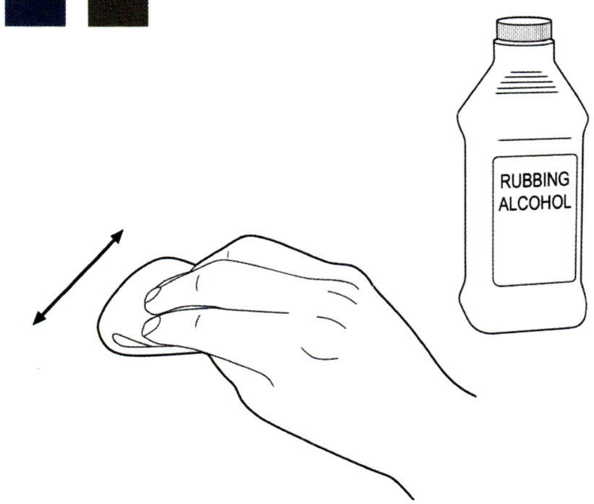

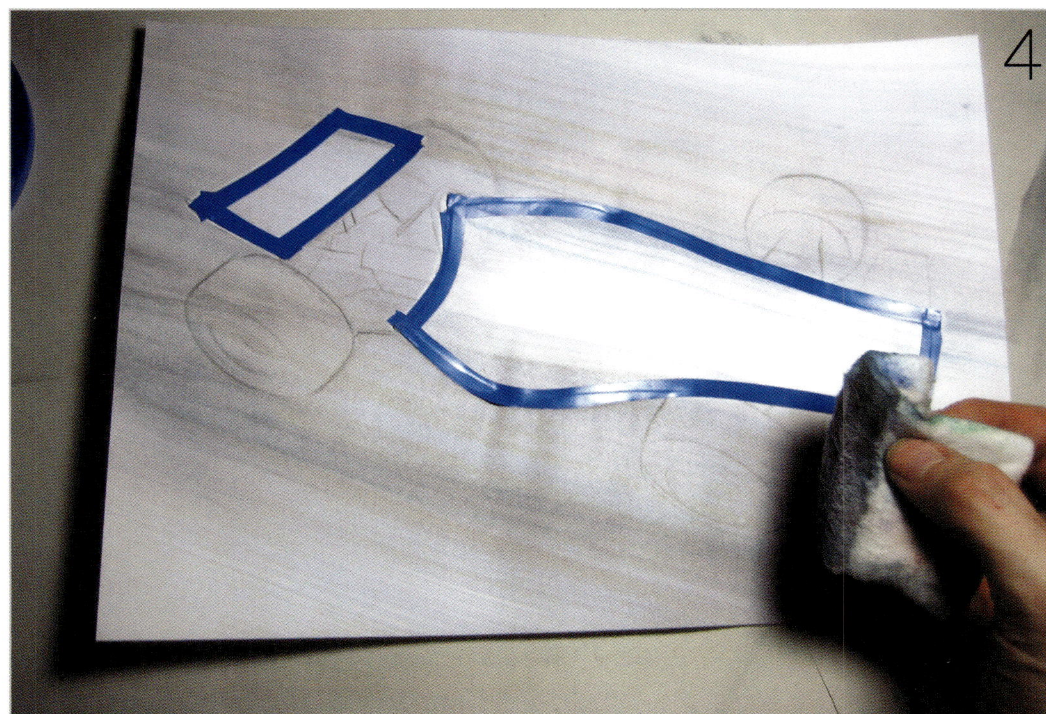

Now, using a Webril pad already stained with blue pastel, apply rubbing alcohol in horizontal strokes. This will blend the brown with the blue and create a dynamic base for the background.

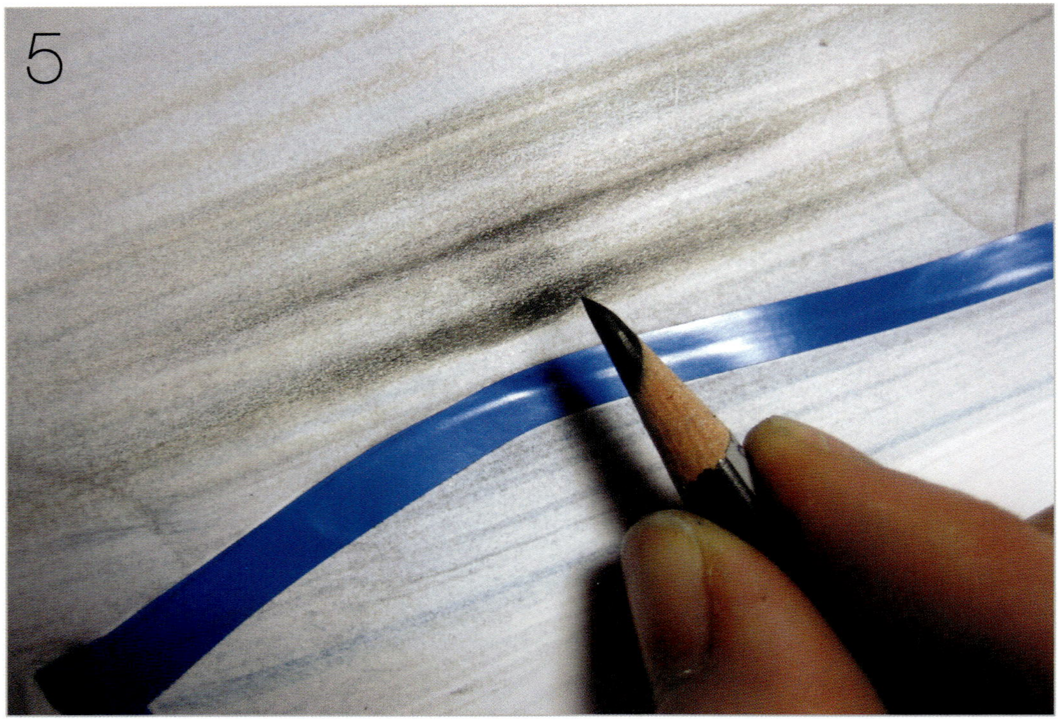

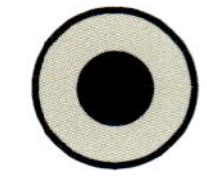

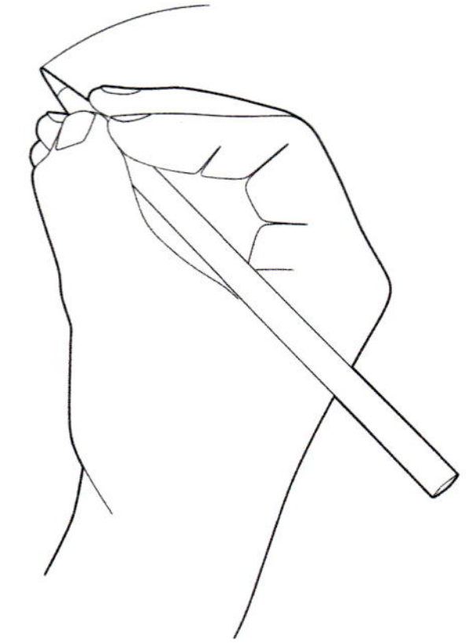

Allow the alcohol to dry. Then, using the side of a black Prismacolor, punctuate certain parts of the moving road surface.

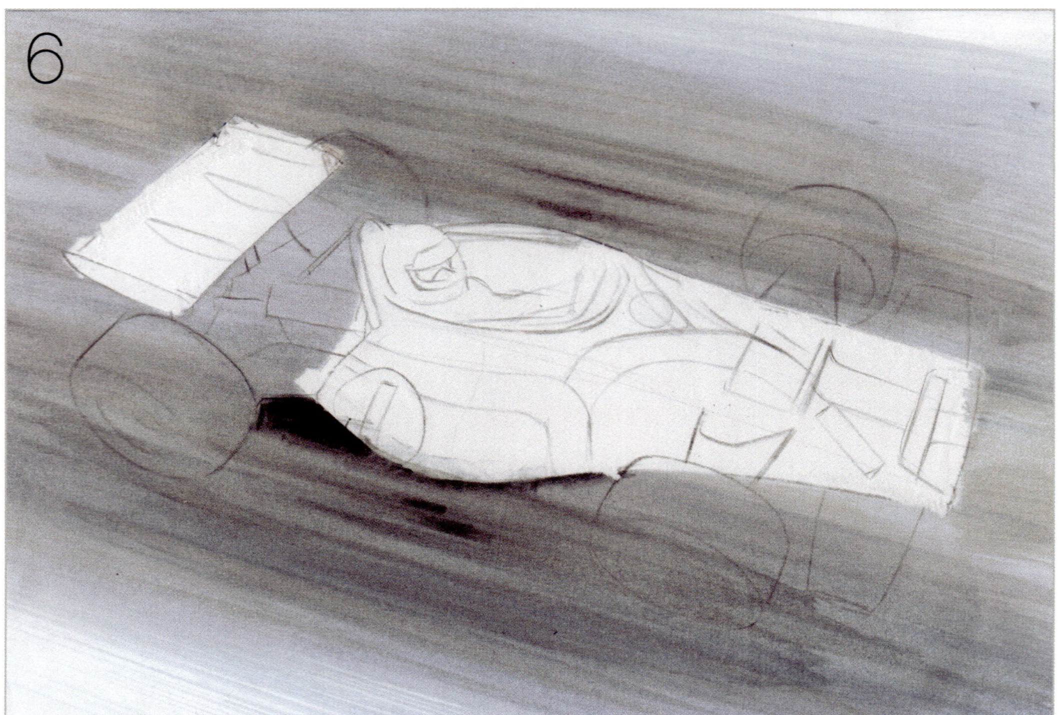

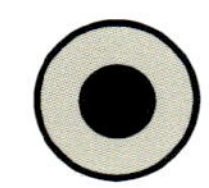

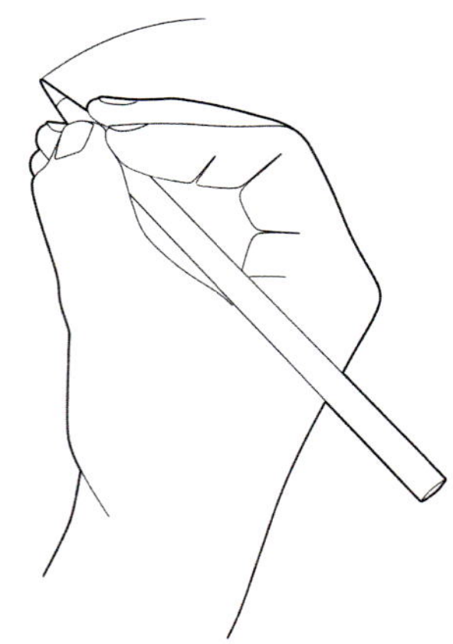

Remove the mylar and assess the contrast of the background. Add more black Prismacolor in the shadow area of the car for definition.

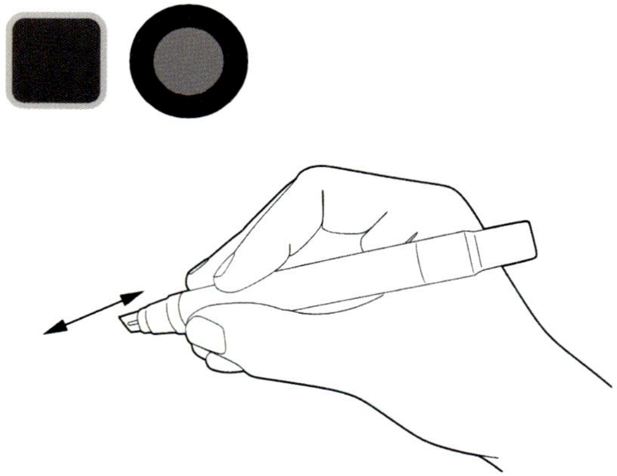

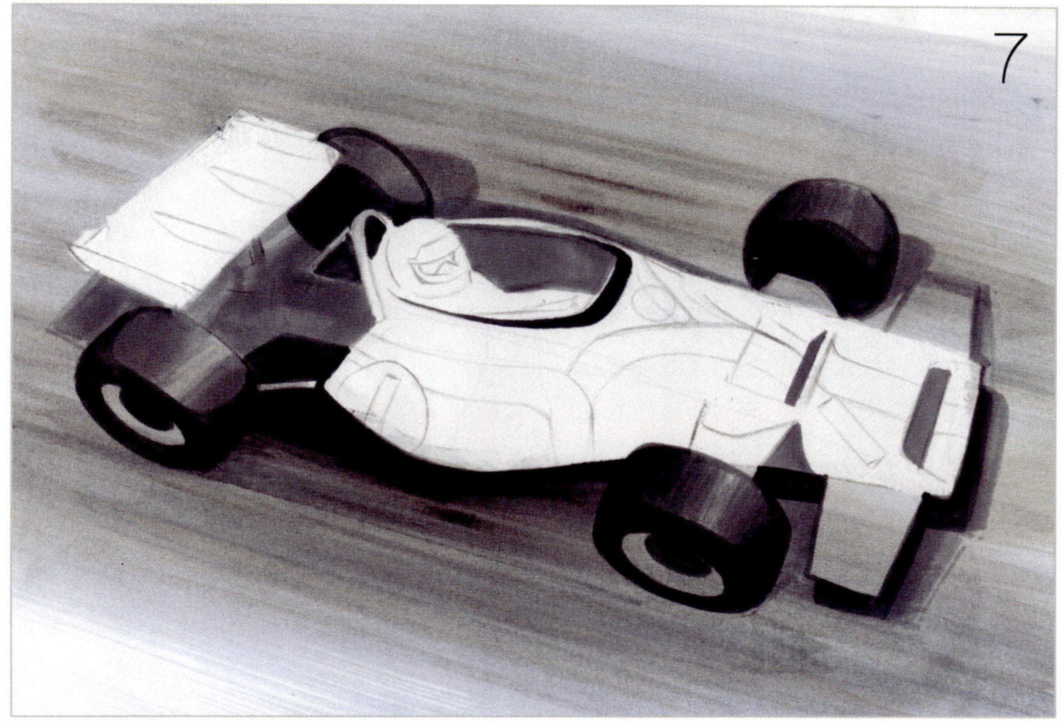

Start laying in the tires and the shadows. Be sure to leave some of the background showing through on the top of each tire (this makes adding highlights to the tires less time-consuming).

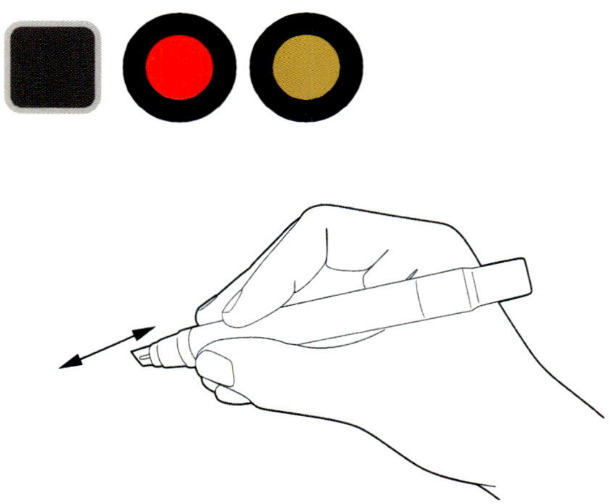

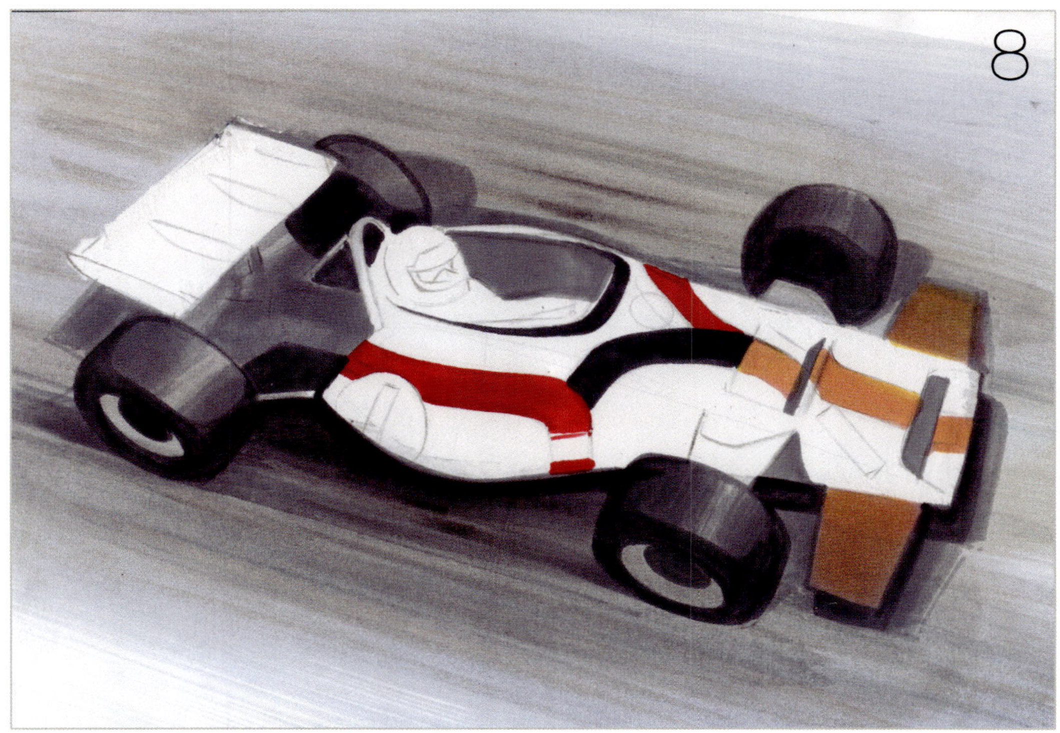

Use Crimson Red and Goldenrod markers to block in the graphics. Again, leave a space blank to aid with highlights.

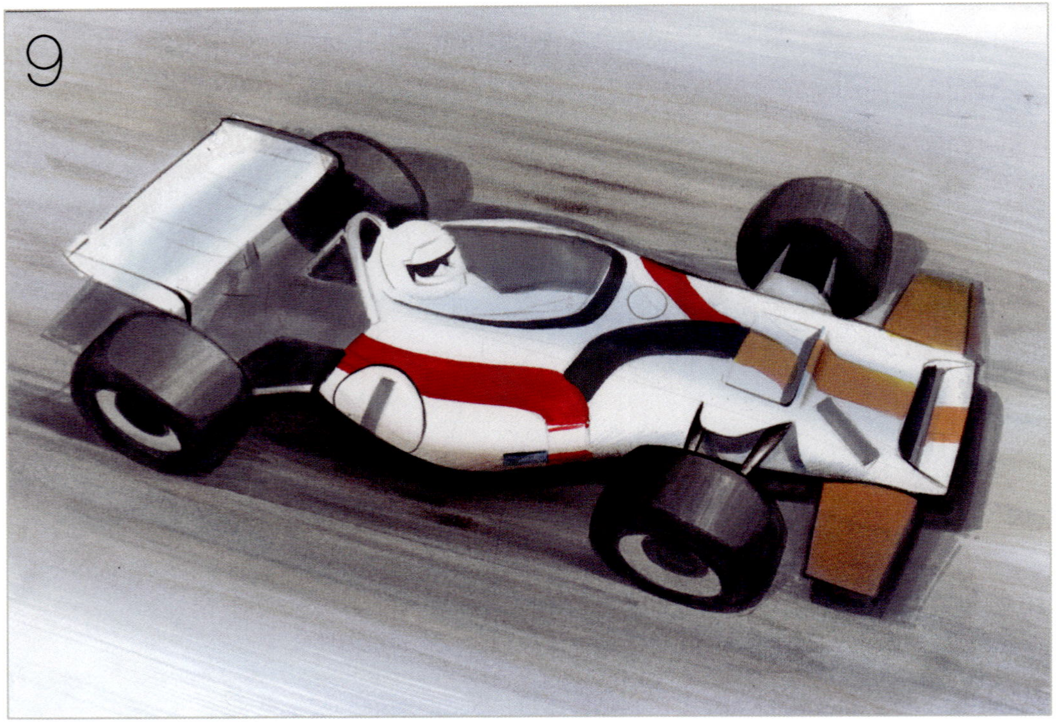

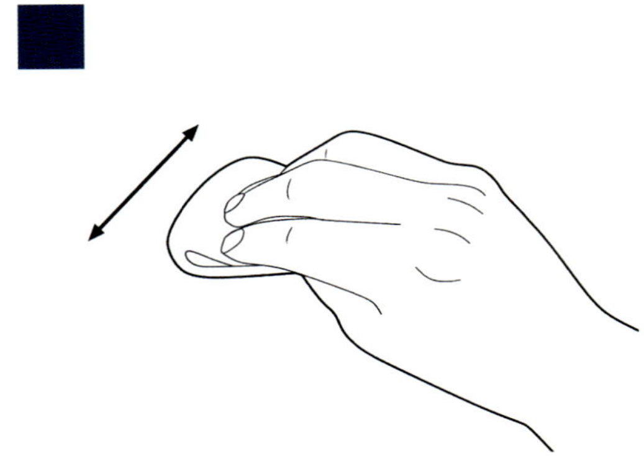

Just because the paint is white, that doesn't mean it has no color. Like other glossy surfaces, it will reflect the blue of the sky tone – but don't apply too much.

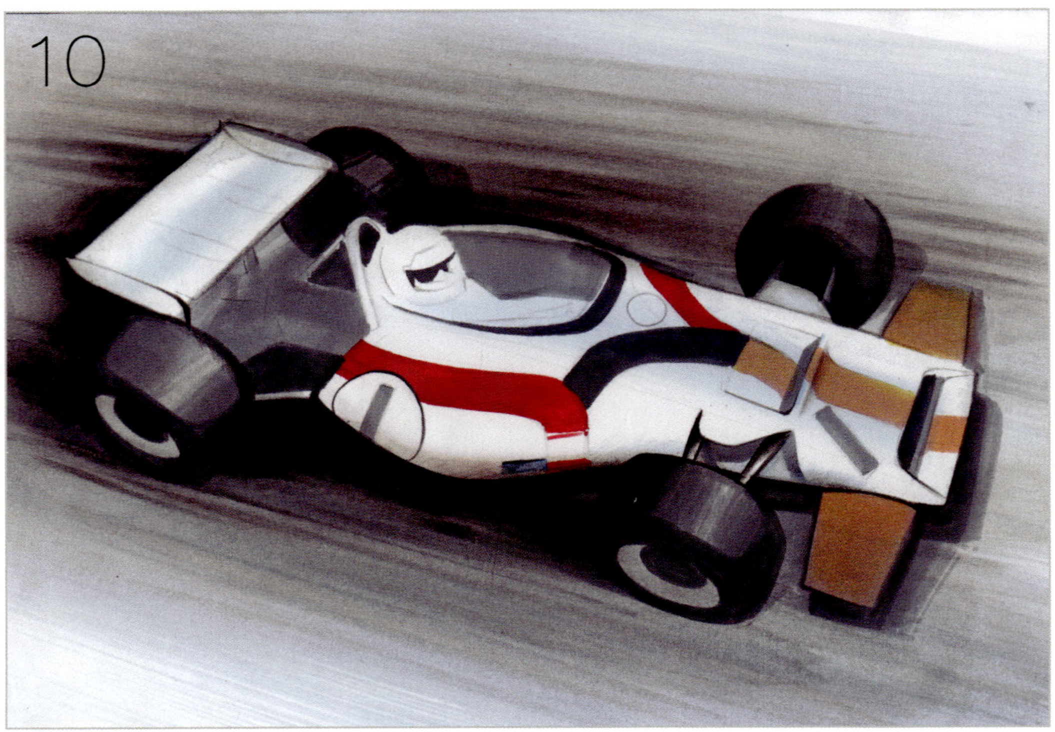

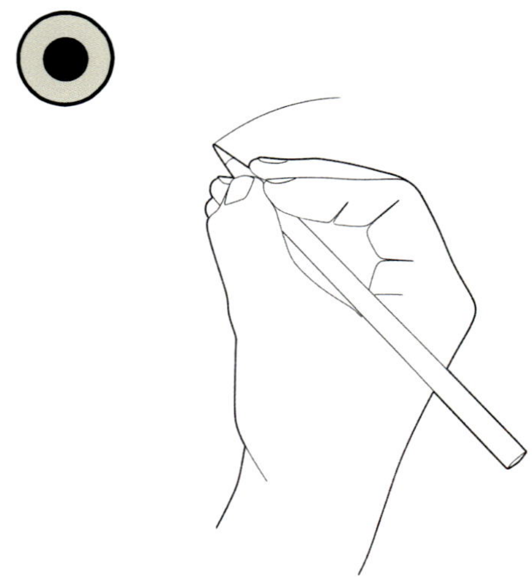

Now, hit the background again with black Prismacolor, emphasizing the shadows and the sense of motion.

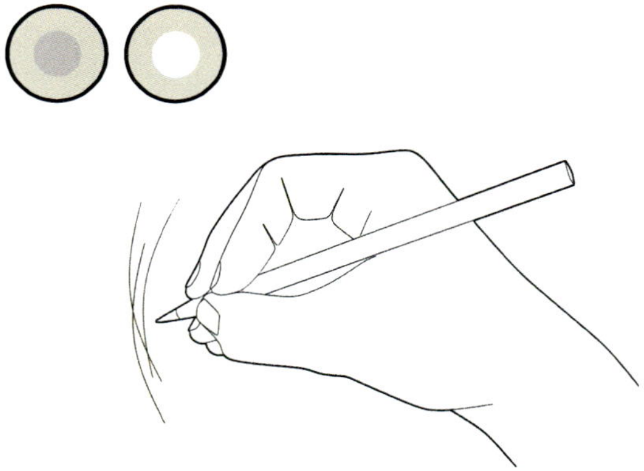

Begin laying in the tire highlights with white and Cool Grey Prismacolor. You can try using circular motions to smooth out gradations.

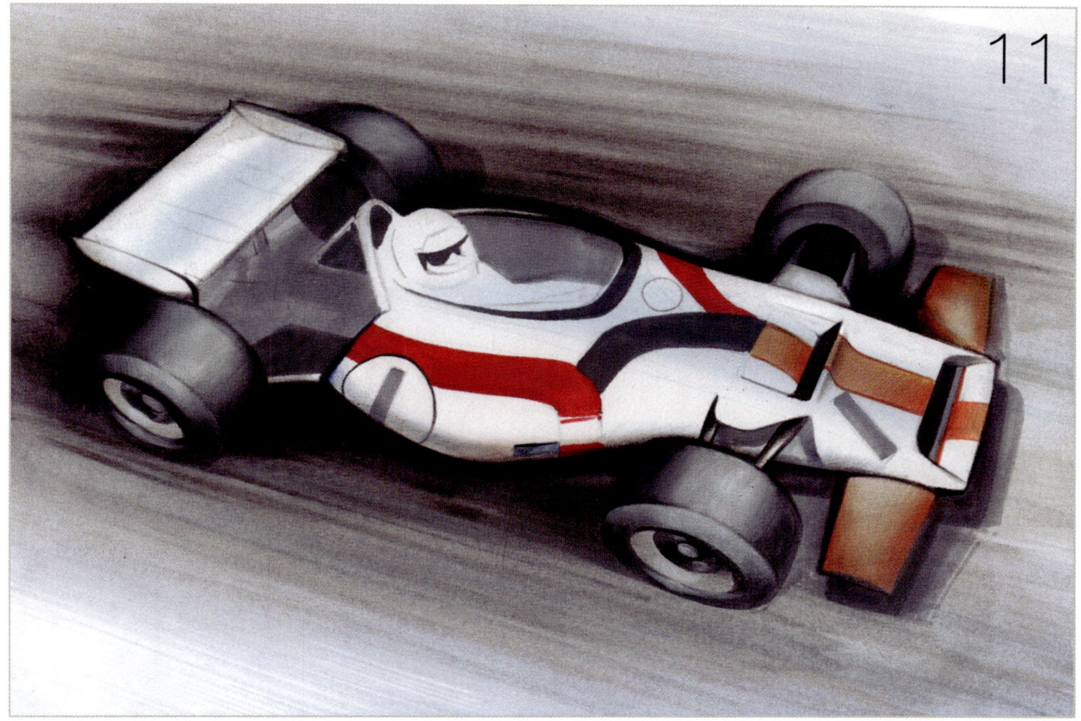

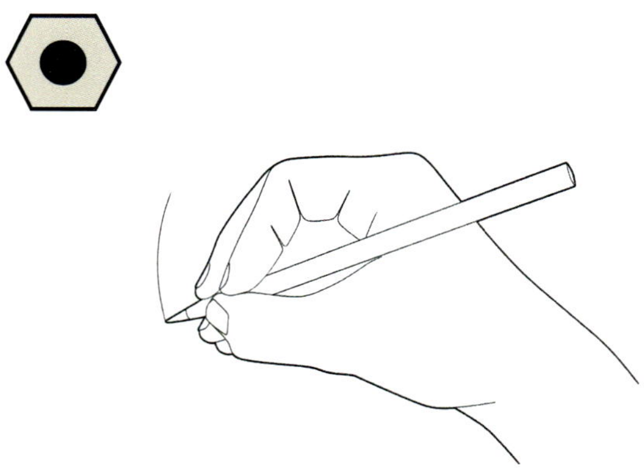

Using a very sharp, black Verithin, carve in the tire treads. Since the car is in motion, you need only worry about the longitudinal grooves.

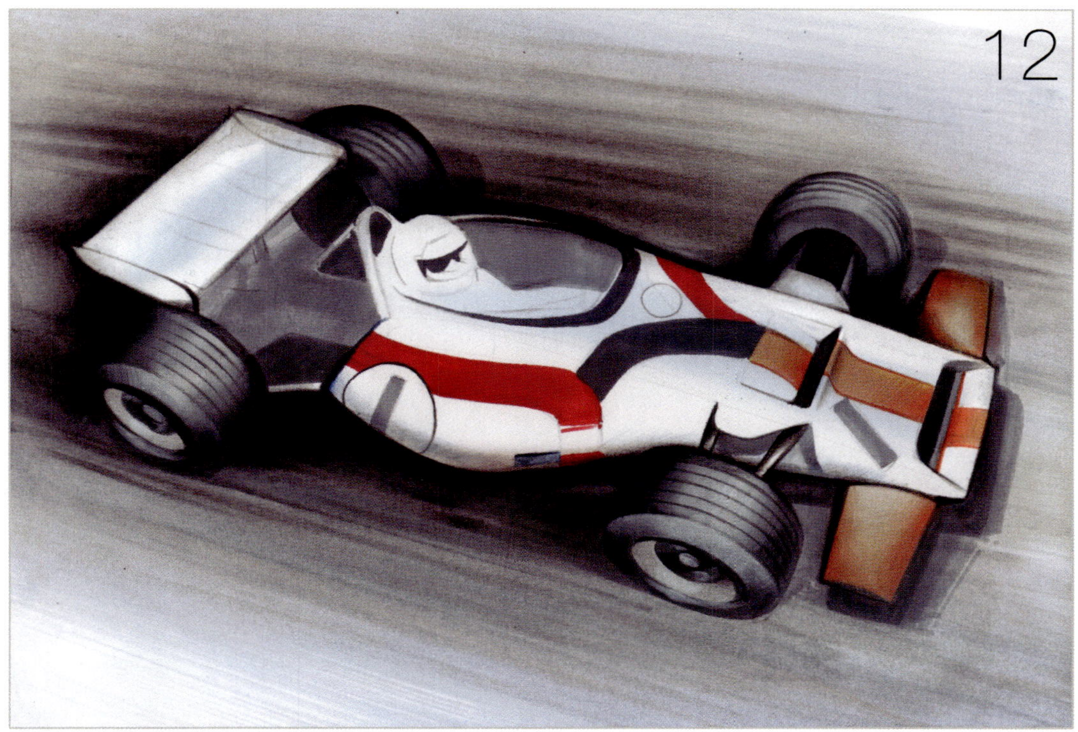

13

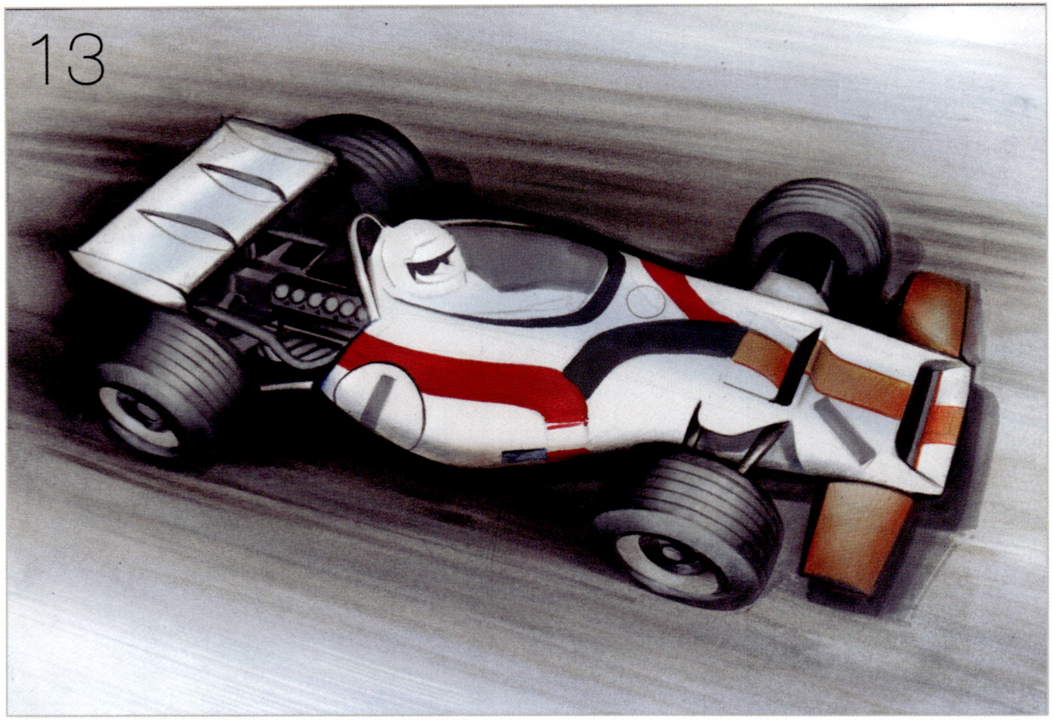

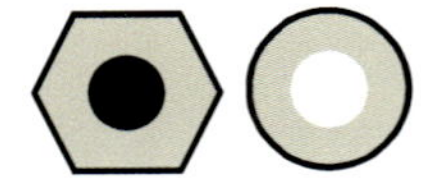

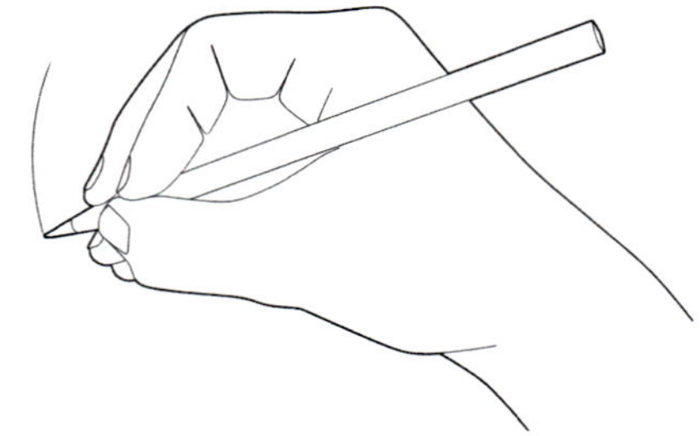

Now, carefully begin laying in the more detailed linework, like the engine parts and the wing sections.

14

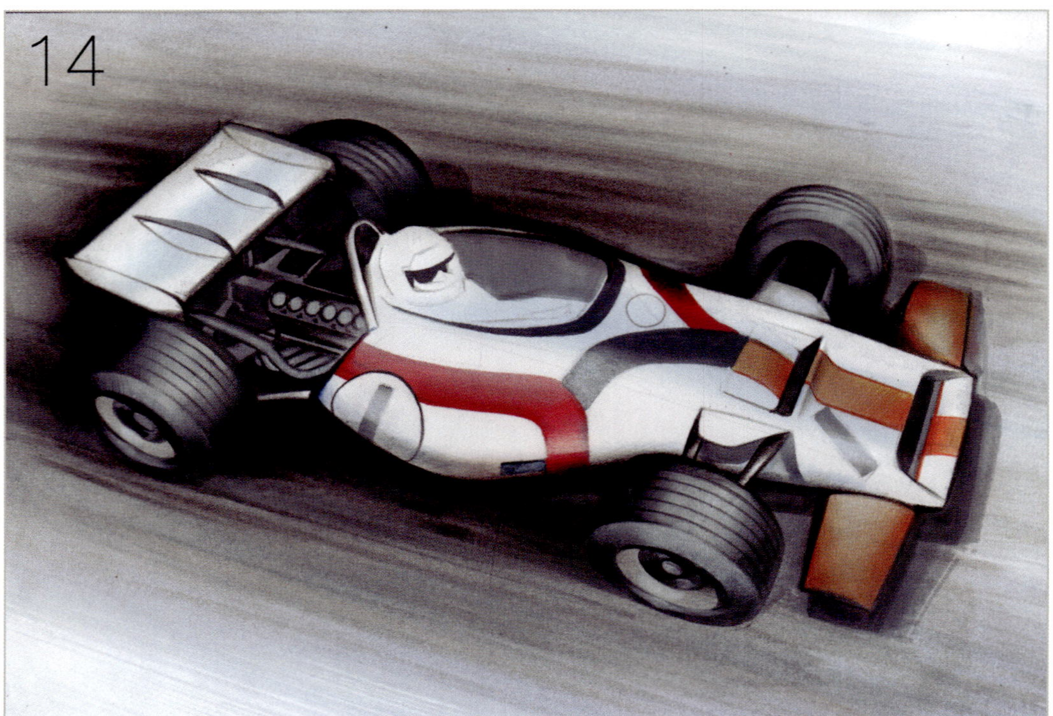

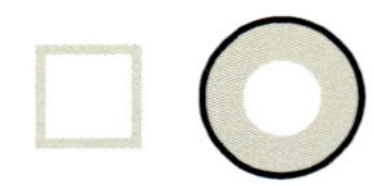

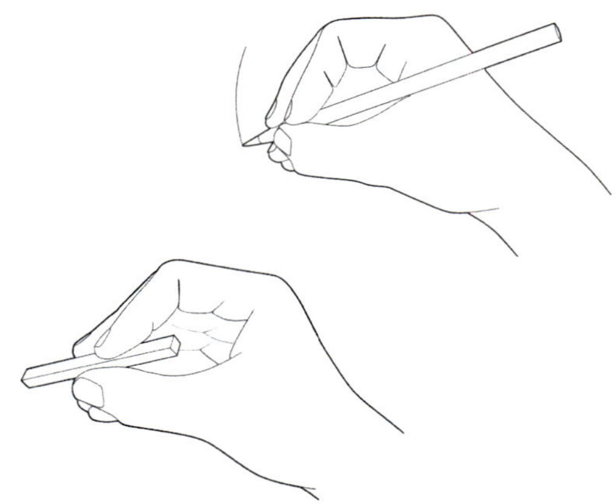

Use white pastel and pencil to continue the body highlights through the red and grey graphics.

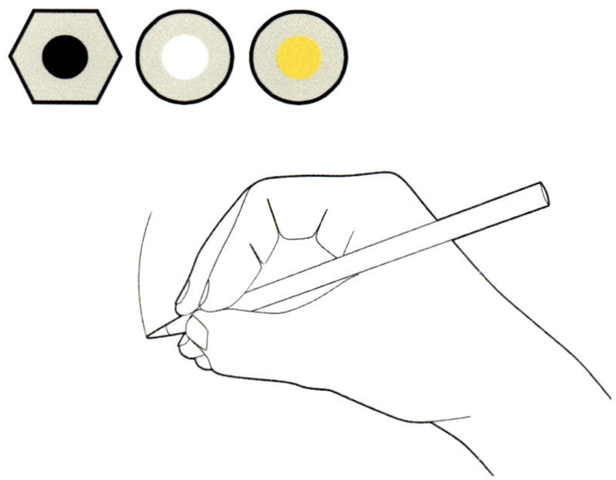

Take time to add details to the driver: His visor, the cut lines on his helmet and the folds in his driving suit. Don't let the details on the driver distract from the overall subject.

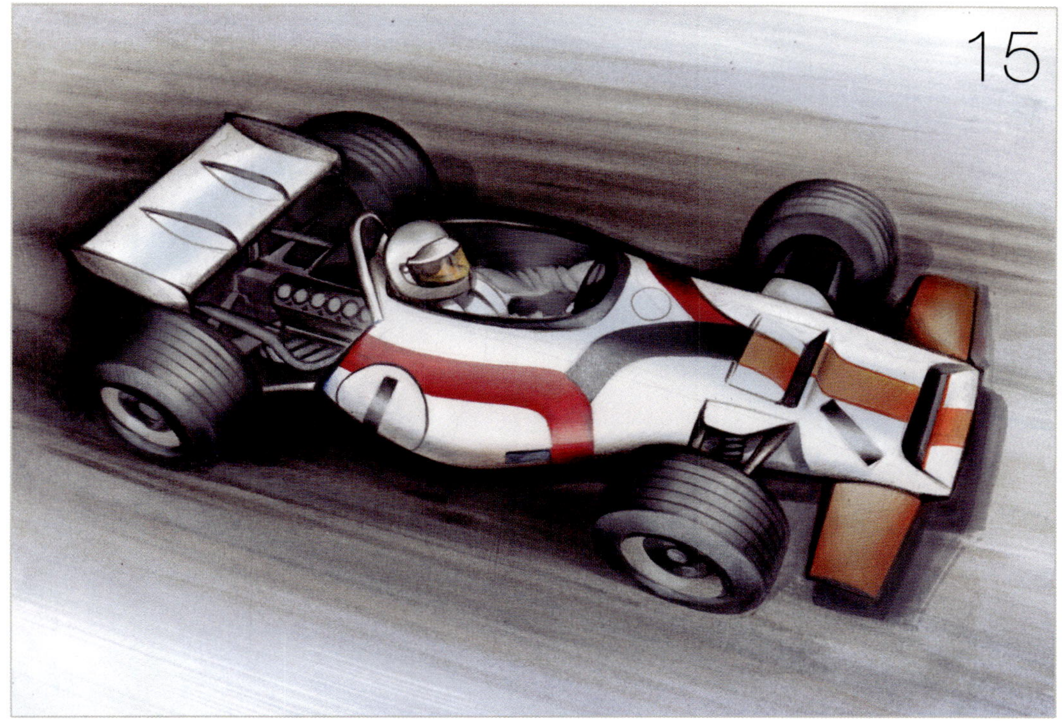

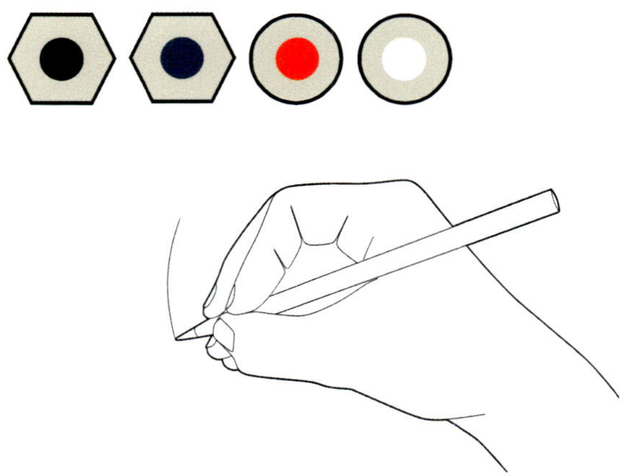

Using an array of pencils, add in the various sponsors and detail graphics. Be sure to slow down, be precise and always keep the pencils sharp. Make a cursory assessment of the details, adding any you may have missed (the mirrors, for example).

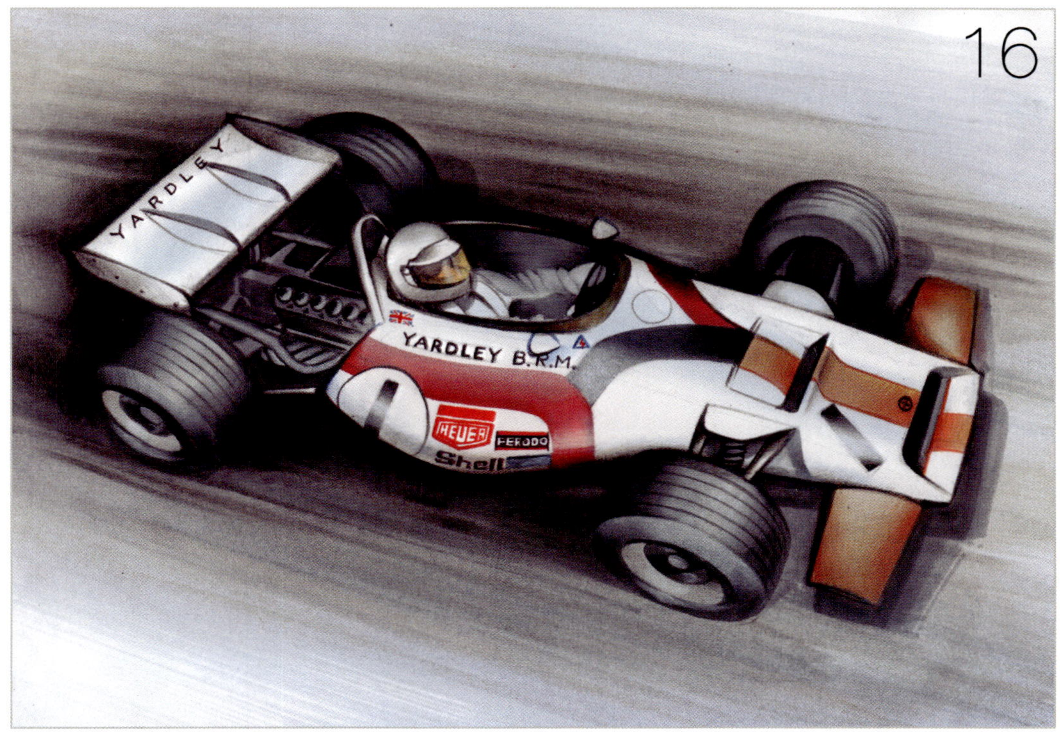

17

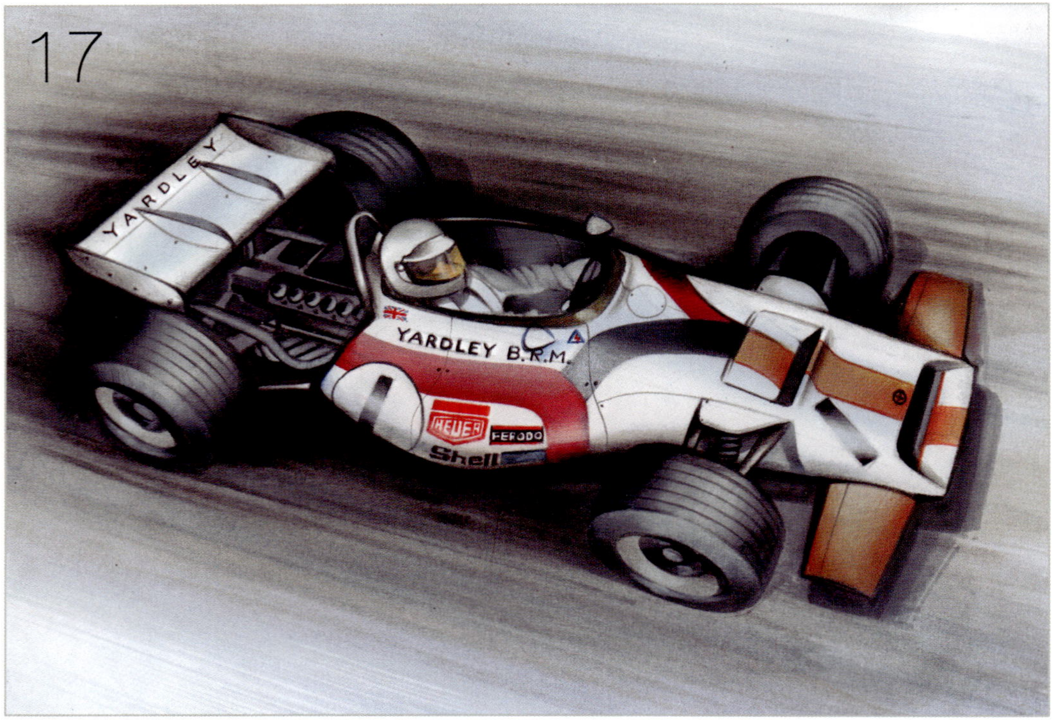

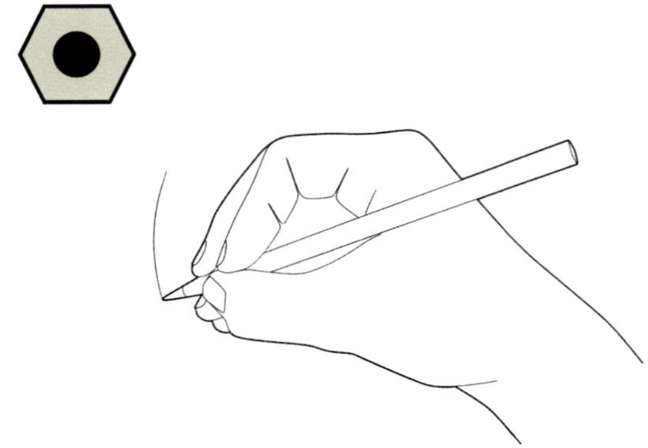

Carefully add part lines and bolts for the body with a black Verithin pencil.

18

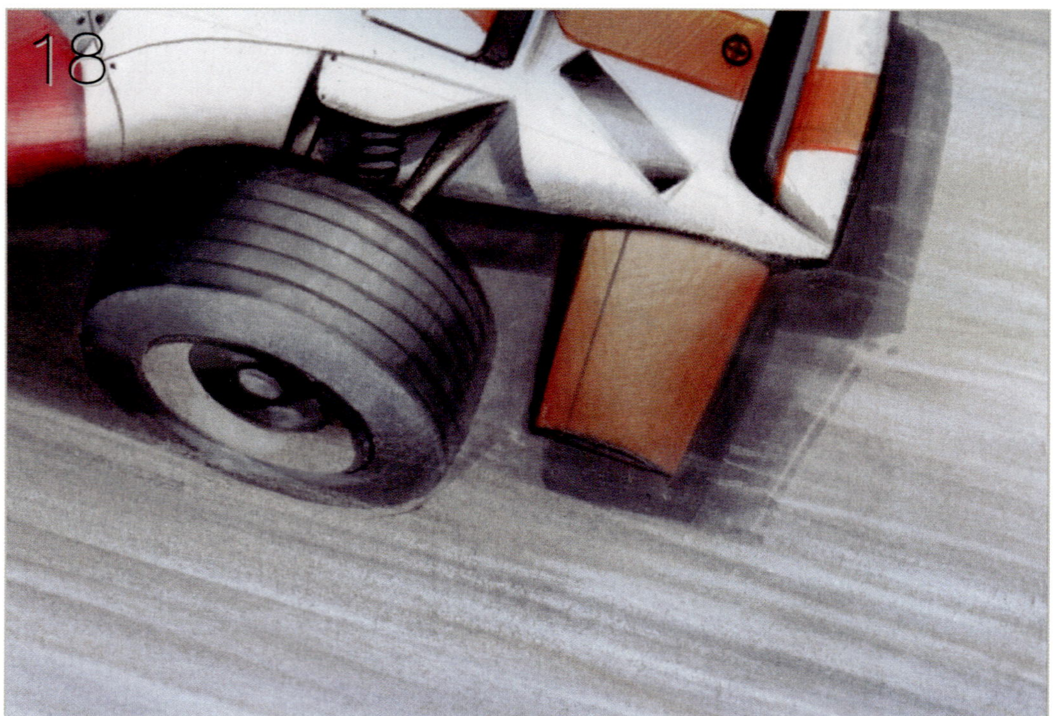

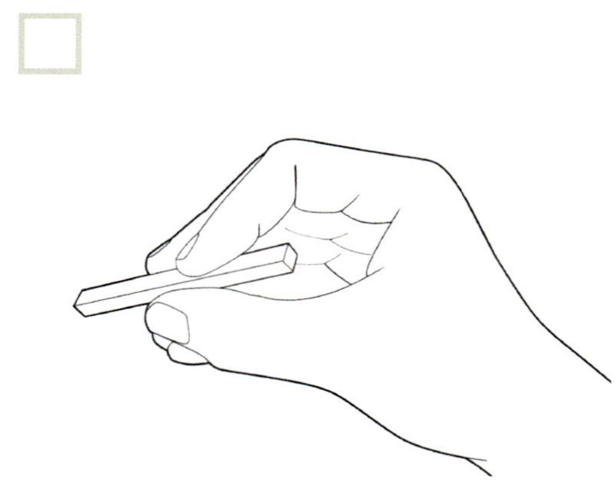

Apply a few extra streaks of white pastel to enhance the sense of motion.

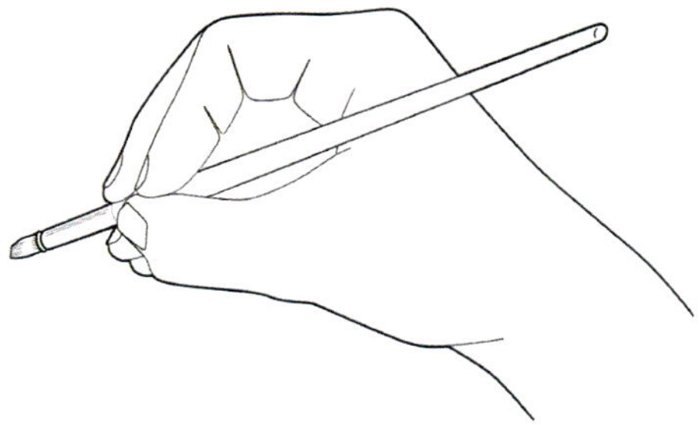

Finally, apply white gouache for the highlights. Remember to keep the highlights on the wheels in line with the wheel hubs.

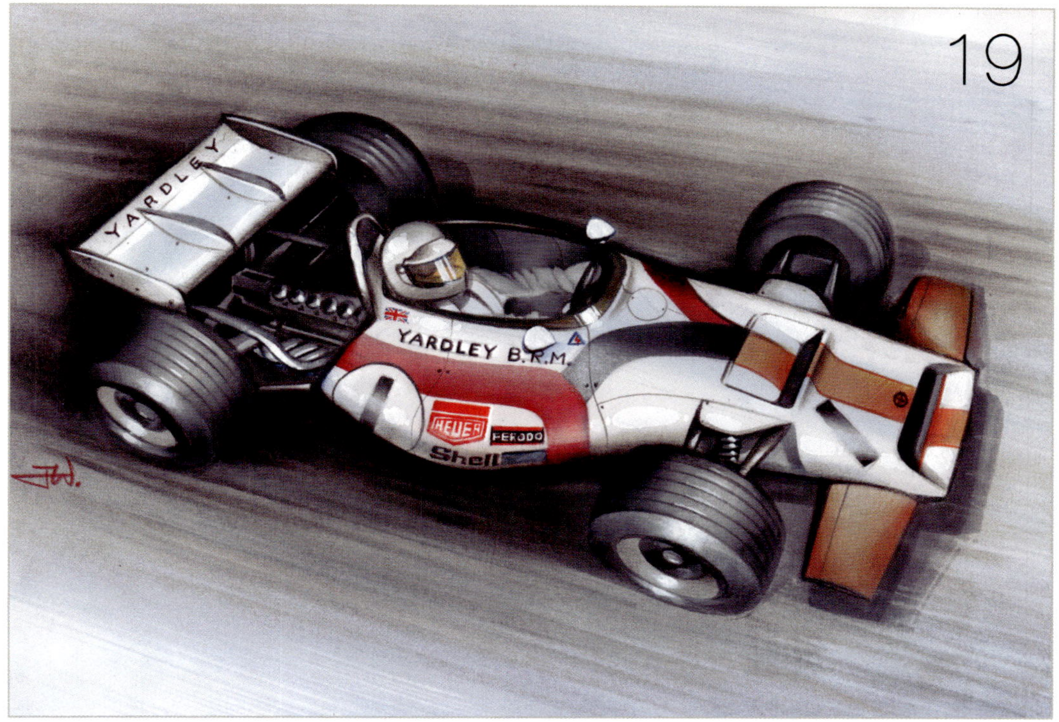

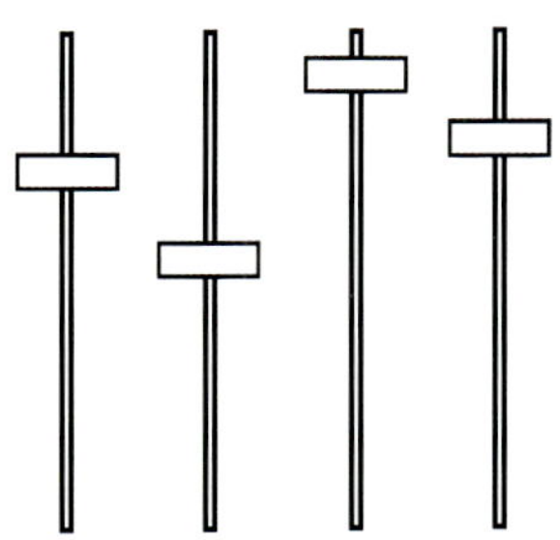

Scan in the rendering and punch up the color saturation in Photoshop. You may choose to apply a blur filter on the road surface to clean up the drawing and focus the eye on the subject.

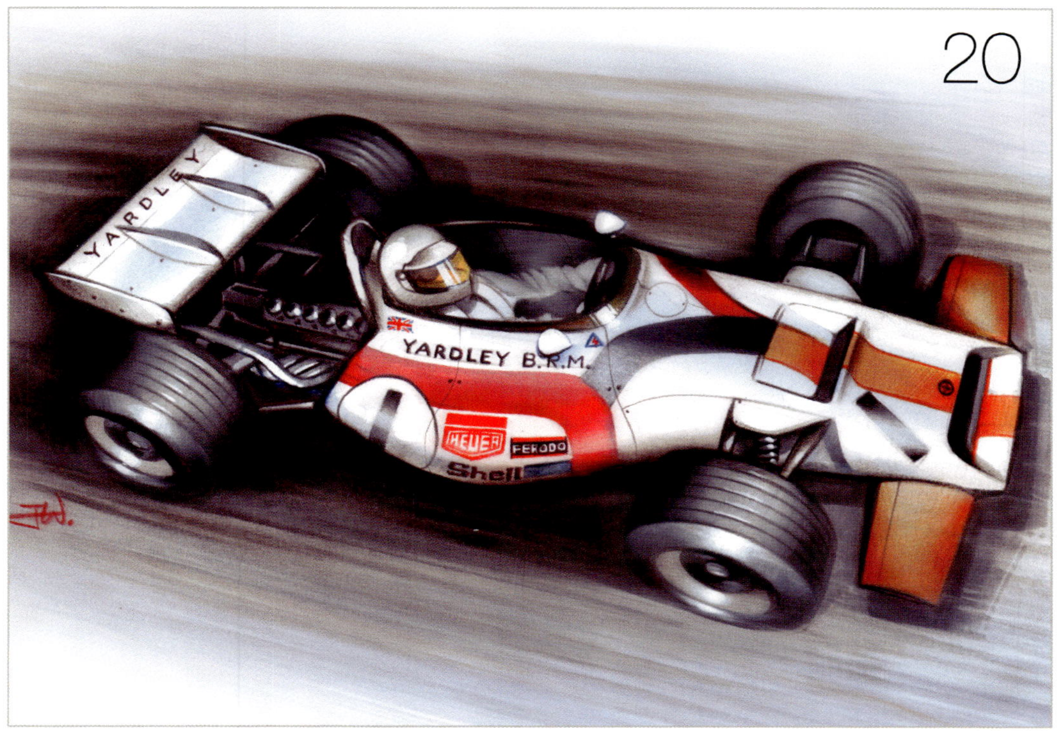

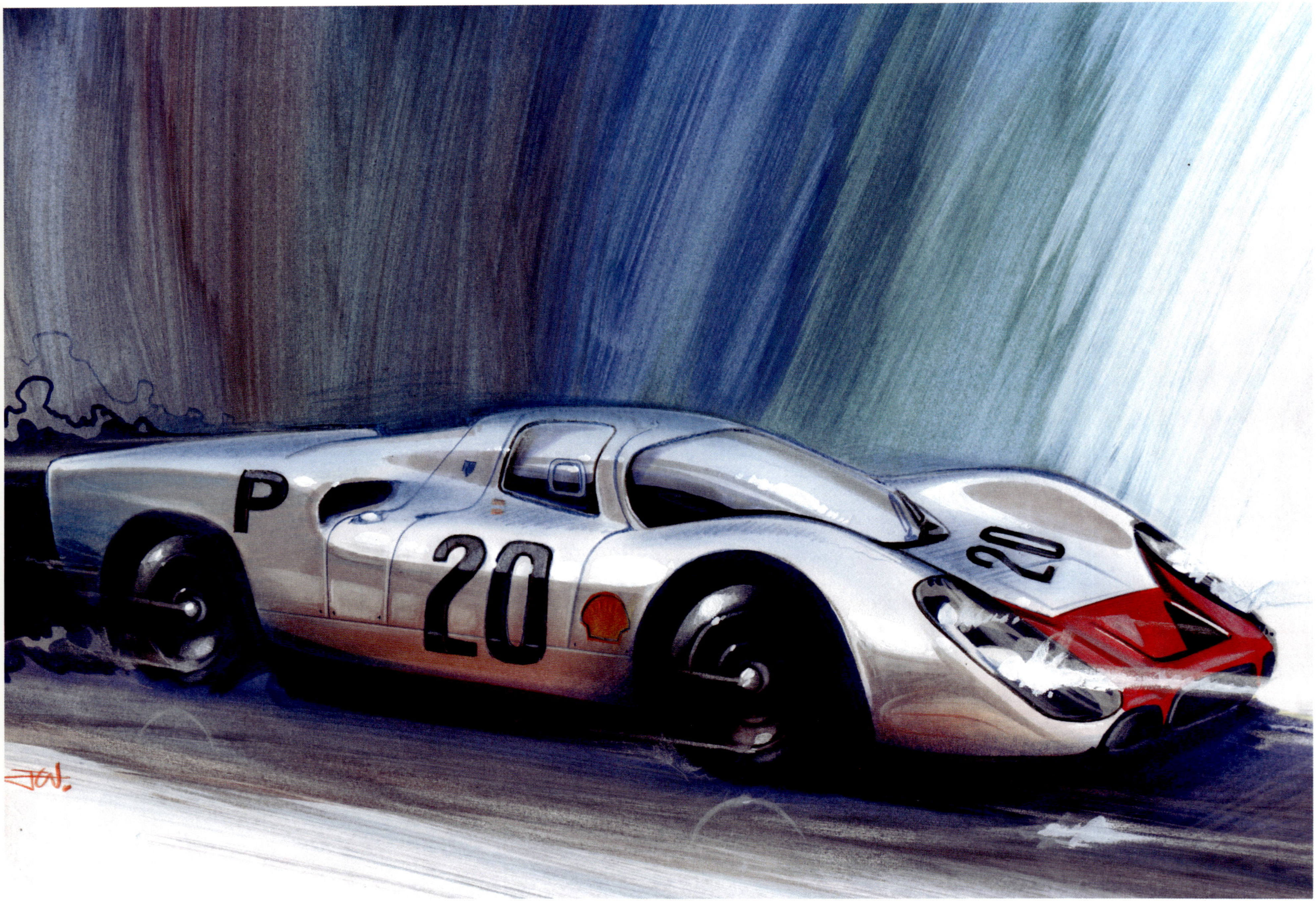
P
20
20

6. Abstract Vignettes

Here's a technique that definitely has its roots in the old school. Abstract vignettes like this one can give a rendering an almost painting-like quality. All it takes is a little planning, a bit of painter's tape, and an adventurous spirit when it comes to color.

The Tools

Prismacolors:
White, Orange.

Nupastels: Orange, Red Violet, Indigo Blue, Sienna Brown, Erin Green, White.

Verithins:
Black,
Indigo Blue.

Gouache:
White.

Prismacolor Markers: Cool Grey 20%, Cool Grey 40%, Crimson Red, Canary Yellow, Indigo Blue.

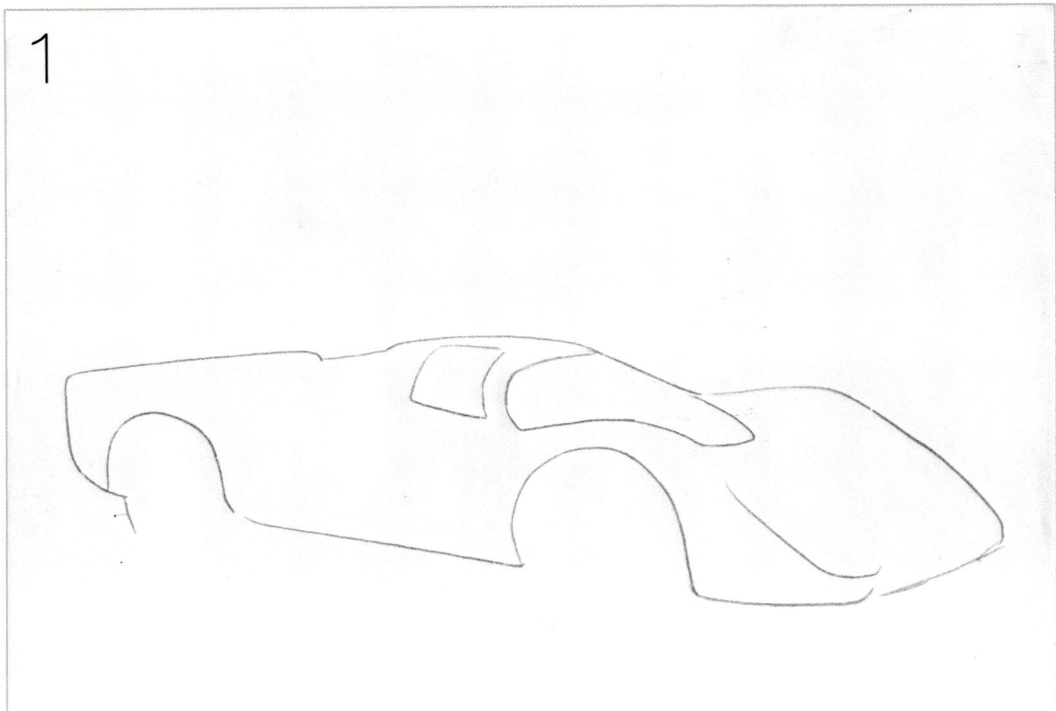

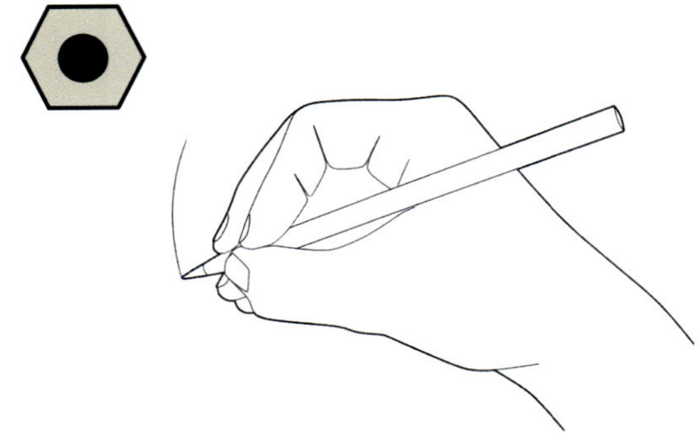

As with the previous tutorial, this racing car is painted white. So try to keep your linework on the marker paper to a bare minimum.

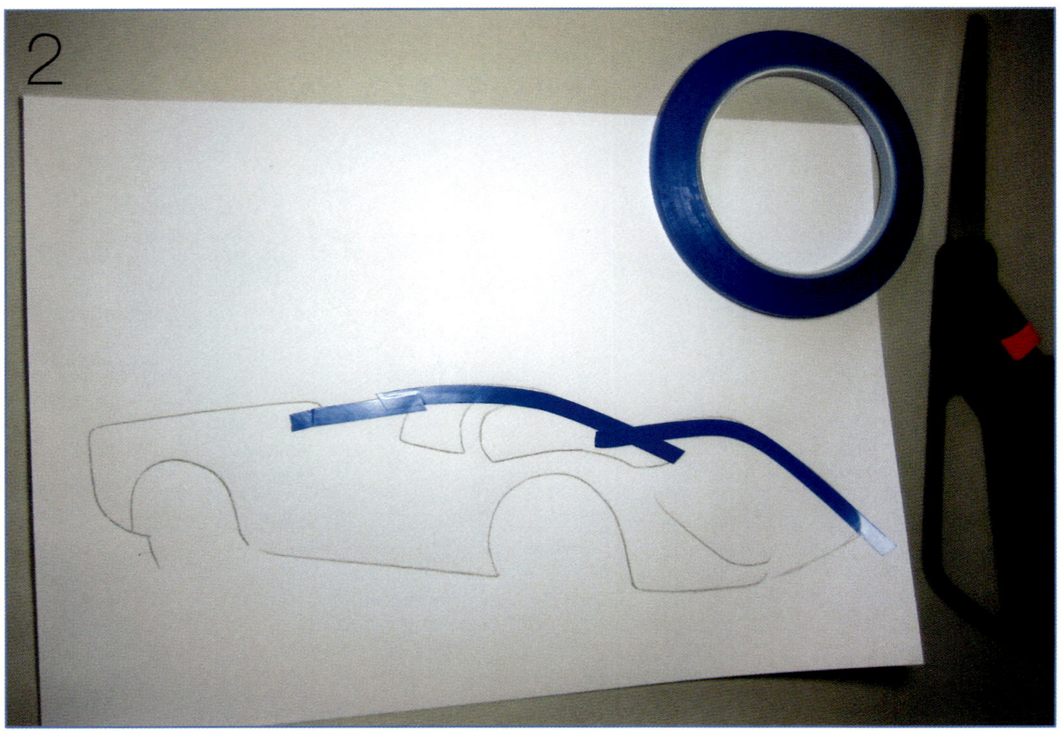

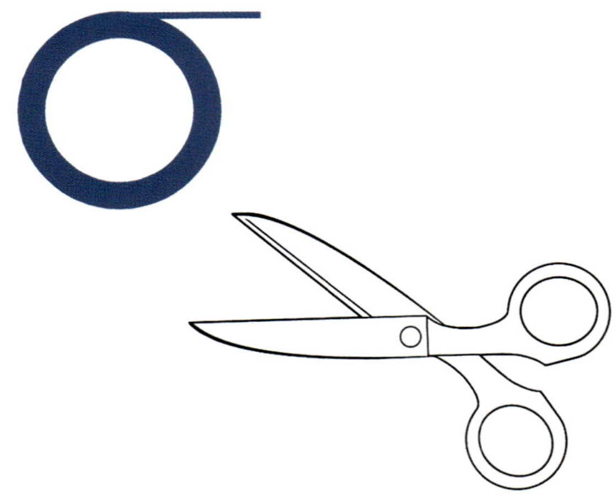

Begin affixing blue painter's tape to the inside edge of the car's profile. It is important for the tape to be smooth, with no crimps or kinks.

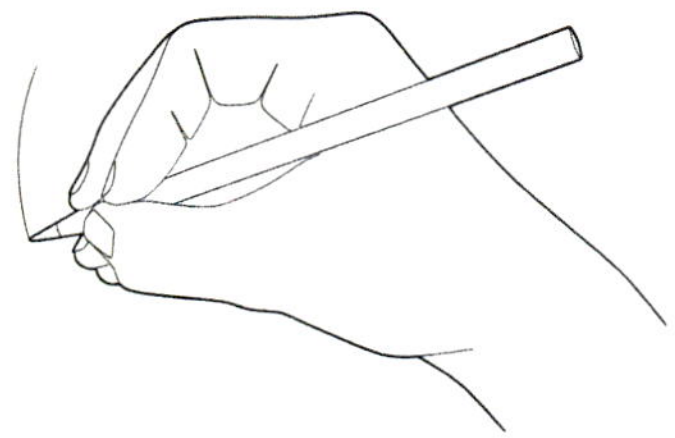

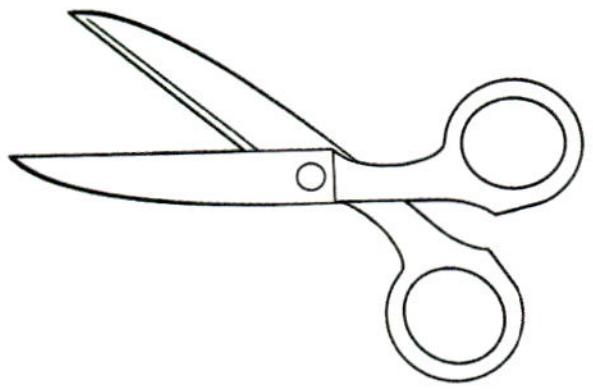

Trace the profile onto a piece of drafting mylar and cut it out with a pair of scissors. Don't worry about capturing the lower portion of the car.

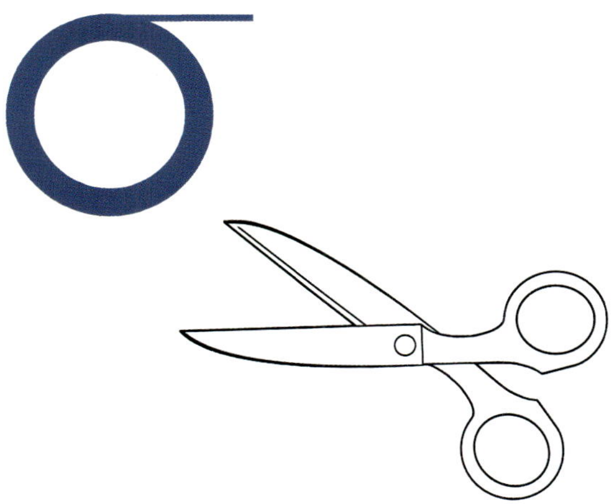

Now attach the mylar to the profile edge, leaving the lower edge loose.

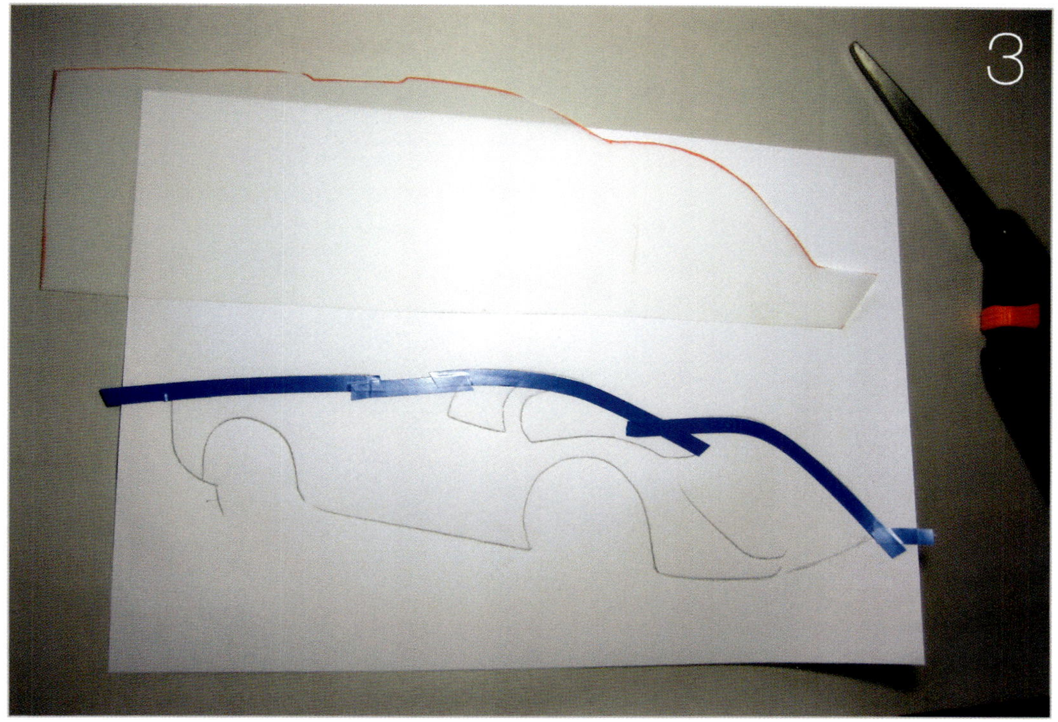

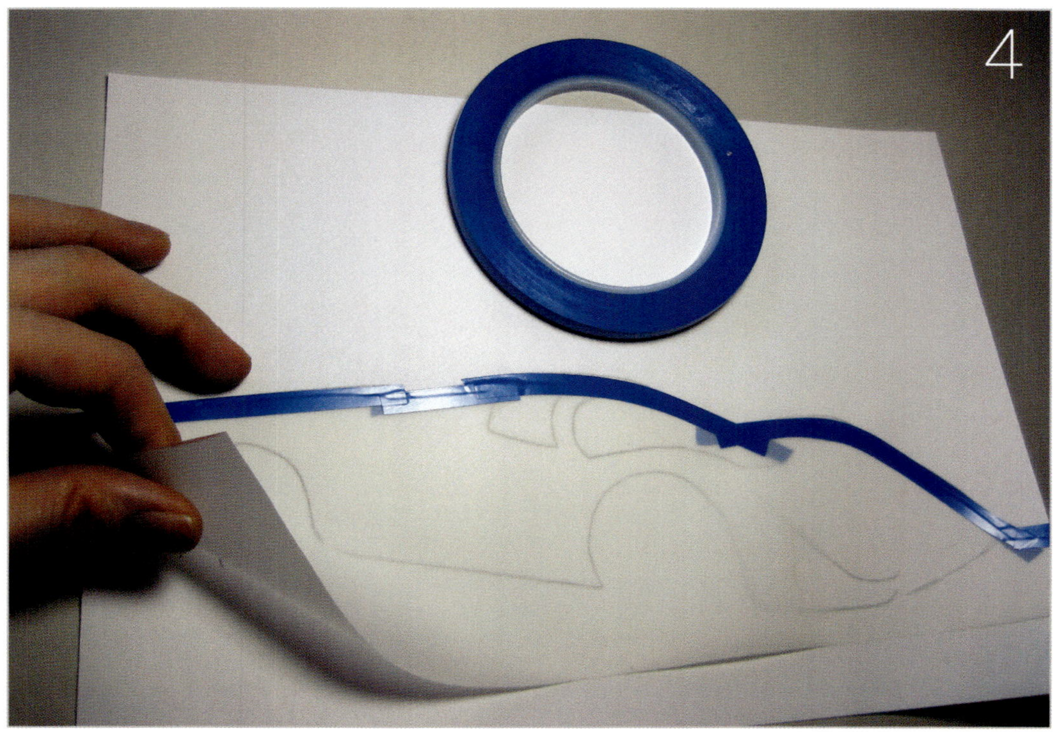

5

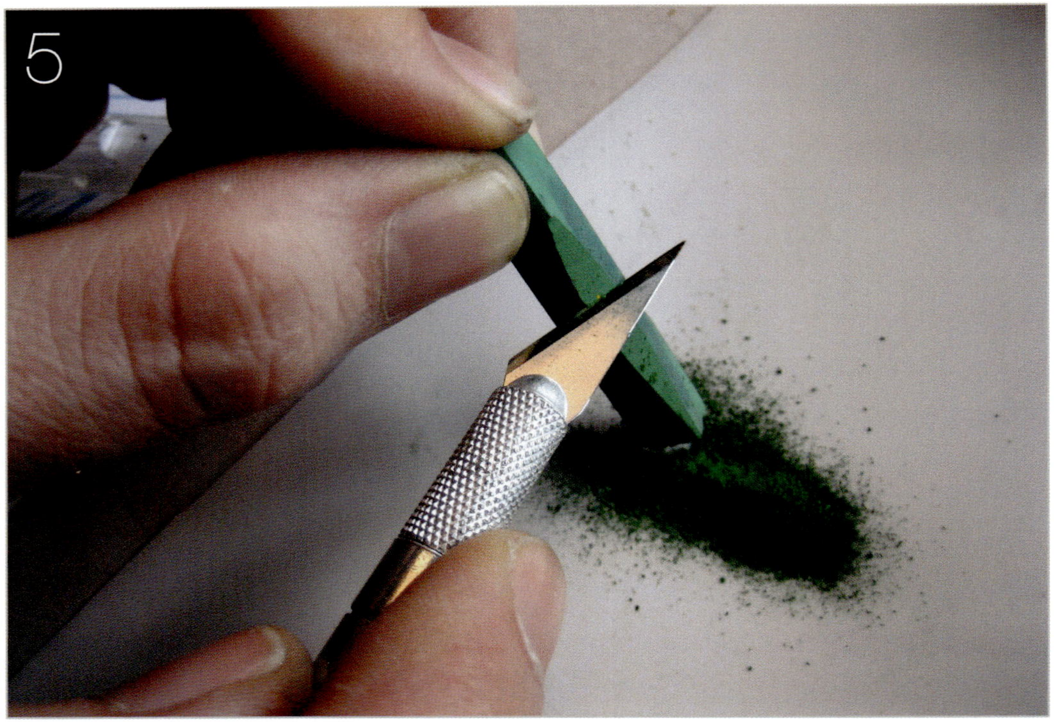

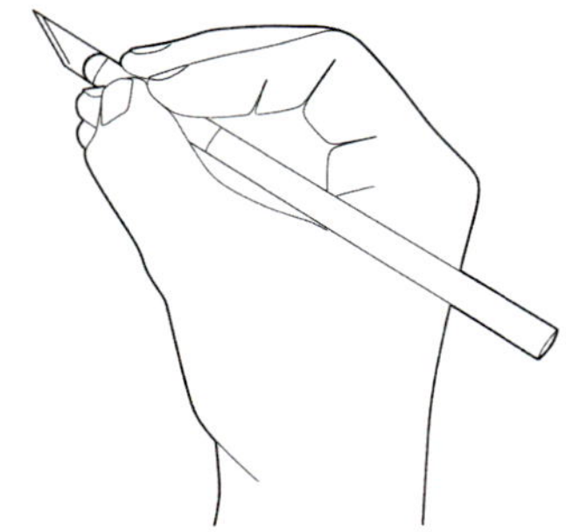

Using an X-acto knife, *carefully* scrape a generous amount of green and blue pastel onto a scrap piece of paper.

6

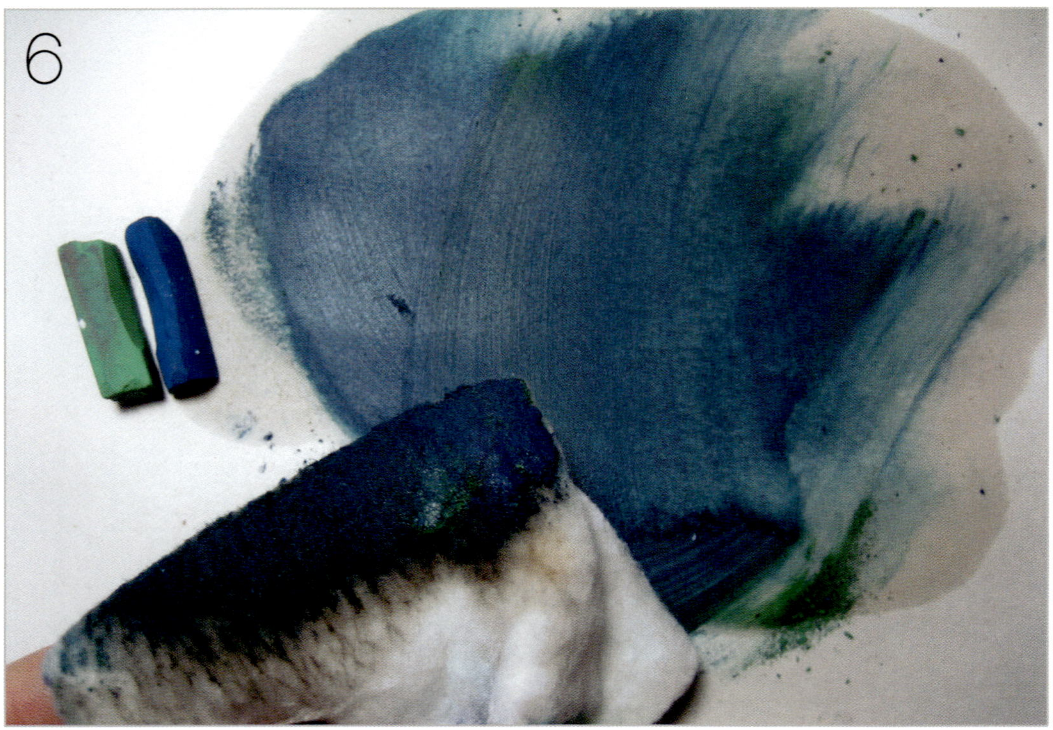

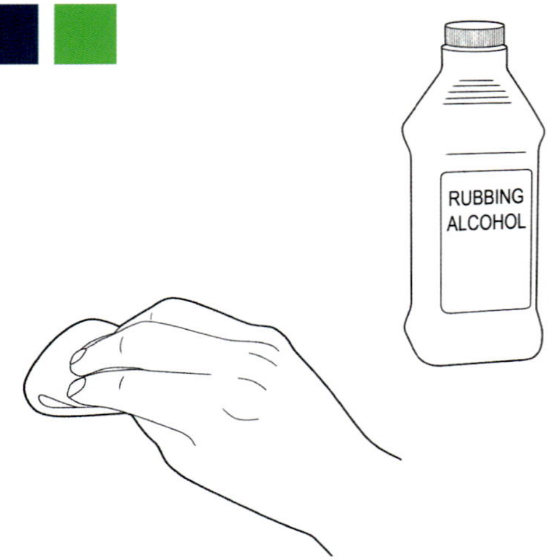

Pour a small amount of rubbing alcohol into the pastels and sop the mixture up with a clean Webril pad.

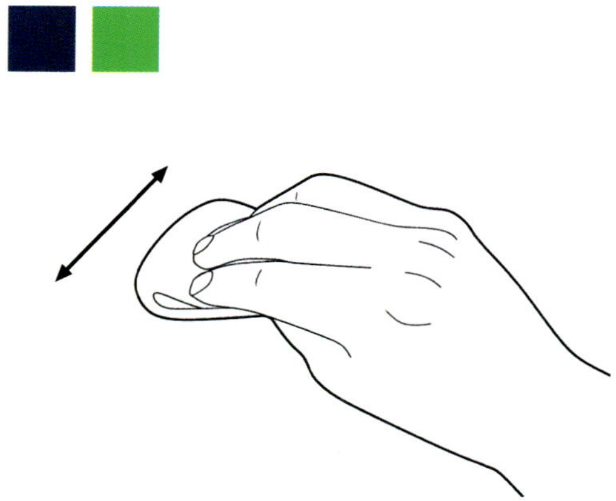

Using north-south strokes, begin brushing in the abstract background. Place a pad of newsprint under the marker paper to absorb the run-off.

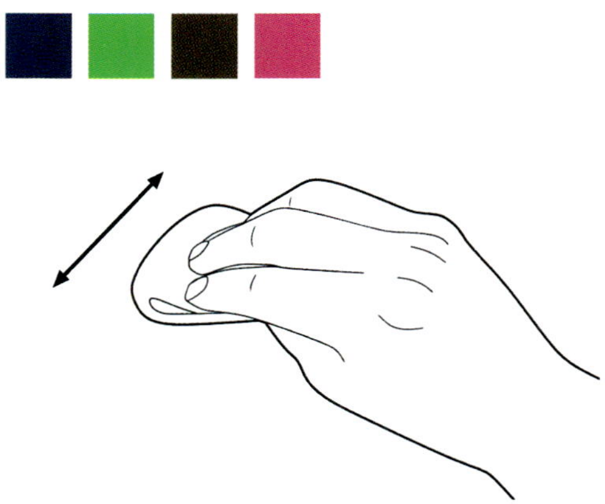

Begin dropping in additional colors in select locations. Blend the colors a little bit, but don't let the background become a muddy mess. Exercise restraint.

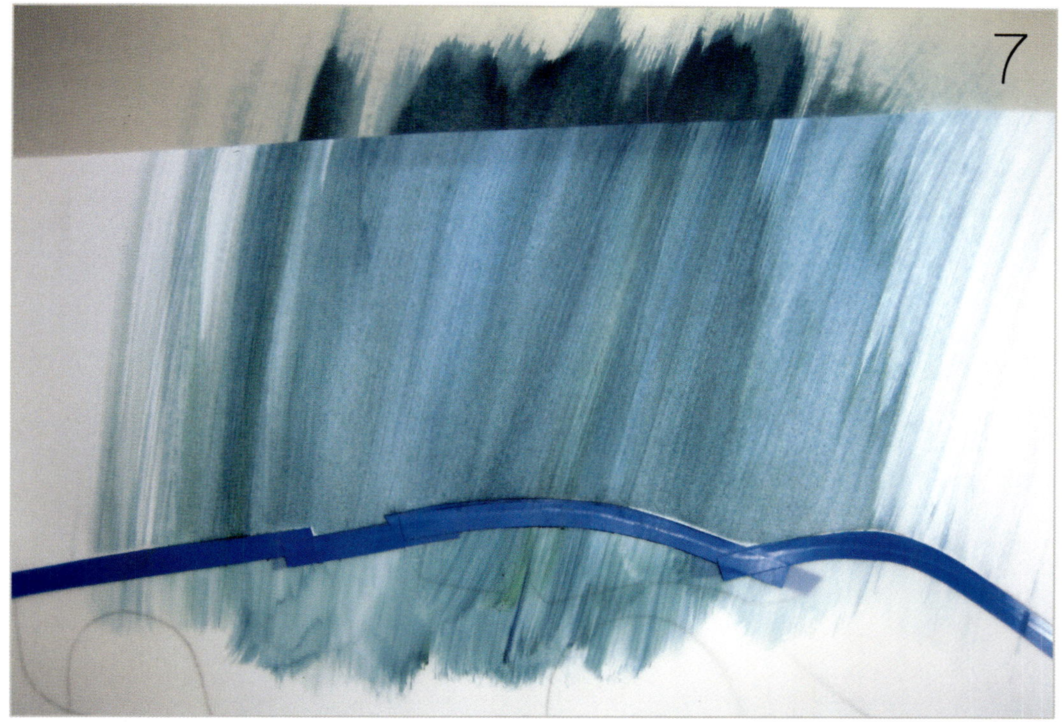

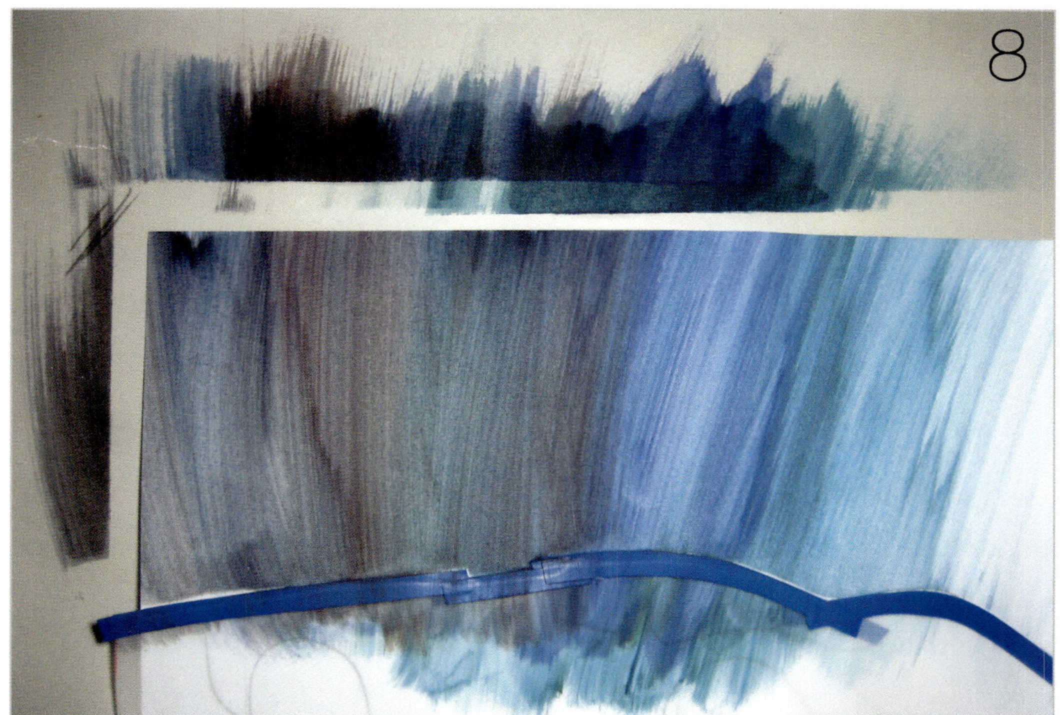

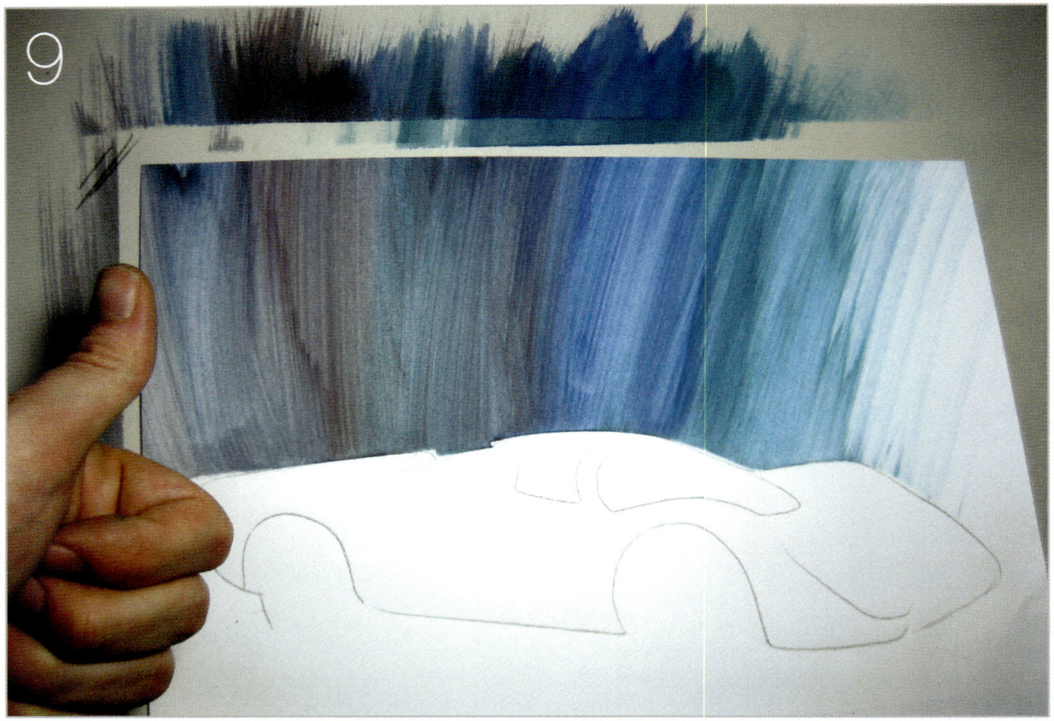

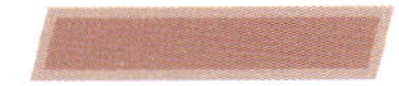

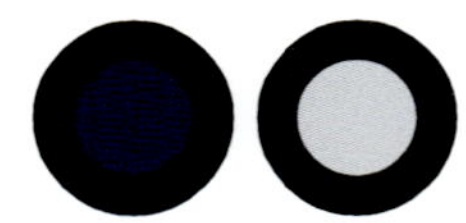

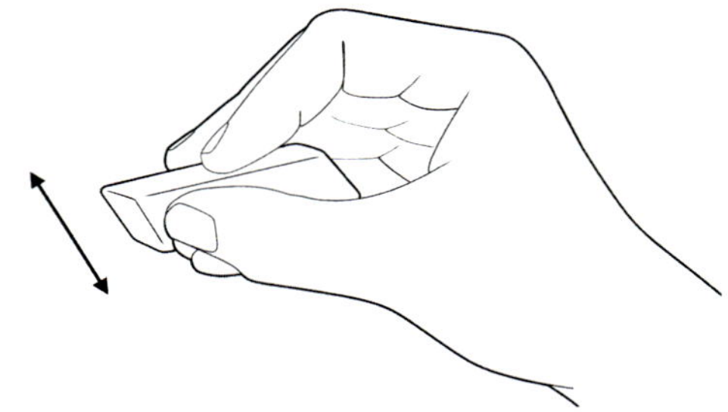

After the alcohol-pastel mixture has dried, remove the painter's tape and the piece of mylar. Erase any pastel strokes that may have eluded the mylar.

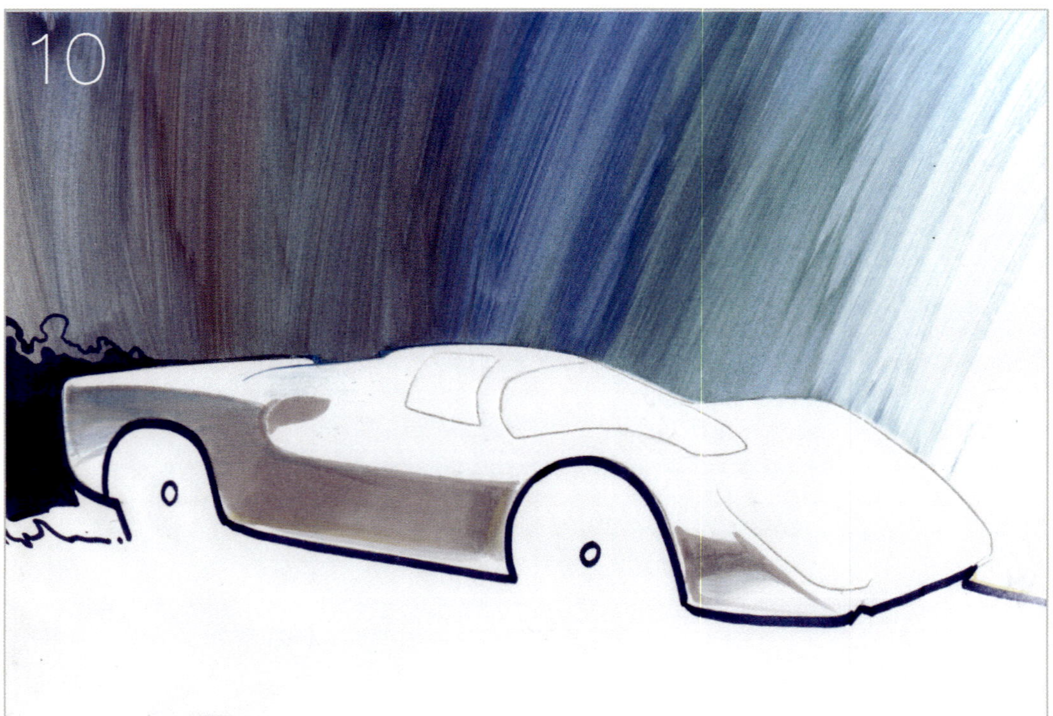

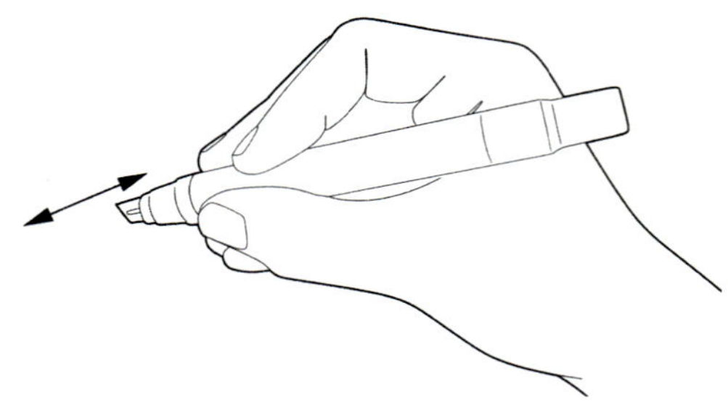

Finally, it is time to proceed to markers. Start off with Cool Grey for the reflections and blue for the wheel wells.

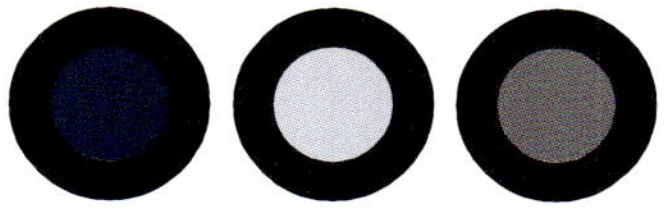

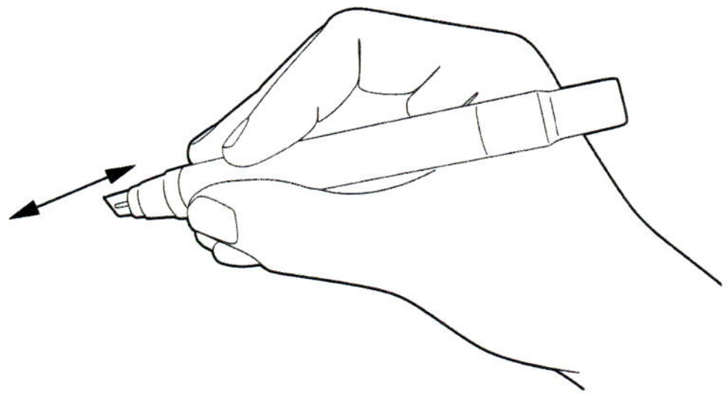

As you fill in the wheelhouses, use the Cool Grey to blend between the edge of the Indigo Blue marker and the white paper; this softens the shadows.

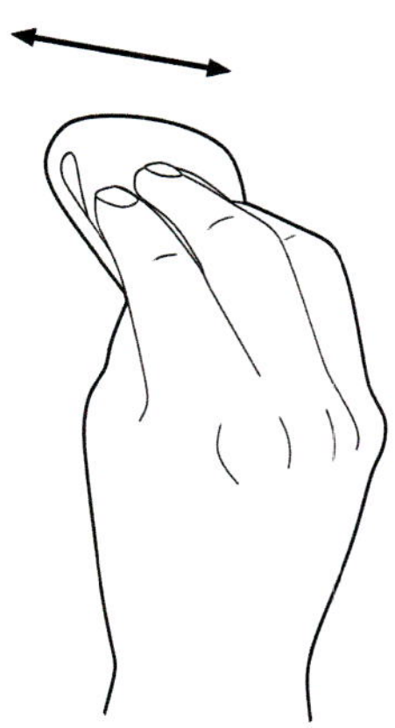

To tie the road surface in with the background, take what is left of the alcohol-pastel mixture and spread it at the base of the car. Again, don't over-do it.

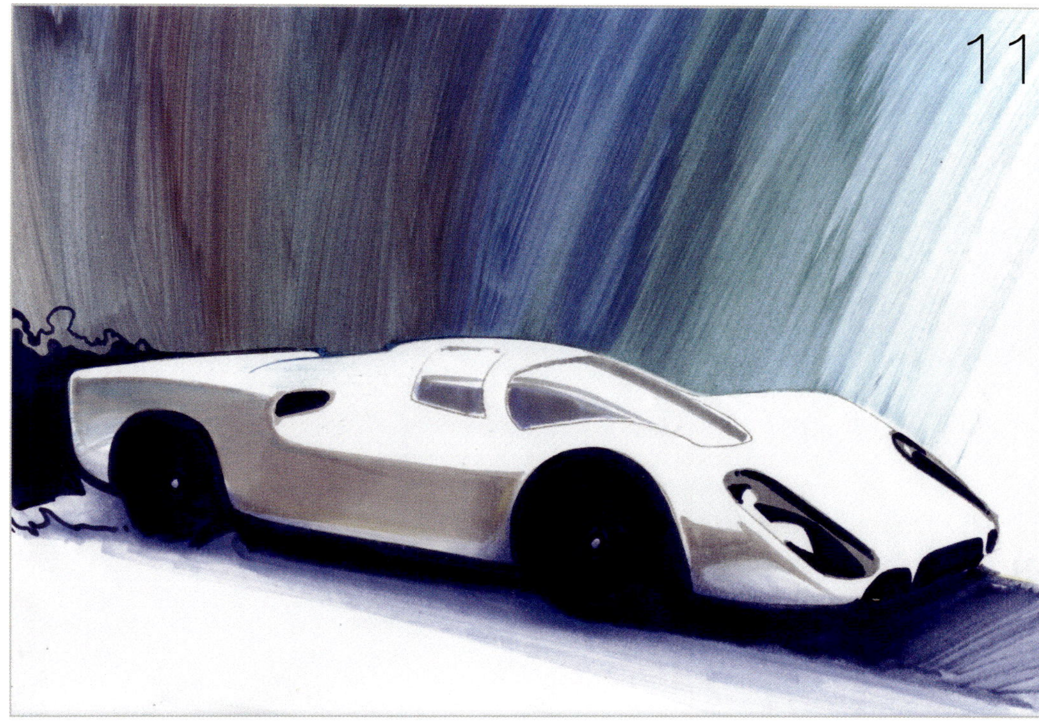

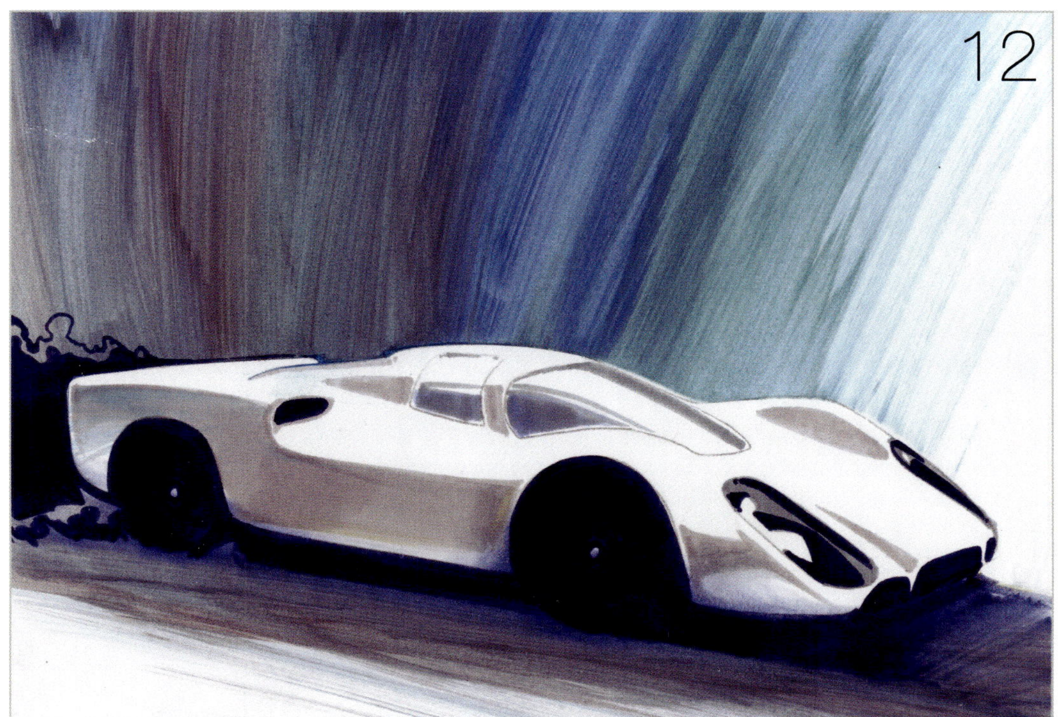

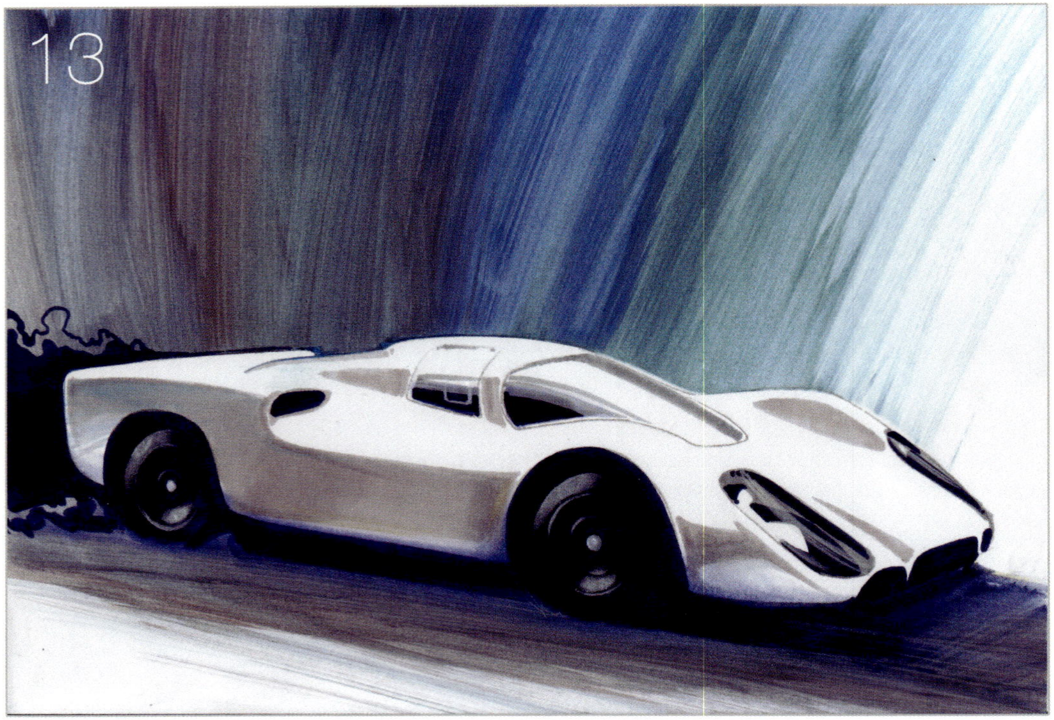

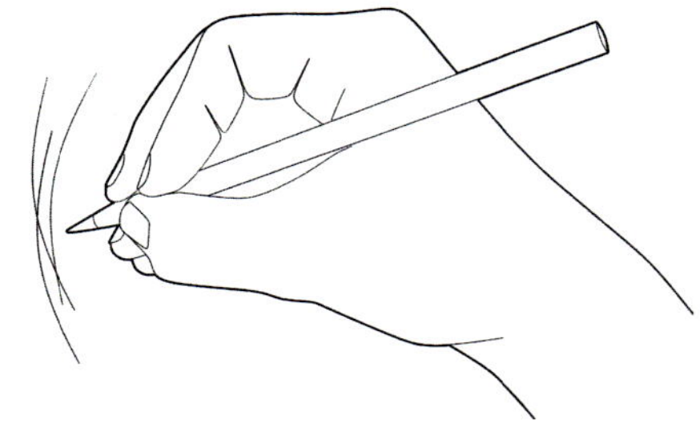

Define the racing tires with white Prismacolor pencil. Make sure that the highlights are in line with the wheel hubs.

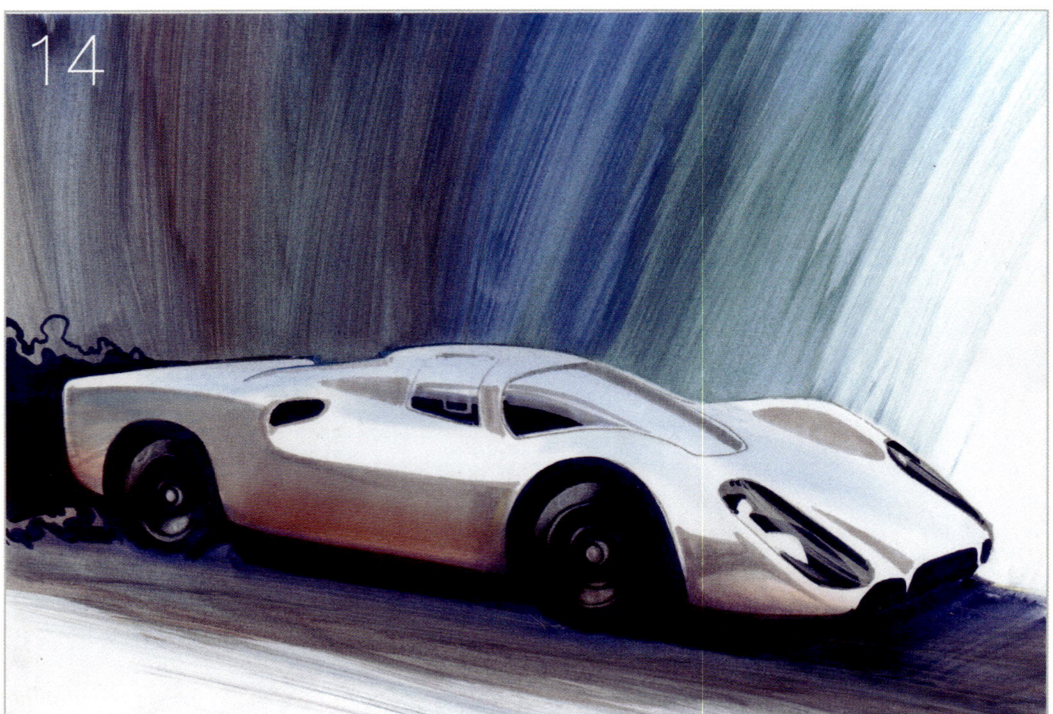

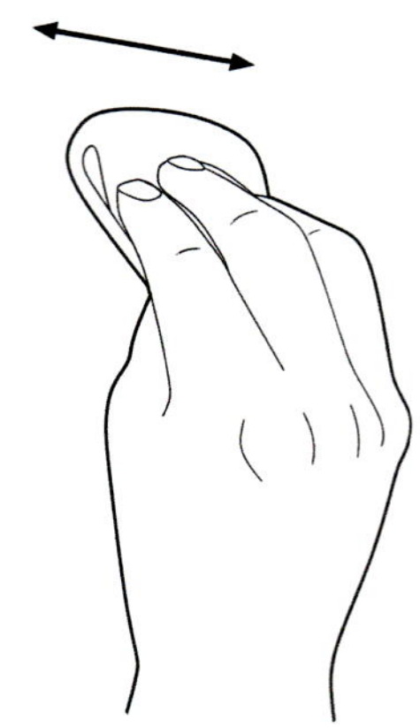

Now add some warm and cool tones to the body with a clean Webril pad (cool on top, warm on the bottom).

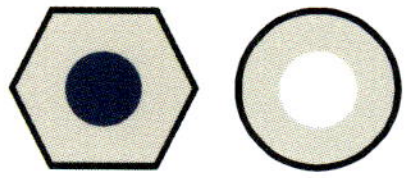

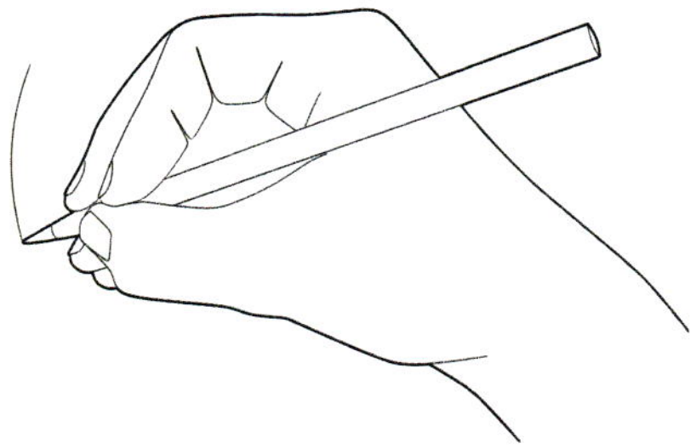

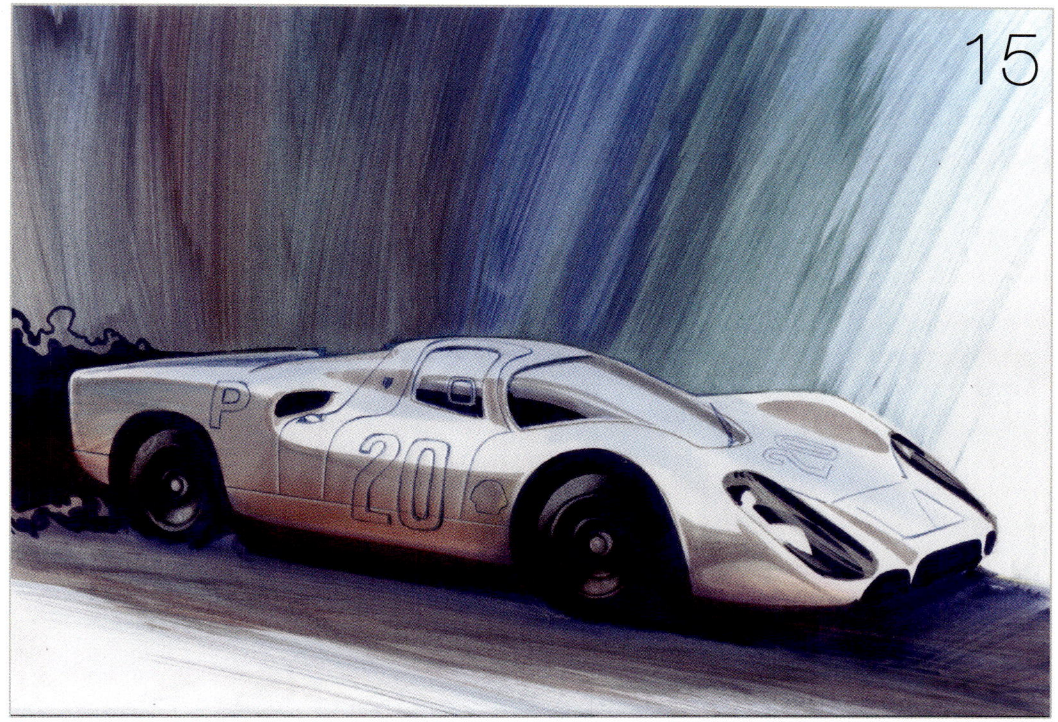

With sharp pencils, begin laying out the sponsors, numbers and shut lines. Make sure you place the shut line highlights on the edges that are facing you, the viewer.

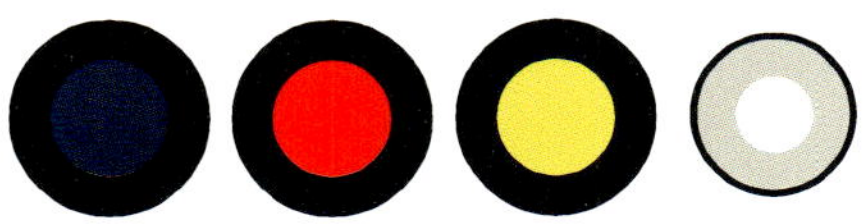

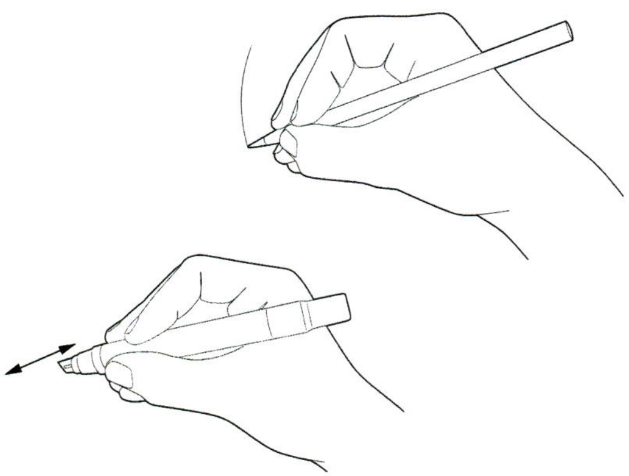

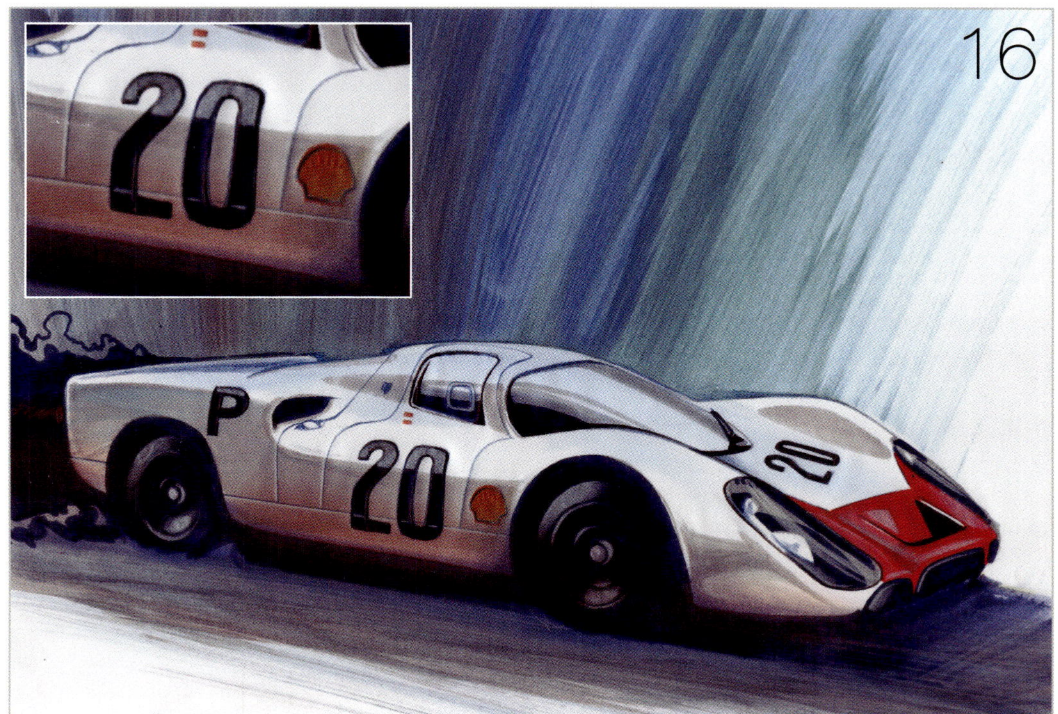

Now fill in the graphics with their corresponding marker colors. Using a white pencil, continue the main bodyside highlight through the "P" and number 20.

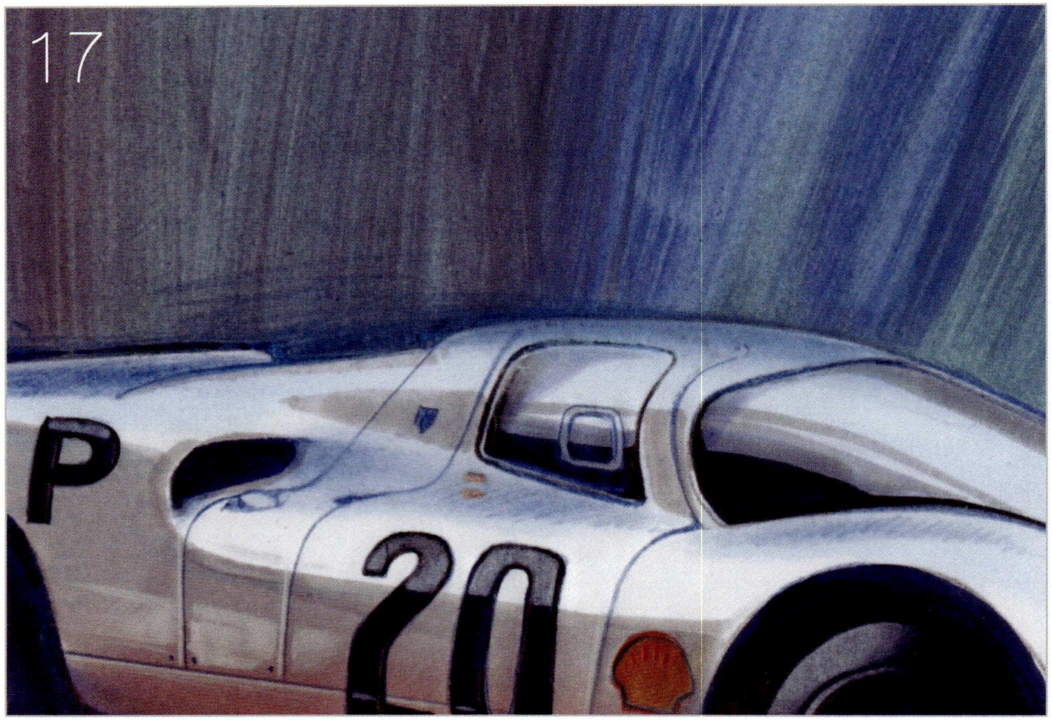

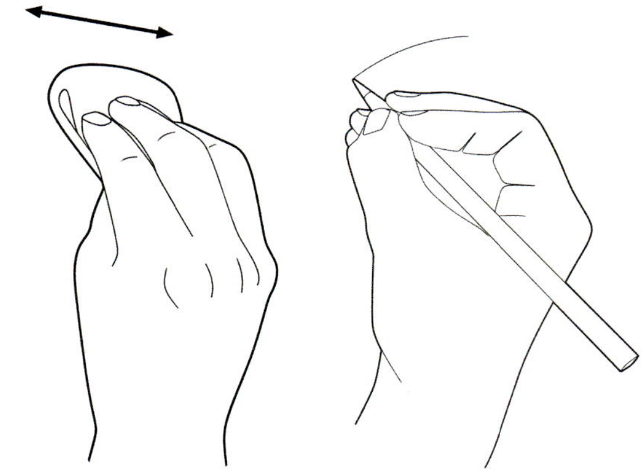

Take a moment to punch up the sky tone with a bit more blue pastel and maybe a bit of blue Verithin.

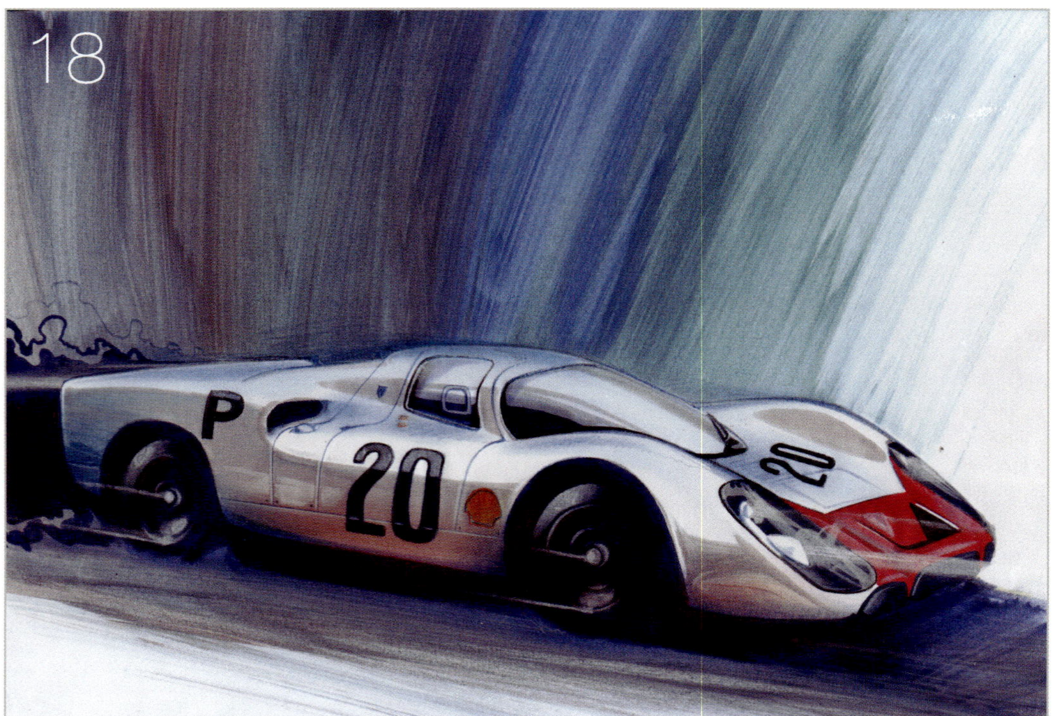

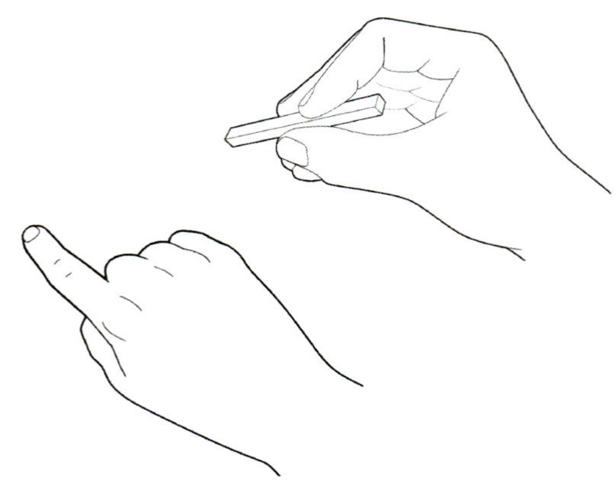

Directly apply white pastel for the headlamp beams and the contrails. Use your fingertip to soften the strokes.

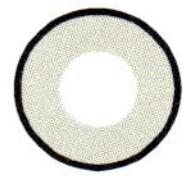

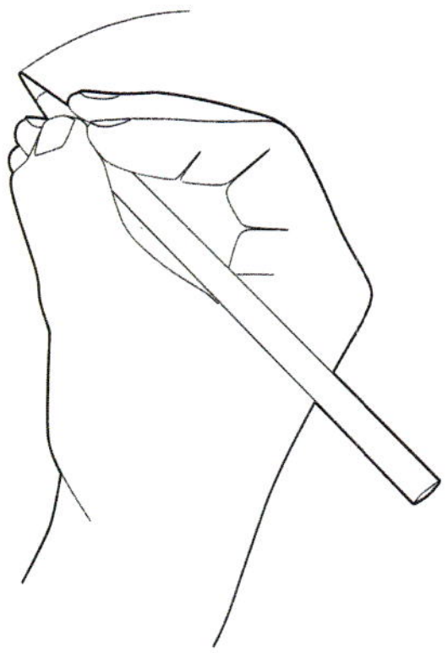

Add a bit more definition to the road surface with a white pencil. For example, reflect the rims into the road to give the suggestion of a damp surface.

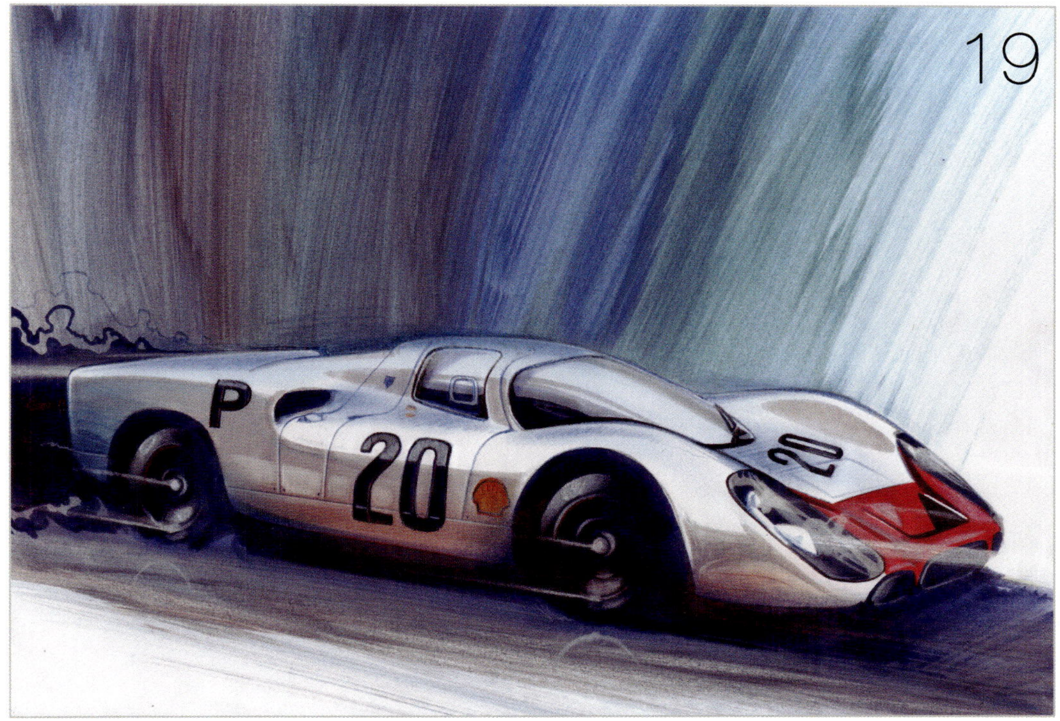

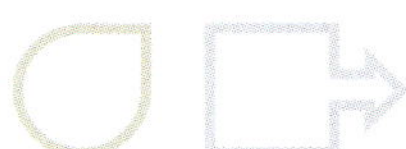

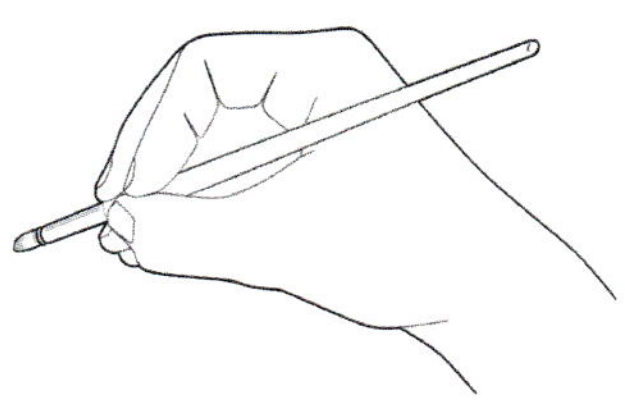

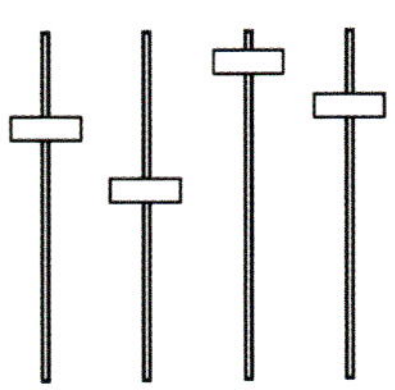

Apply gouache highlights here in a more painting-like manner, especially near the headlamps. Scan in the finished car and raise the color levels in Photoshop.

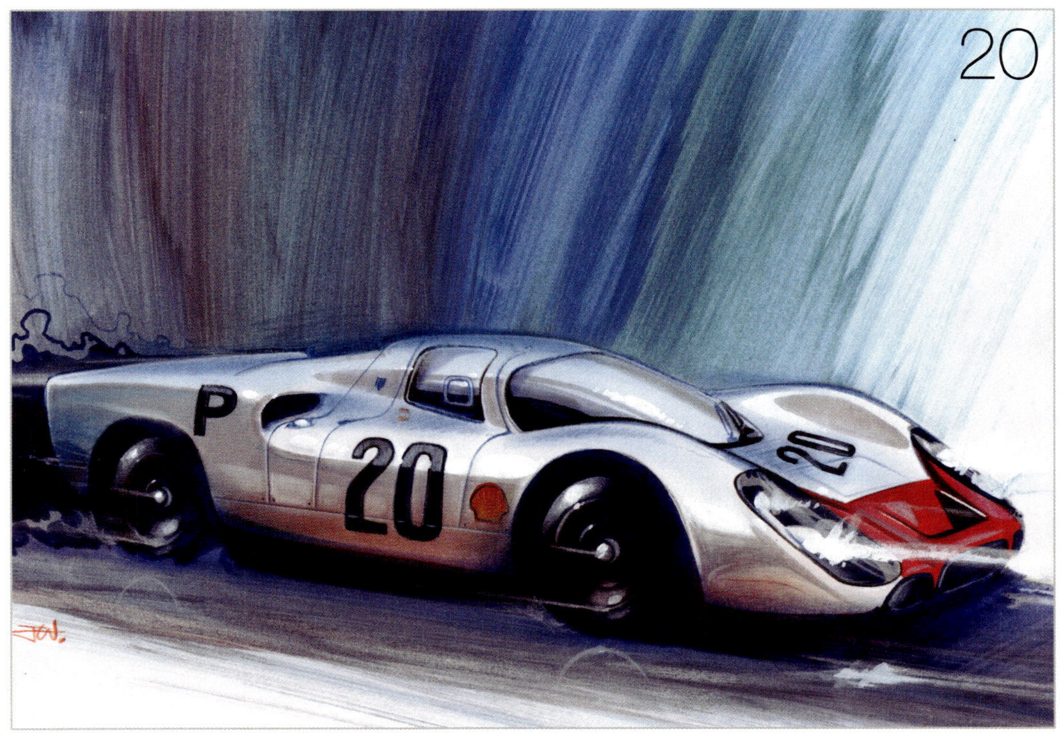

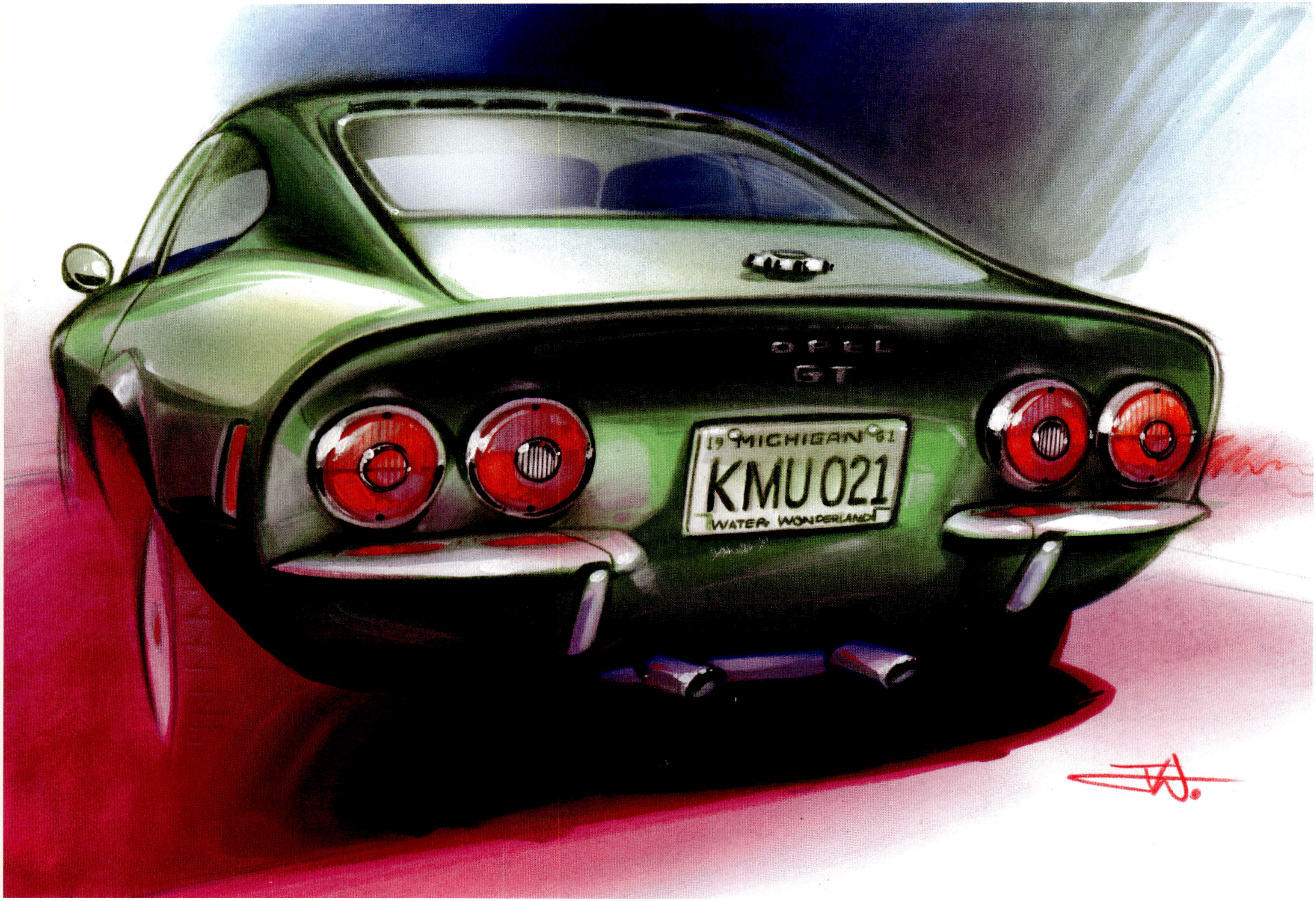
OPEL
GT
19 MICHIGAN 61
KMU 021
WATER WONDERLAND

7. Abstract Shadows

Just because most shadows are black or grey doesn't necessarily mean they have to be rendered that way. Applying abstract colors to your rendering in strategic locations can add some zest that wouldn't be there otherwise.

The Tools

Prismacolors:
White, Dark Green, Crimson Red.

Nupastels: Red Violet, Indigo Blue, Crimson Red, Erin Green, Lemon Yellow.

Verithins:
Black.

Gouache:
White, Red.

Prismacolor Markers: Celadon Green, True Green, Grass Green, Olive Green, Mulberry, Cool Grey 20%, Crimson Red, Orange, Indigo Blue, Black.

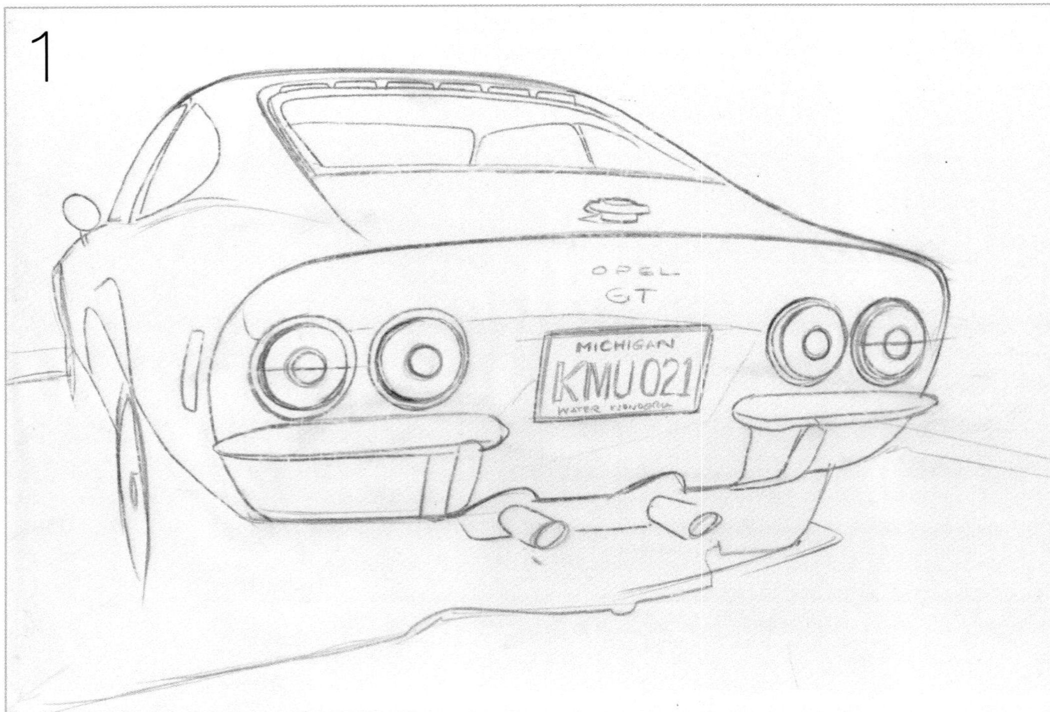

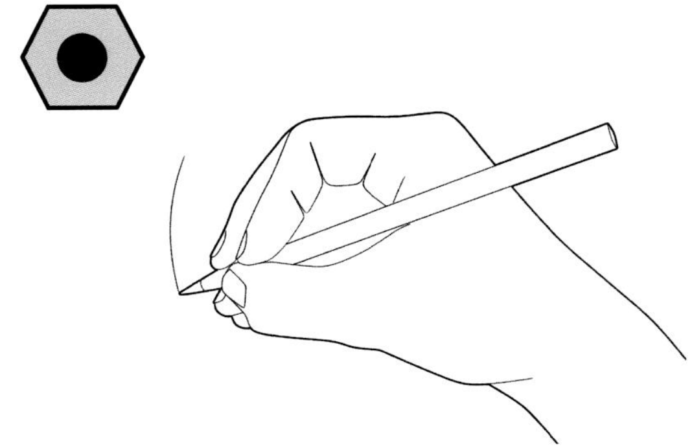

Start by laying out the linework of the car. Take time to describe a few details as well, but only in a cursory way.

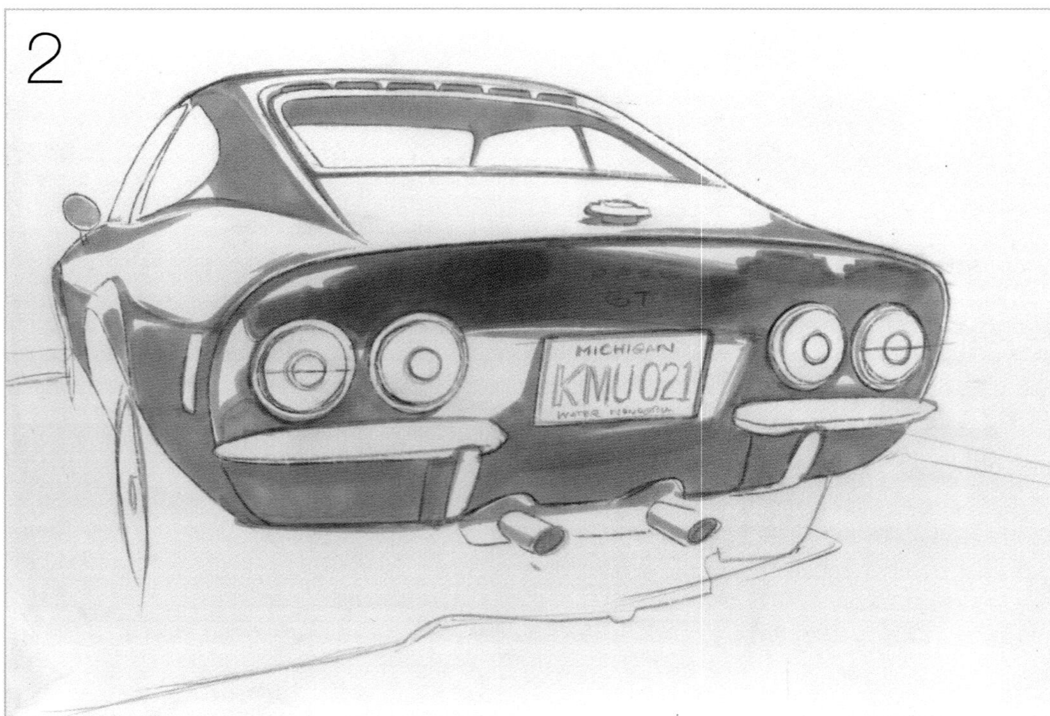

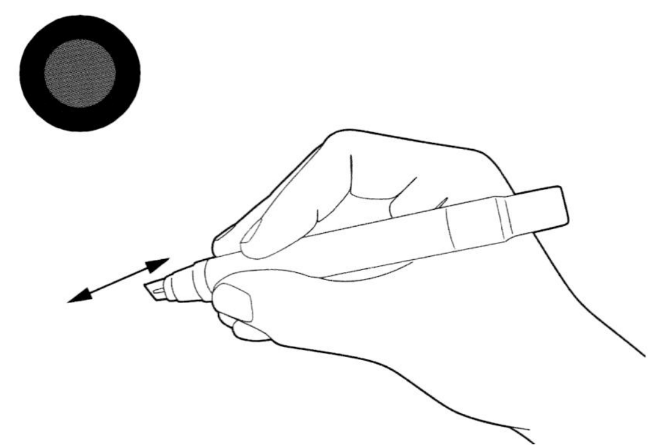

Apply a layer of Celadon Green marker in all the dark areas of the car – except for the shadow. Leave some isolated patches of white paper on the underside edge of the deck lid; these will represent light reflected up from the taillamp bezels.

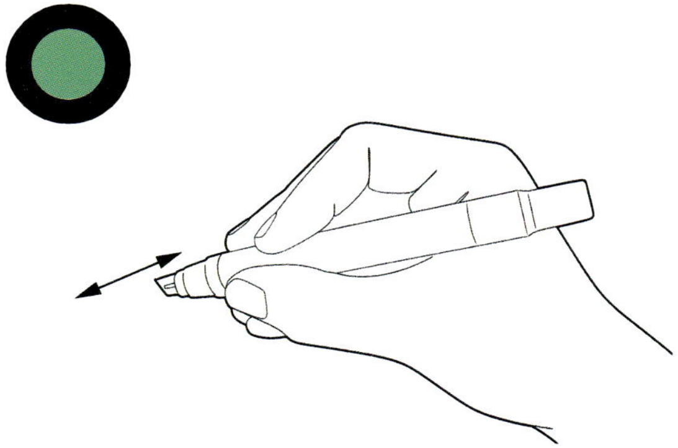

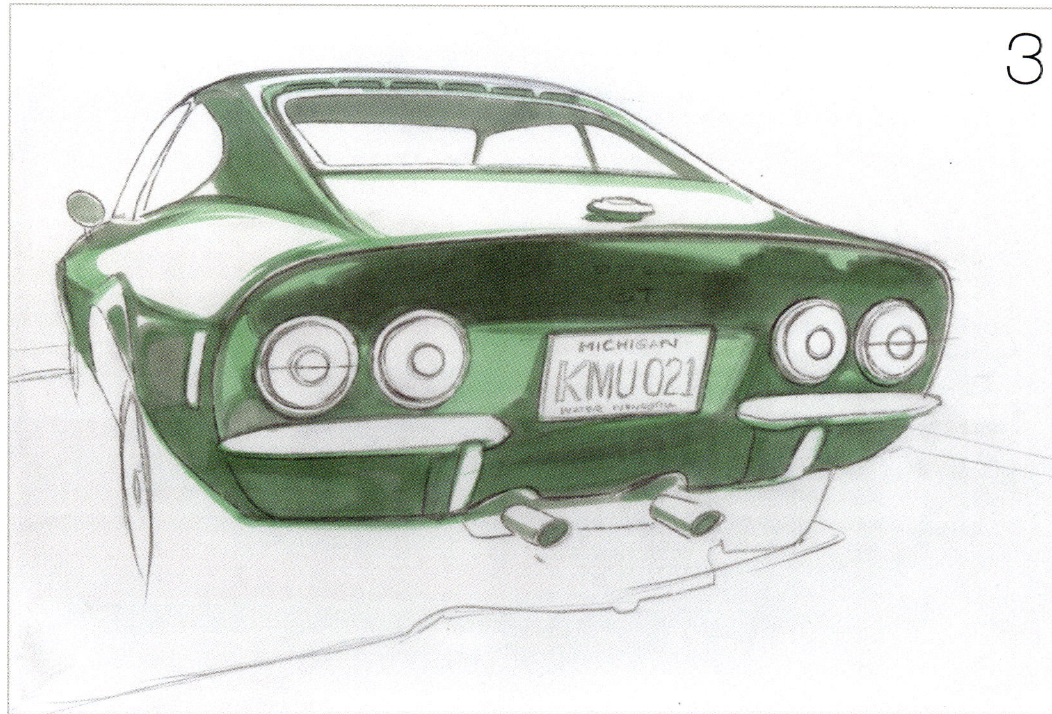

Next, lay in a second layer of marker, this time True Green. Here, you can see how layering one marker over the other adds depth to the rendering.

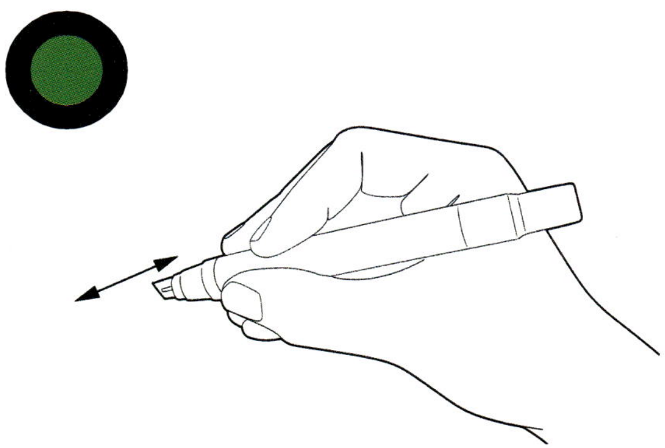

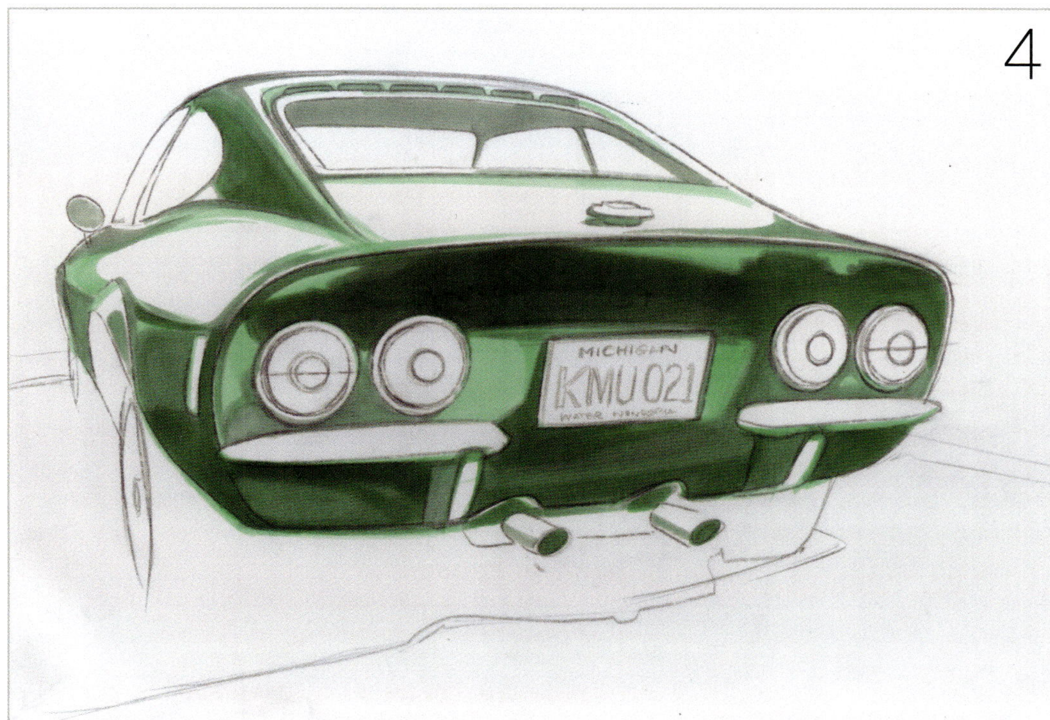

Now try punching up some of the darker reflections with Grass Green.

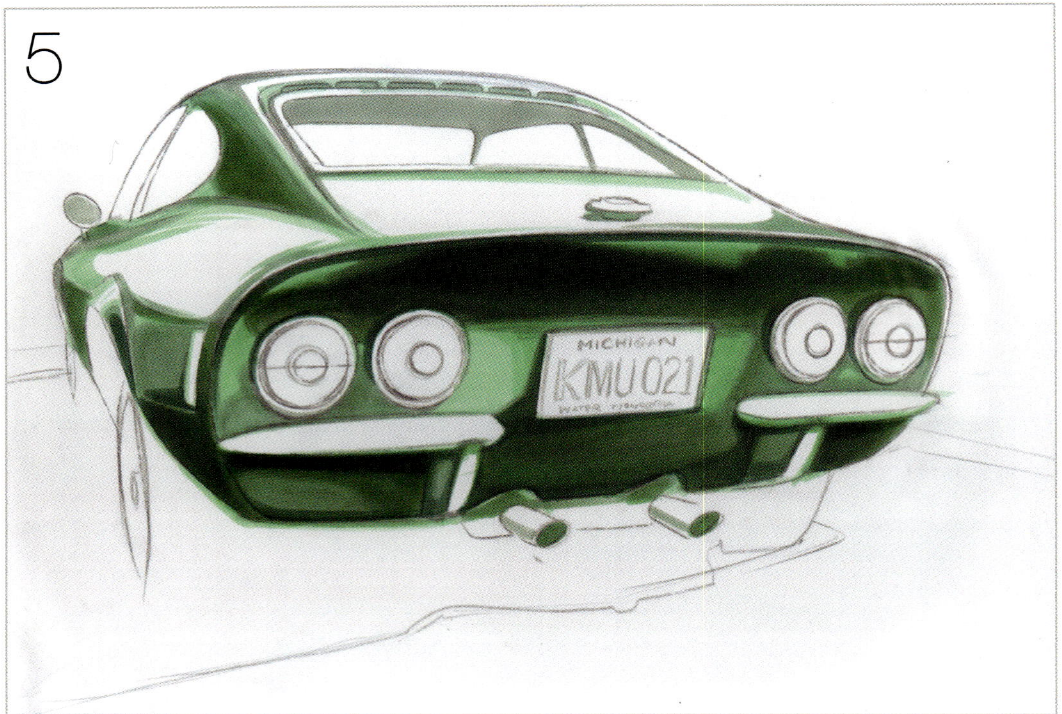

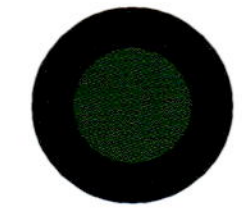

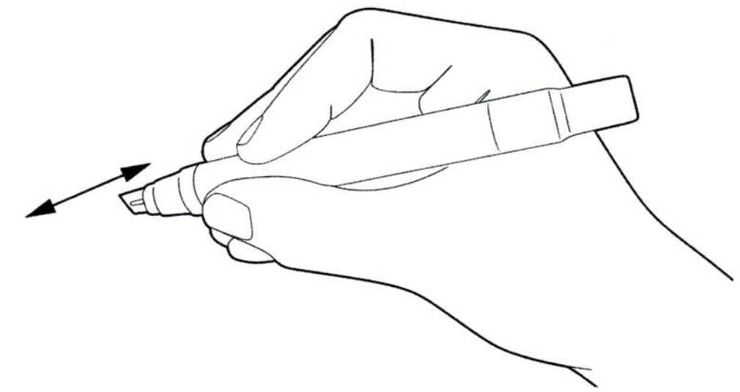

To take the marker depth even further, hit the underside of the deck lid with a thin layer of Dark Olive Green.

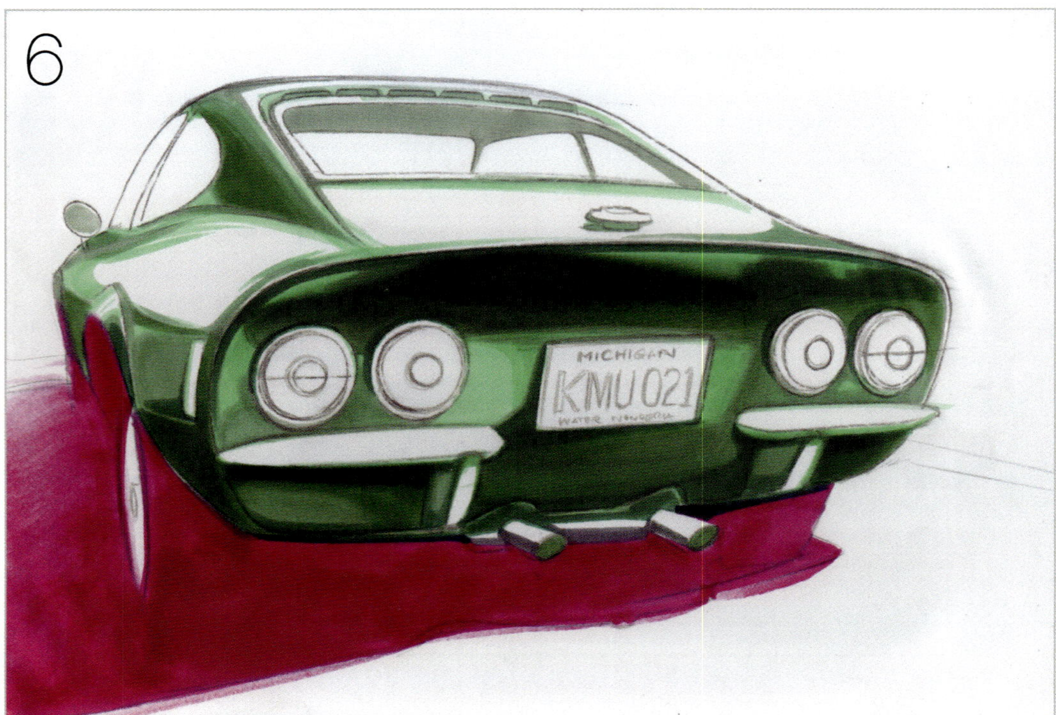

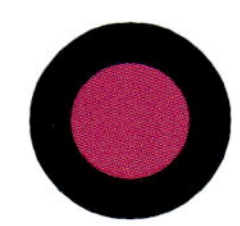

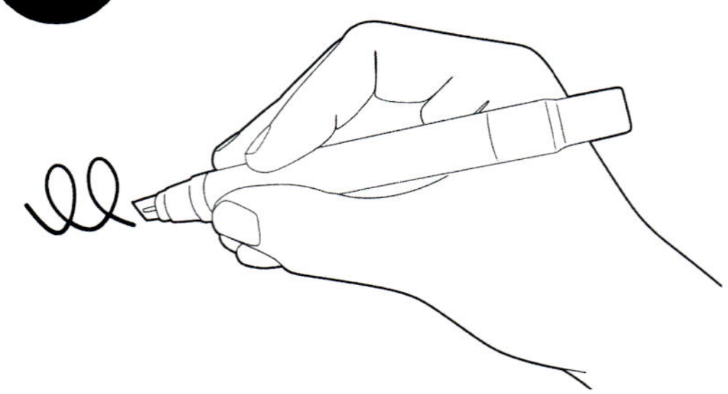

Now it's time for the abstract shadow. Be sure to apply the Mulberry marker in circular motions to minimize the visible strokes.

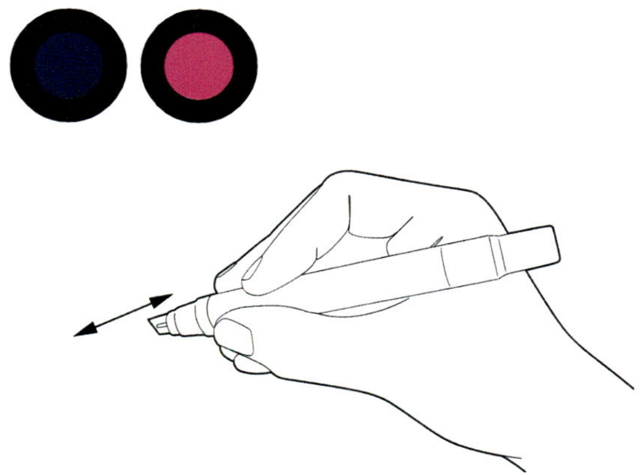

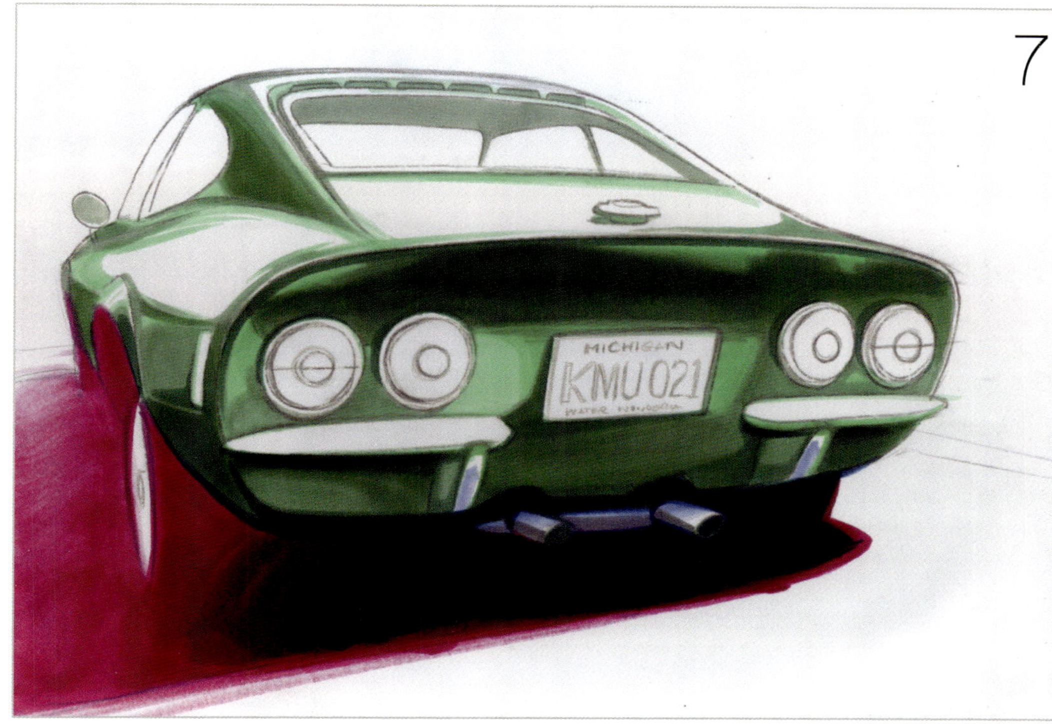

For the dark side of the shadow, brush in some Indigo Blue from right to left. Return with a bit of Mulberry to smooth out the color transition.

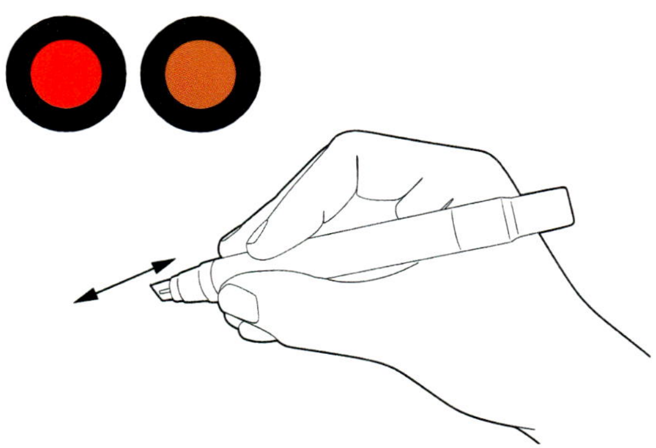

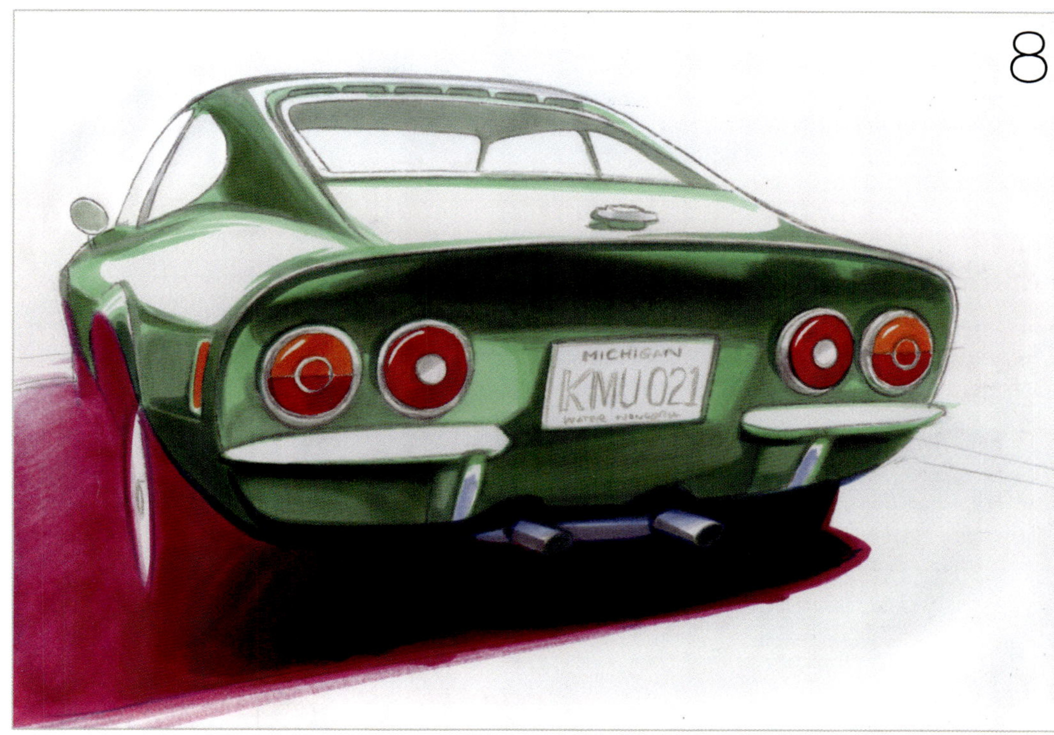

Take a moment to block in the taillamps. Leave small slivers of paper showing through on each lamp's top-left edge; this will represent a highlight later.

9

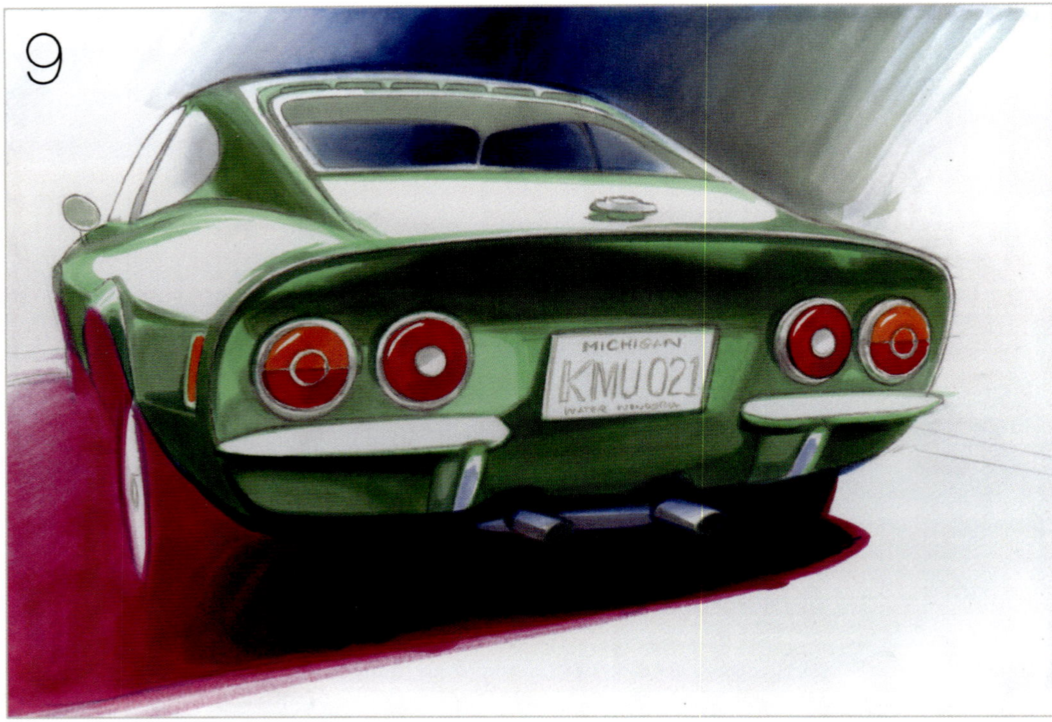

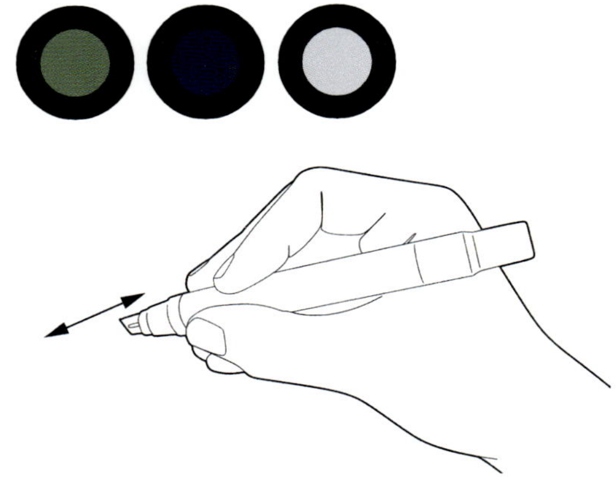

For the background, try a mixture of Celadon Green and Indigo Blue. Add Cool Grey to smooth out the edges.

10

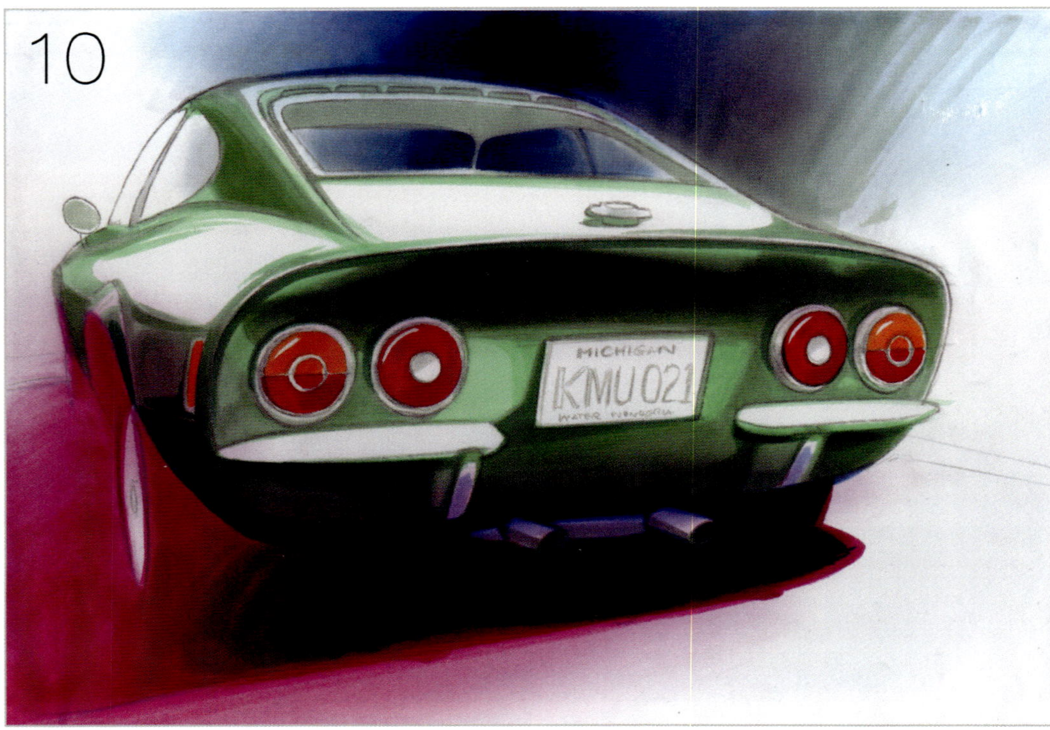

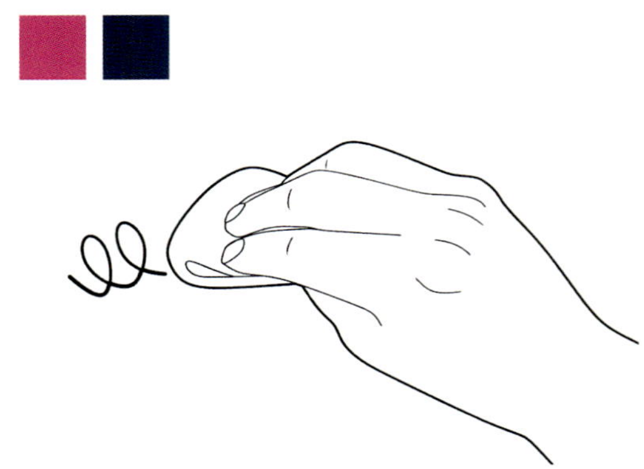

Now, using a Webril pad, take the background and the shadow one step further with a modest amount of powdered violet and blue pastel.

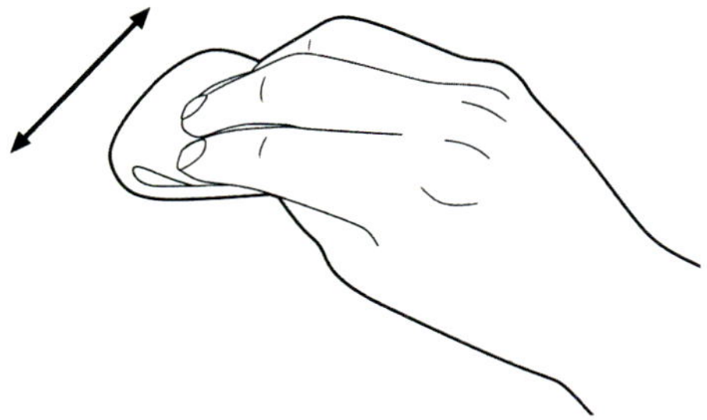

Since the marker work in this piece is the dominant feature, the sky tone should be comparatively subtle. Use a mixture of green and yellow on the driver's side and a blue-green combination on the far side of the roof.

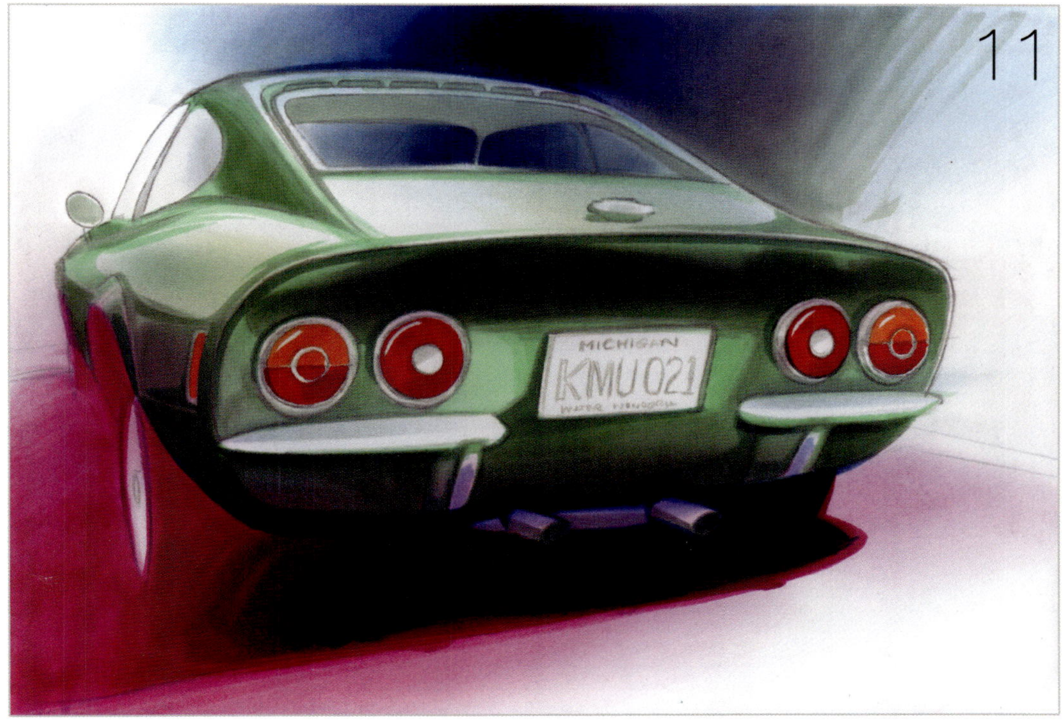

11

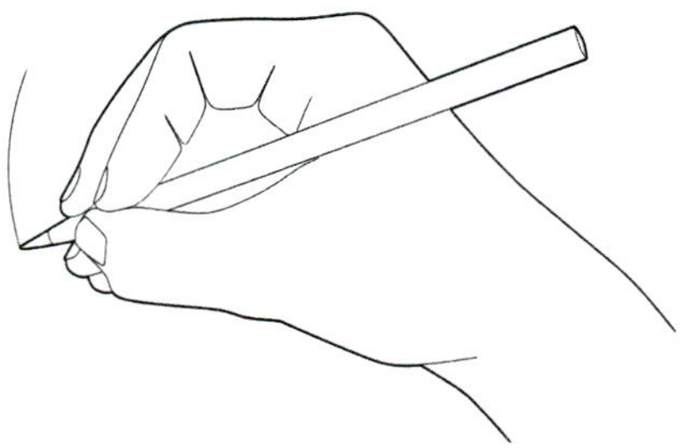

For the outlines, instead of using black Prismacolor pencils, try green ones. They will give the sketch a richer feel.

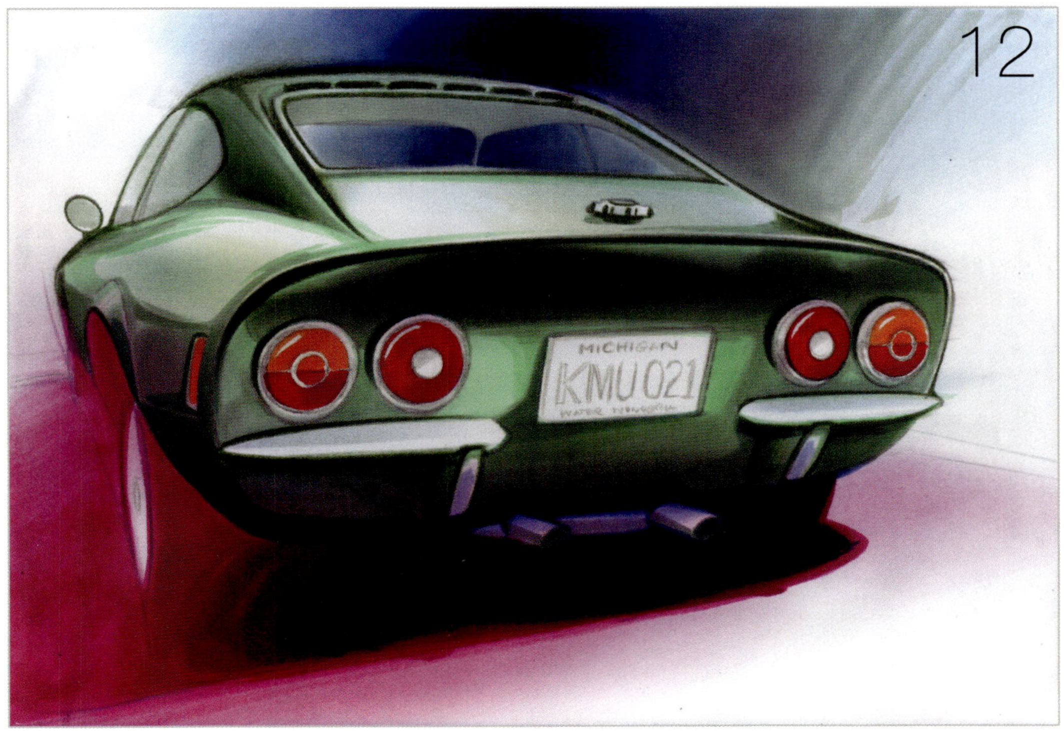

12

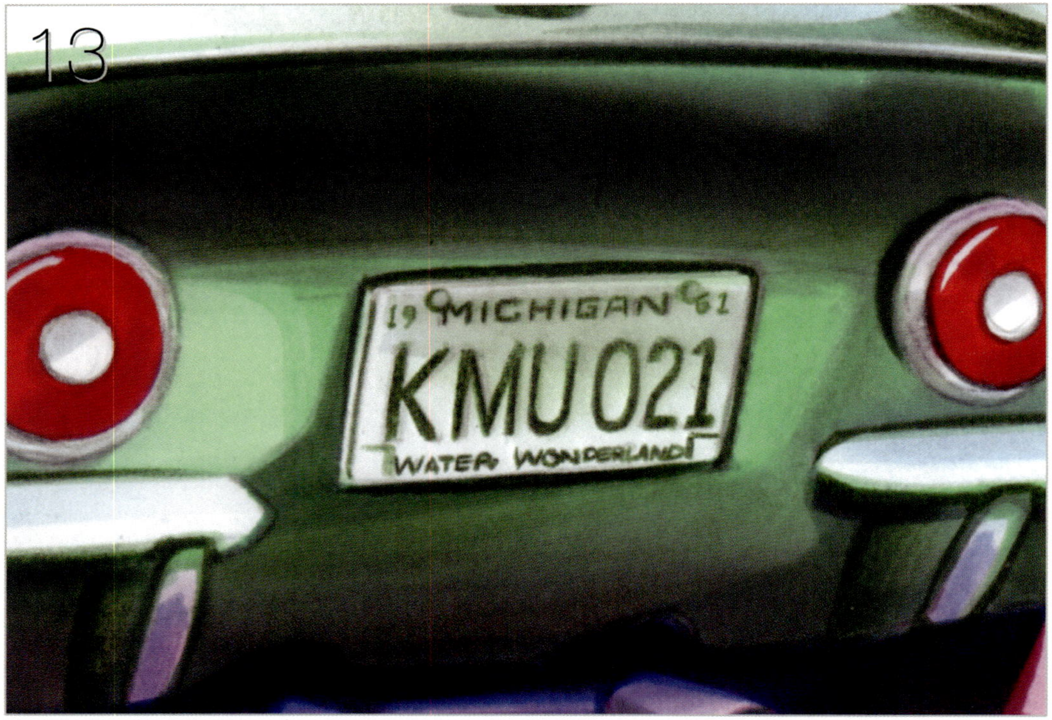

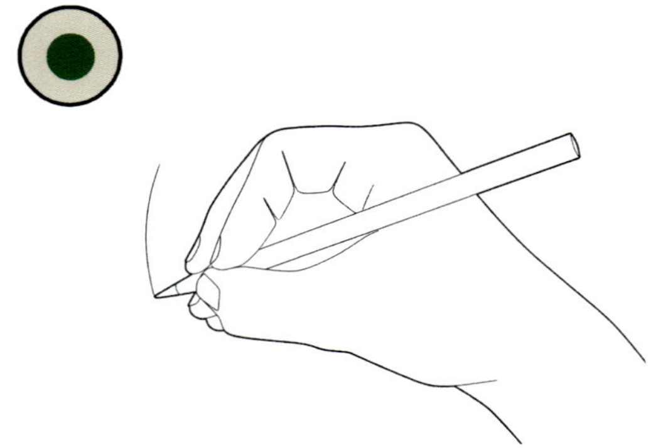

For the license plate lettering, be sure your green pencil is extra sharp.

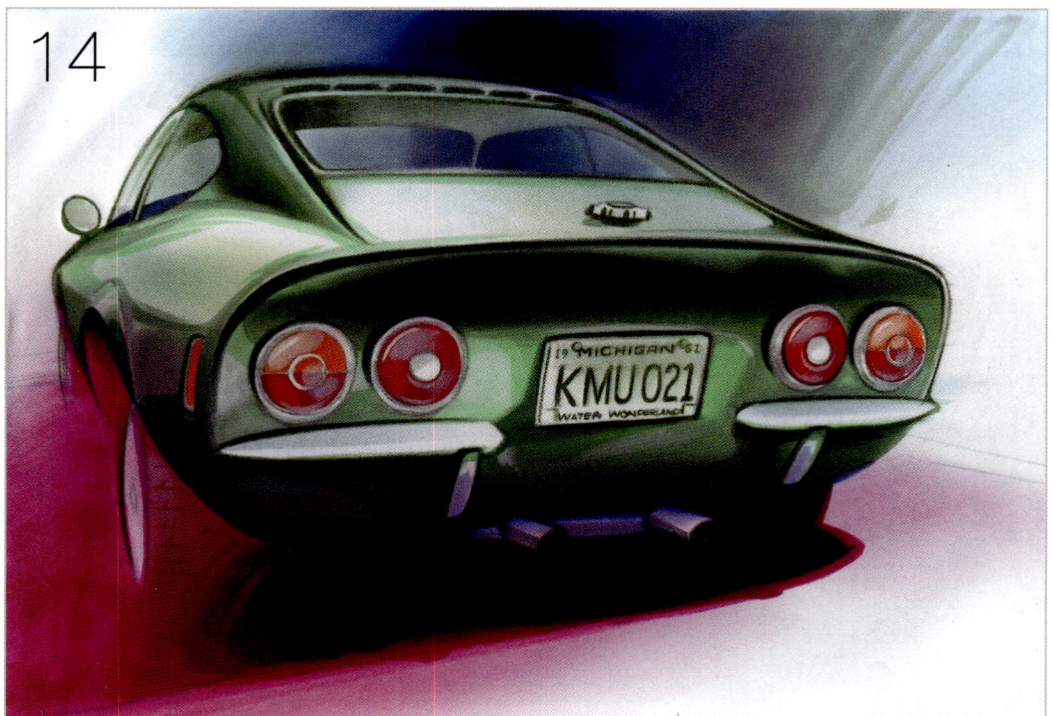

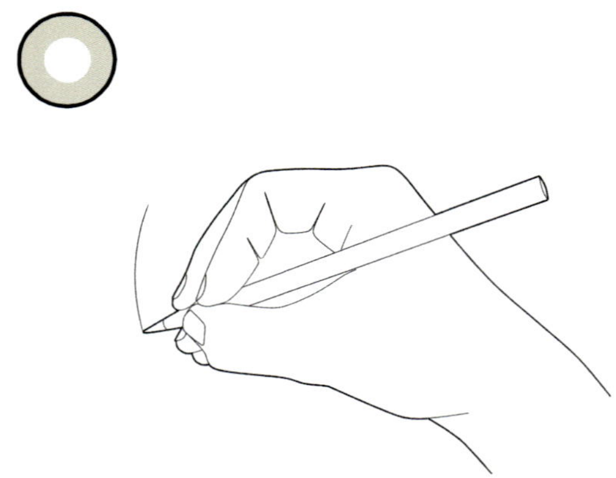

Begin adding some shape to the taillamps with subtle highlights. Similar to the tire highlights shown in previous tutorials, make sure the highlights are in line with the center of the lamp.

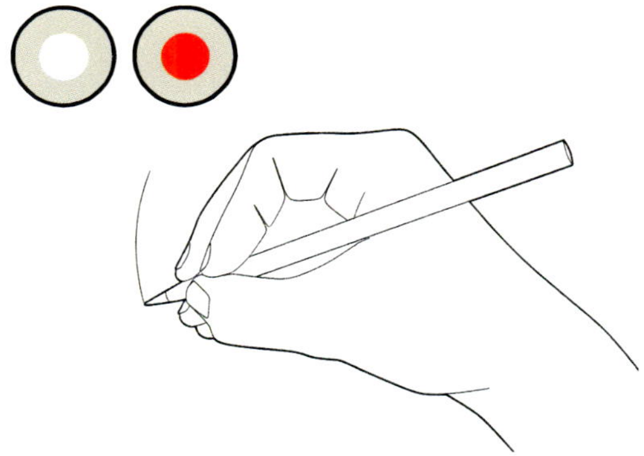

Continue to define various details with white pencil. With red pencil, add a pair of thin taillamps reflected into both chrome bumpers.

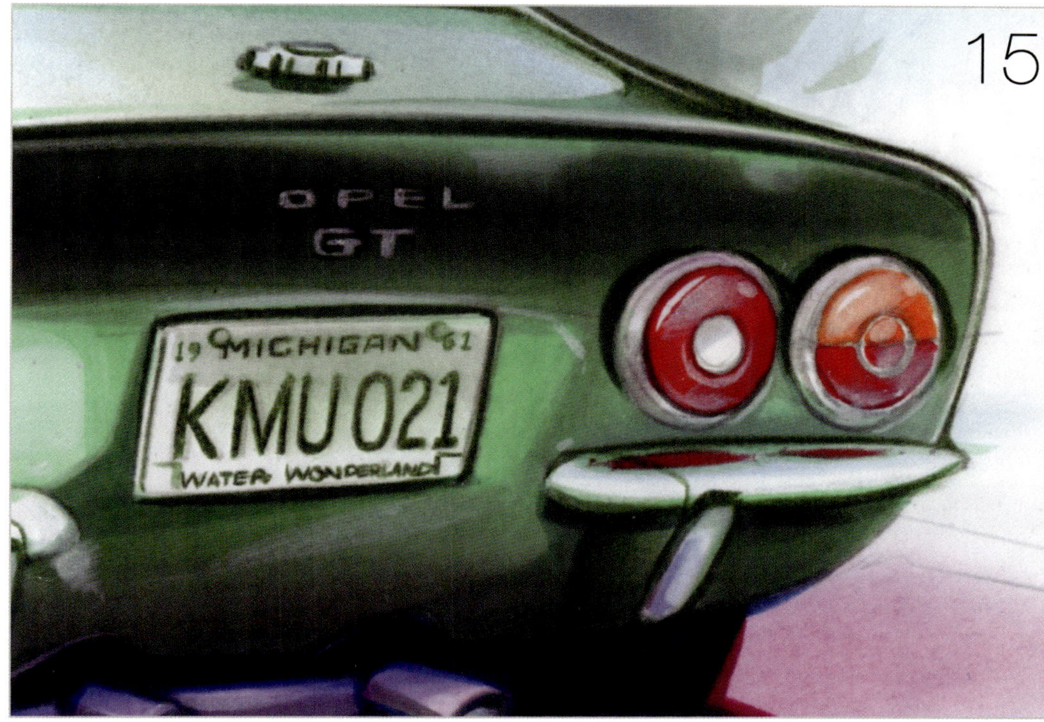

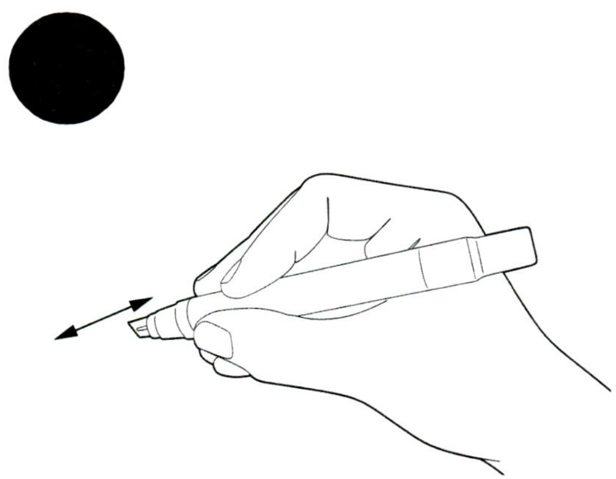

With the fine tip of a black marker, add the dark, sudden reflections which are appropriate for chrome bezels.

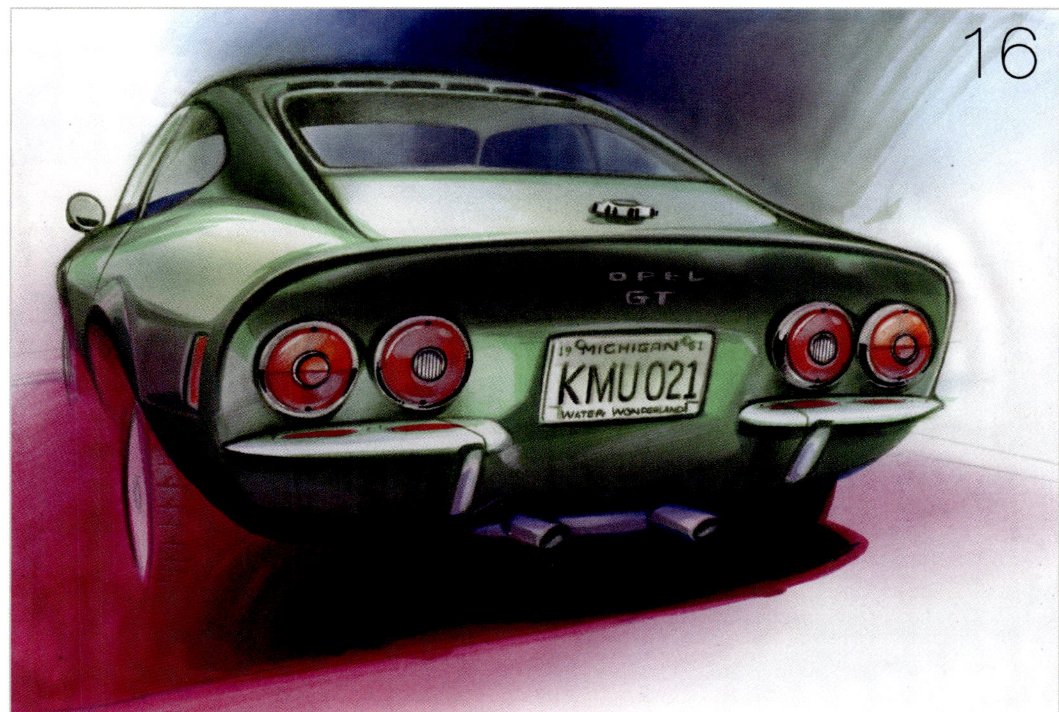

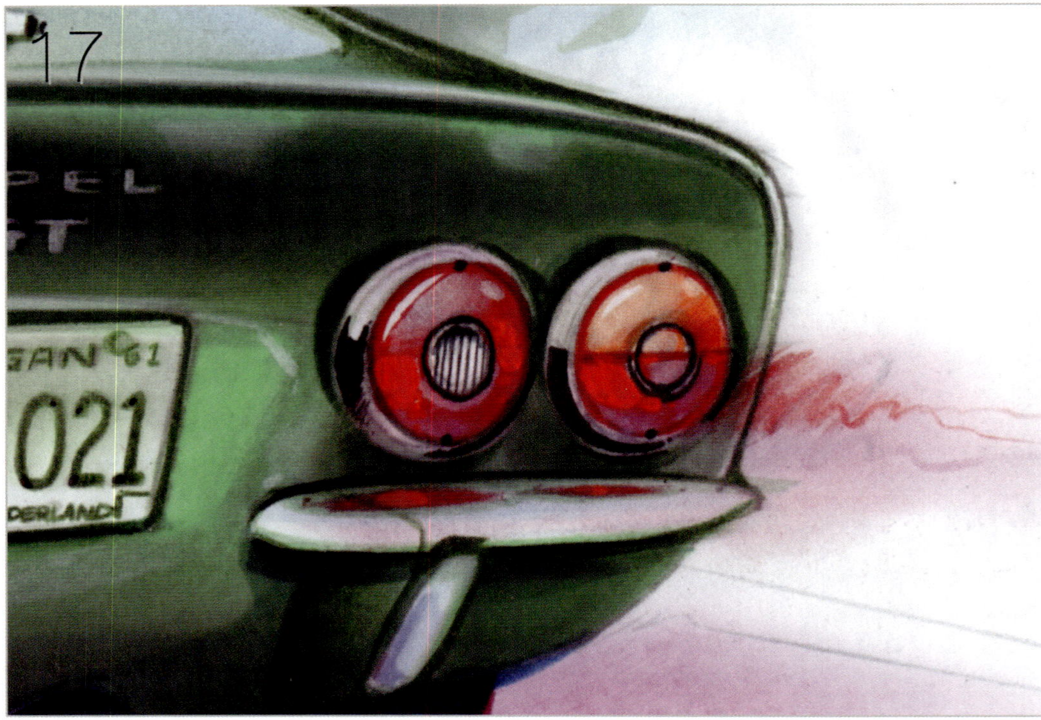

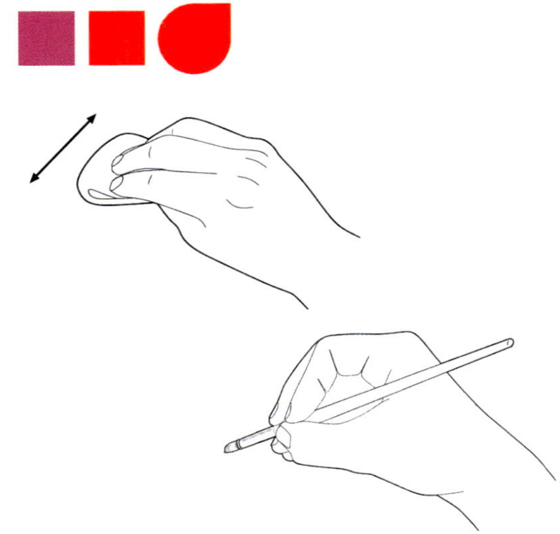

For the taillamp details, add a bit of red pastel directly to the page and blend it out with the Webril pad you used in the Mulberry shadow. Light the lamps with a bit of red gouache.

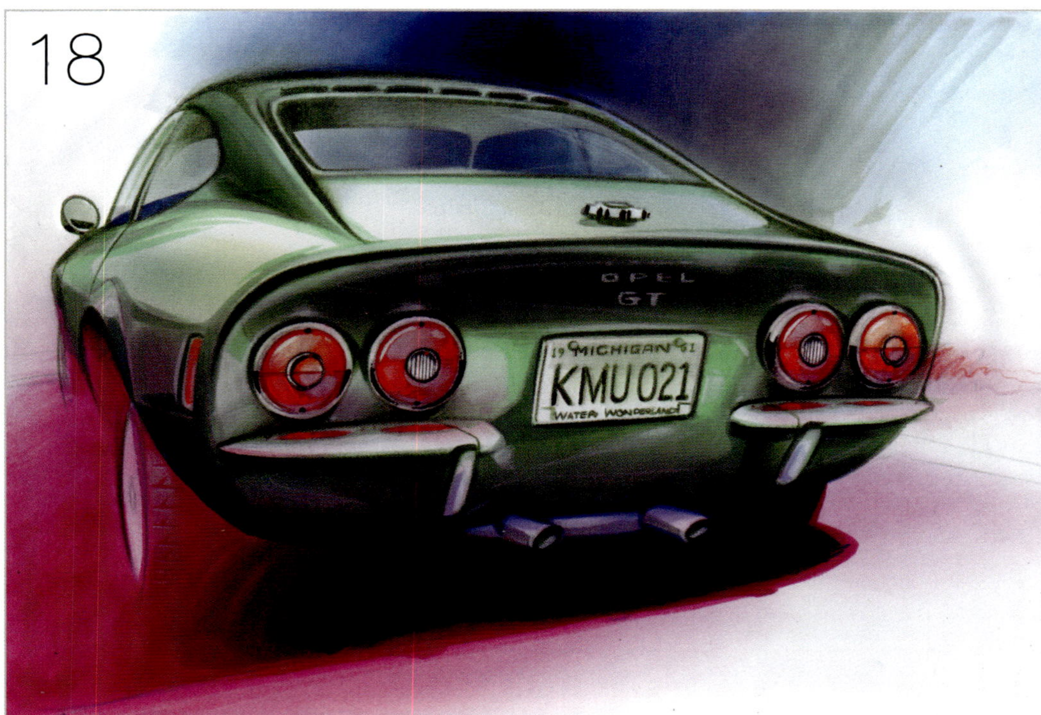

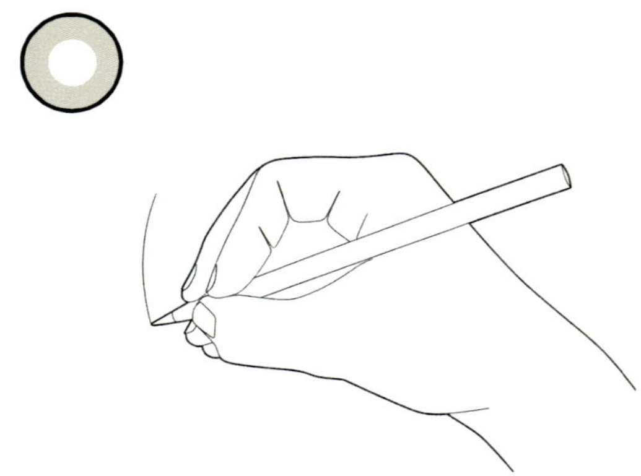

Add some subtle highlights and abstract swirls to the reflections. Some are coming from the taillamp bezels, others are simply artistic license.

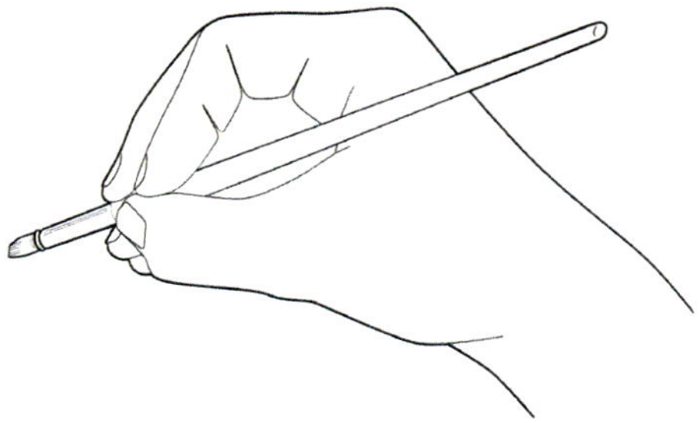

For the white gouache highlights, concentrate on areas that are undeniably shiny, like the bezels.

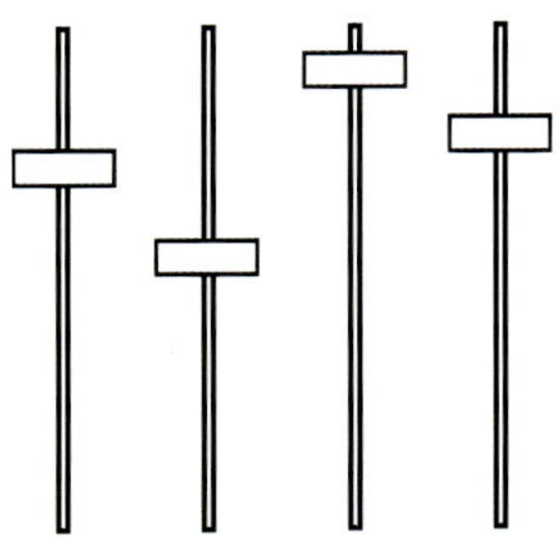

Scan in the finished car and turn up the color saturation in Photoshop. Take a moment to add emphasis on the lit roof; this will increase the contrast.

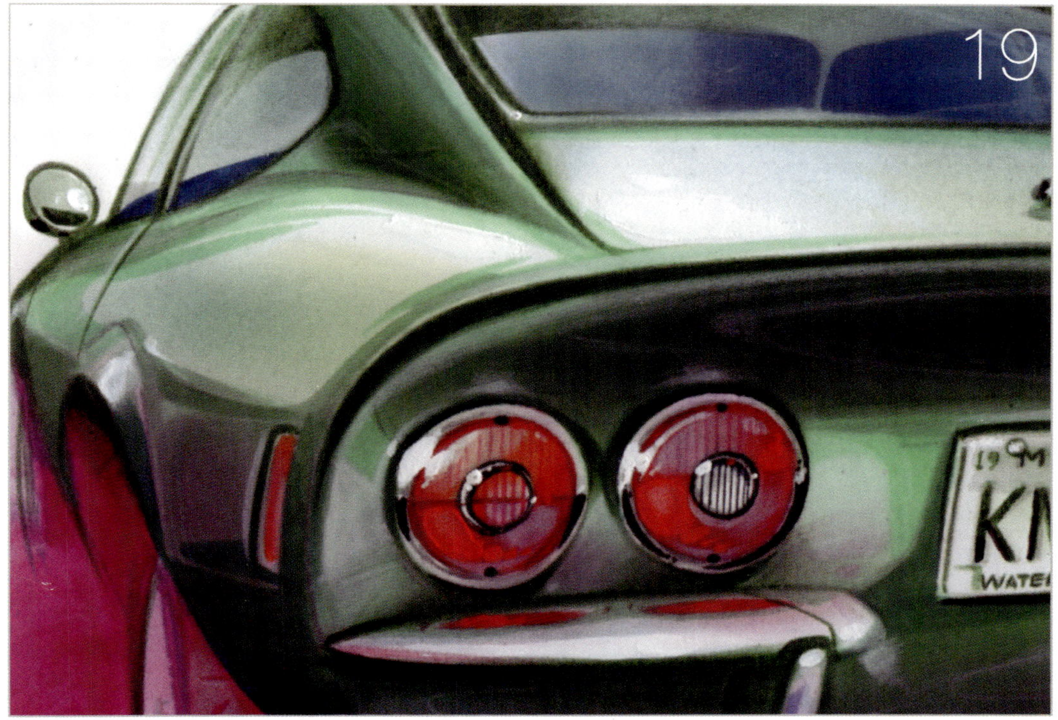

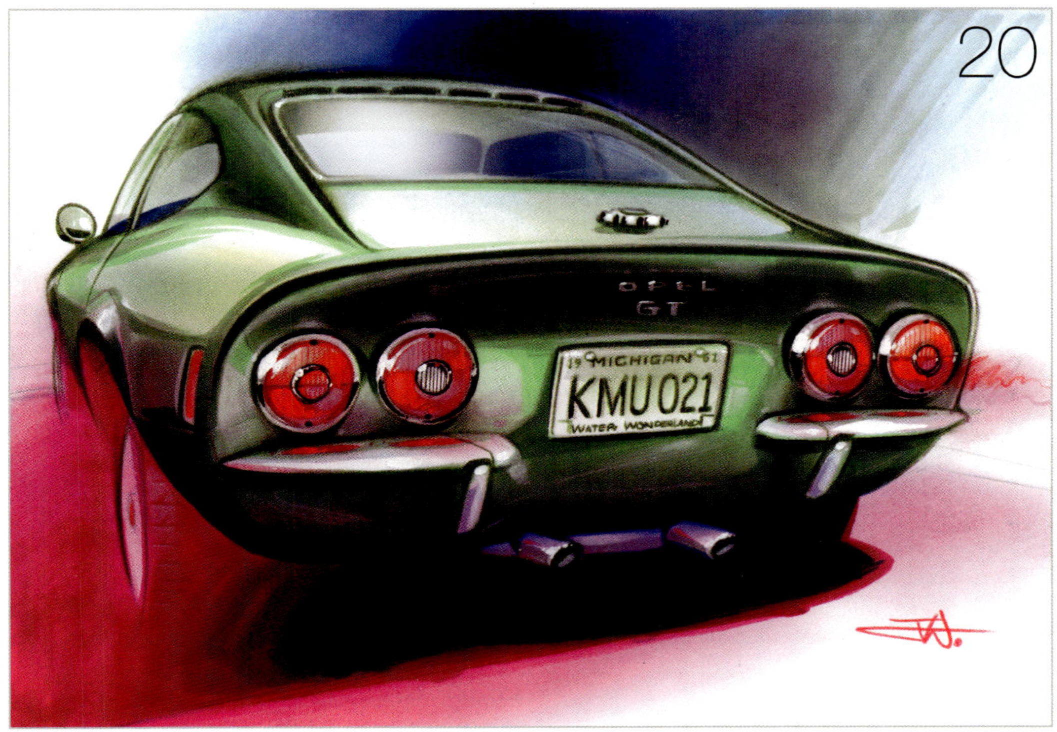

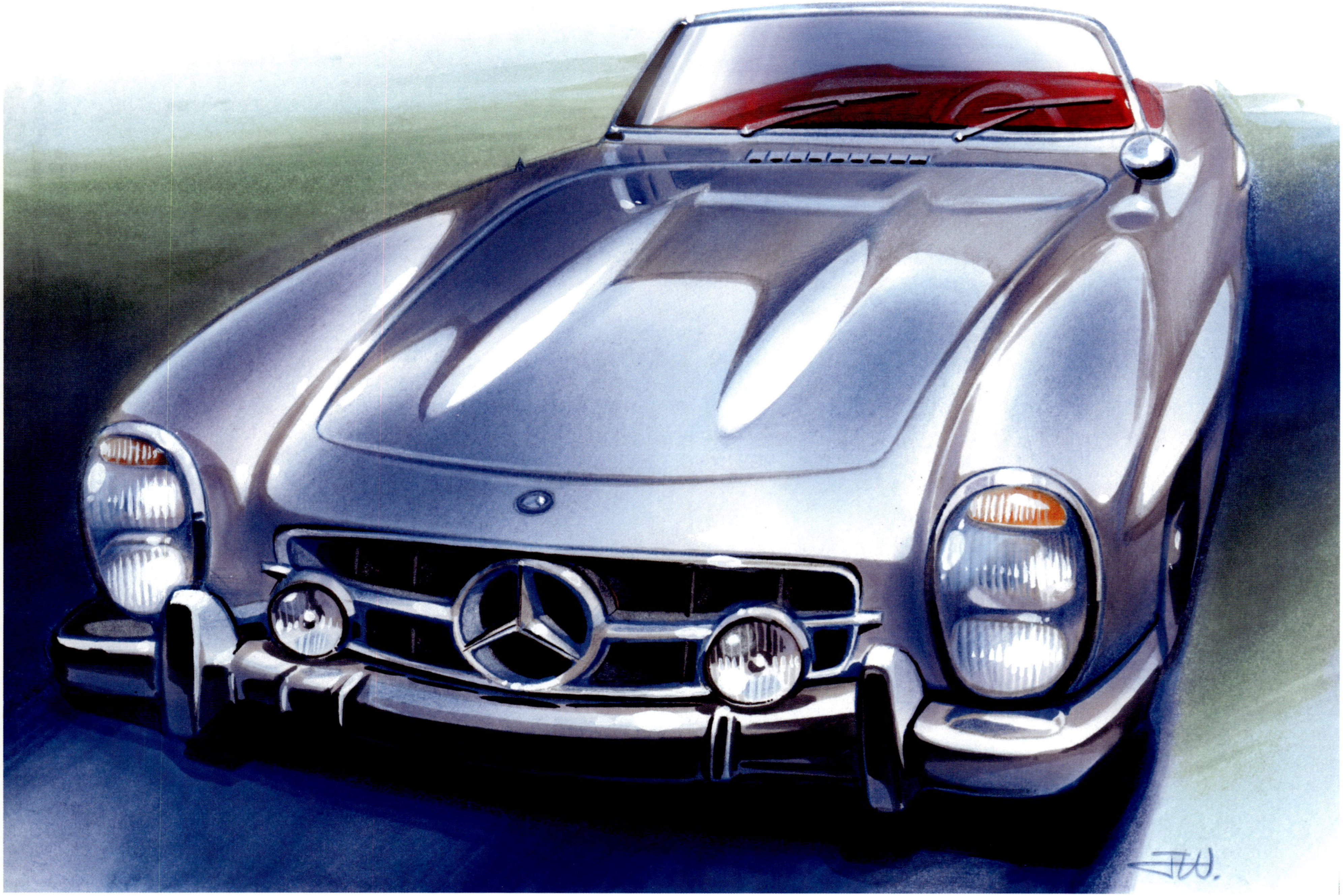

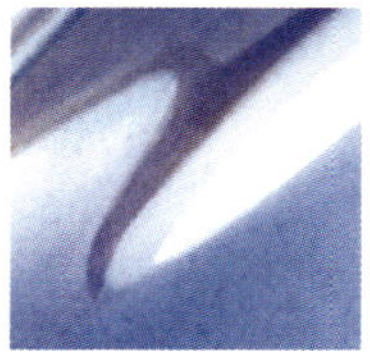

8. Cross-Car Reflections

For subjects featuring dramatic hoods and fenders, cross-car reflections can be an excellent Viscom tool. Picture an awning or a full, summertime tree hanging over this Mercedes-Benz. The reflections cast by these items can be far more descriptive than the standard pastel sky tones.

The Tools

Prismacolors:
White.

Nupastels:
Indigo Blue, Violet.

Verithins:
Black,
Indigo Blue.

Gouache:
White.

Prismacolor Markers: Celadon Green, Cool Grey 40%, Cerulean Blue, Black, Cool Grey 70%, Crimson Red, Orange, Indigo Blue.

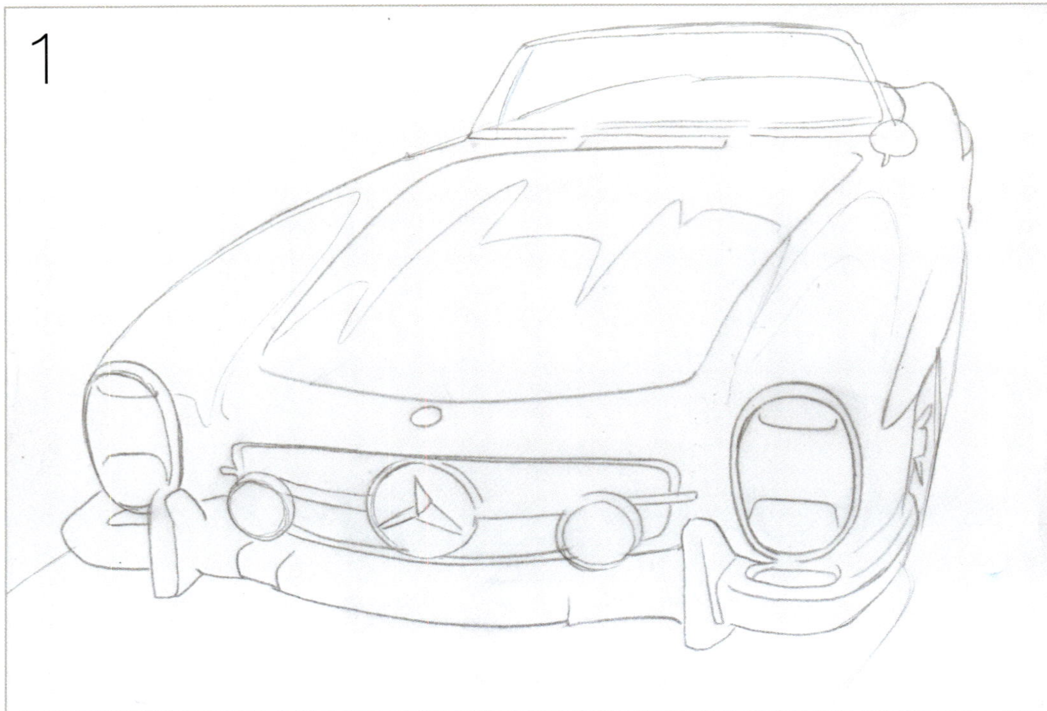

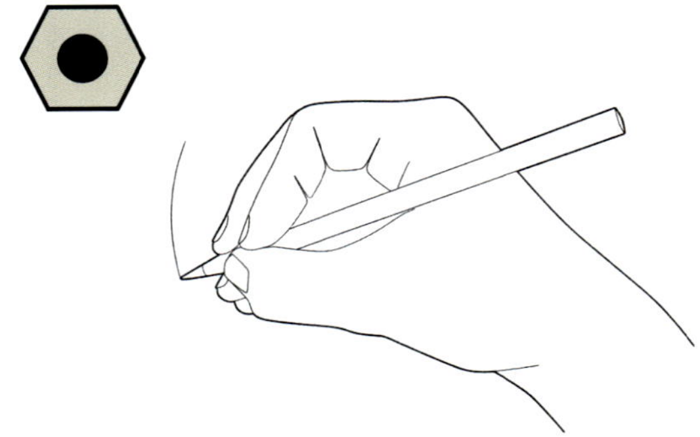

In addition to laying out the main theme lines of the car, map out the main boundary for the cross-car reflections.

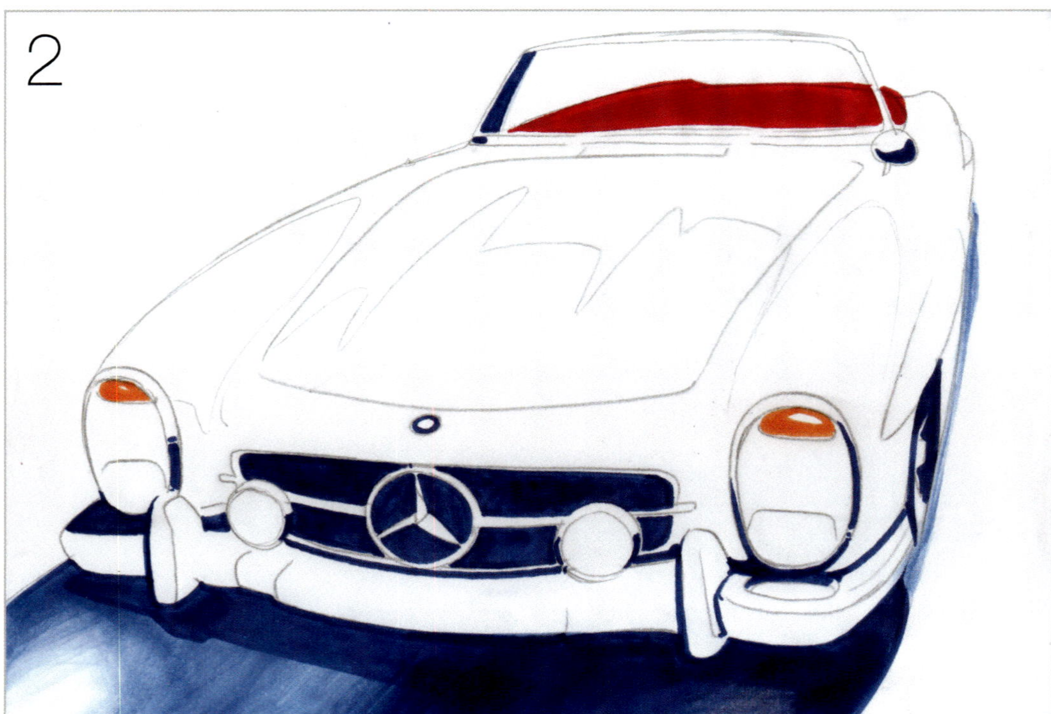

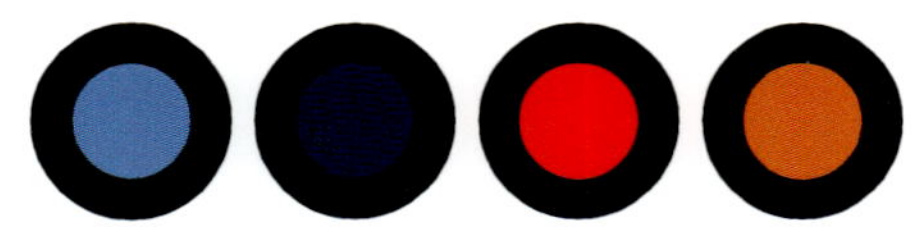

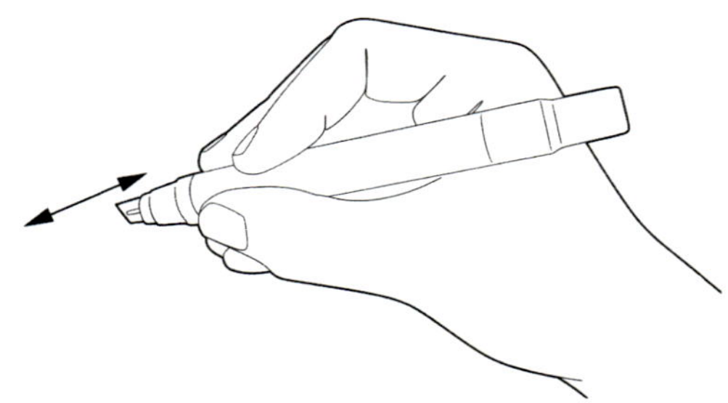

Block in the grille openings and the main shadow with the Cerulean and Indigo Blue markers. Try leaving some paper showing through on the far side of the shadow to make it more dynamic. Block in the interior with red and the turn signals with orange.

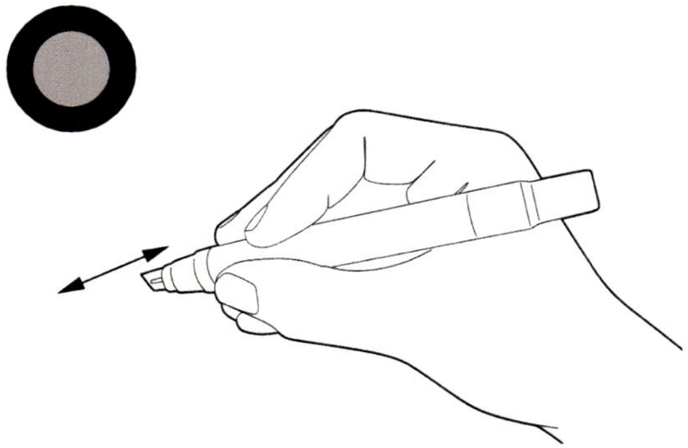

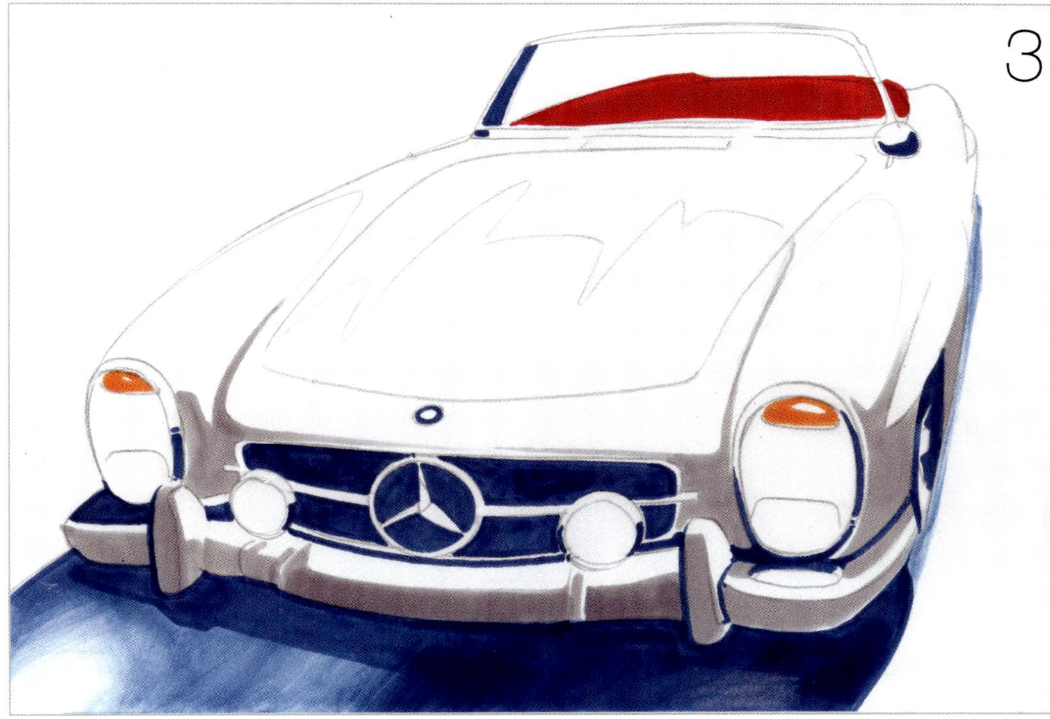

Now define the horizon reflection with Cool Grey 40%. Since the cross-car reflections will be dramatic, keep these reflections clean and simple.

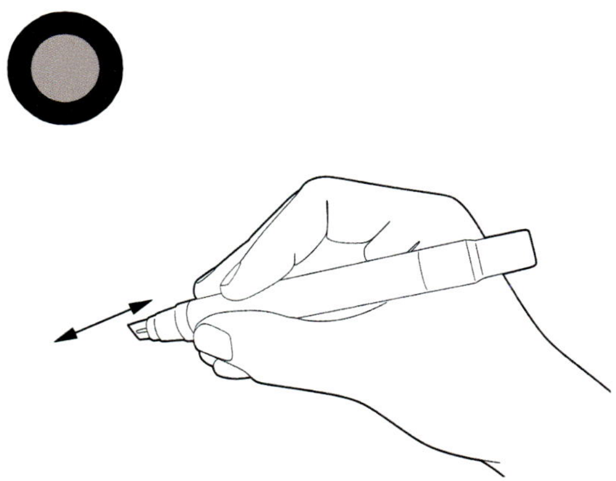

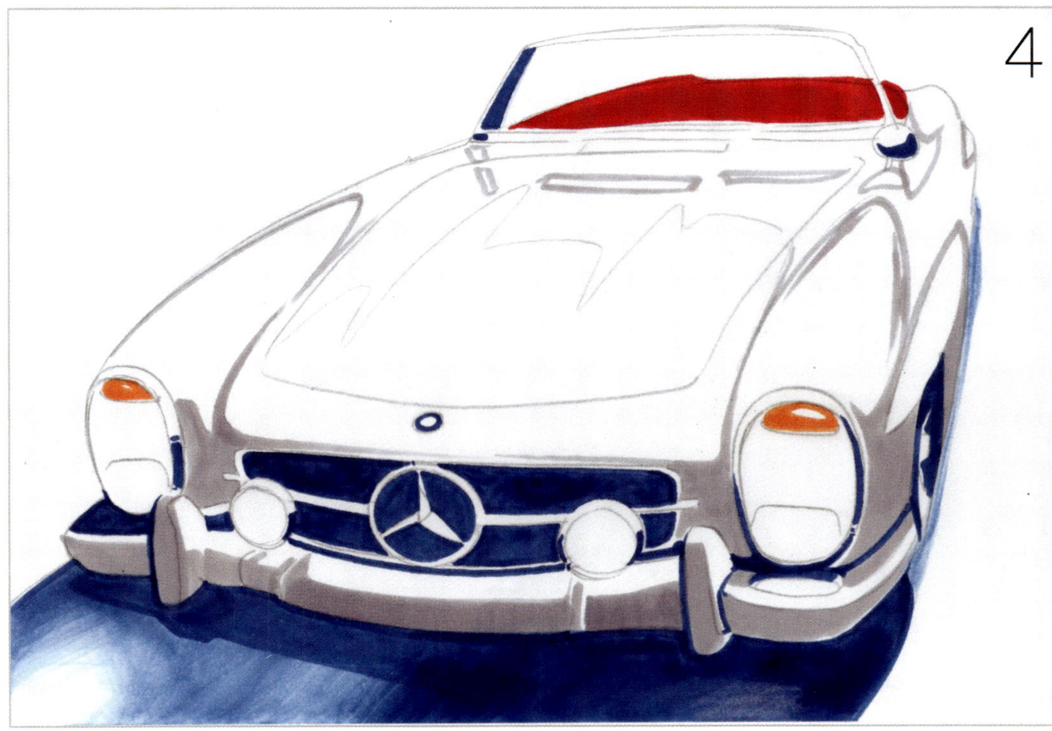

Use the chisel point of the same marker to outline the fender reflections.

5

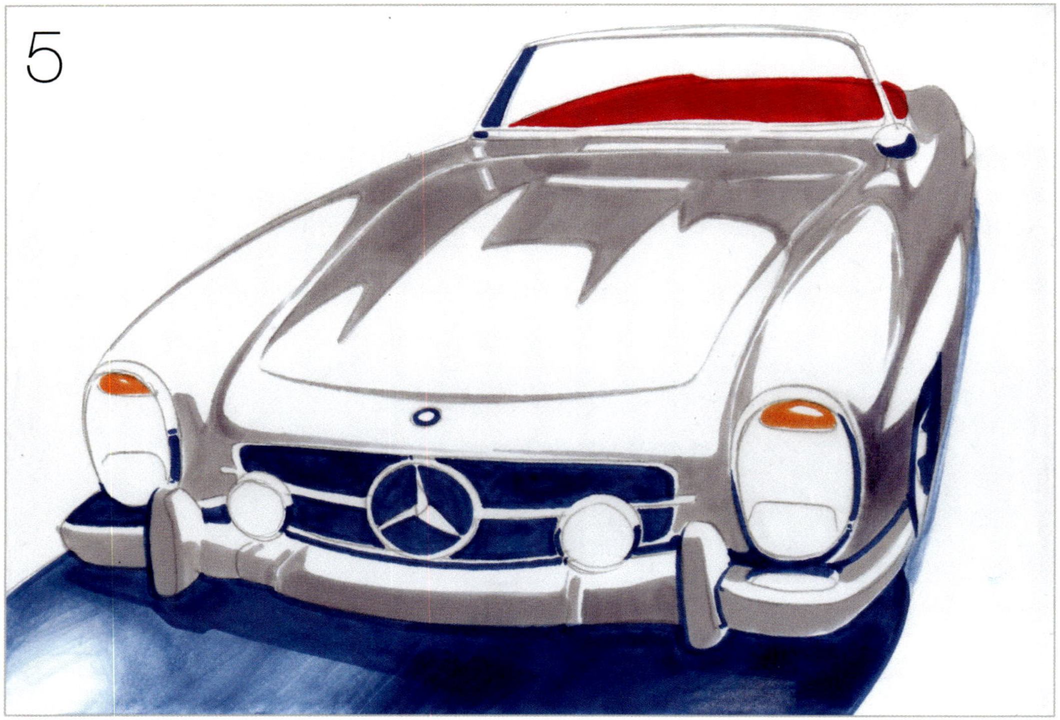

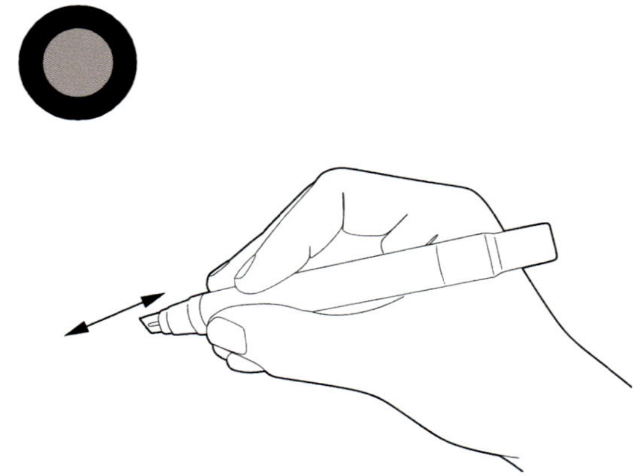

When blocking in the cross-car reflections, be sure to leave some slivers of white just under the windshield; these will represent light being reflected off the chrome pillars and crossbar.

6

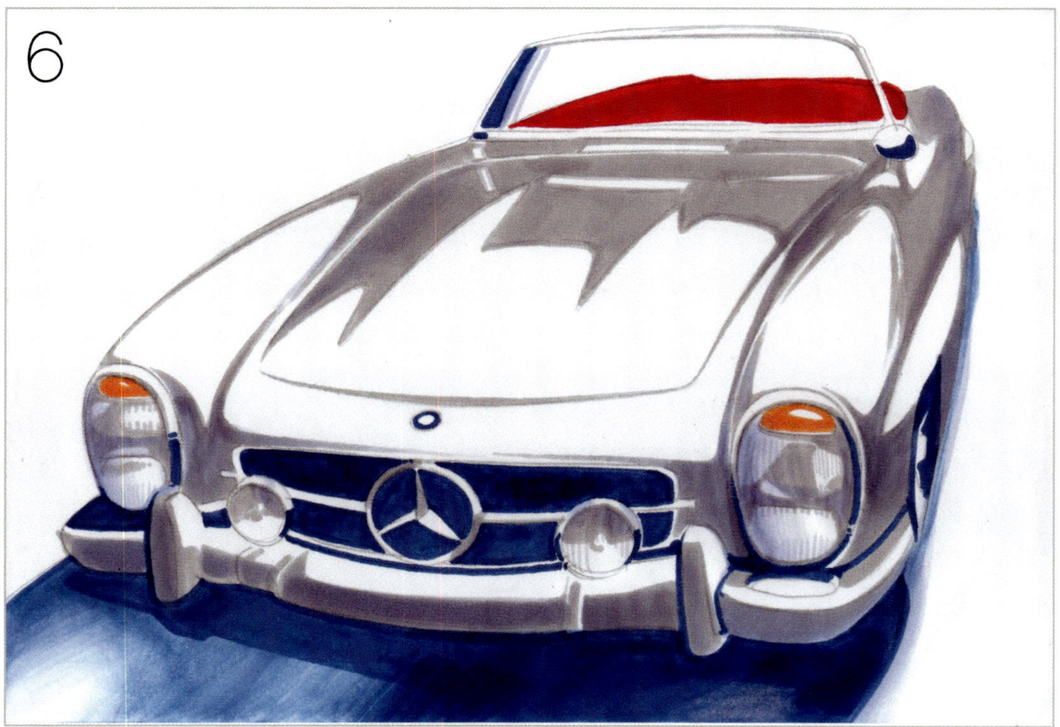

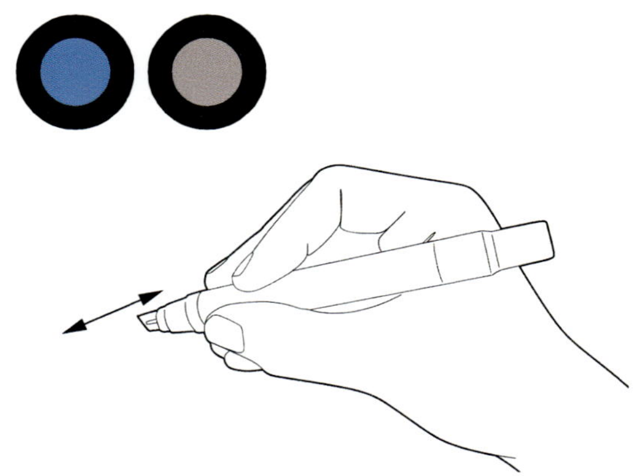

Define the inside surface of the headlamps with Cool Grey. Ghost in a bit of Cerulean Blue on one side of the lozenge-shaped lamps; this will give the lenses some crown.

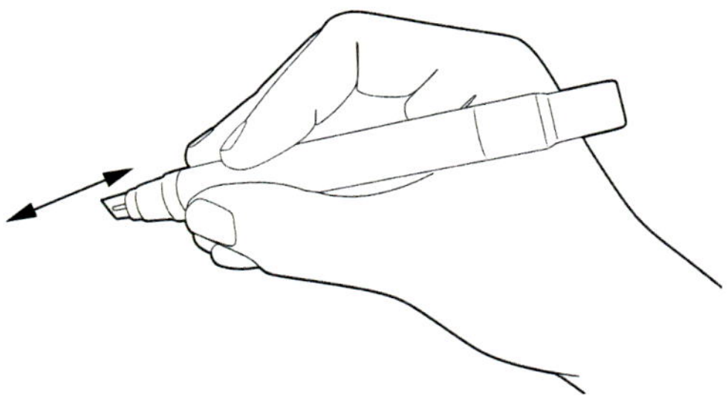

Using a Cool Grey 70% marker, punch up the reflections in the chrome bumper and the foglamps; this will help set them apart from the body.

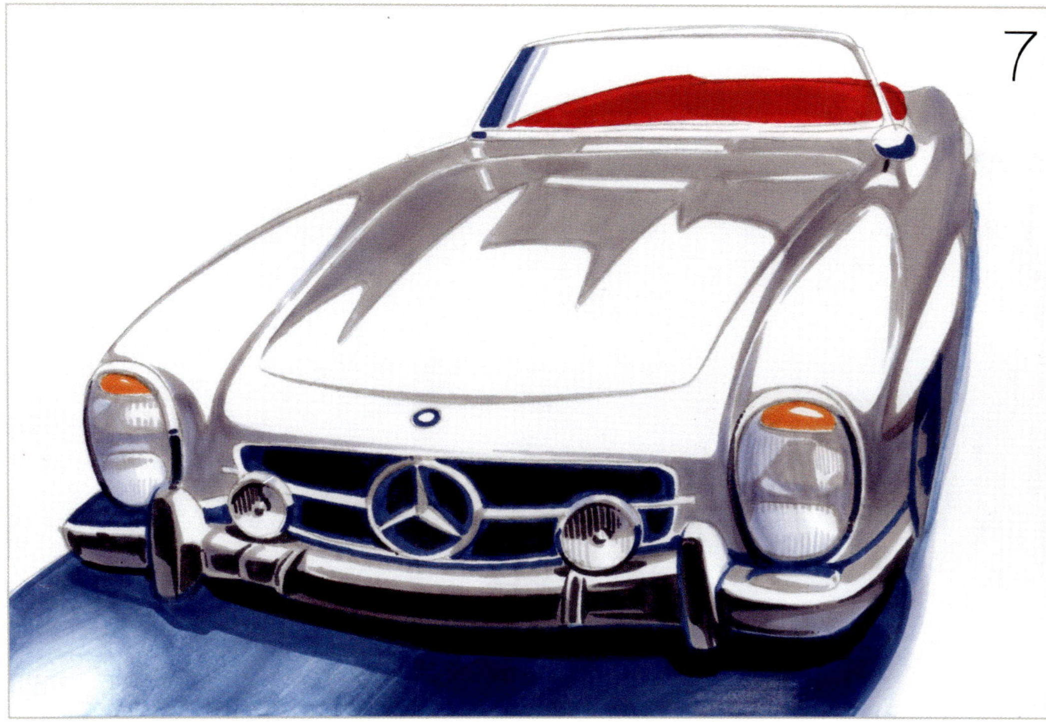

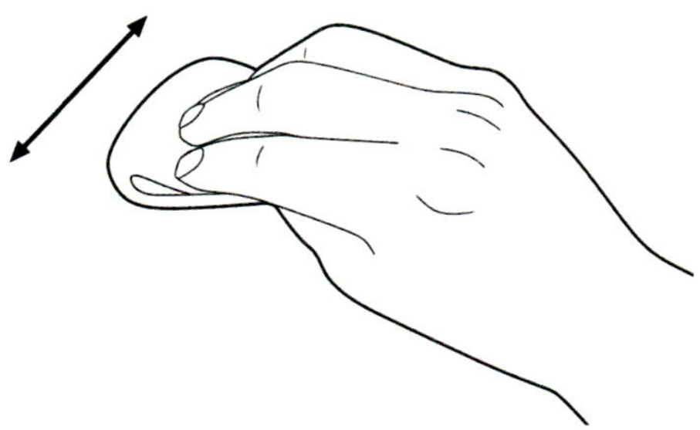

For the sky tone, start by gently brushing in some blue pastel powder.

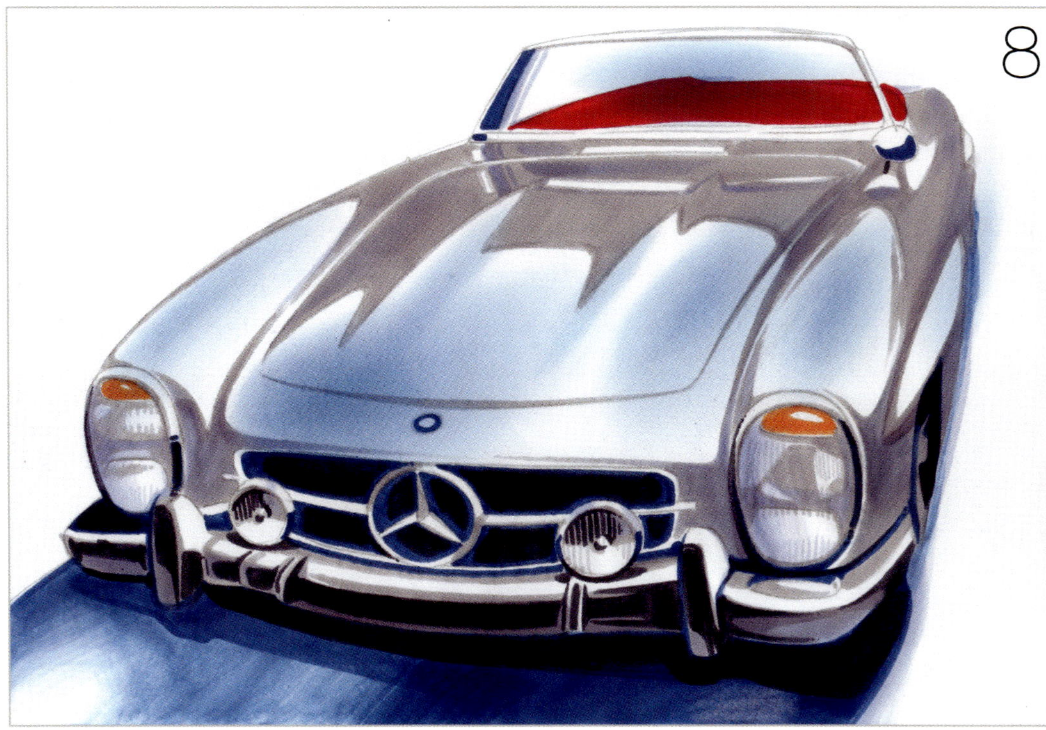

9

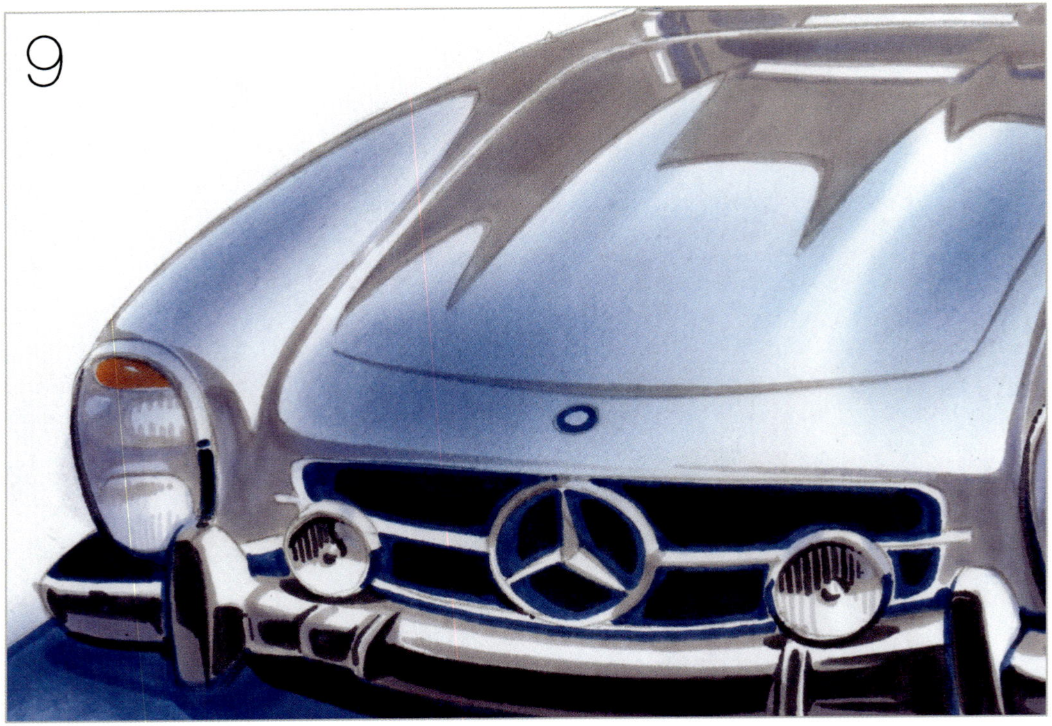

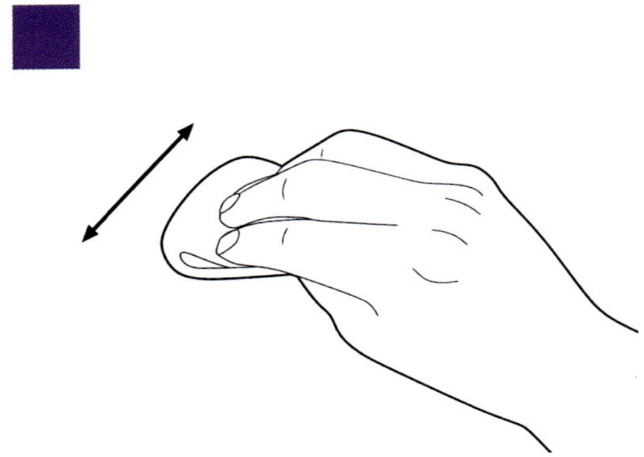

Now blend a small amount of violet pastel into the middle of the sky tone. This will add a further dimension to the color.

10

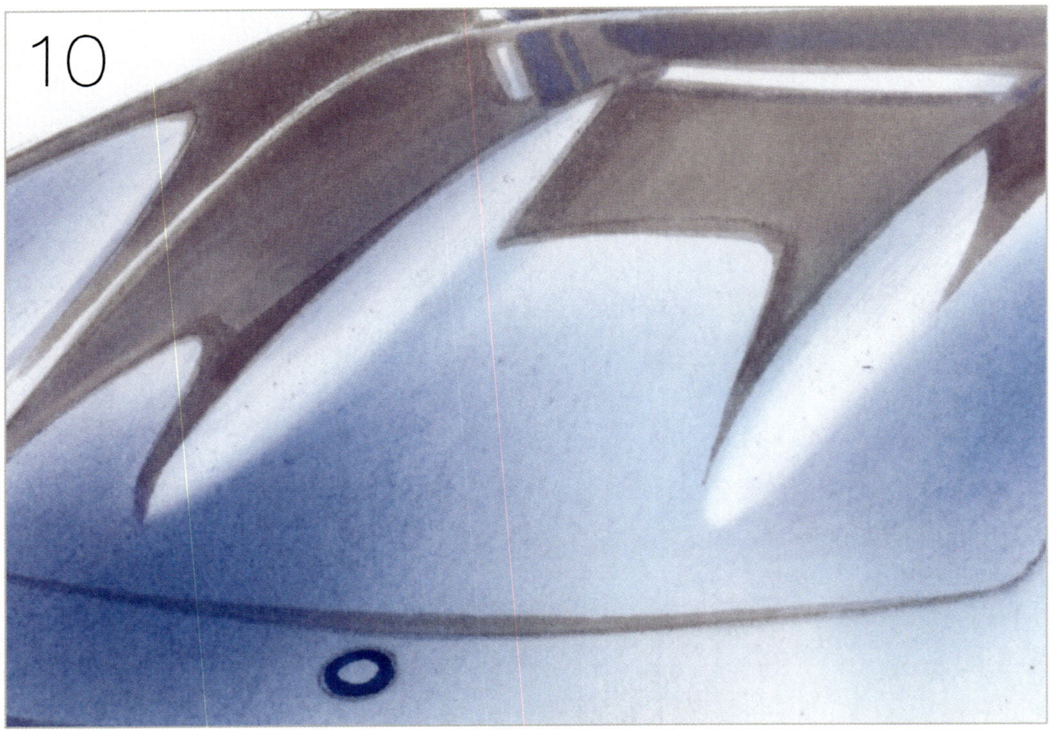

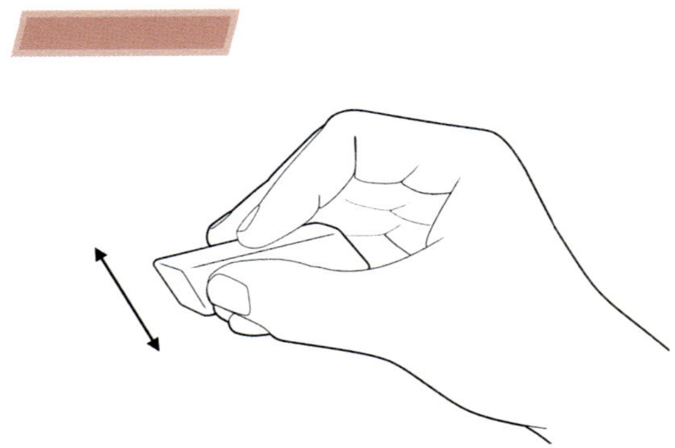

With a clean eraser, emphasize the highlights defining the pill shapes in the hood surface.

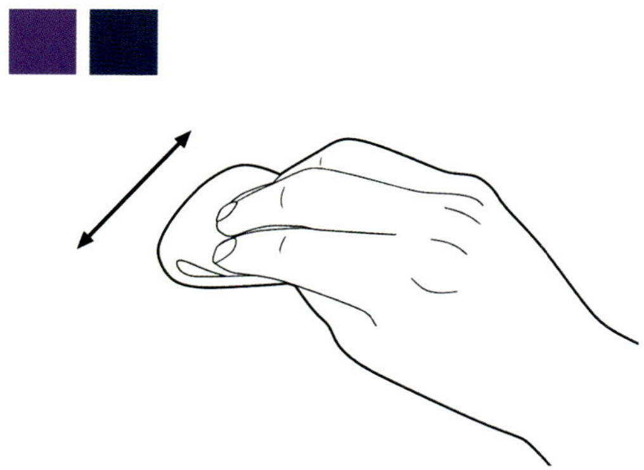

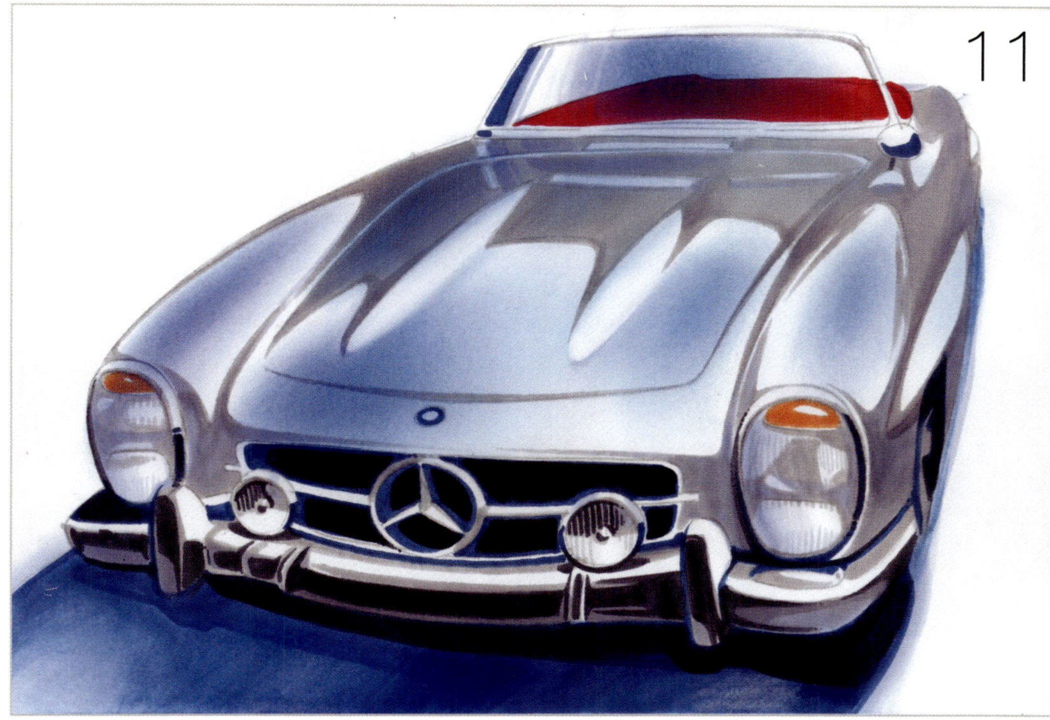

Add a mixture of blue and violet pastel to the far side of the cross-car reflection so it doesn't appear flat next to the sky tone.

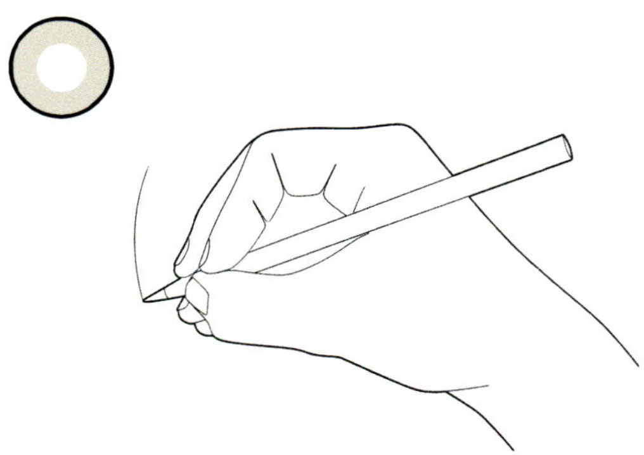

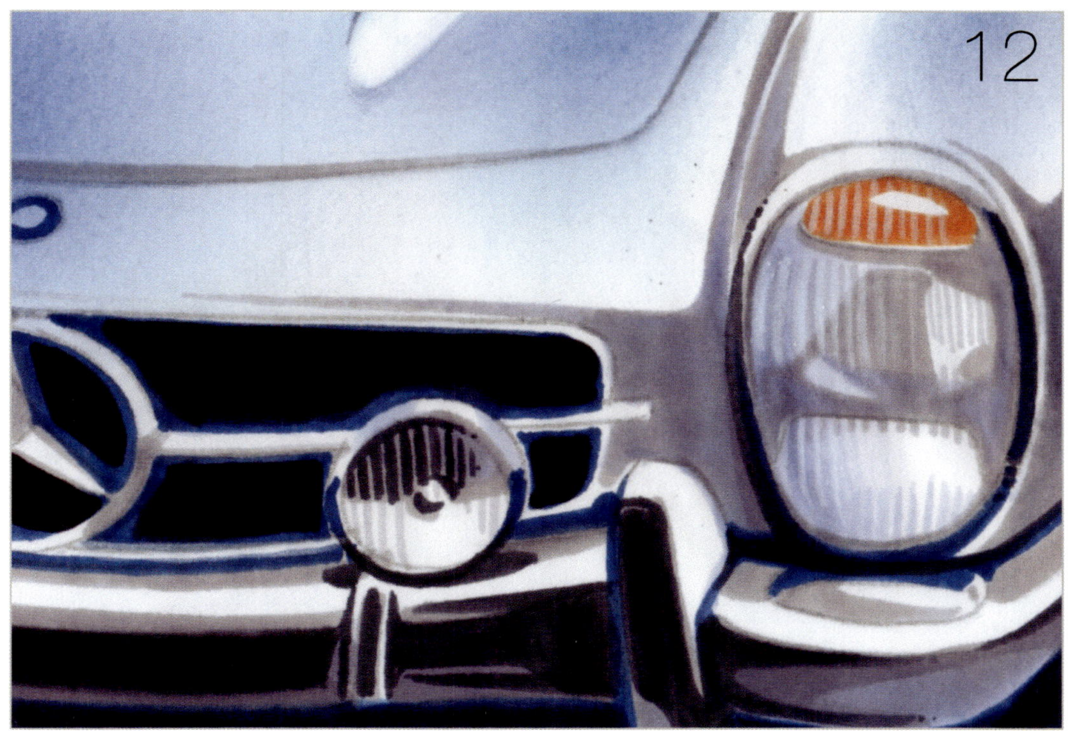

Define the vertical fluting of the headlamp lenses with a white pencil.

13

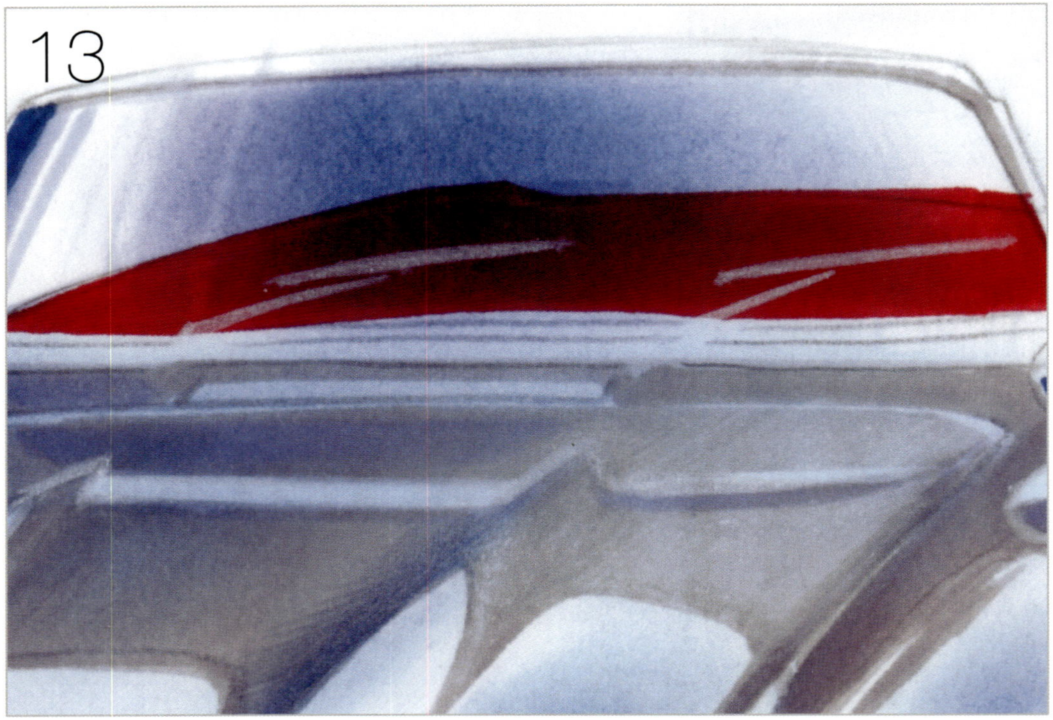

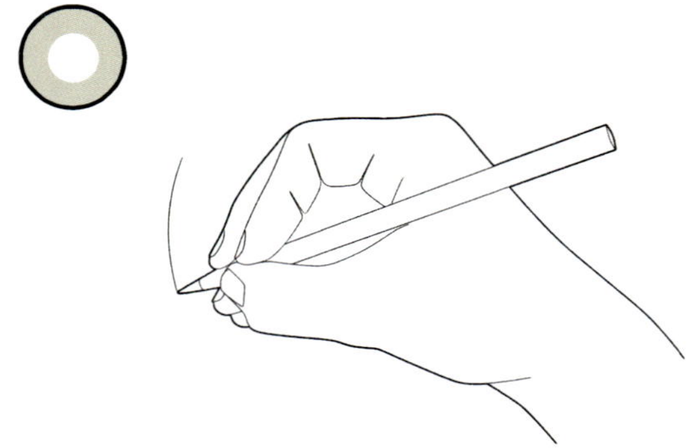

Continuing with white pencil, add details like the windshield wipers and the thinner parts of the crossbar reflection.

14

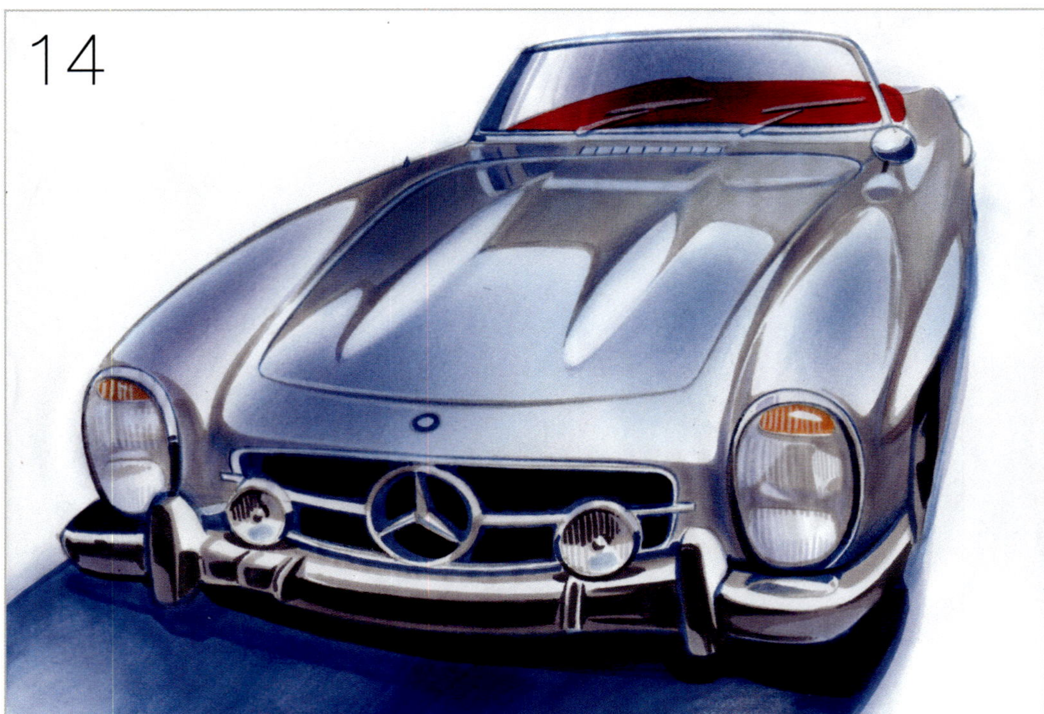

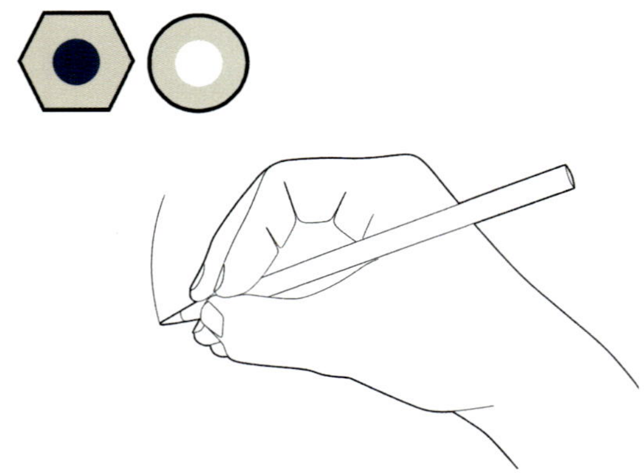

Now begin tightening up the line-work throughout the drawing with blue Verithin and white Prismacolor pencils.

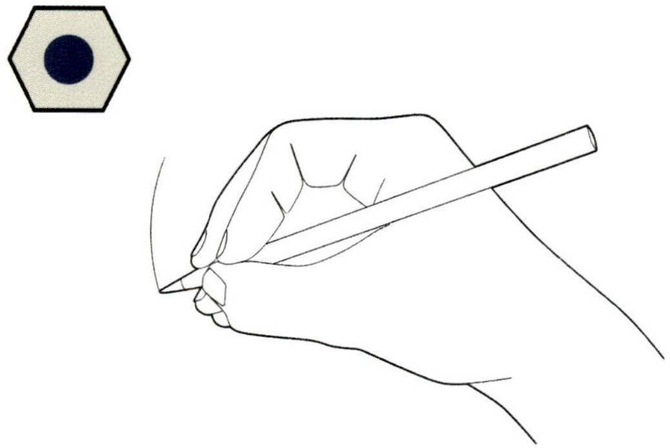

Add more detail to the headlamp lenses with a blue Verithin, this will make the white fluting even more prominent.

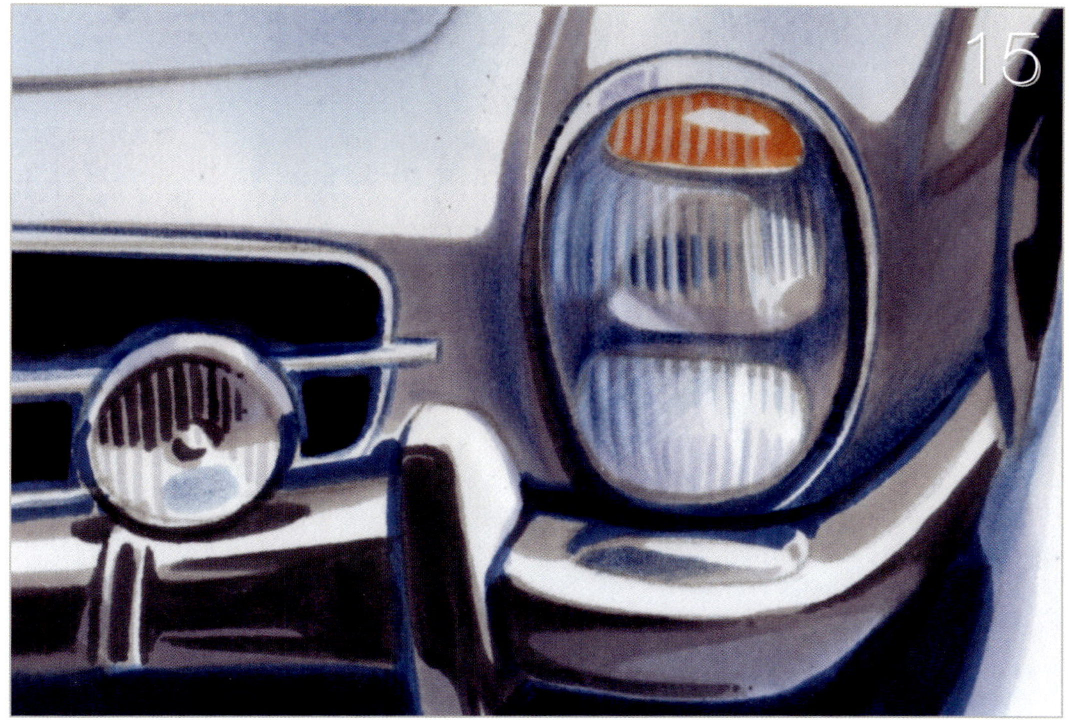

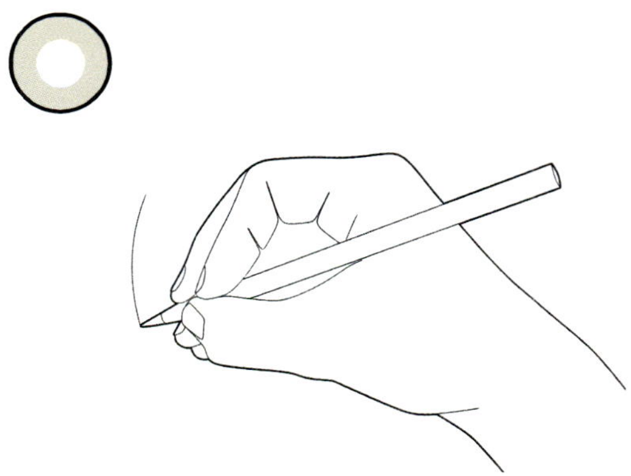

Define the grille strakes with a white Prismacolor. Don't forget to leave an angled black shadow on each one.

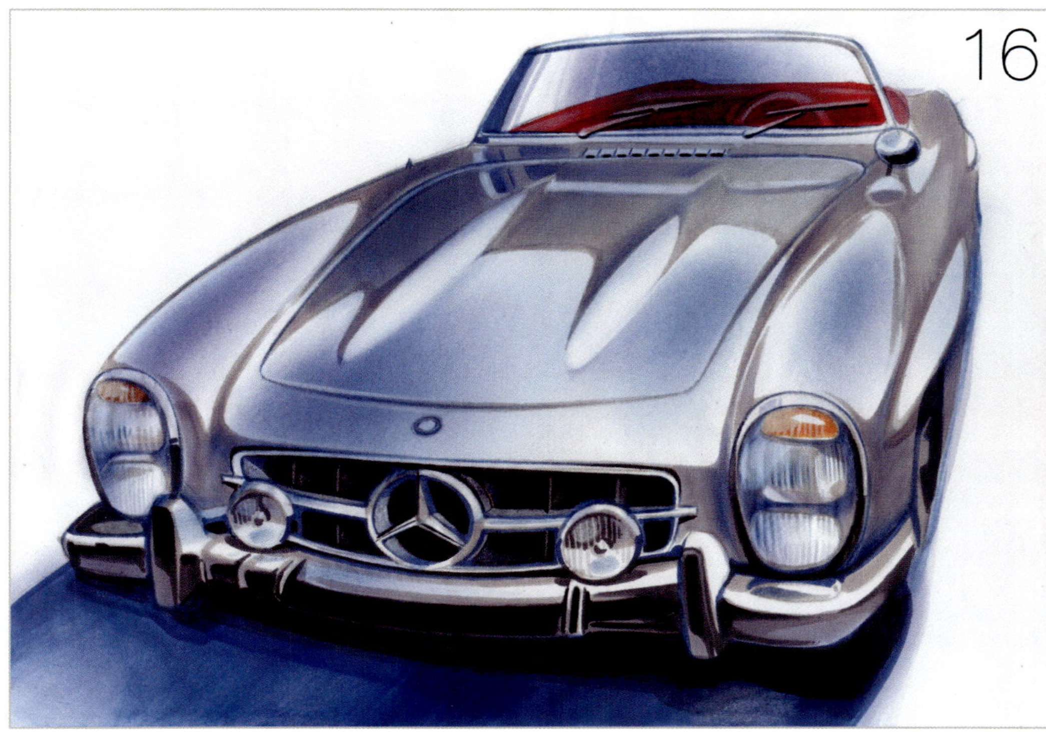

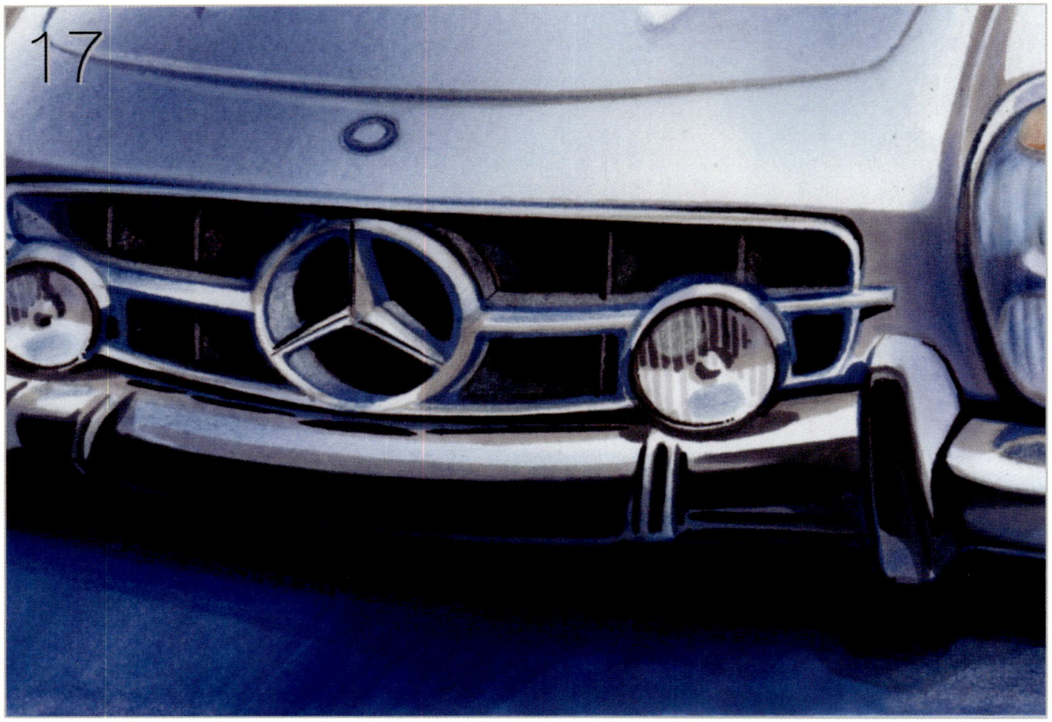

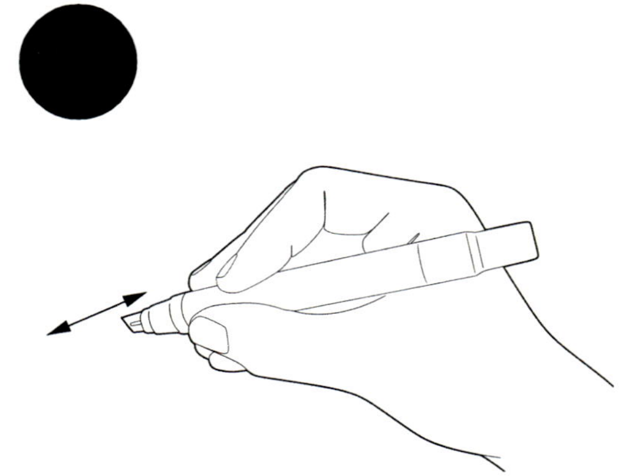

Add one final marker layer to the chrome reflections, this time black. Be sure to leave some of the lighter tones peeking through here and there.

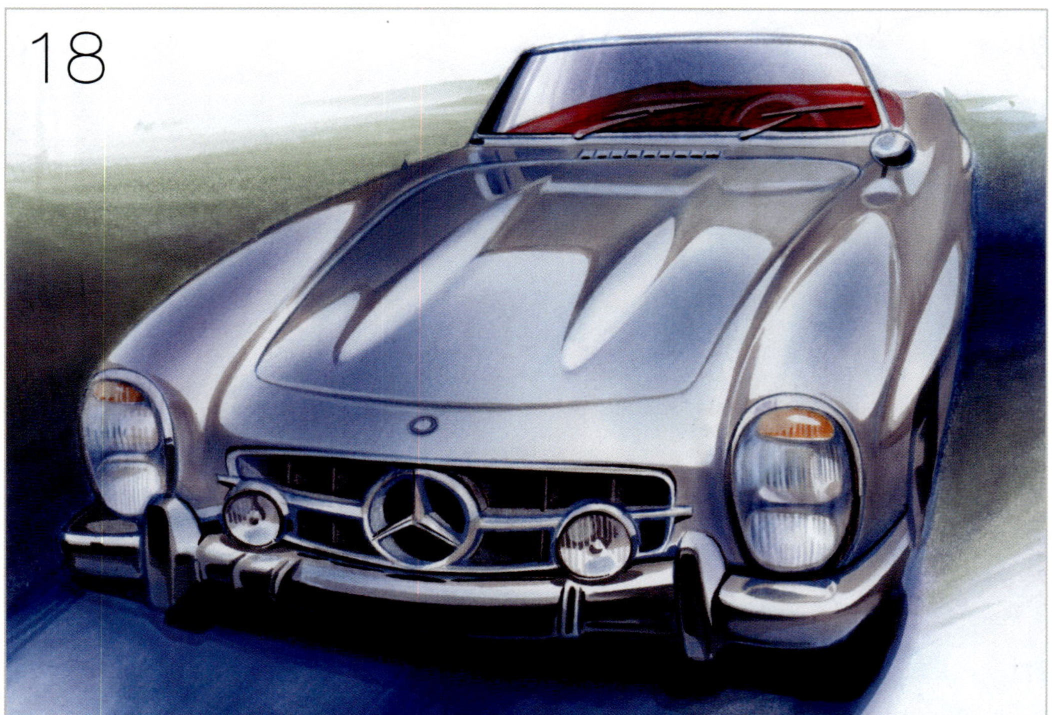

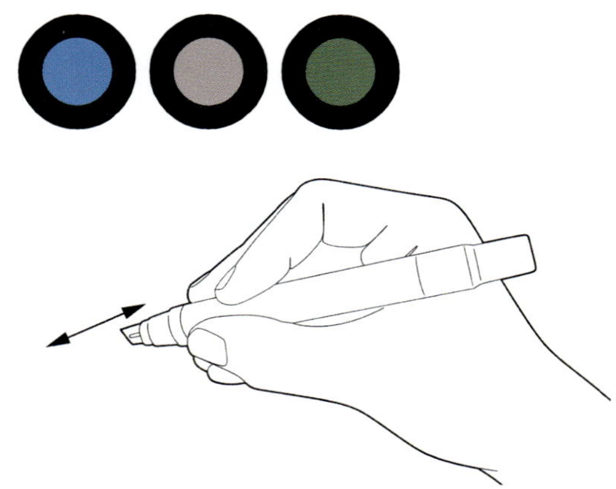

For the background colors, try employing a mixture of Cerulean Blue, Cool Grey and Celadon Green.

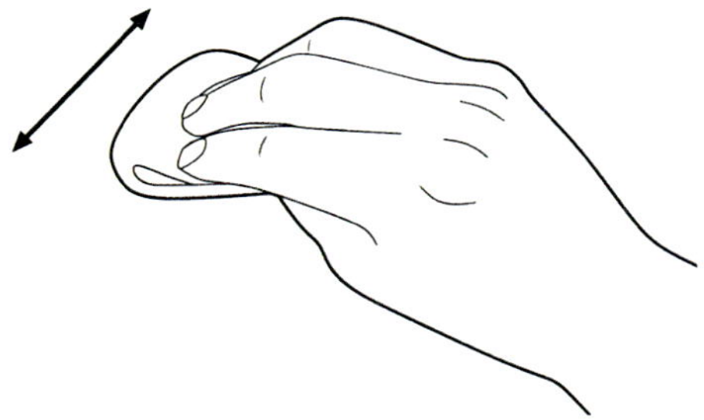

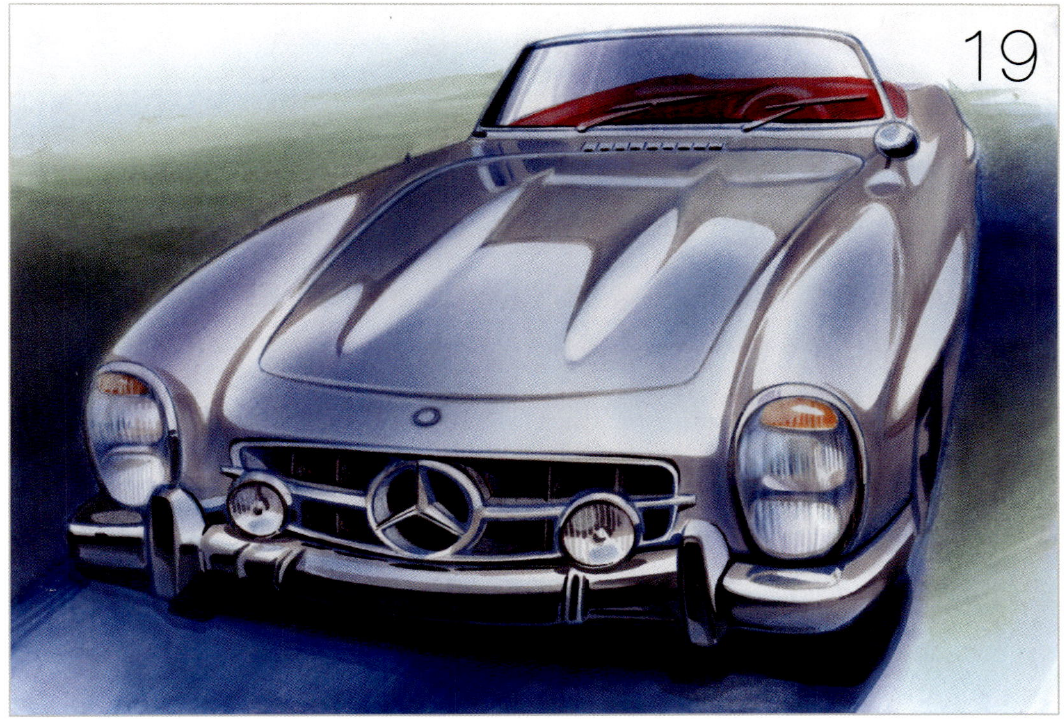

Soften the background colors by applying a bit of blue pastel powder.

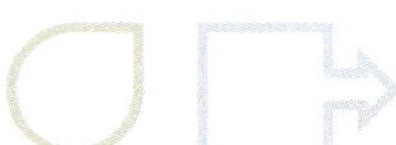

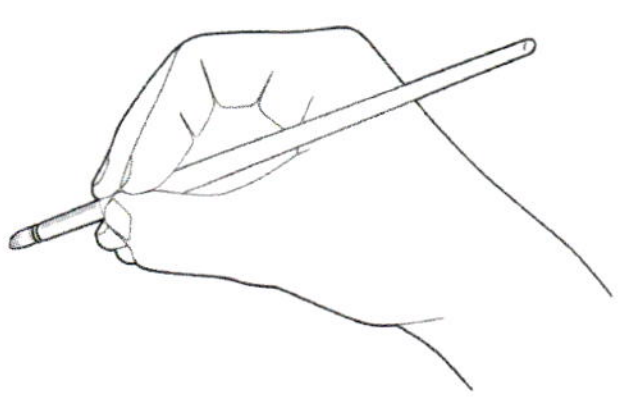

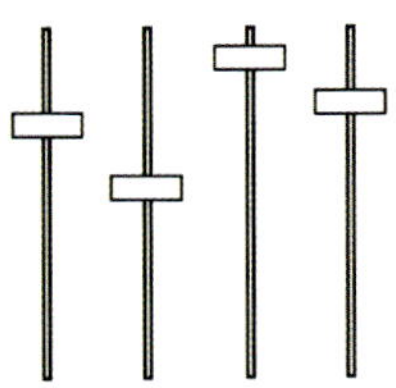

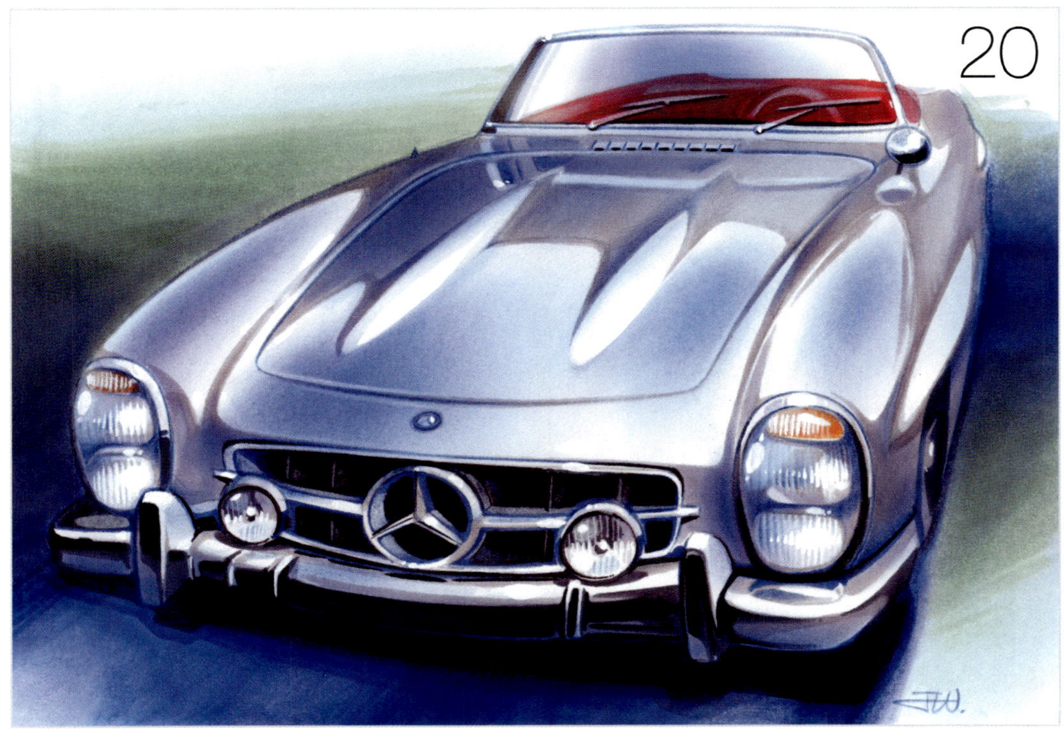

After the careful application of gouache highlights, scan in the finished rendering. Turn up the contrast and color saturation in Photoshop.

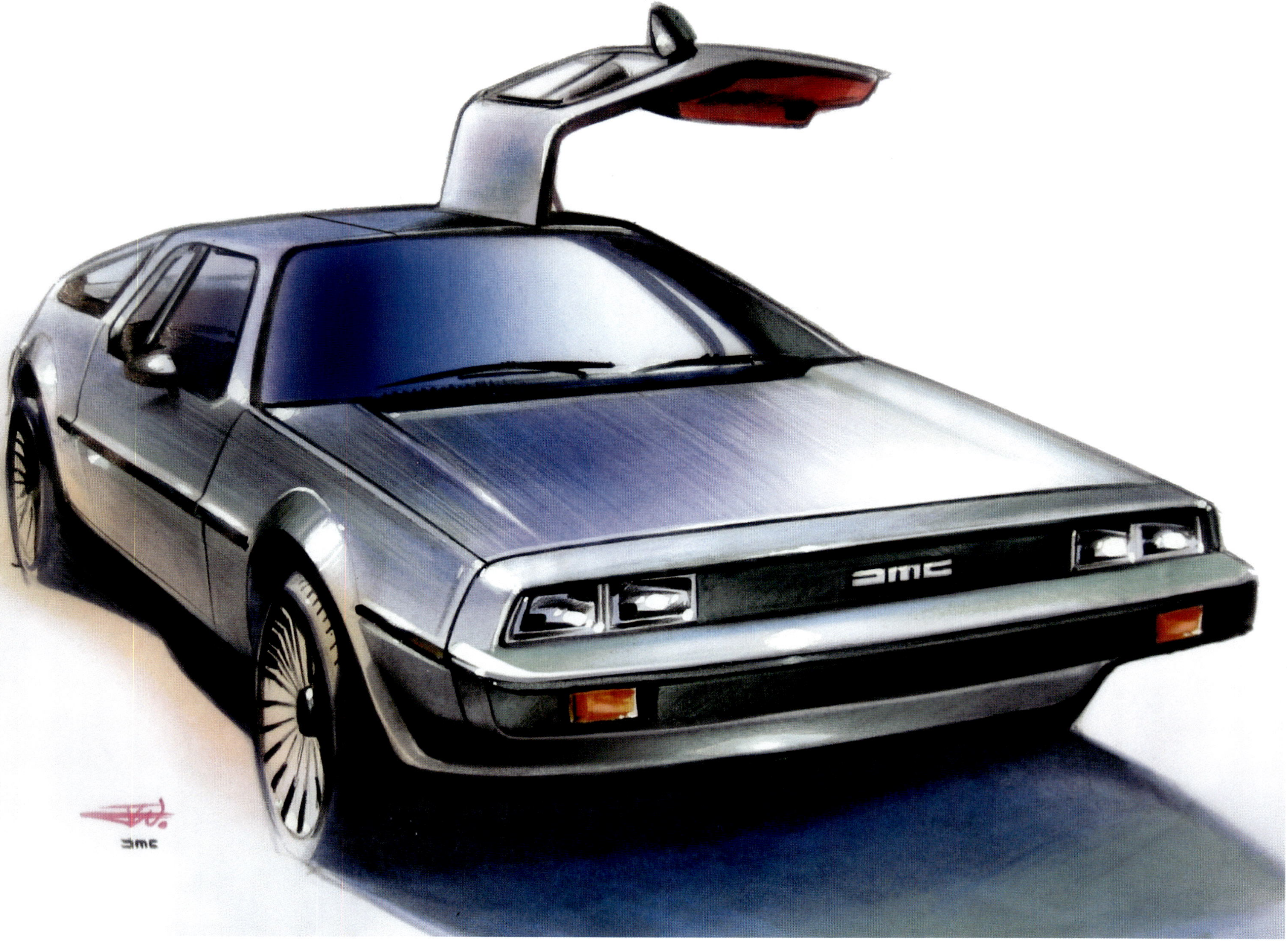

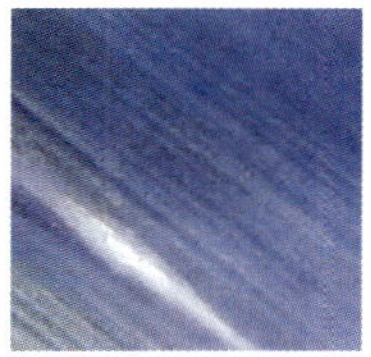

9. Stainless Steel

Rendering stainless steel presents some unique challenges. It doesn't have hard reflections like painted metal, and there's the grain to consider as well. Here, we present a simple, step-by-step process for representing this material – in the guise of the De Lorean DMC-12.

The Tools

Prismacolors: White, Celadon Green, Black.

Nupastels: Indigo Blue, Violet, Orange.

Verithins: Black.

Gouache: White.

Prismacolor Markers: Indigo Blue, Cool Grey 20%, Cerulean Blue, Black, Cool Grey 70%, Crimson Red, Orange.

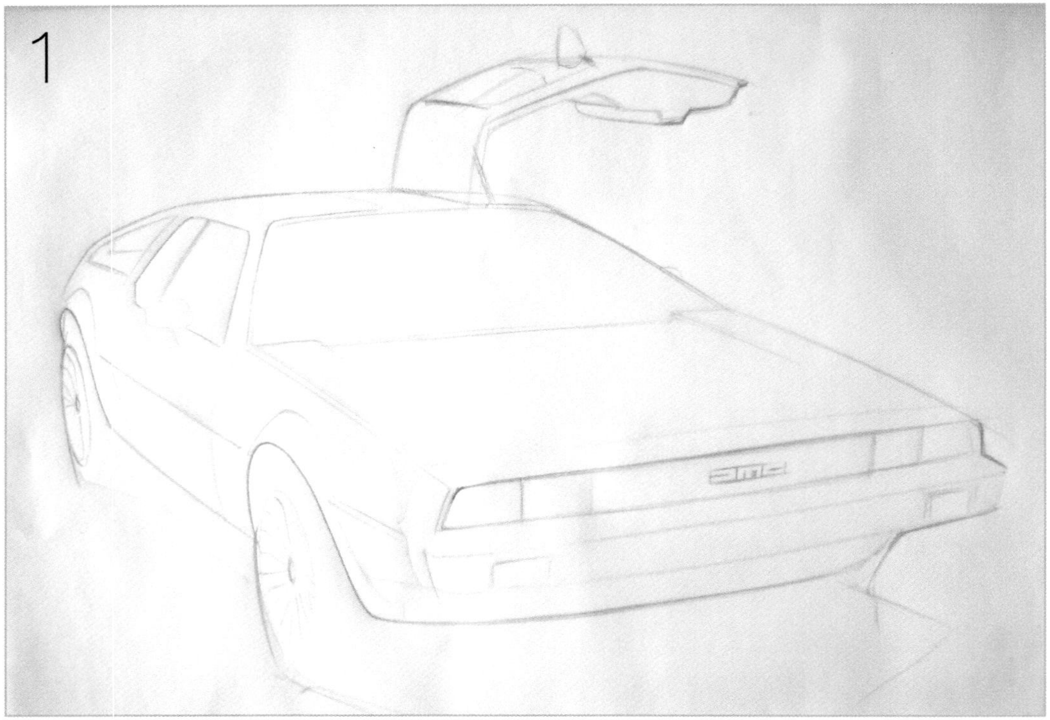

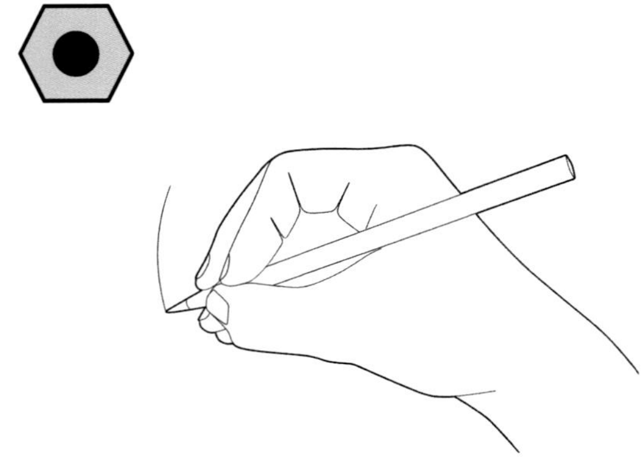

For this particular rendering, try switching to George Vincent Vellum. On vellum, you apply medium to both sides.

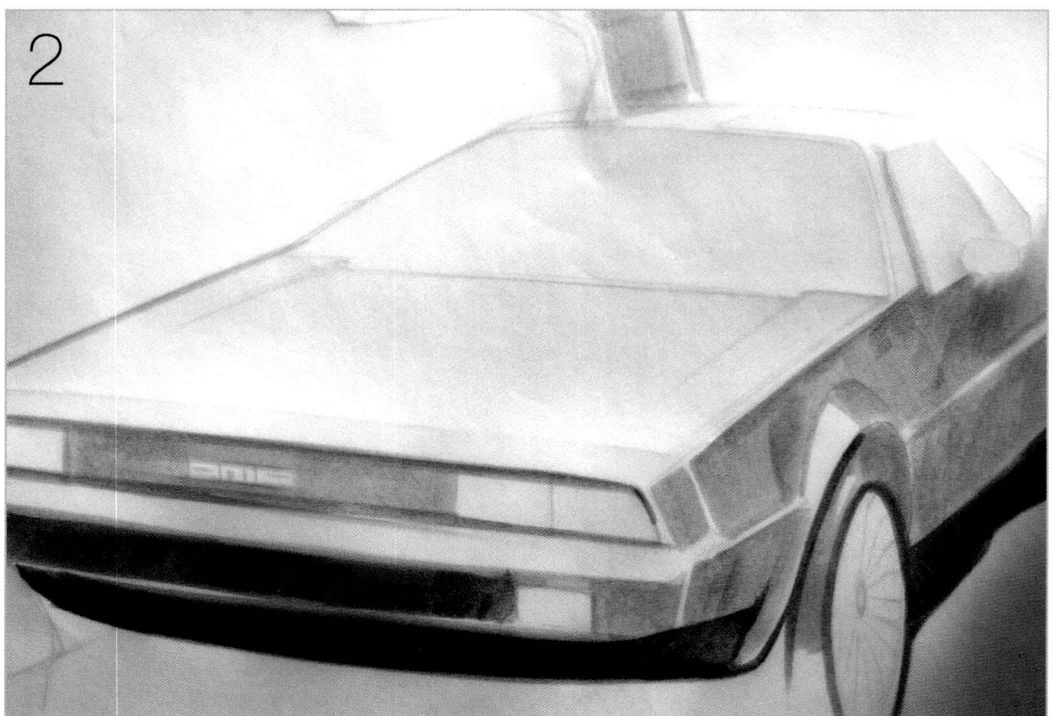

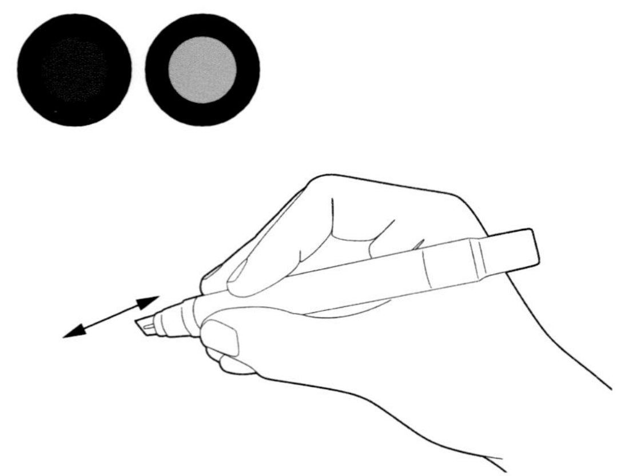

Begin by laying in the bodyside color and black plastic on the back with Cool Grey 20% and 70% markers.

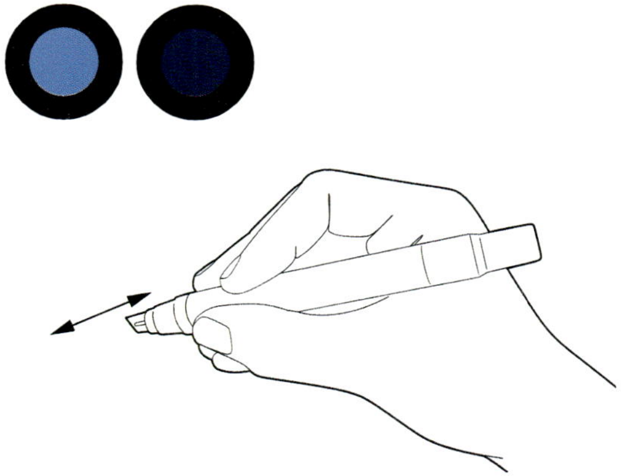

Switching to the front, apply a mixture of Cerulean and Indigo Blue in the wheel wells and the main shadow. Notice how working on both sides of the paper adds depth to the drawing.

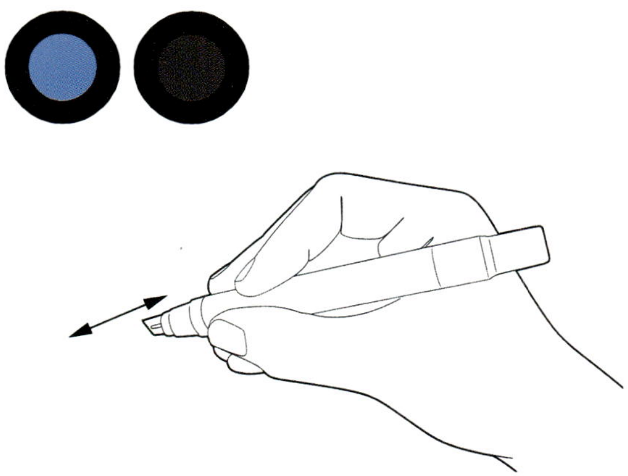

Staying on the front, layer the grey tones from the back with Cerulean Blue; emphasize the bodyside crease with Cool Grey 70%.

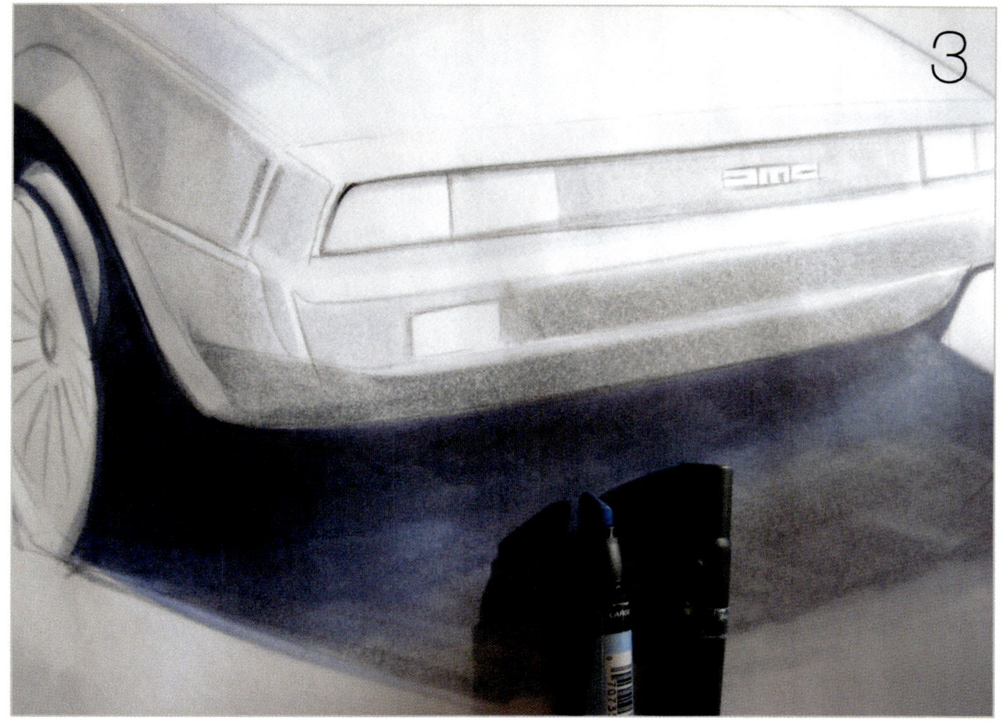

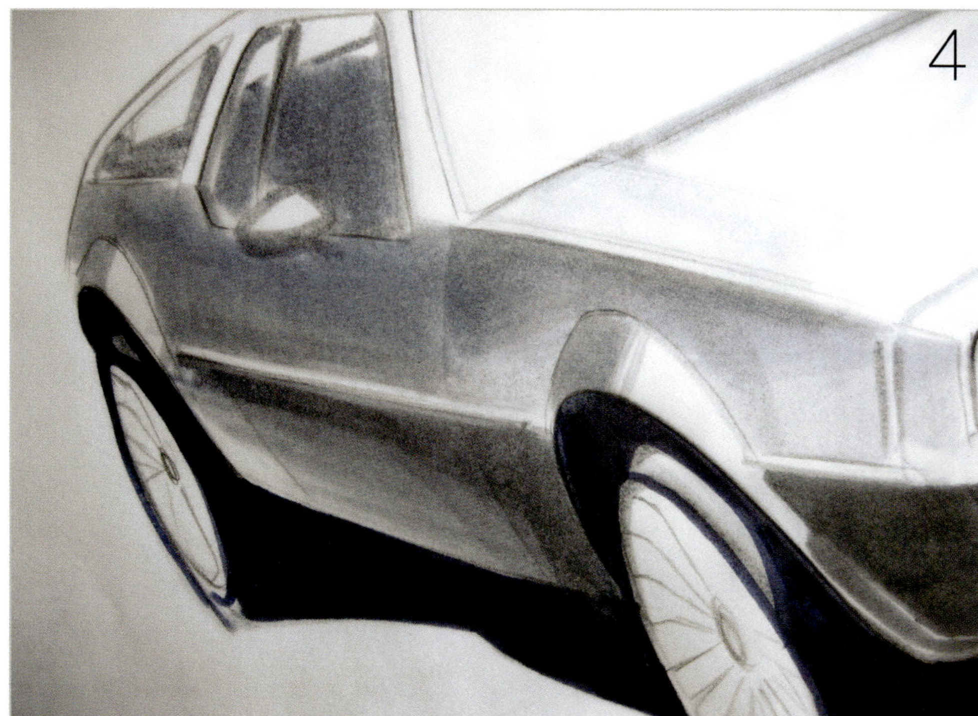

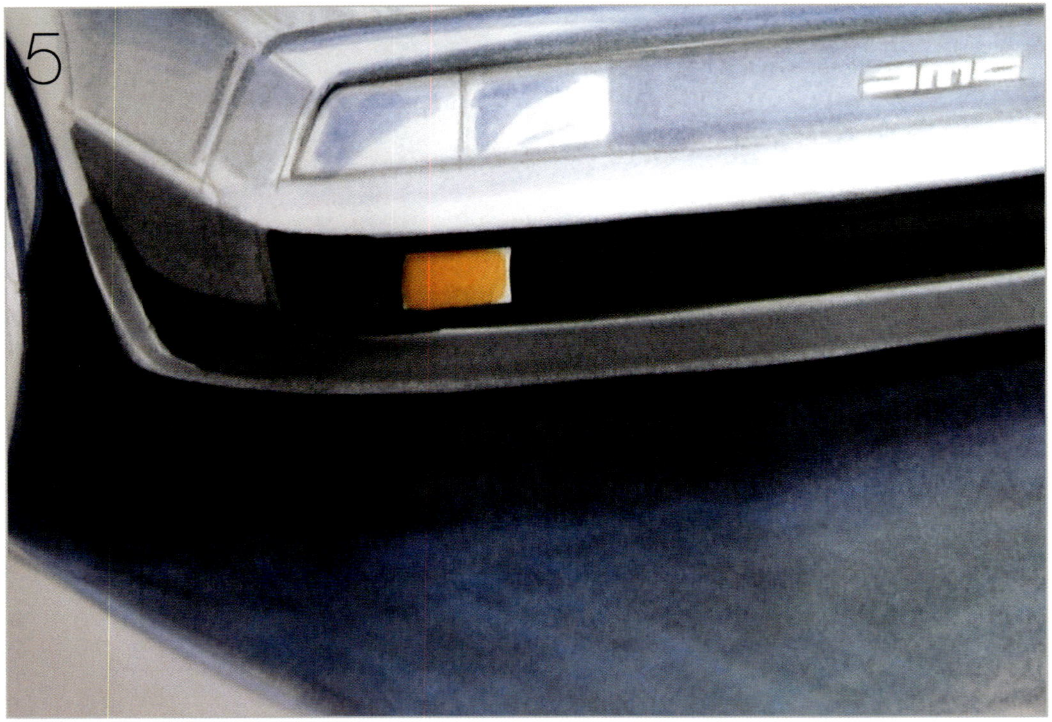

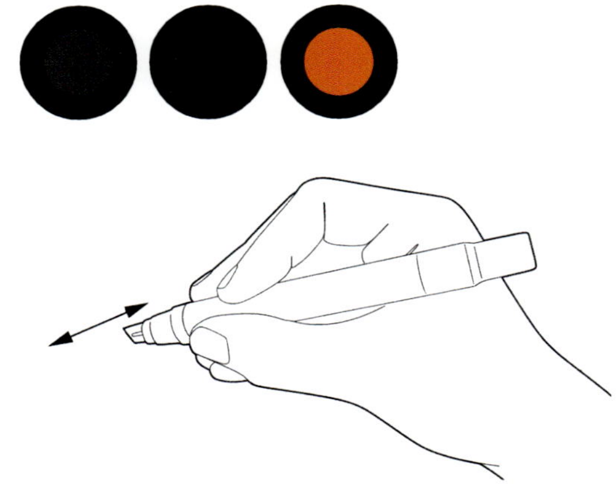

Reinforce the front bumper and shadow border with Cool Grey 70% and black. Block in the turn signals.

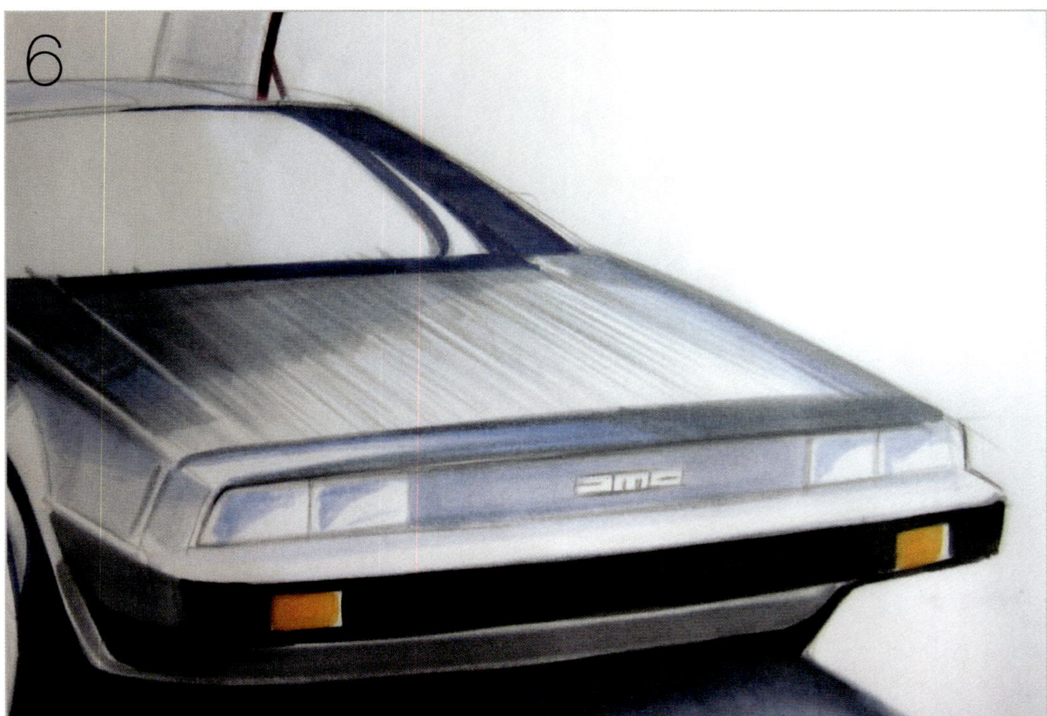

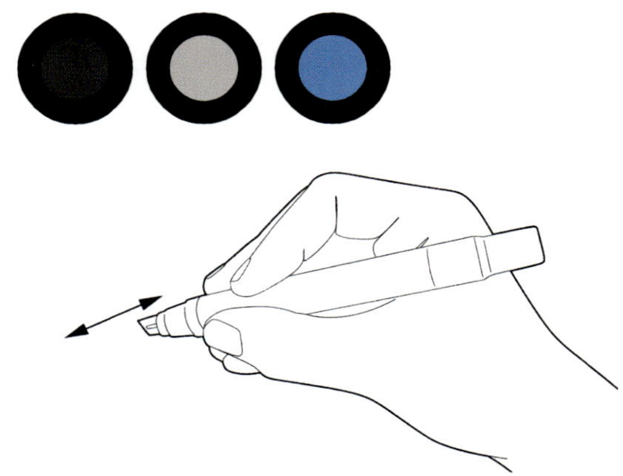

For the base portion of the car's stainless steel grain, apply swift, pointed strokes of blue and grey marker.

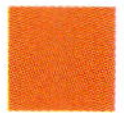

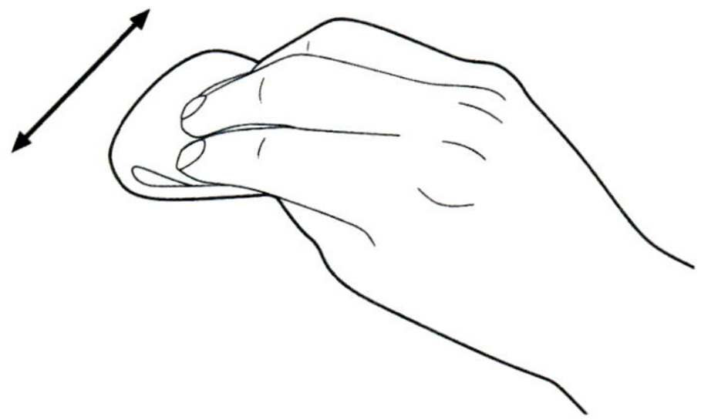

On the back of the page, lay in some powdered orange pastel on the lower half of the bodyside and next to the shadow. This will be the "warmth" of the ground countering the cool sky.

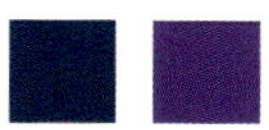

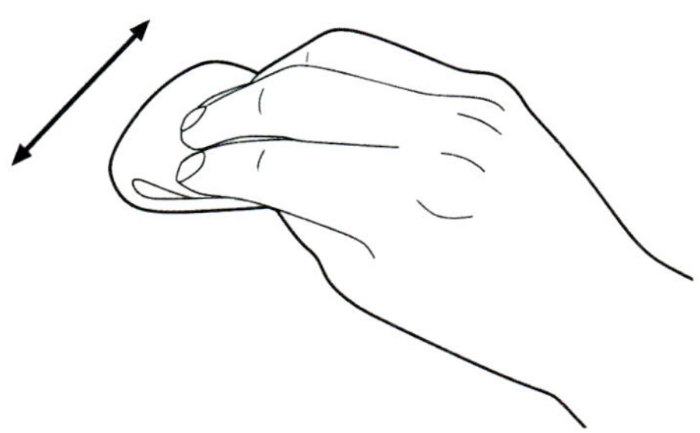

On the front, apply powdered pastel as a sky tone that transitions from violet to blue. Be sure to put the violet side of the sky tone near the warm side of the car.

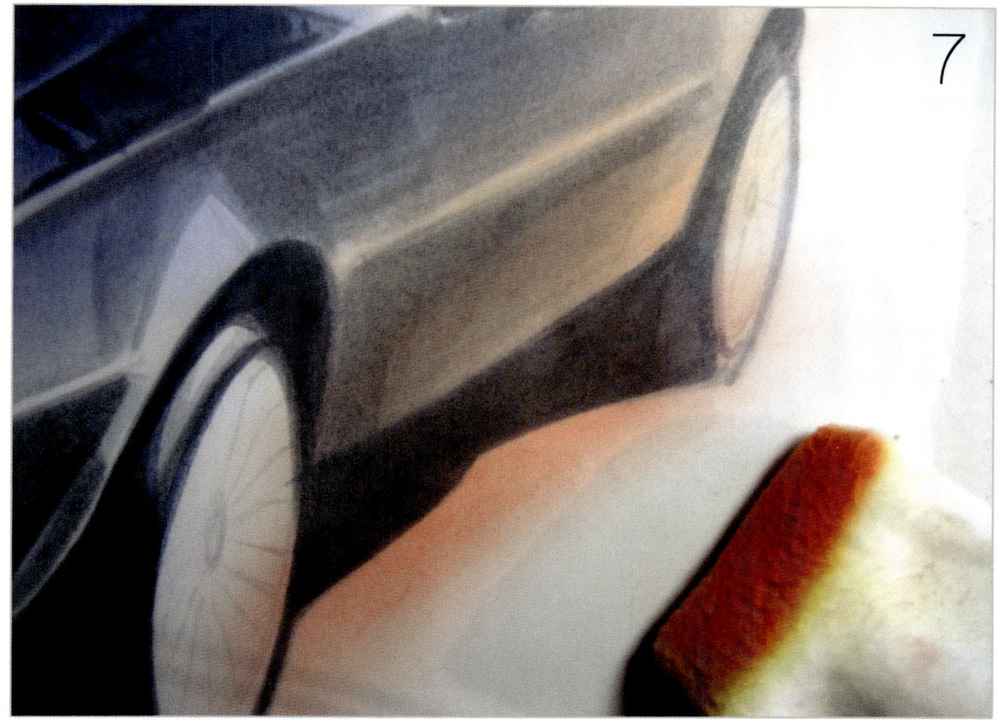

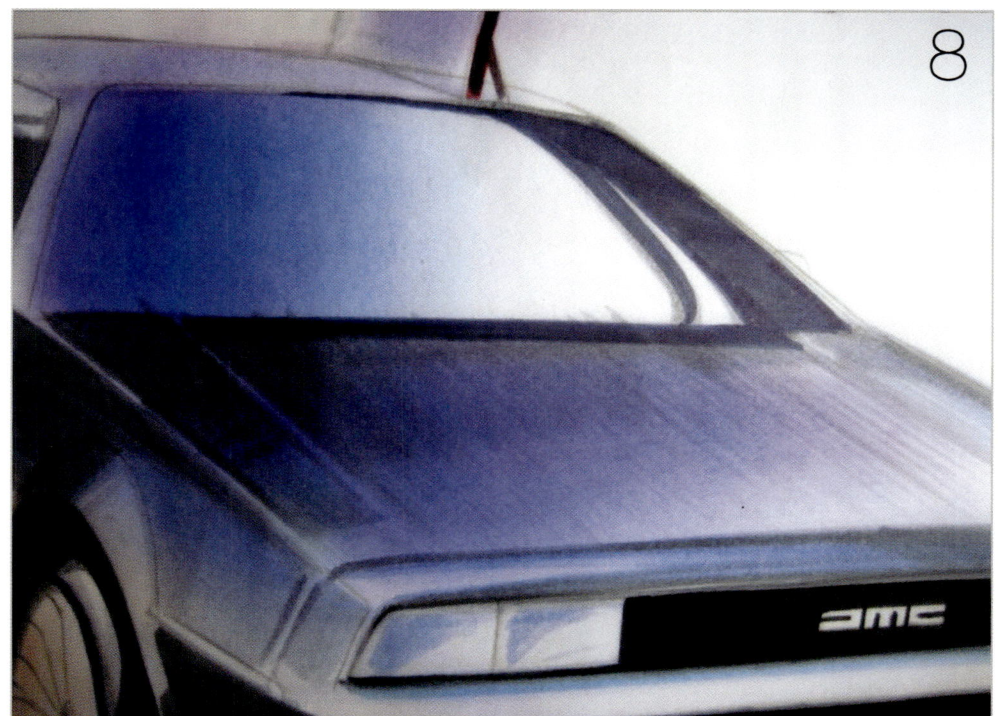

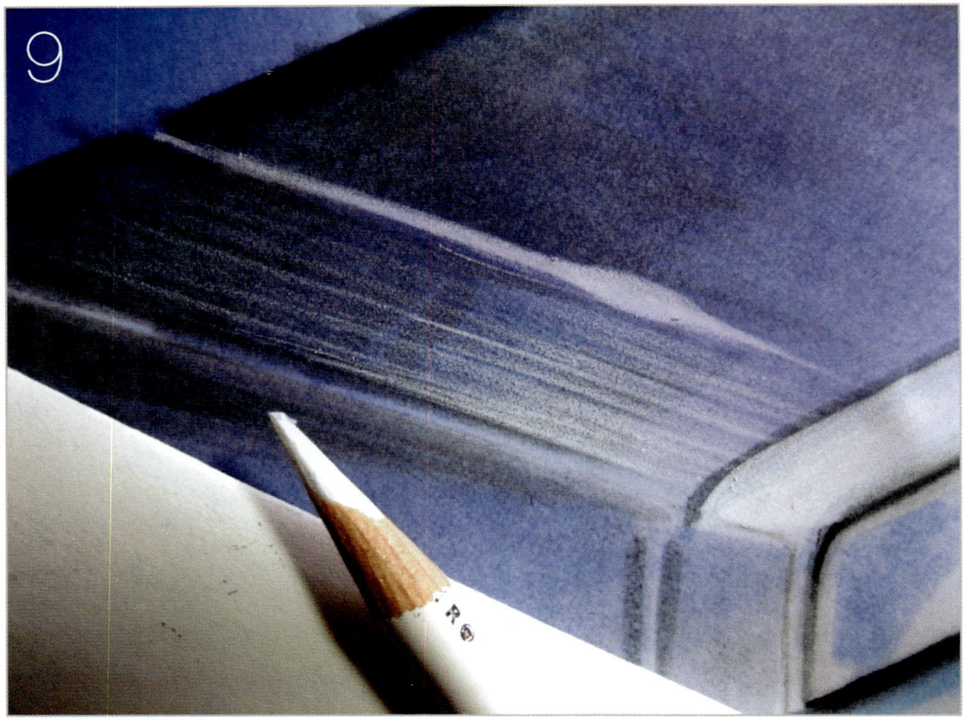

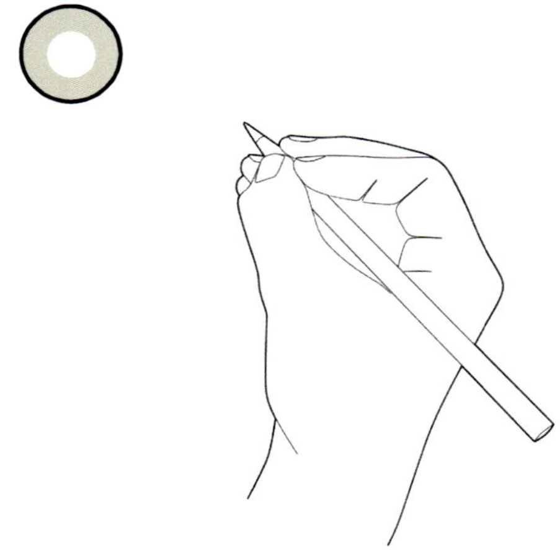

White Prismacolor pencil lines will provide the "peaks" of the stainless steel grain. Use swift strokes, but be sure to keep the direction consistent.

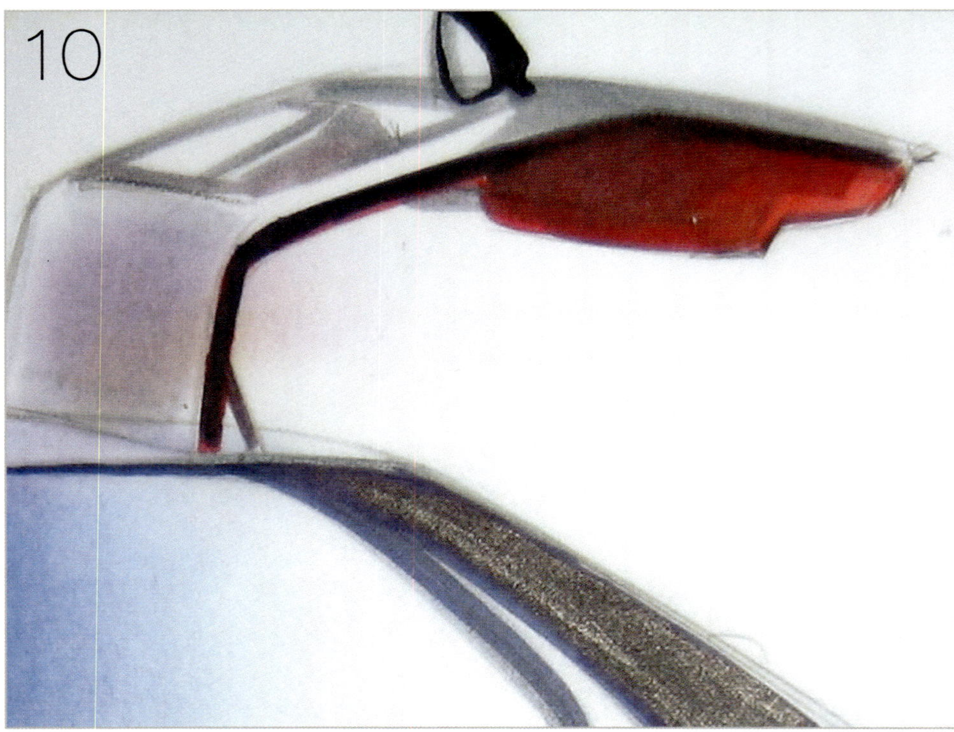

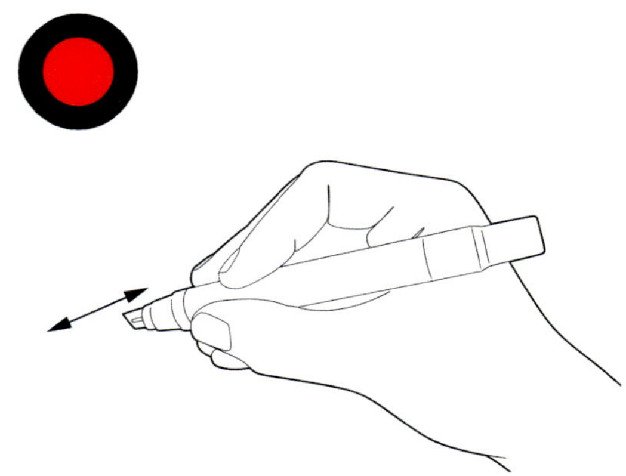

Not much of the interior is visible in this rendering, so make it count. The red on the front is made richer by the grey on the back.

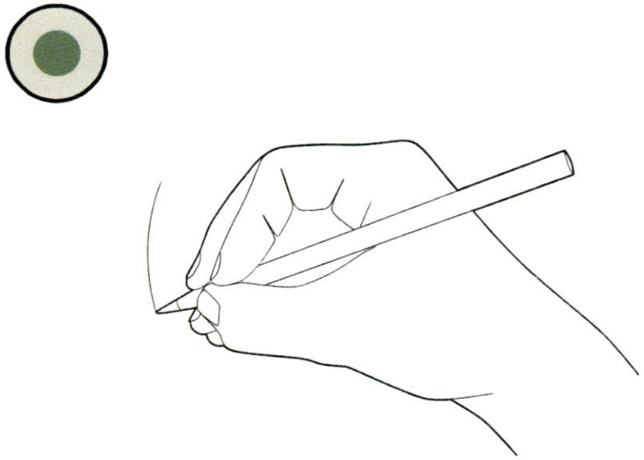

To give the black bumper and the front fascia some added character, lay in a bit of Celadon Green pencil. Like the pastels above, the green should fade progressively across the front.

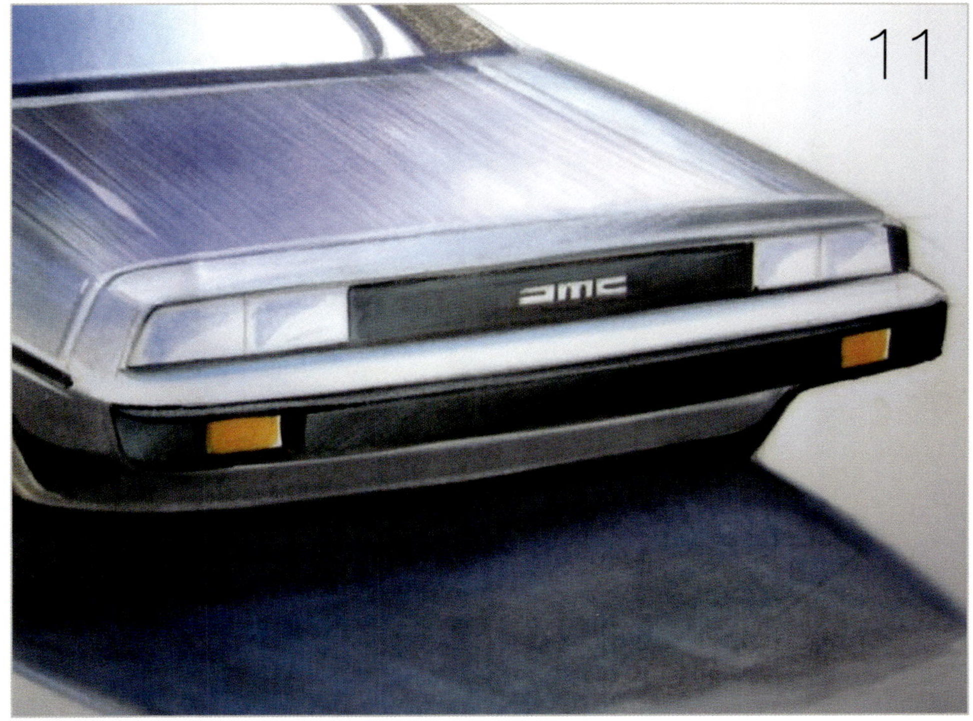

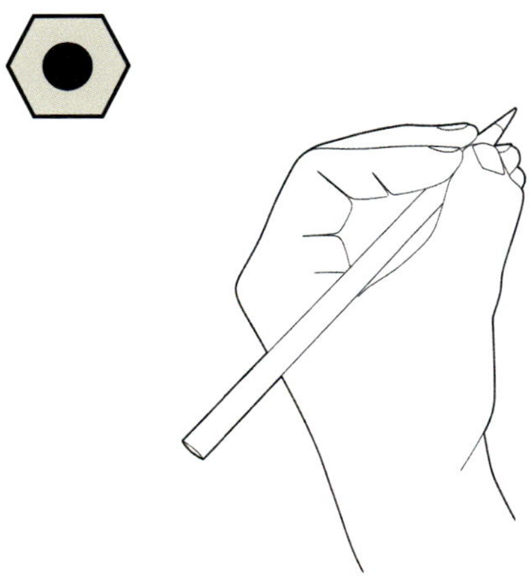

Black Verithin pencil will provide the "valleys" of the stainless steel grain. As with the white pencil, make sure the strokes are well-aligned.

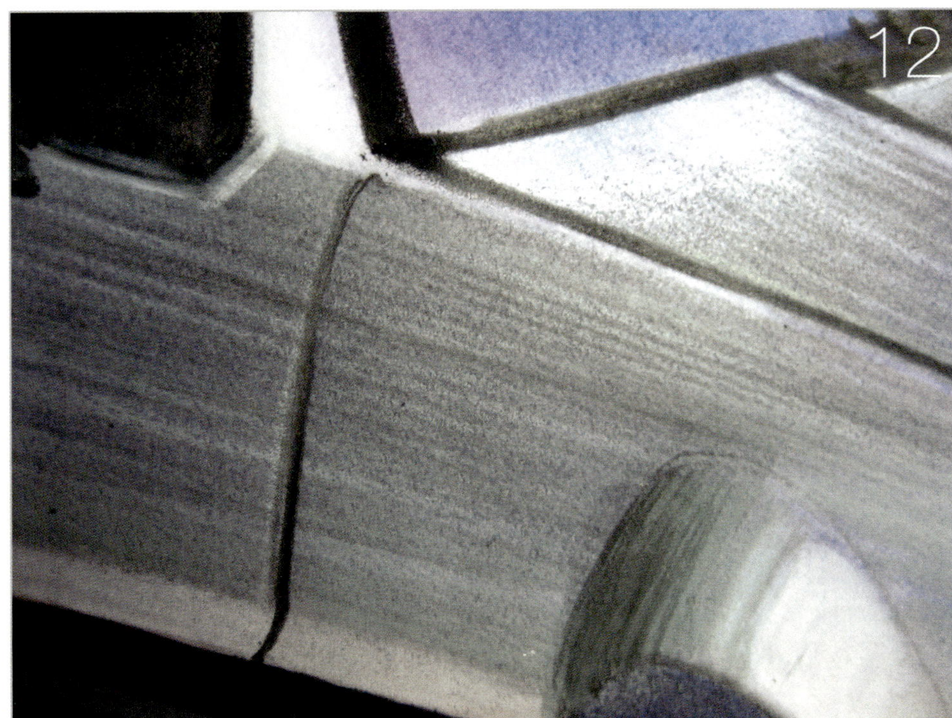

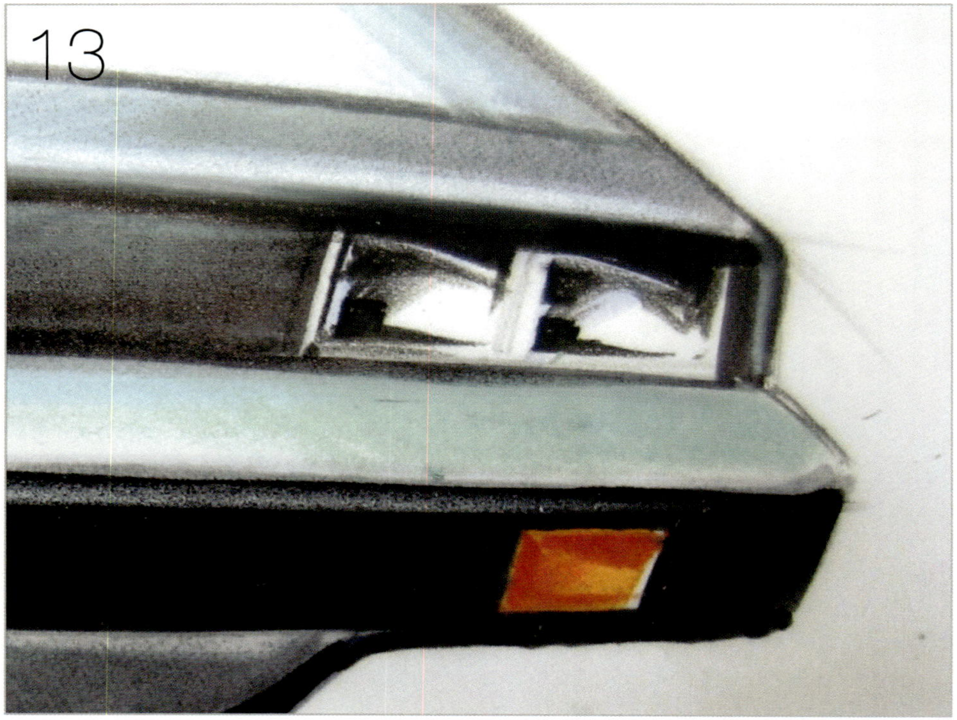

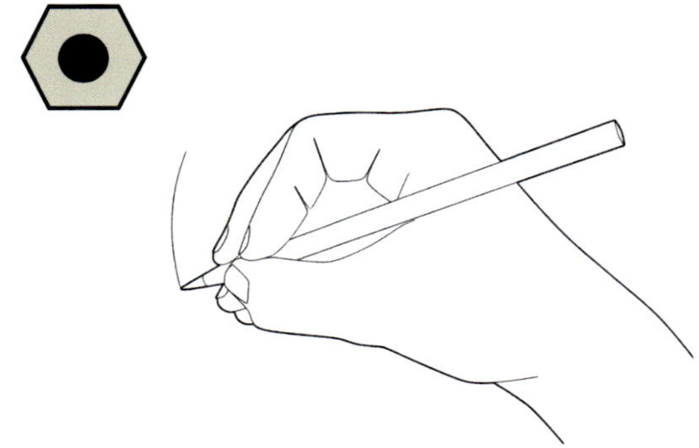

Staying with the black Verithin, take a moment to detail the headlamp cans. Remember the rule about lamps: Dark on the top, light on the bottom.

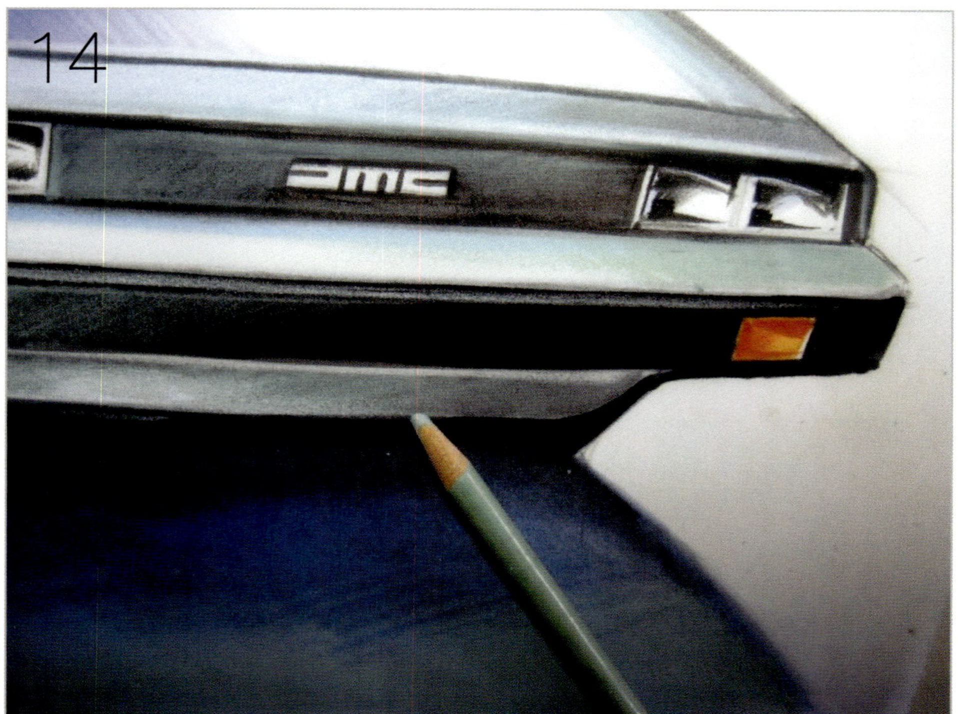

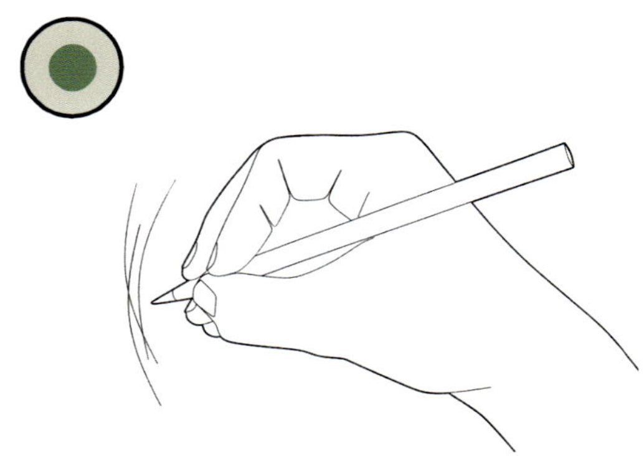

To give the shadow a bit more character, feather in a layer of Celadon Green pencil on the lower edge.

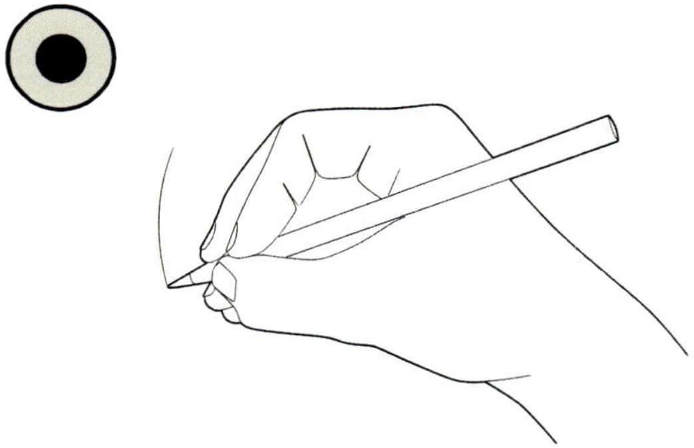

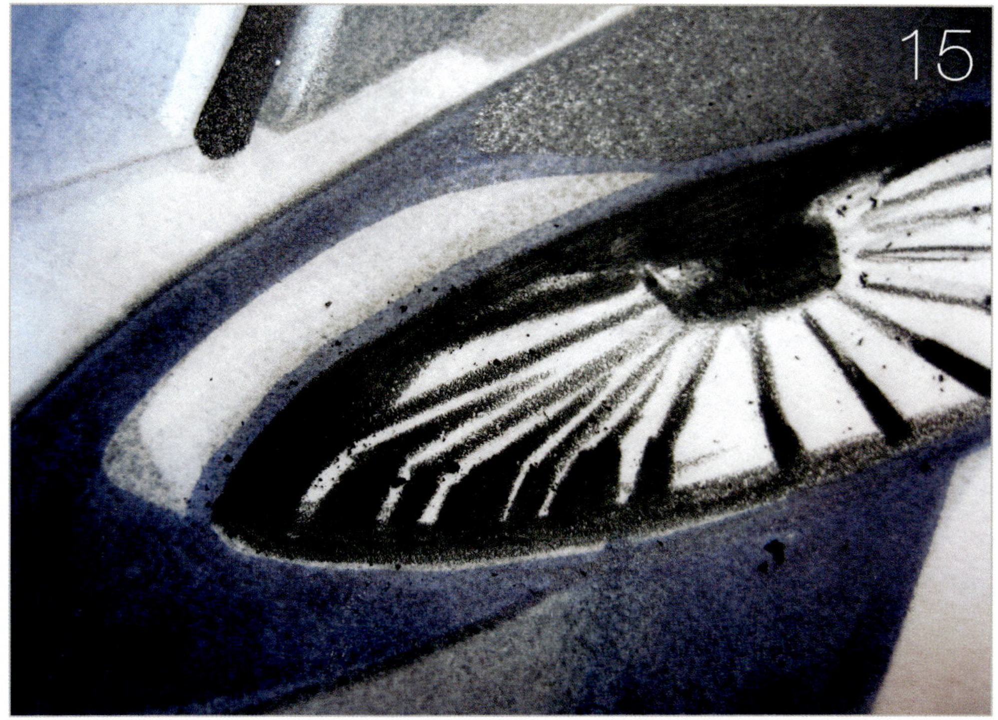

Apply deep shadows in the milled wheel ribs using a sharp, black Prismacolor pencil.

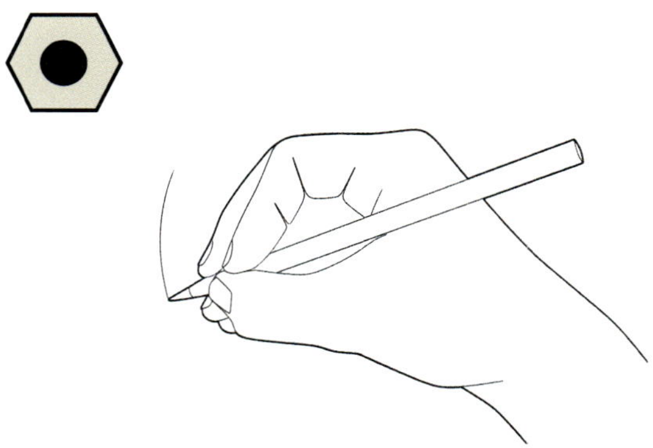

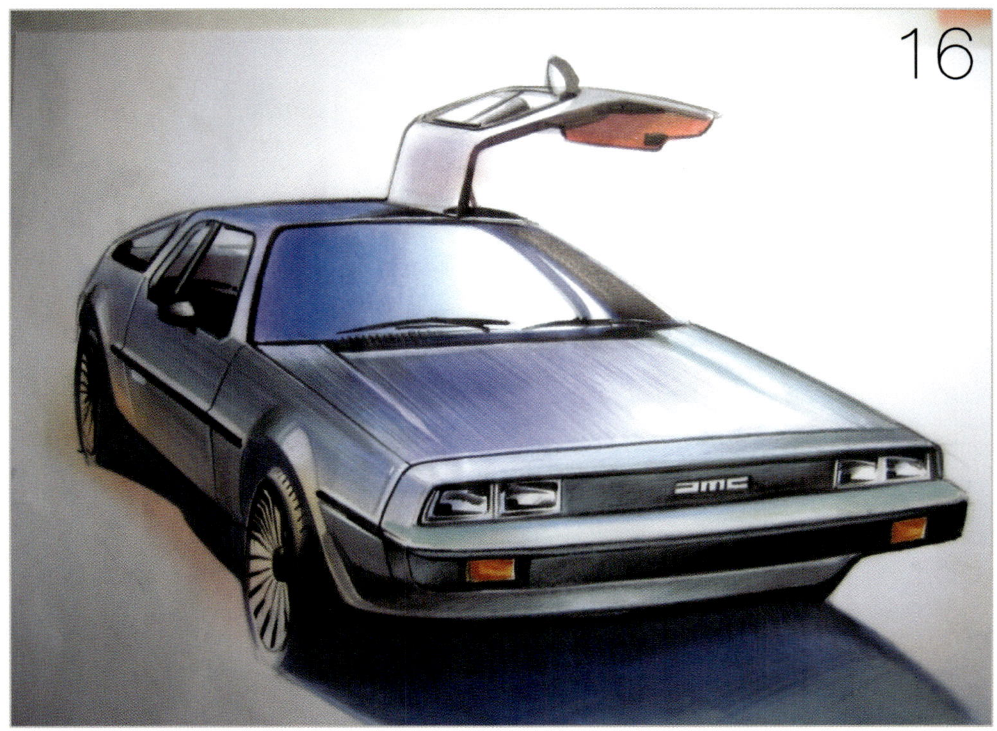

Now go around the car with a sharp, black Verithin and carve in details: Shut lines, wipers, vent louvers, etc.

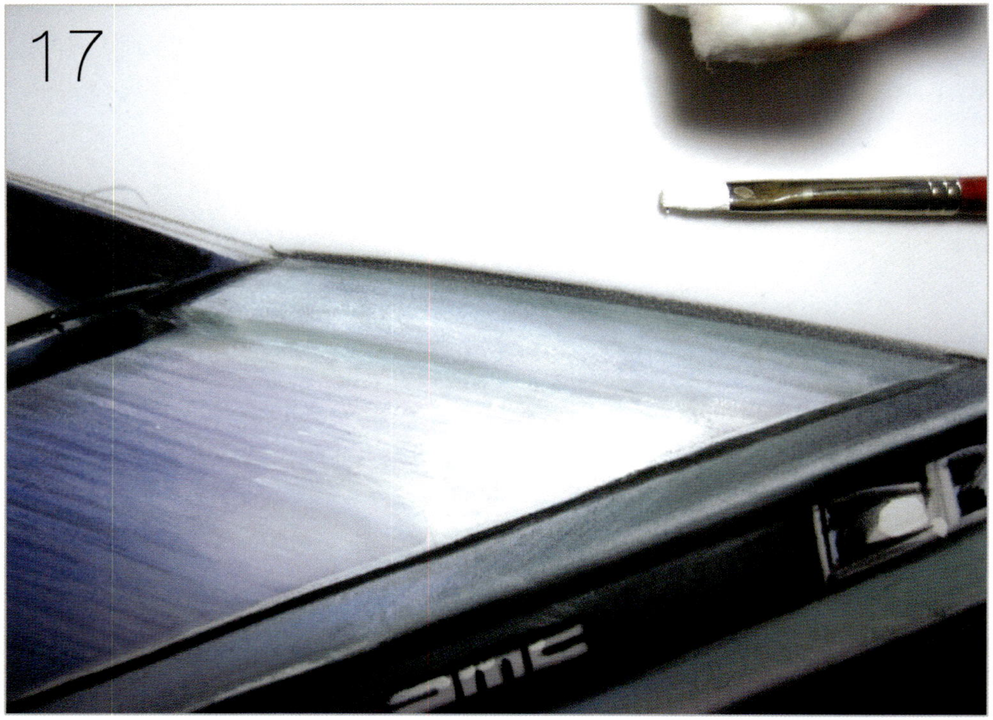

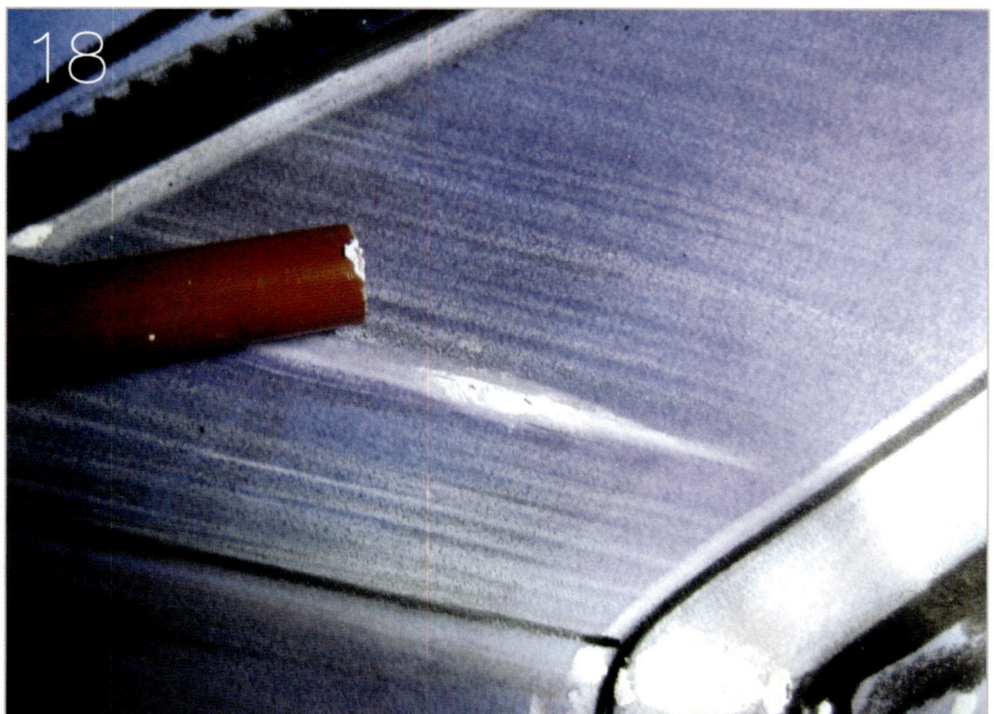

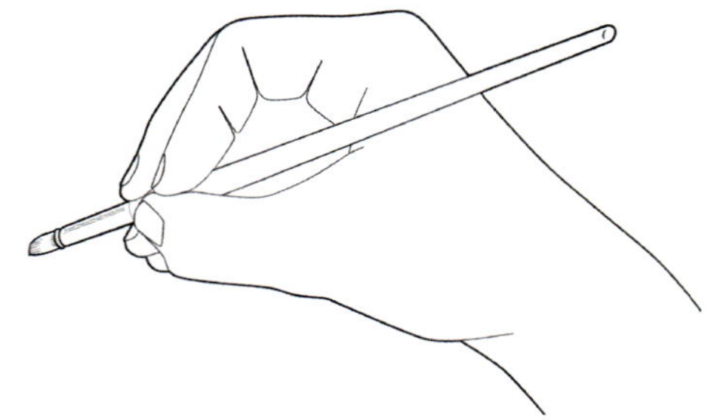

Stainless steel highlights follow the direction of the grain and also tend to be far more broad. Take these factors into account when applying the white gouache. Remember that the fascia is painted plastic, so it won't light the same as the stainless steel.

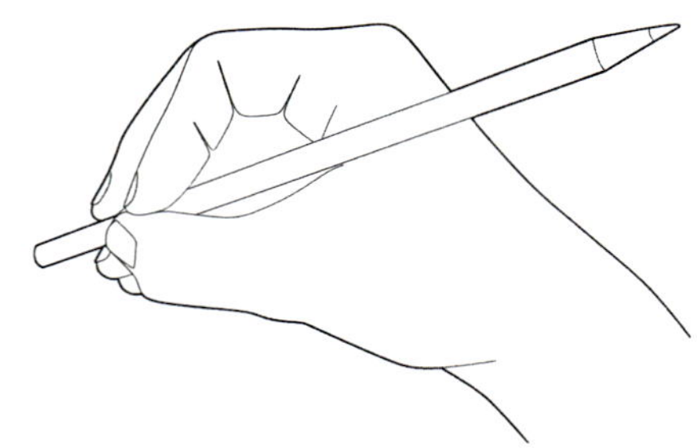

If you apply too much gouache, wait for it to dry and scrape away the excess with the blunt end of a pencil.

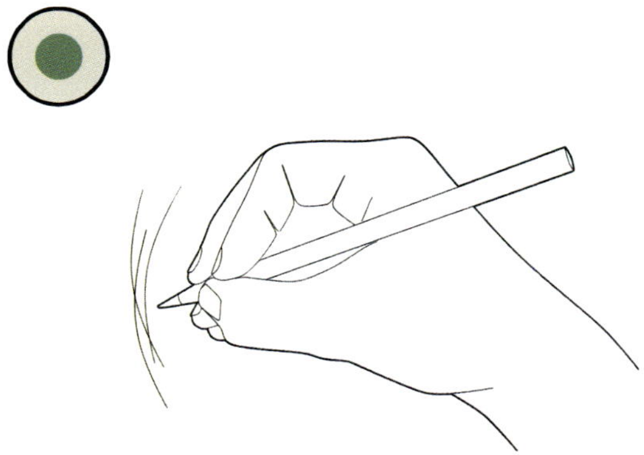

Make one more cursory pass with a Celadon Green pencil, adding some subtle shading in the reflections.

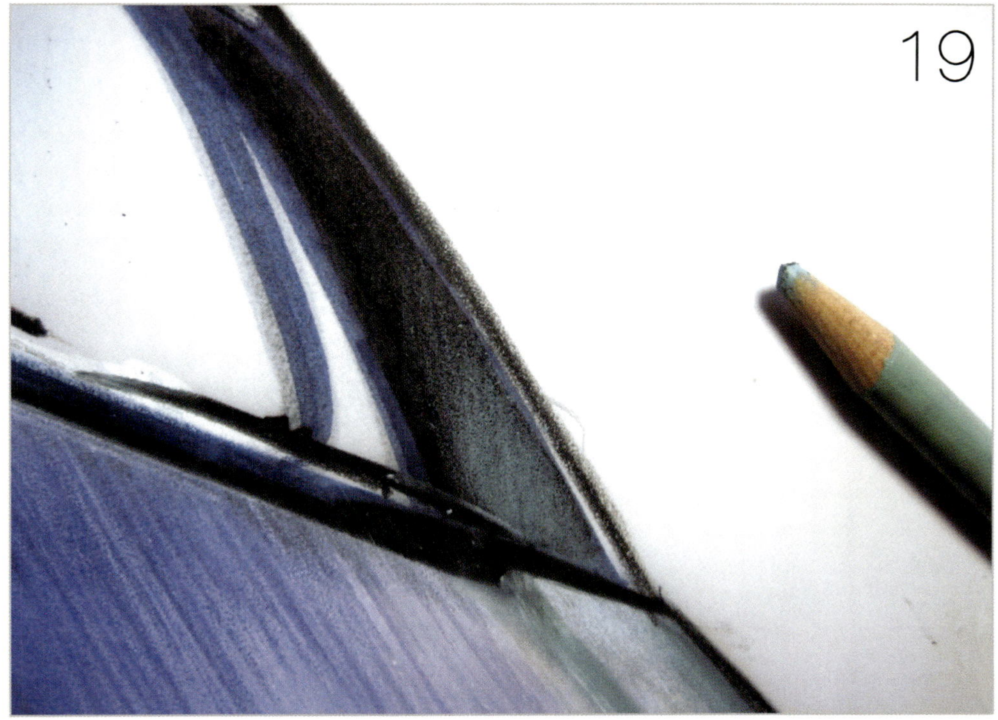

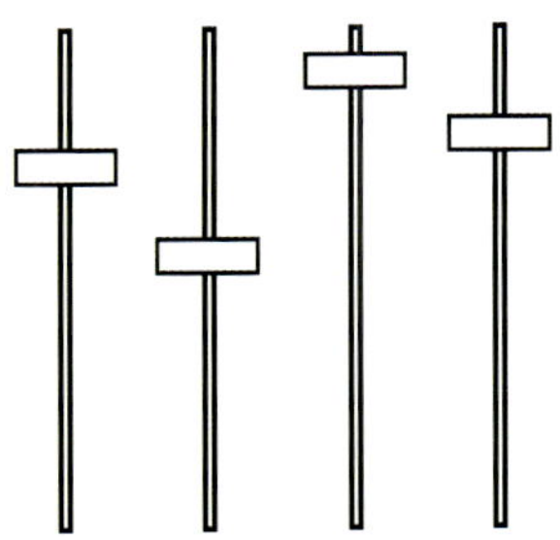

Scan in the finished rendering. Clean up the highlights and adjust the levels to your liking in Photoshop.

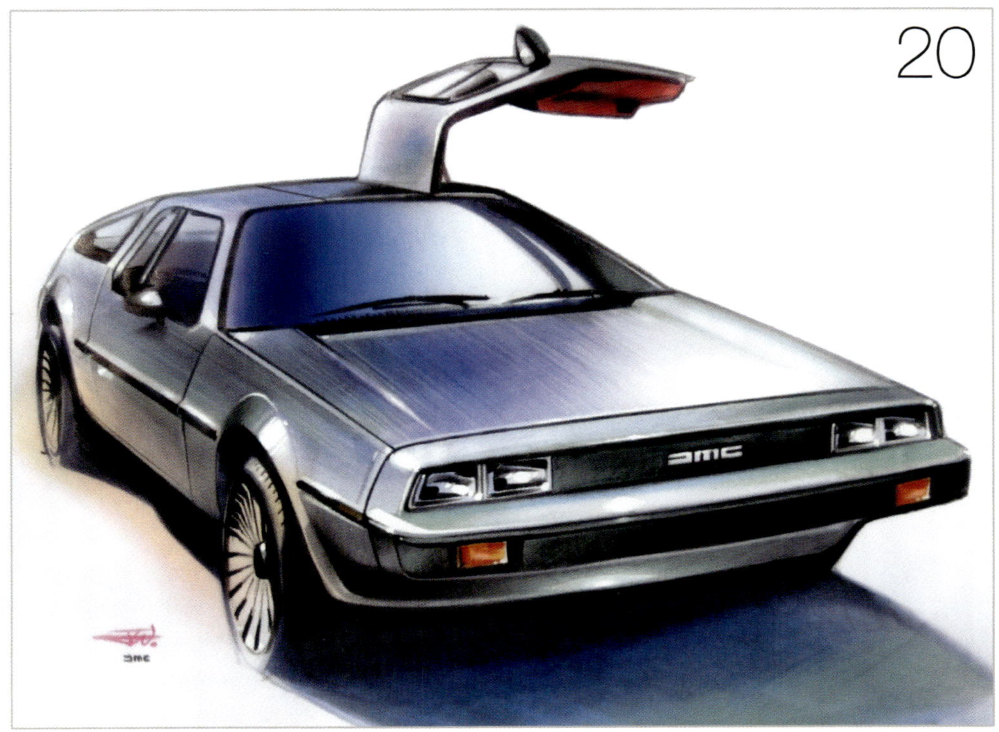

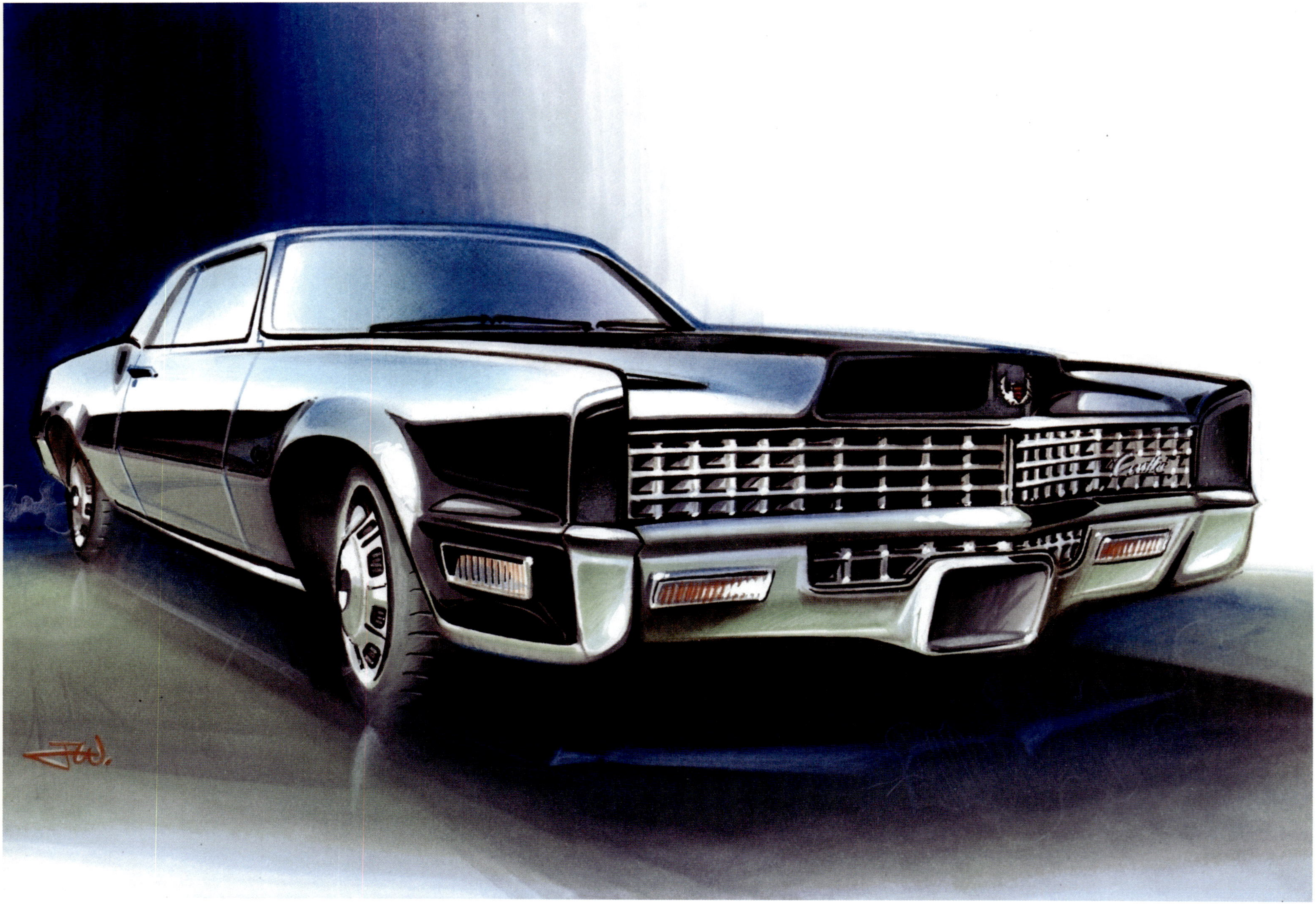

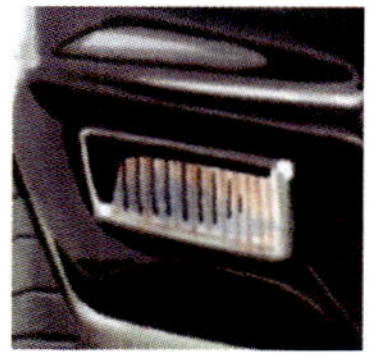

10. Gloss Black 101

When rendering gloss black paint on marker paper or vellum, you have to pretend it's the hottest, brightest day of the year. The alternative is to fill in the whole car with black marker, which would leave the subject looking leaden. What follows is a simple gloss black technique, using the floor color as a key descriptor.

The Tools

Prismacolors: White, Orange.

Nupastels: Indigo Blue.

Verithins: Black, Indigo Blue.

Gouache: White.

Prismacolor Markers: Indigo Blue, Cool Grey 20%, Cerulean Blue, Cool Grey 70%, Celadon Green, Black.

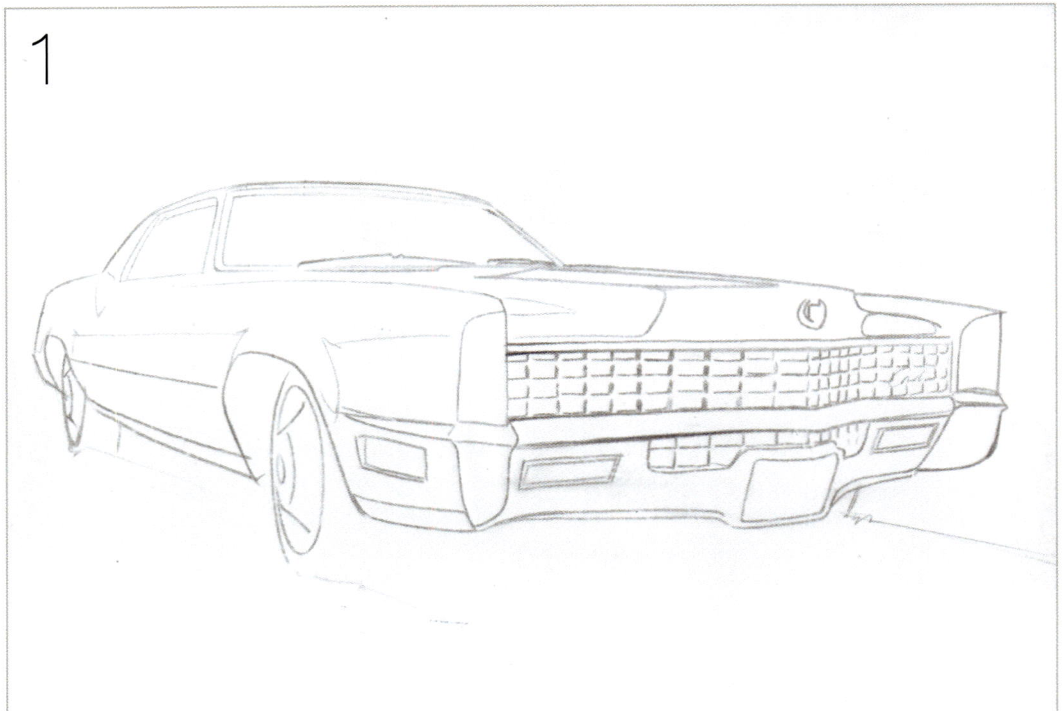

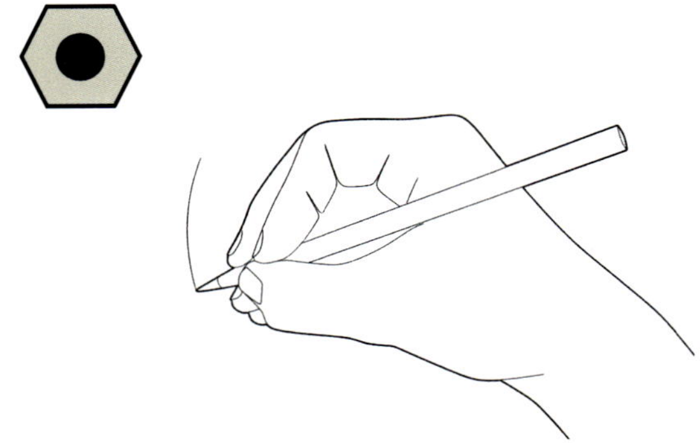

When laying out the drawing on marker paper, take the time to pencil in every strake and cross member in the grille. Resist the urge to rush things.

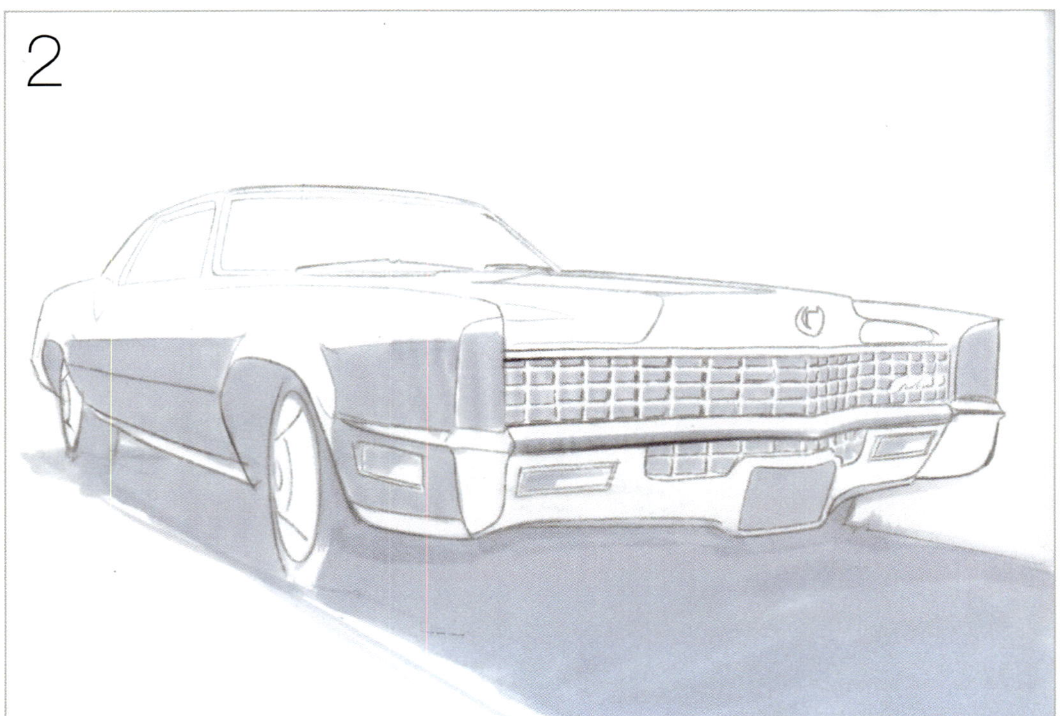

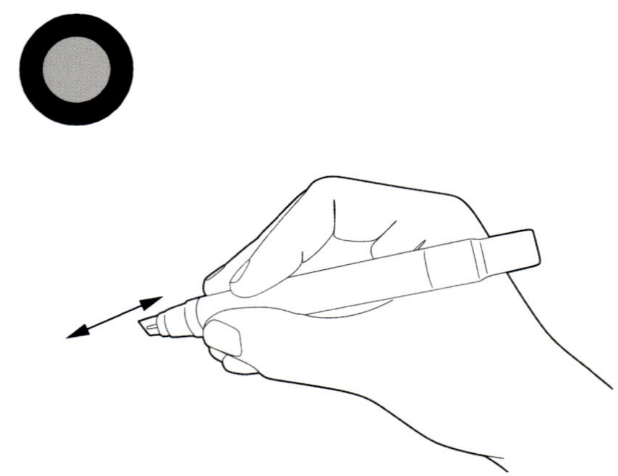

Begin by laying in the bodyside reflection, the grille holes and shadow with Cool Grey 20%.

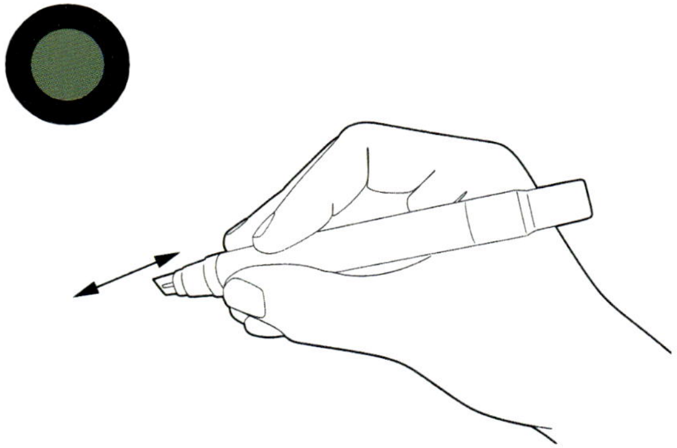

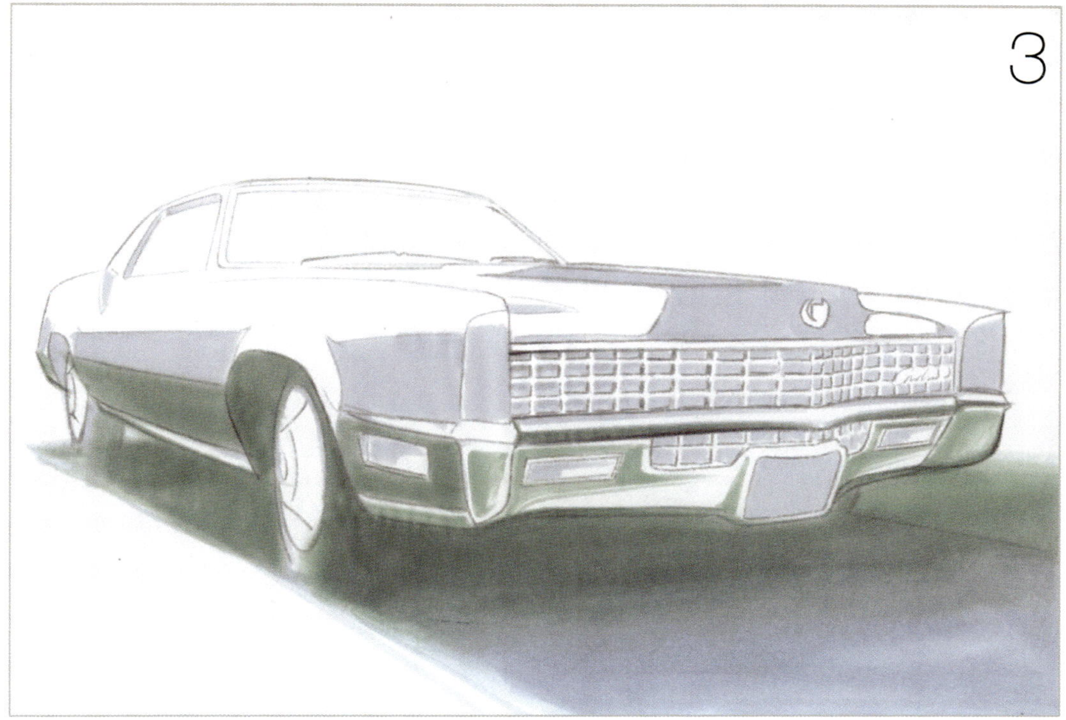

Now switch to a Celadon Green marker for the darker portion of the shadow, the lower part of the bodyside and the chrome bumper reflections.

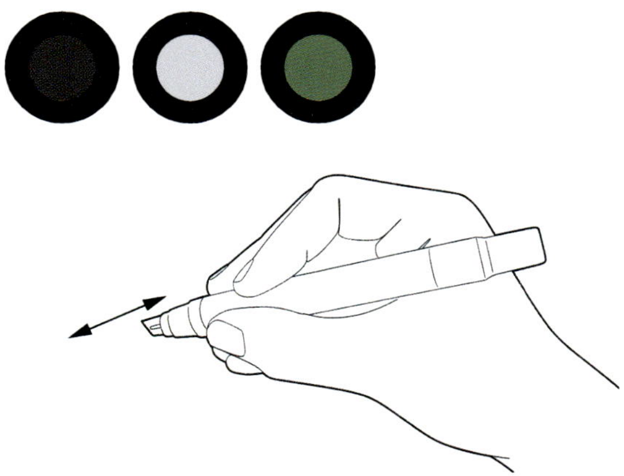

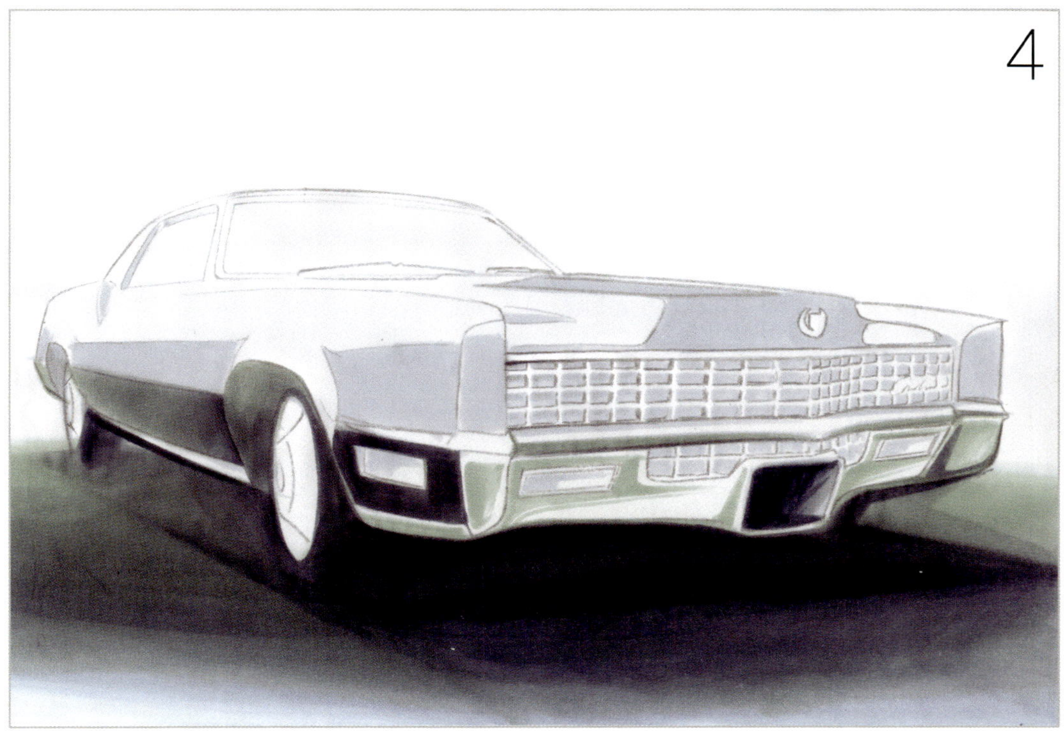

Take things one step further by including Cool Grey 70% in the mix.

5

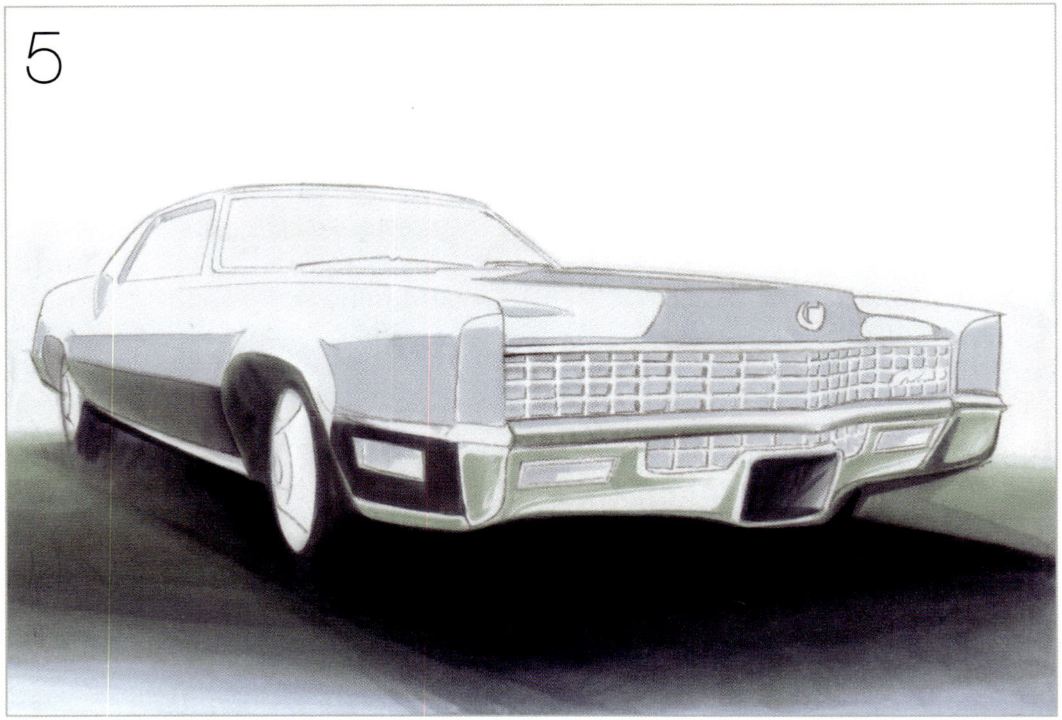

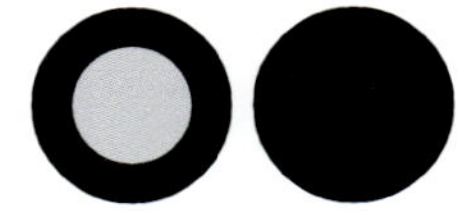

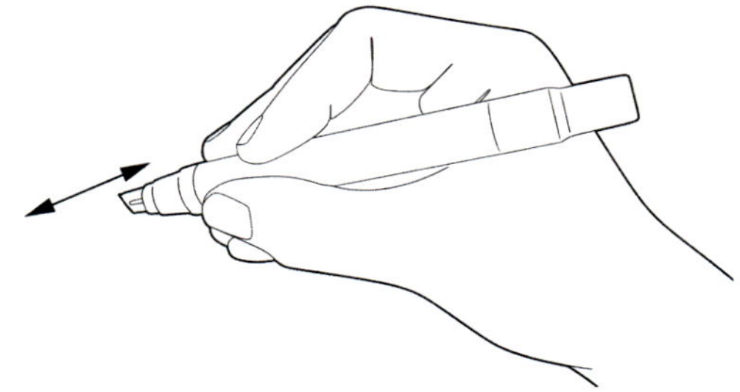

Using Cool Grey 20%, smooth out the shadow's transition from dark to light. Add some black marker in extreme dark areas for a little extra punch.

6

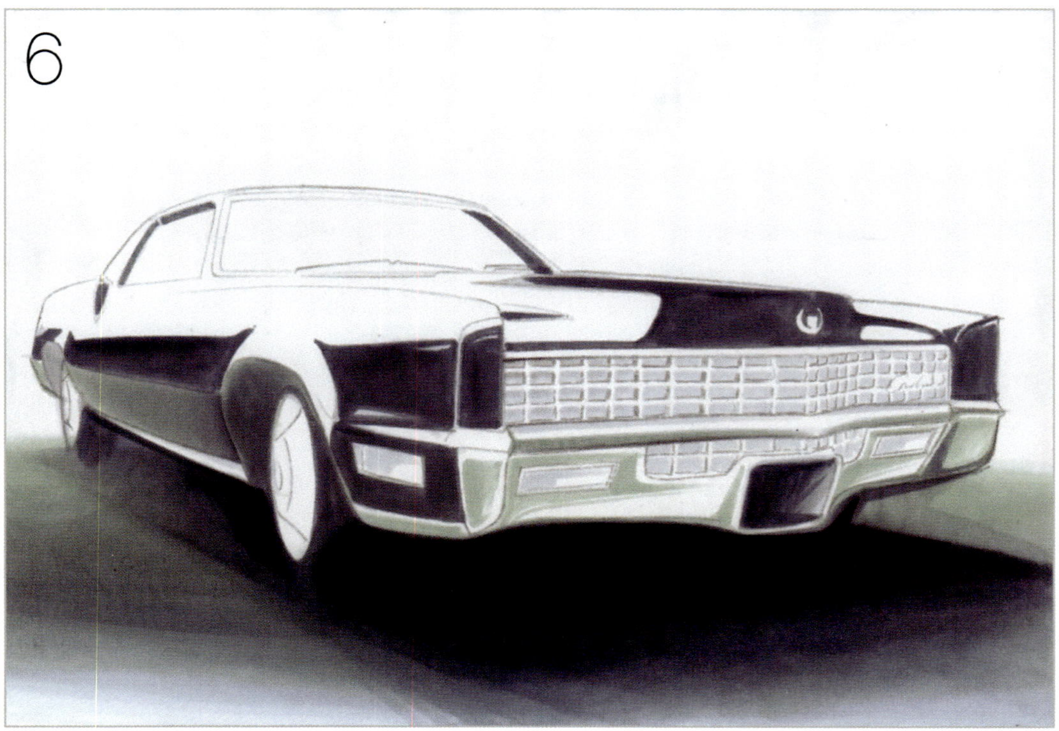

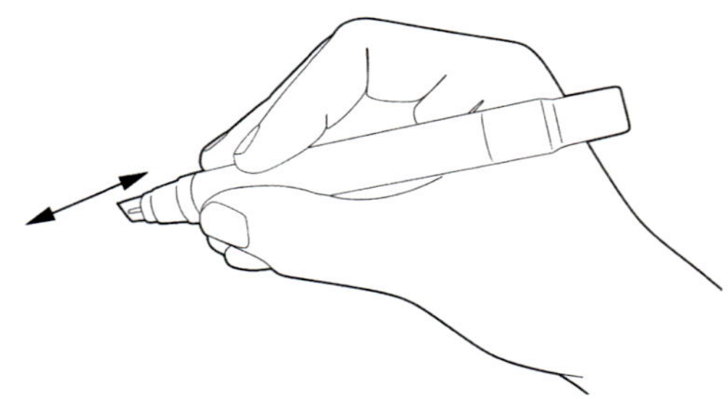

Add a second layer of marker to the main body reflections; this time use Cool Grey 70%.

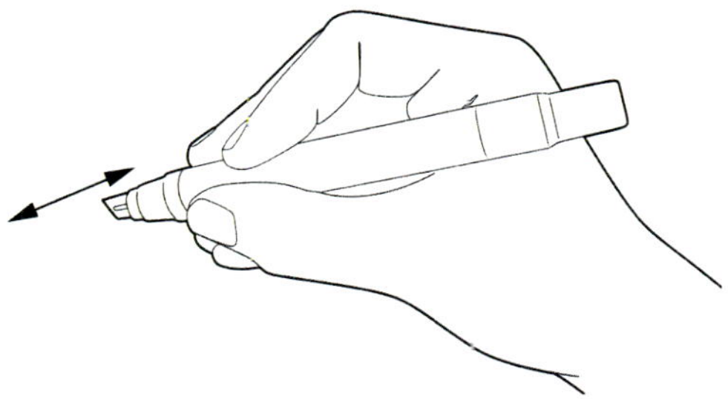

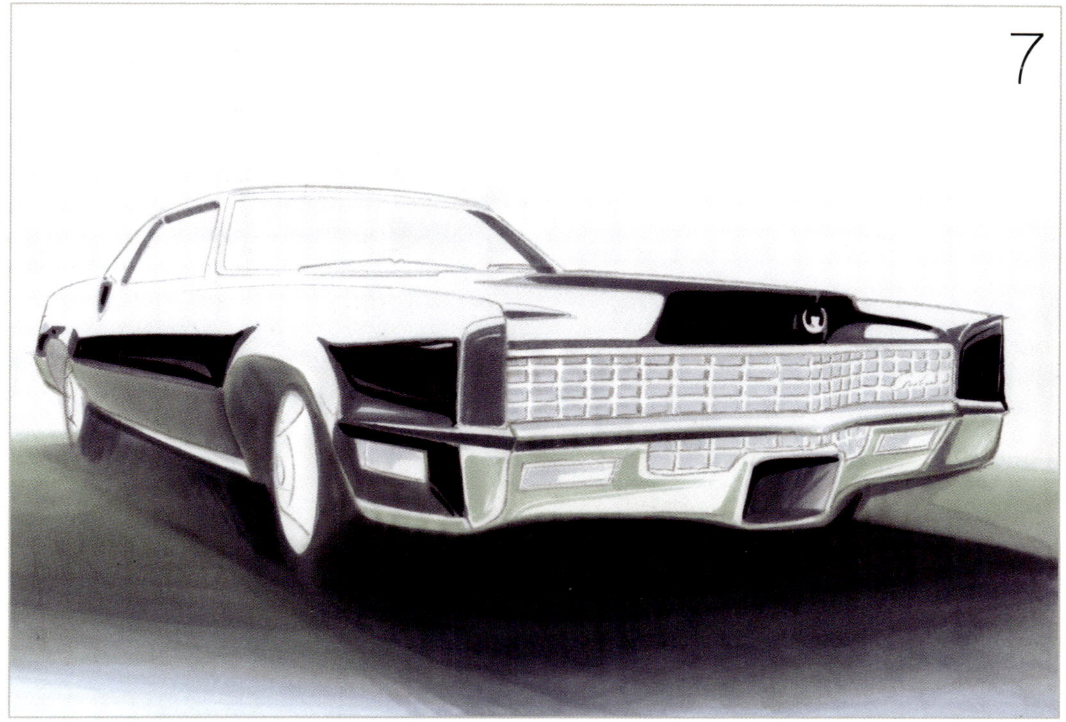

Fill in the darkest part of the main reflections with a black marker. Notice how the sheer, under-turned part of the bodyside is reflecting light from the floor.

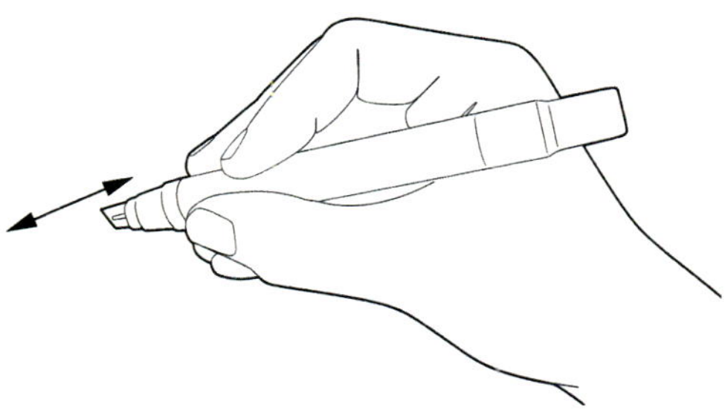

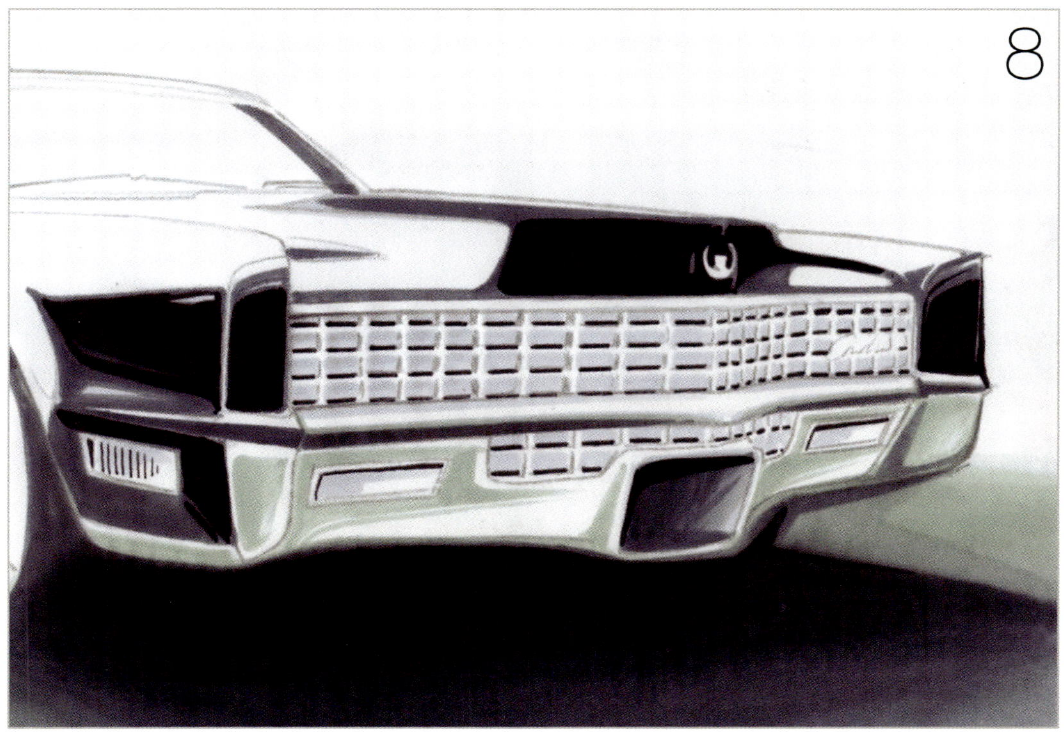

Now comes the tedious part: The grille openings. Begin with the lower edges; you may want to use a straight edge as a guide.

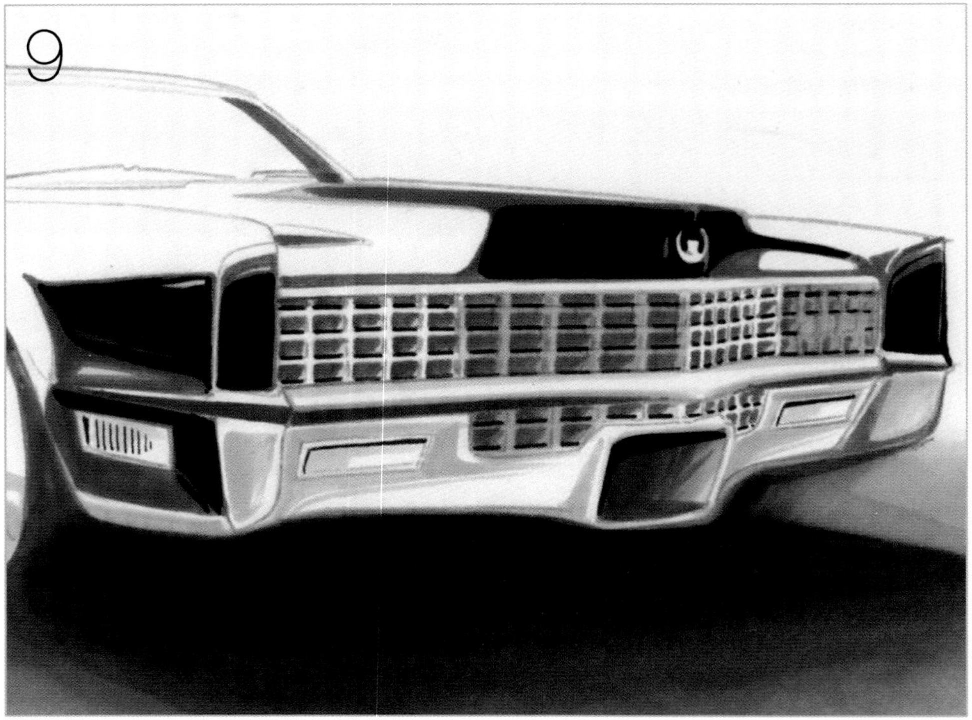

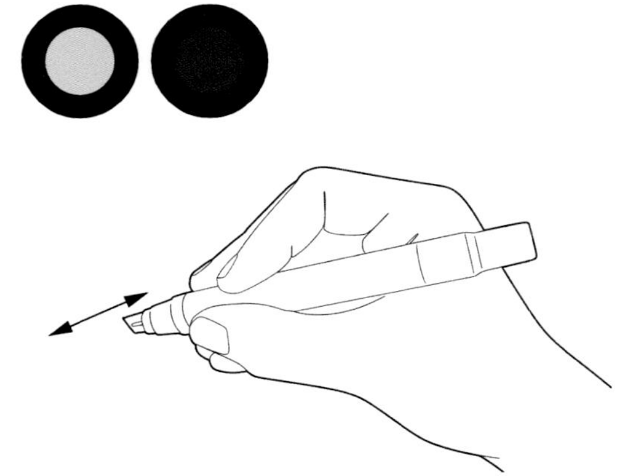

Fill in the holes themselves with Cool Grey 70%. To show the faceted nature of the grille, shade one angle a bit darker with Cool Grey 20%.

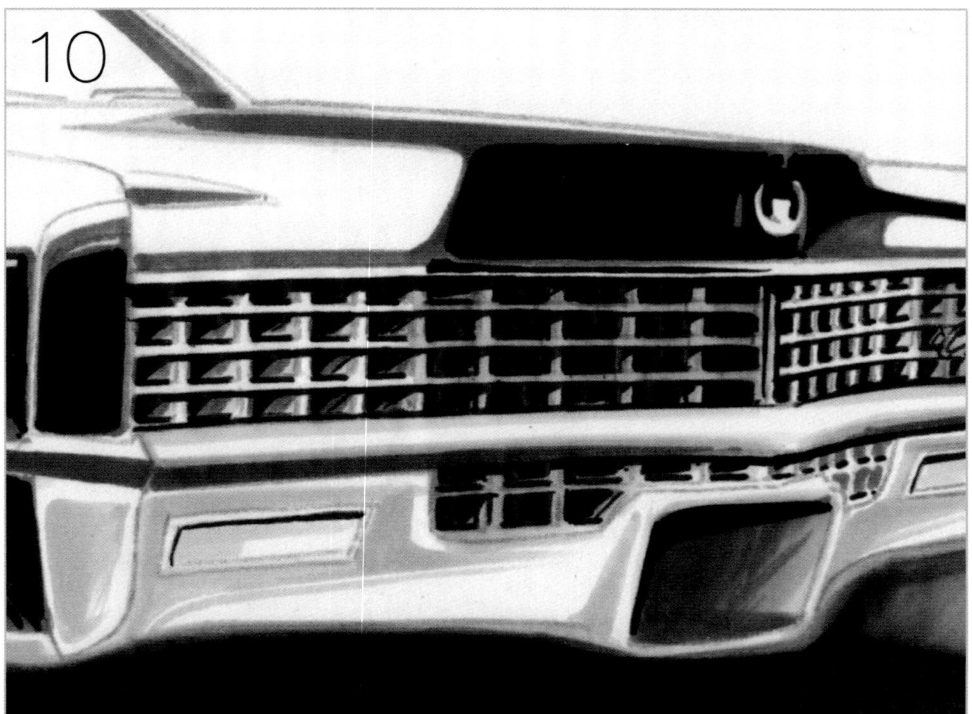

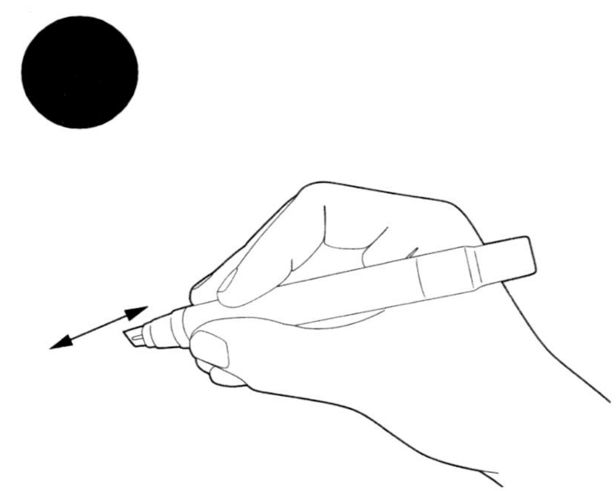

Using the fine point of a black marker or a felt tip pen, add the diagonal shadows cast by the cross members.

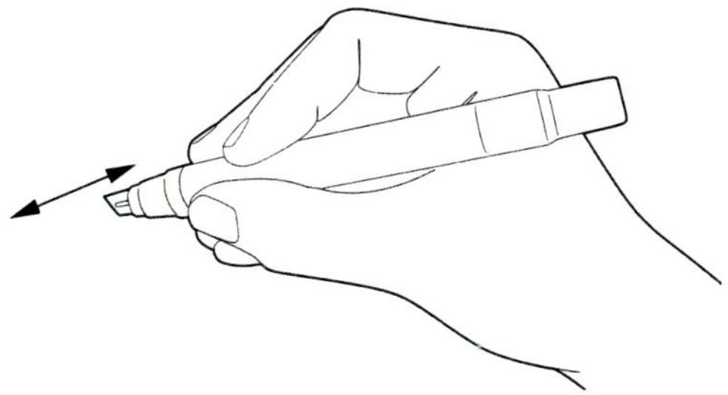

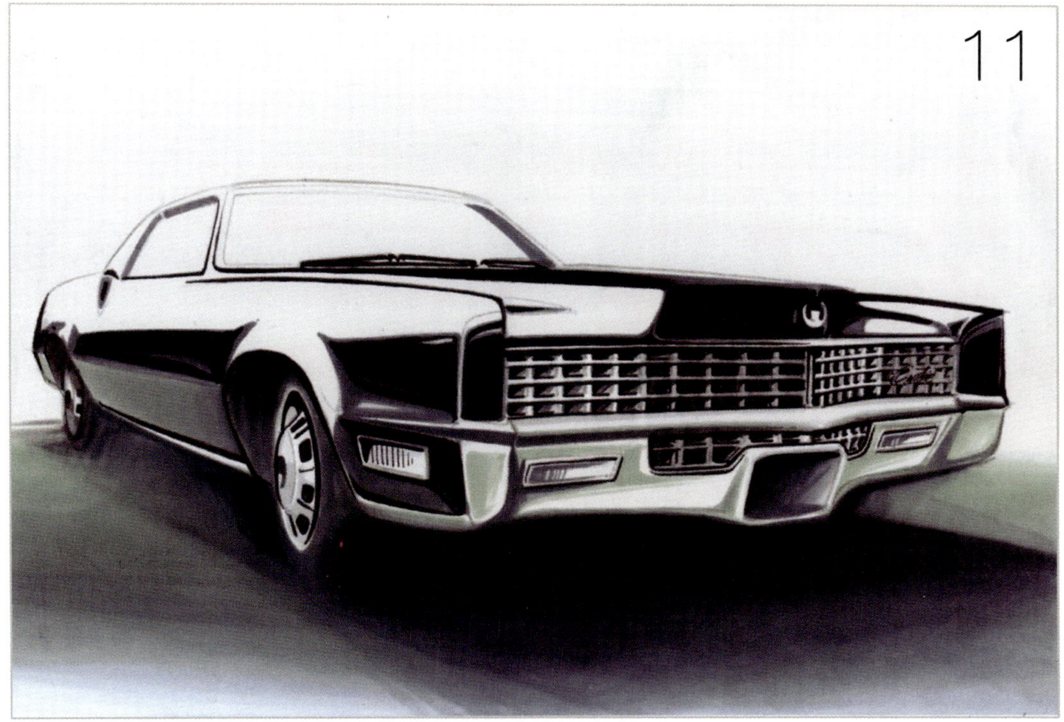

11

Using the same black marker or felt tip pen, go around the car and add in the secondary reflections, like those on the edges of the chrome moldings.

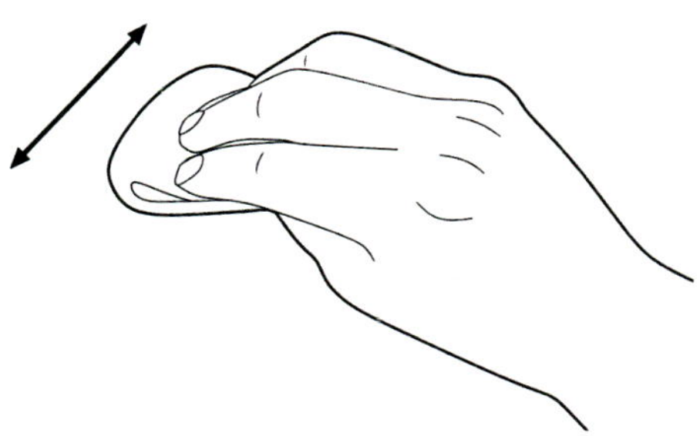

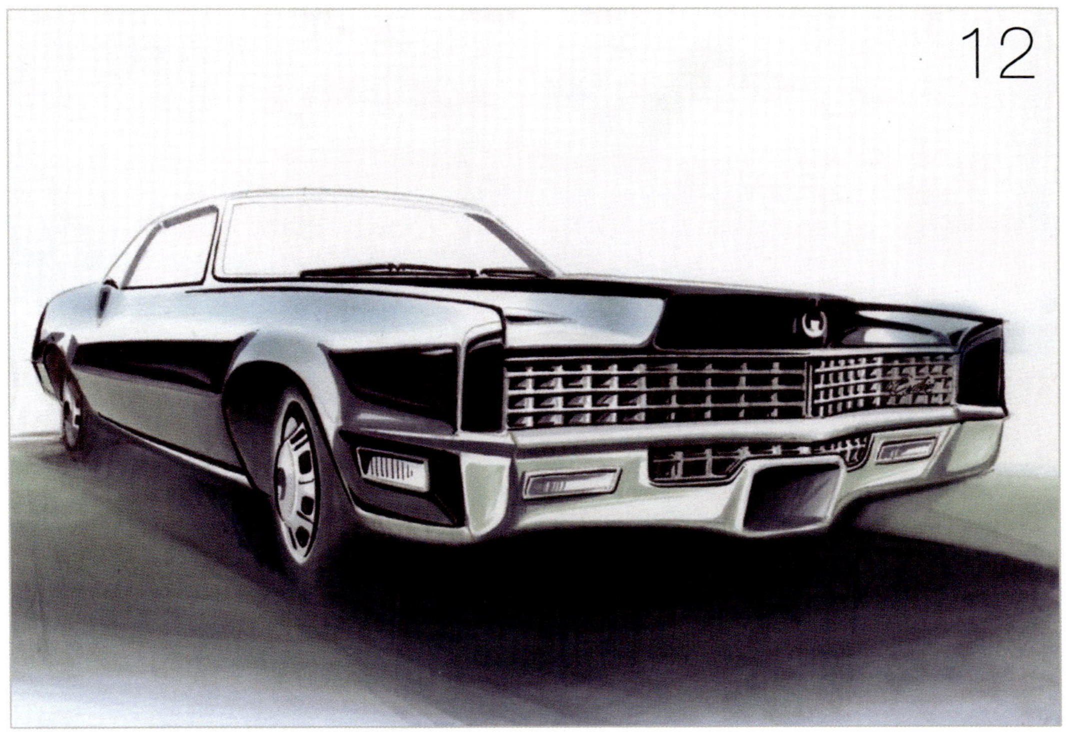

12

Now, ghost in a very modest amount of blue paste' powder. Given there is minimal paper showing through, you won't want to over-do the sky tone.

13

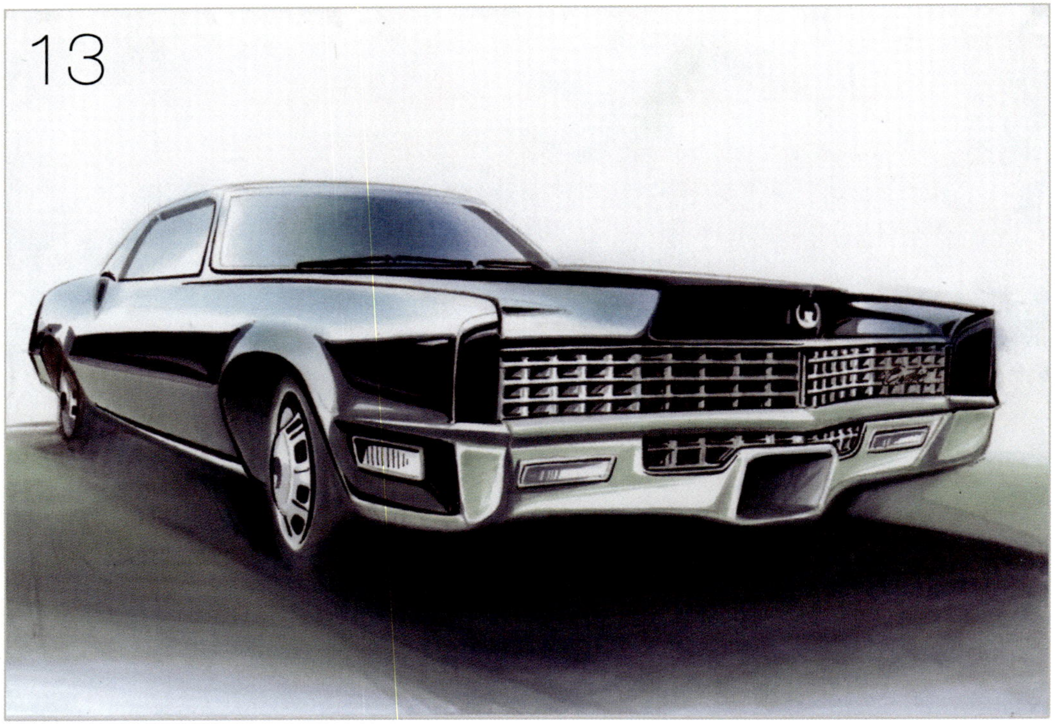

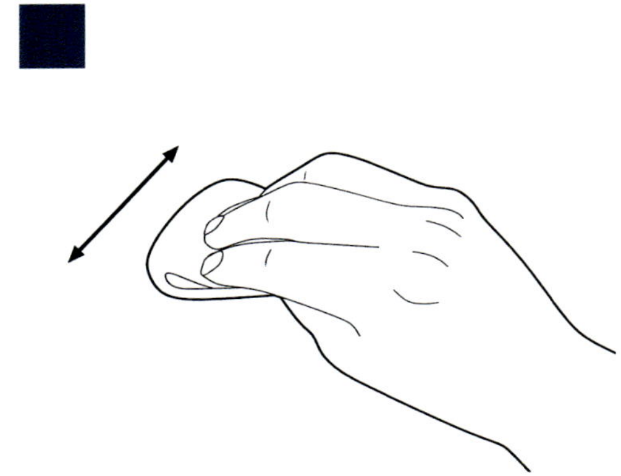

For the windshield, concentrate the blue pastel in the middle of the glass; use a clean Webril pad to blend out the edges of the color.

14

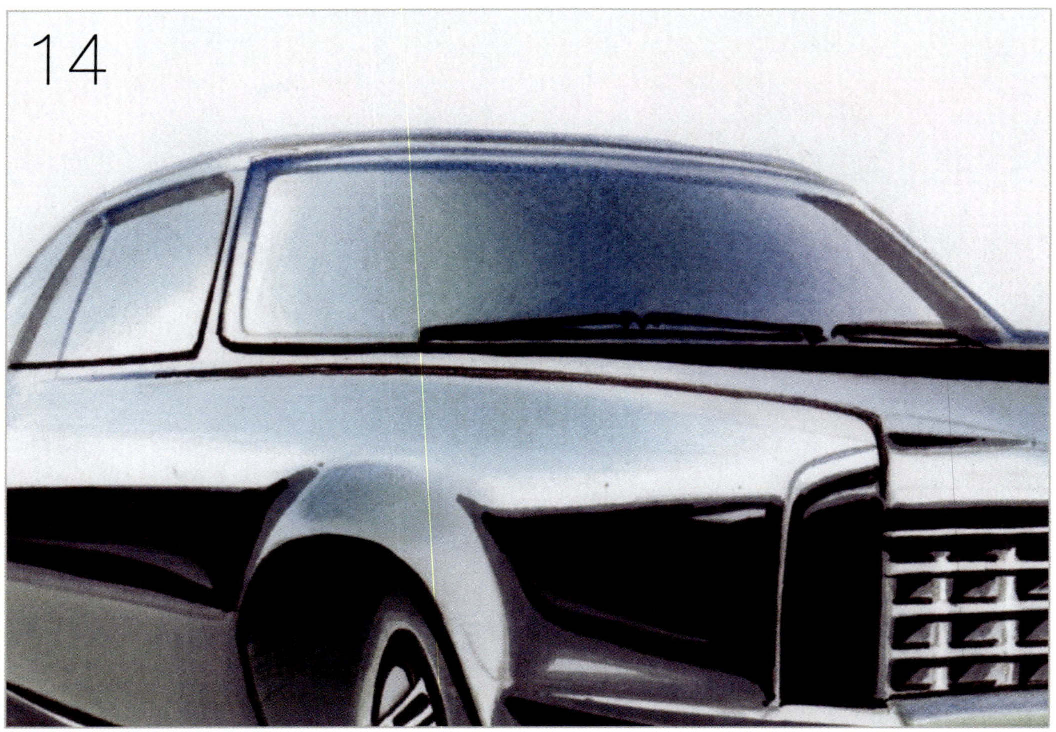

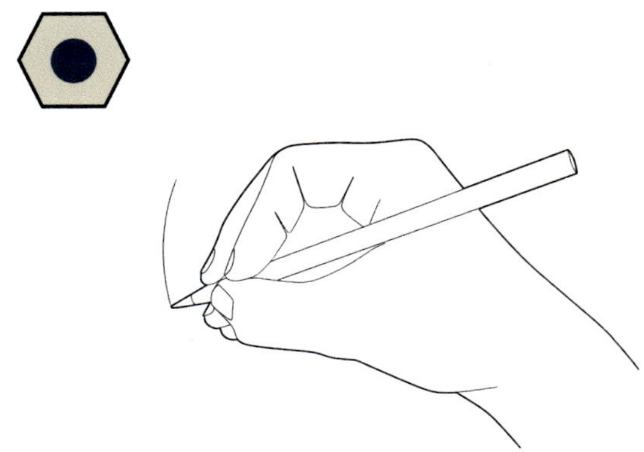

Begin tightening up the drawing with Verithin pencils. Instead of black, try using blue near the sky tone.

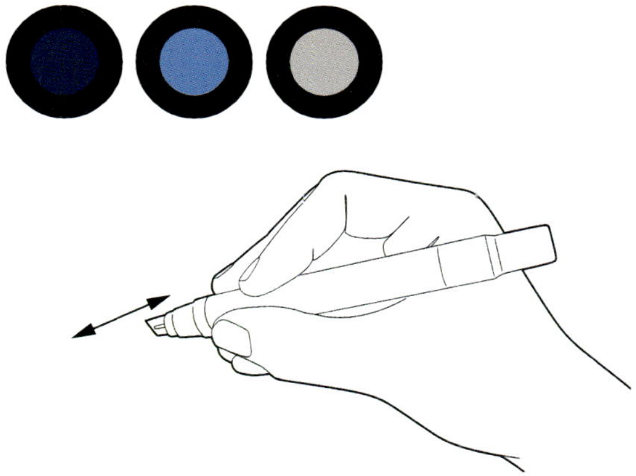

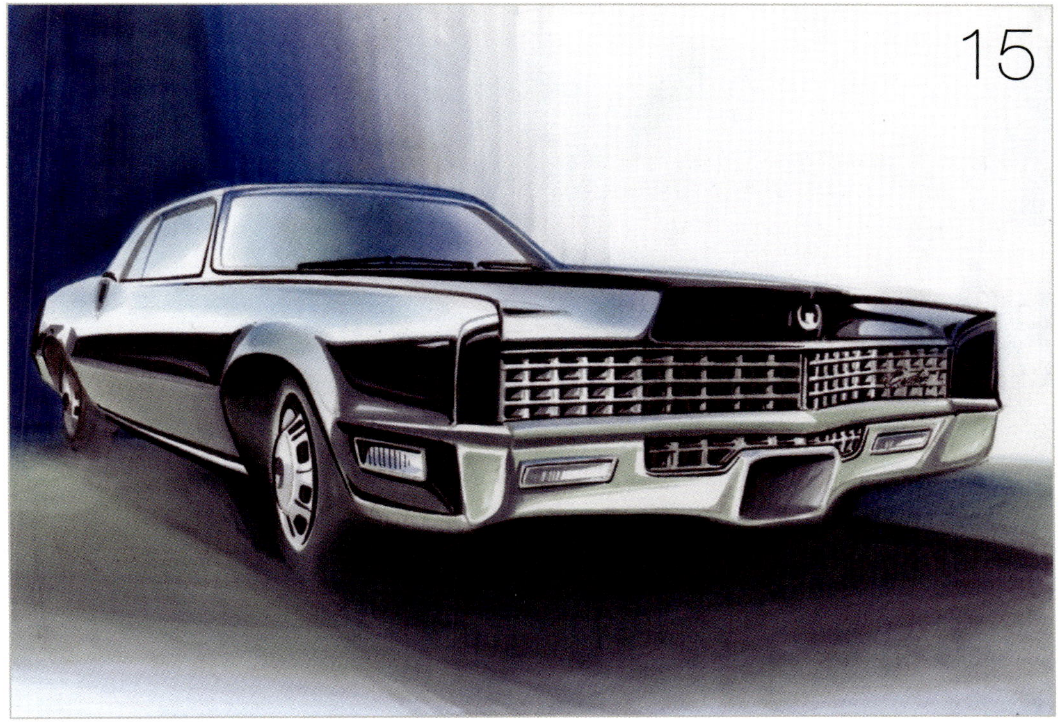

Using an array of cool markers, apply a vignette in the background. Put the darkest part of the vignette behind the lightest part of the car. Use upward strokes to avoid getting any marker on the car itself.

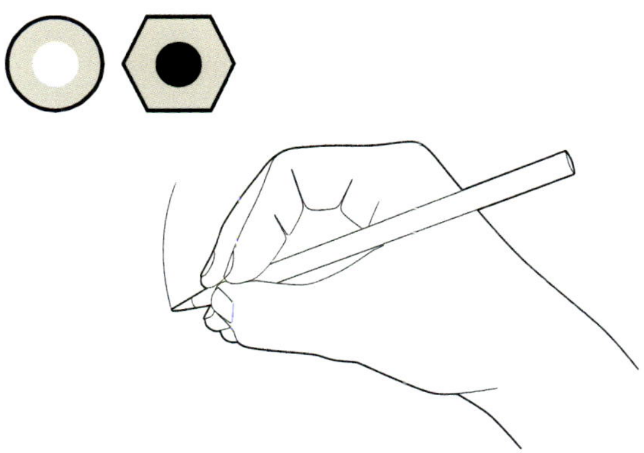

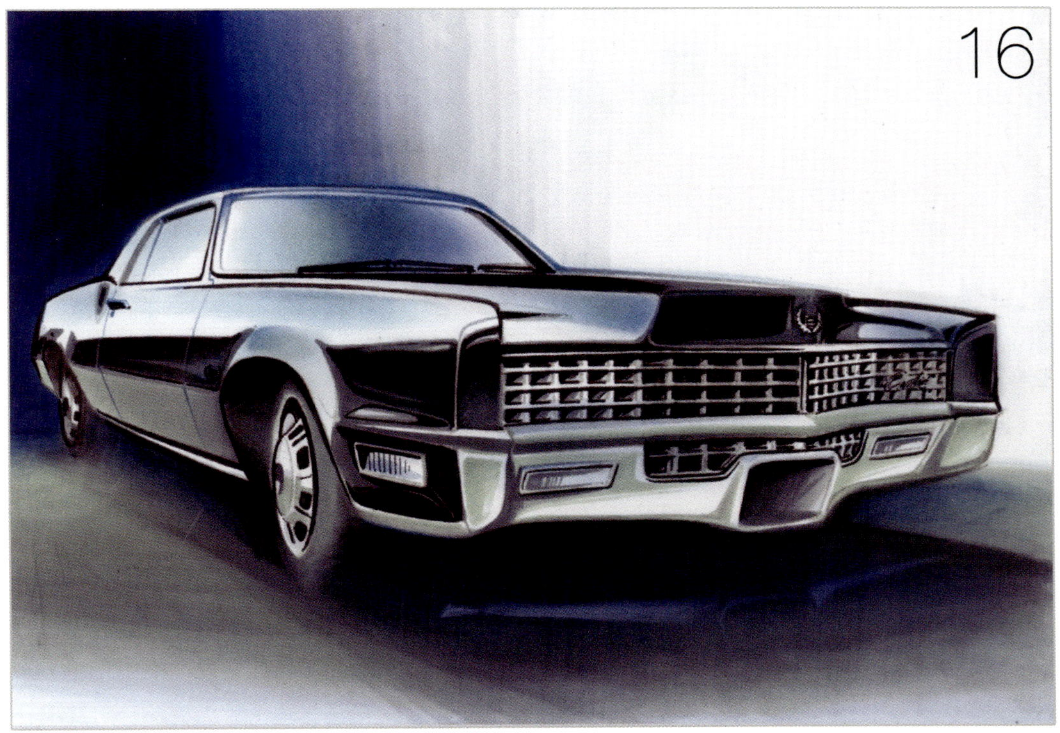

Add subtle gradation and detail to the black reflections using white and black pencils.

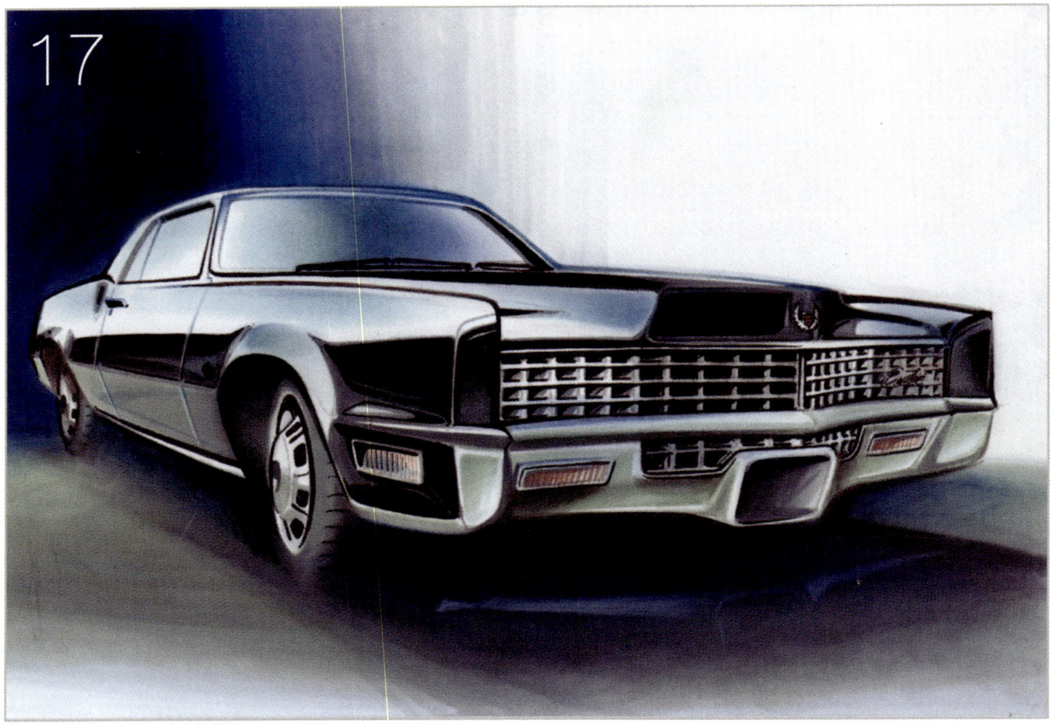

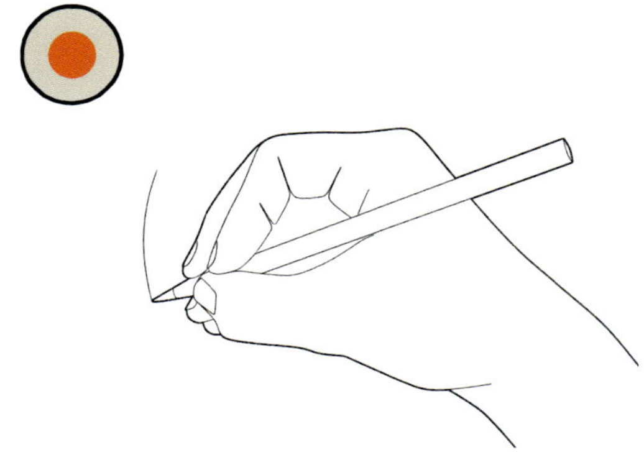

With a sharp pencil, add orange linework in the turn signal lenses.

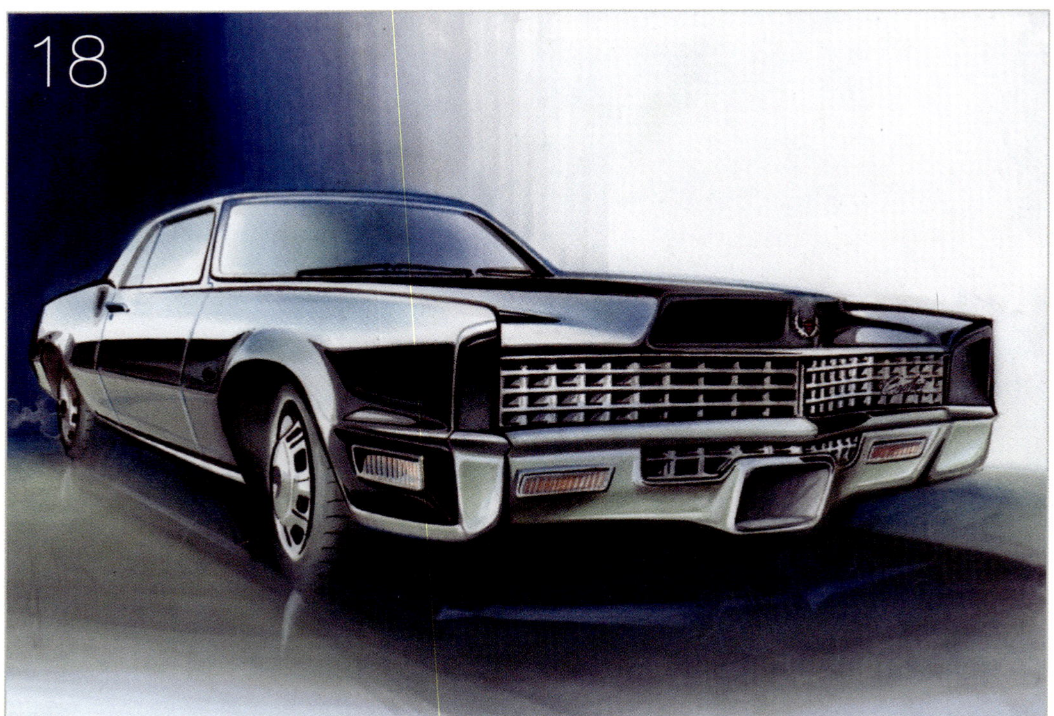

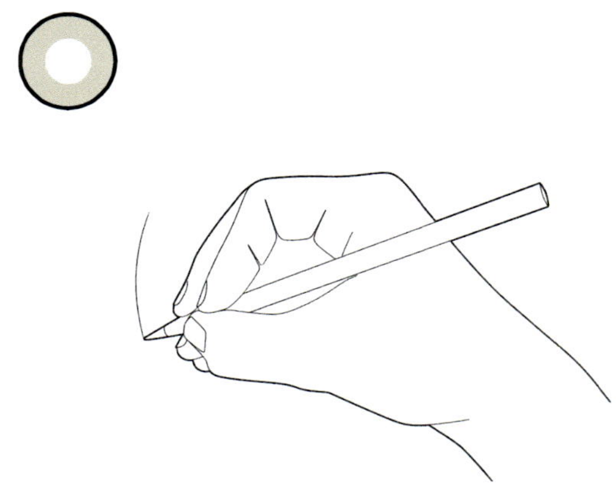

Add one more layer of detail with white pencil: Wheel cover reflections, the rocker reflection and the exhaust.

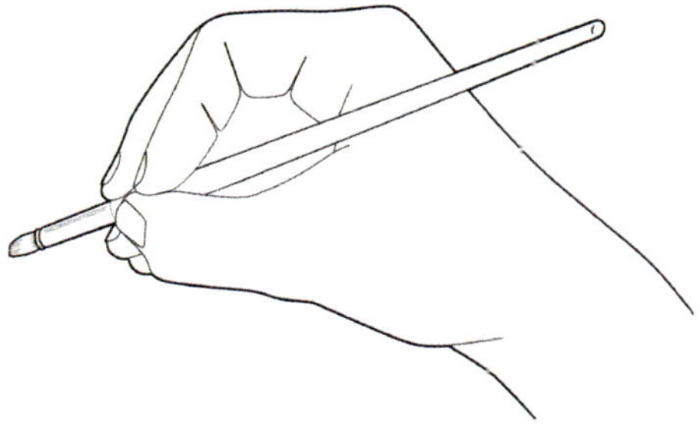

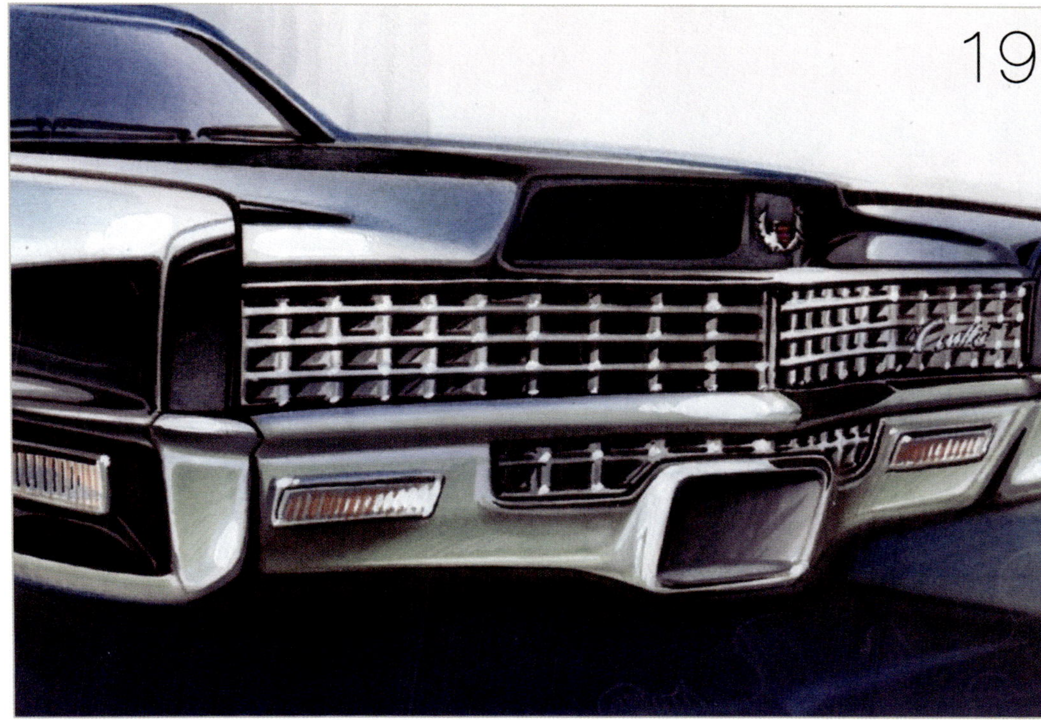

19

Carefully apply white gouache for the highlights. For the grille, apply small dots of gouache at the intersections.

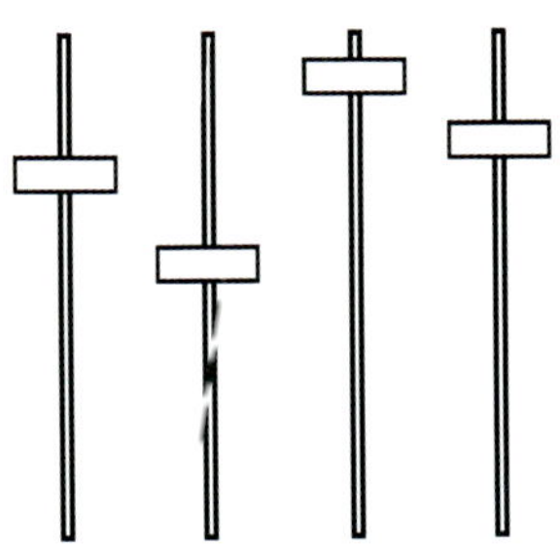

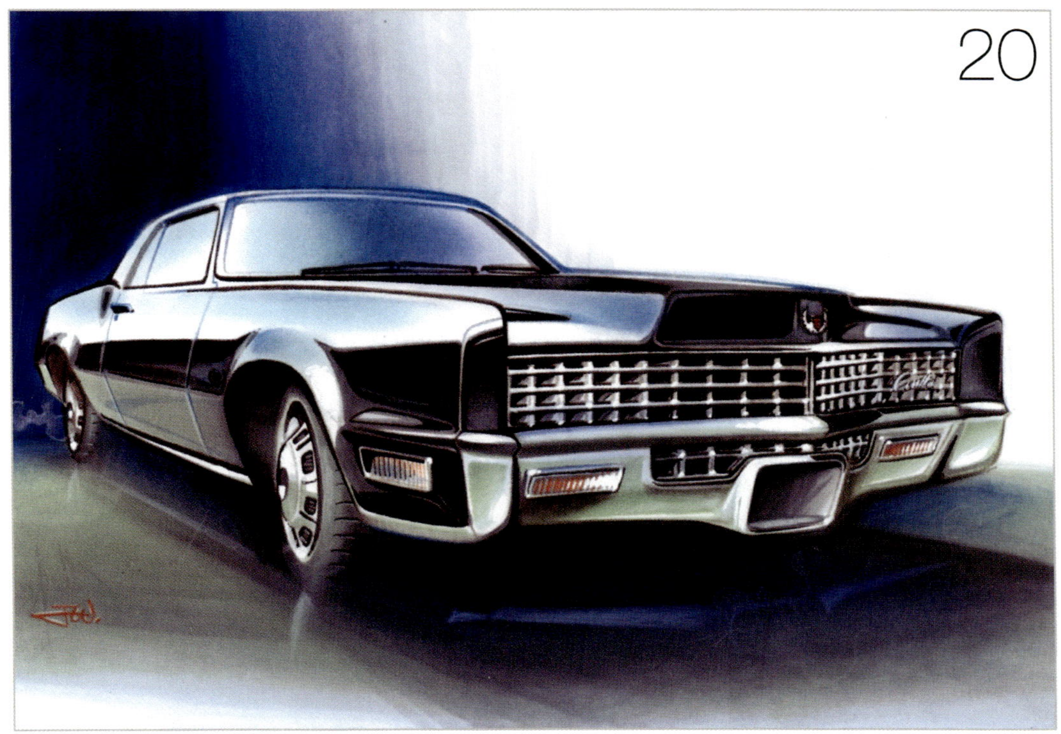

20

Scan in the rendering and turn up the color saturation in Photoshop.

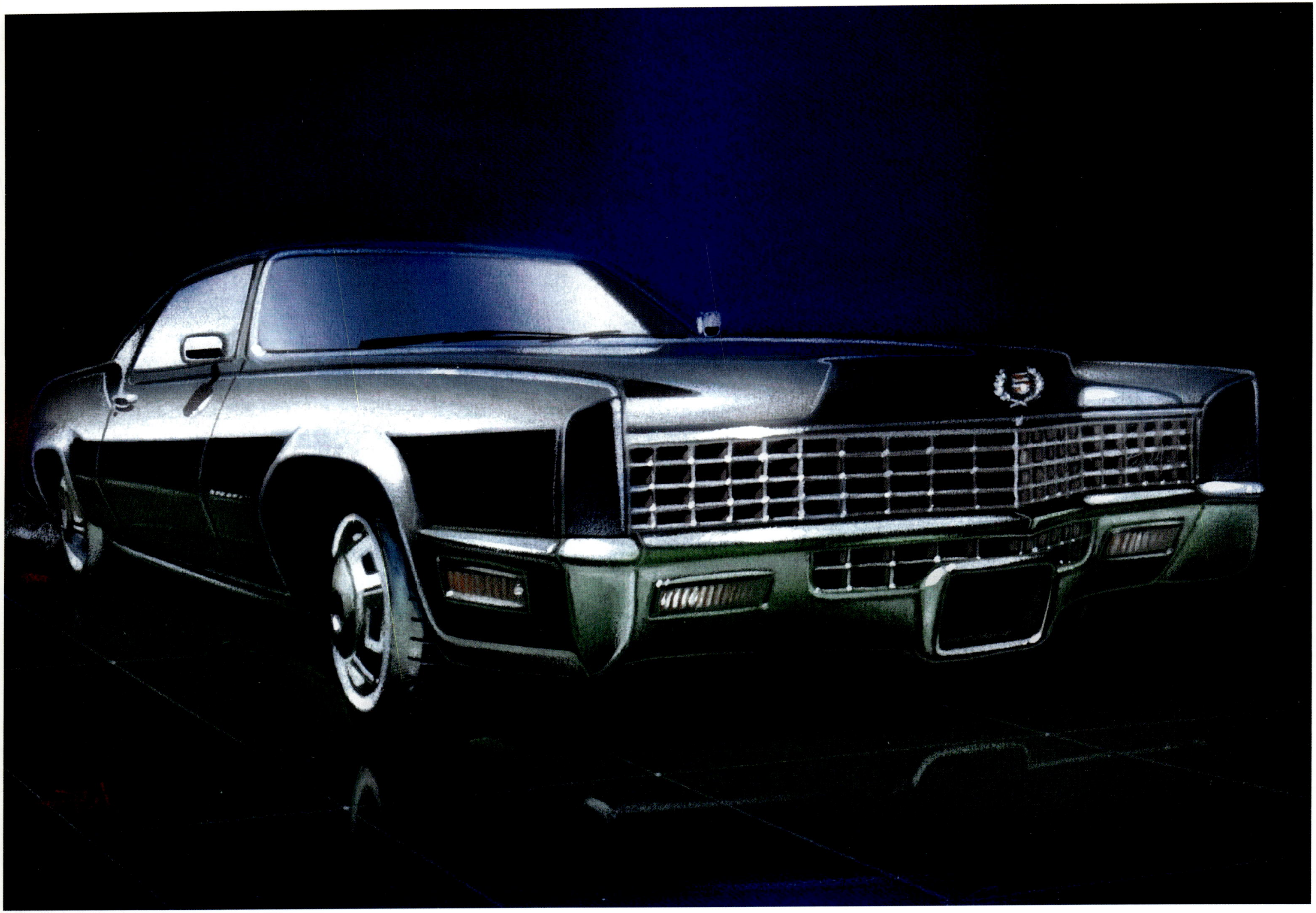

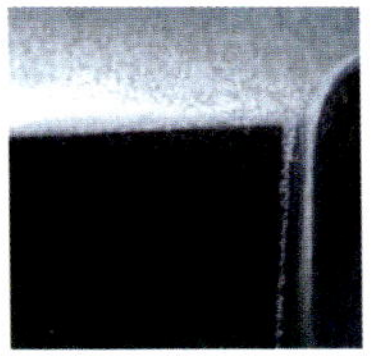

11. Black Magic

Now let's try the same subject again – only this time, with black Canson paper. Here, the darkest dark is given to us rather than the lightest light. Using floor reflections and background colors, you can create a mysterious, almost velvet-like atmosphere that draws the viewer in.

The Tools

Nupastels: White, Blue, Green, Violet, Red.

Prismacolors:
White, Crimson Red, Celadon Green, Orange, Cool Grey 20%, Black.

Verithins:
Black,
Indigo Blue.

Gouache:
White.

Prismacolor Markers: Black, Cool Grey 20%.

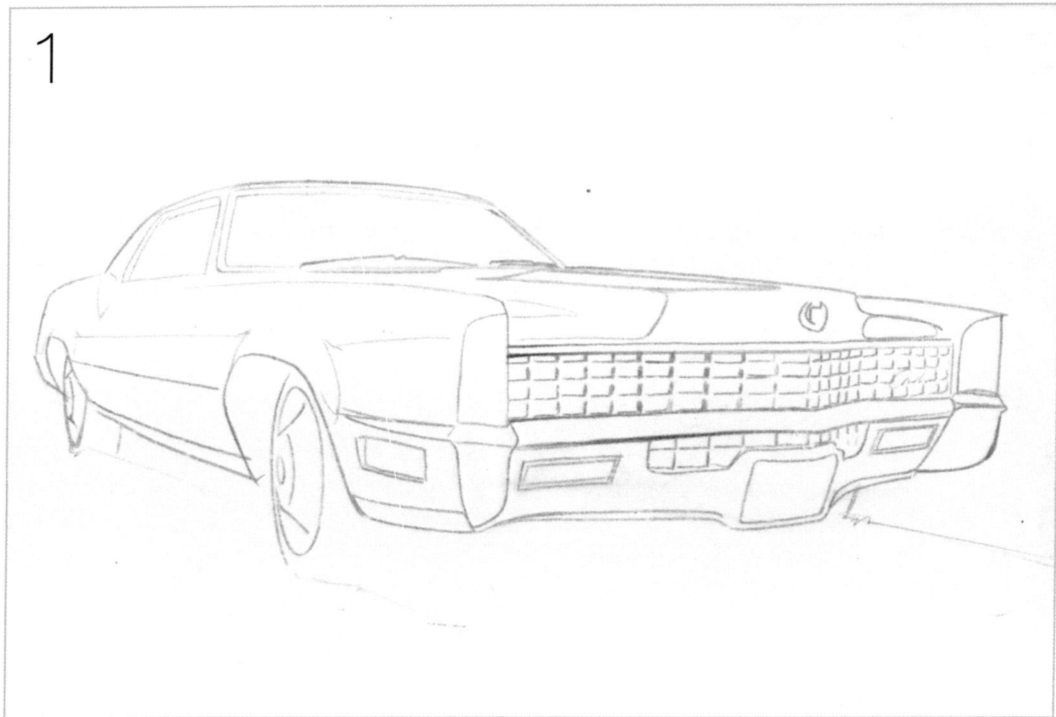

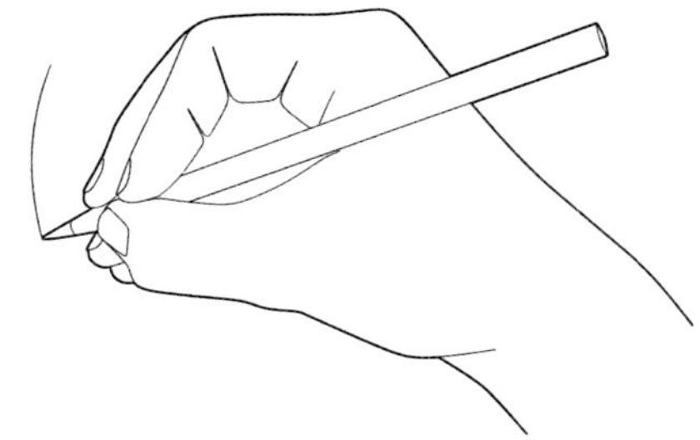

Start this rendering the same way you started the previous one: Laying out the car on a piece of white paper.

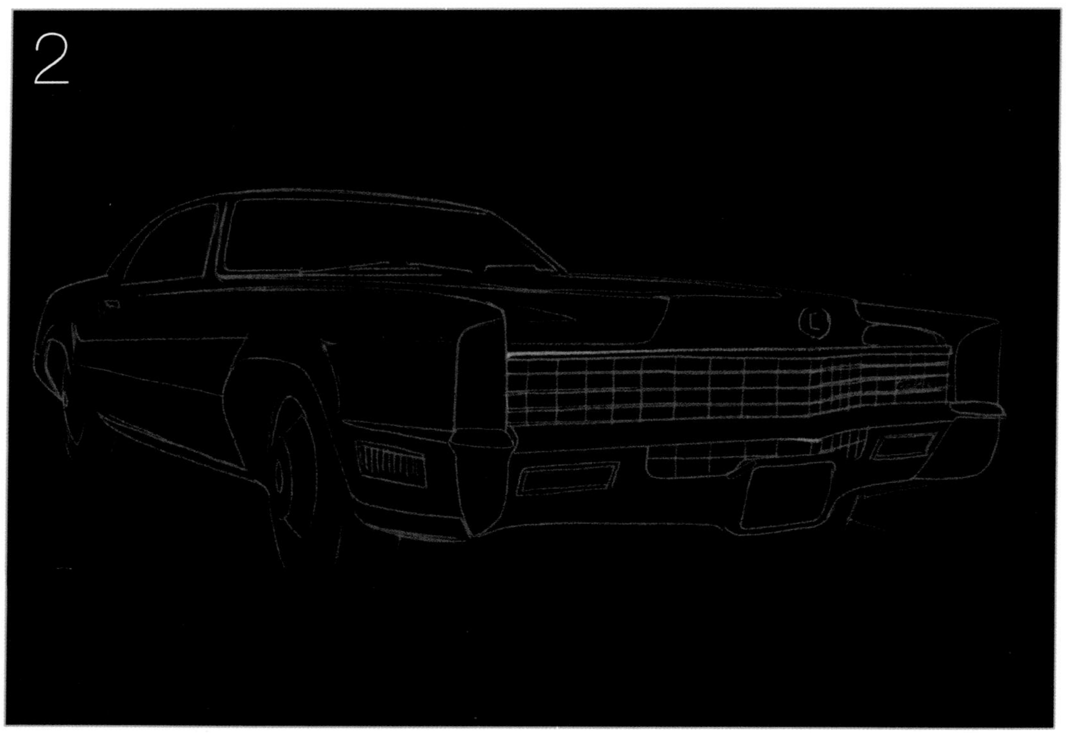

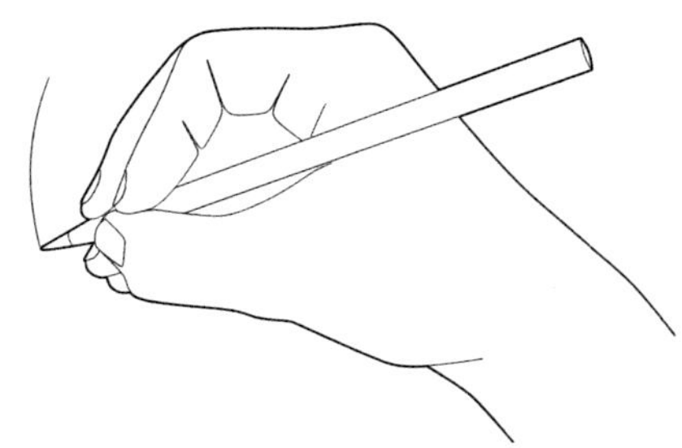

Now, using a sheet of white transfer paper, carefully apply the outline to the smooth side of the black Canson.

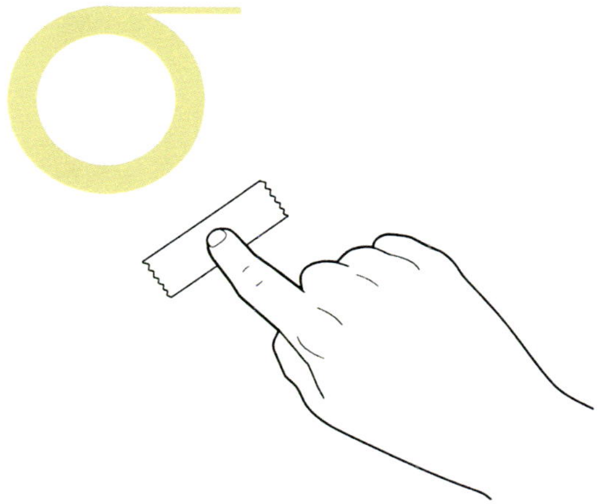

Place a medium blue piece of Canson under the black piece. With a sharp X-acto, cut straight through both pieces around the glass. Replace the black bits with the blue bits and secure them on the back with masking tape.

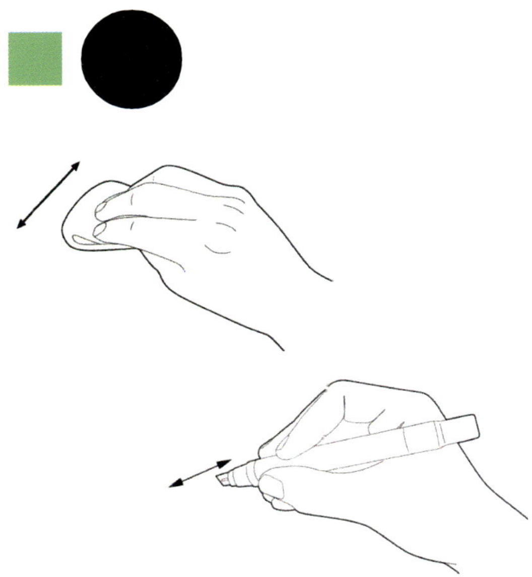

With a light touch, ghost in green pastel powder on the floor. Use a black marker to clean up the shadow's edge.

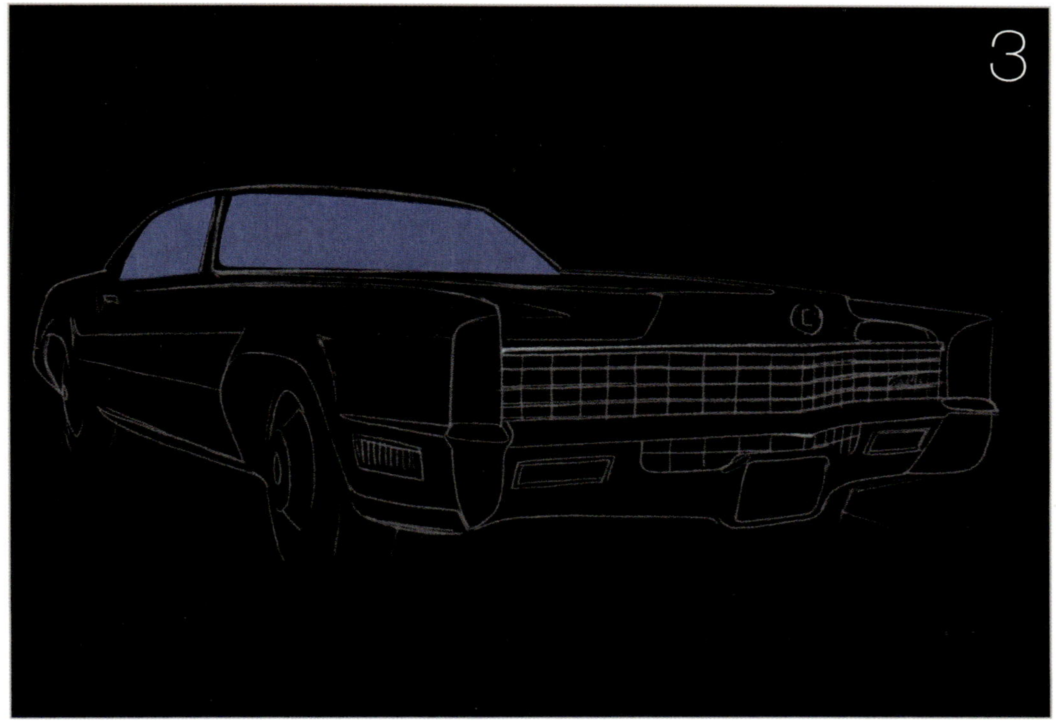

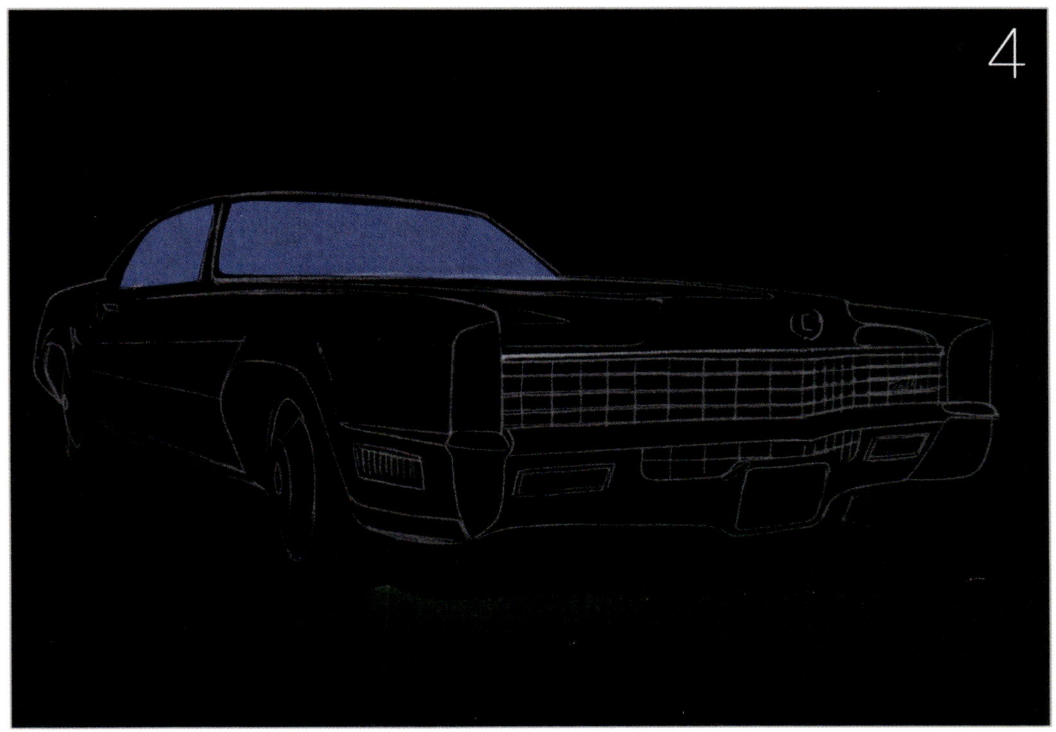

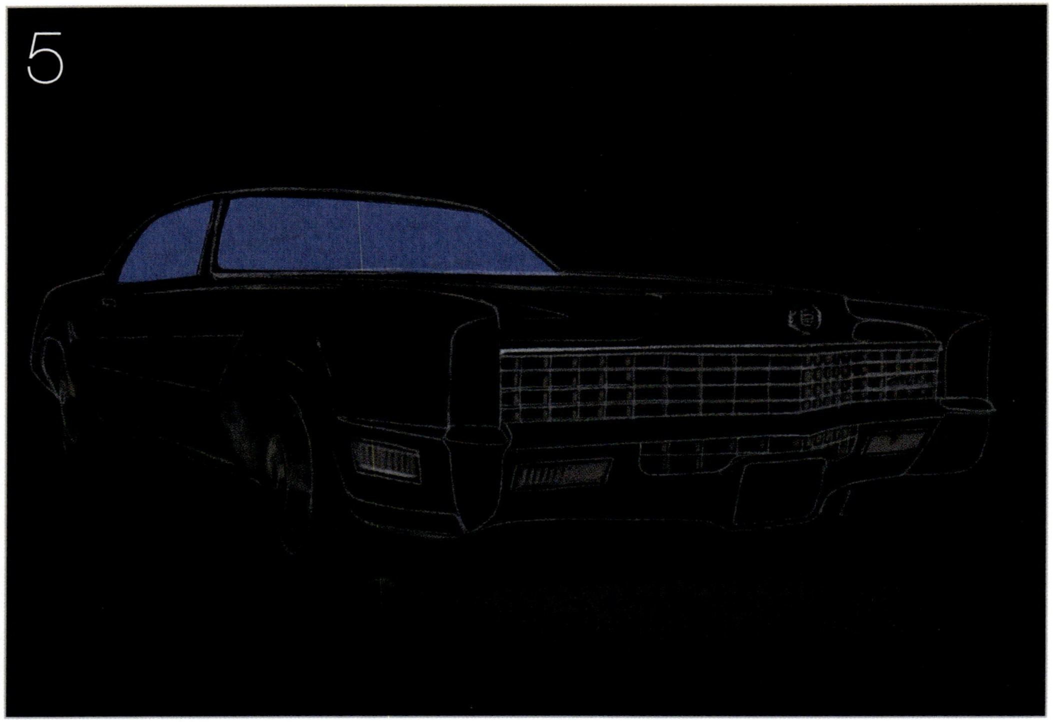

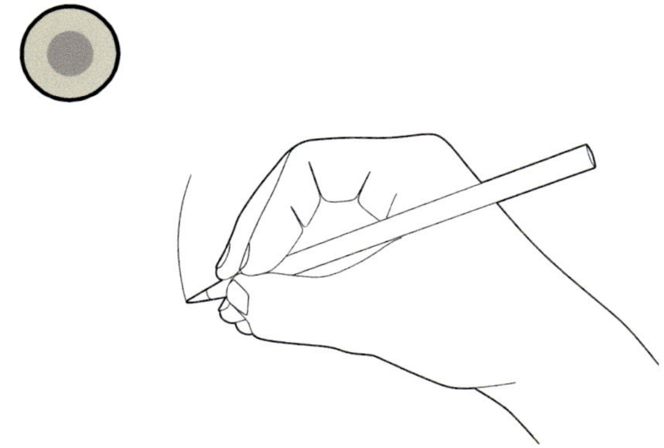

With a Cool Grey pencil, begin adding some subtle highlights in select places: Wheel wells, turn signals, etc.

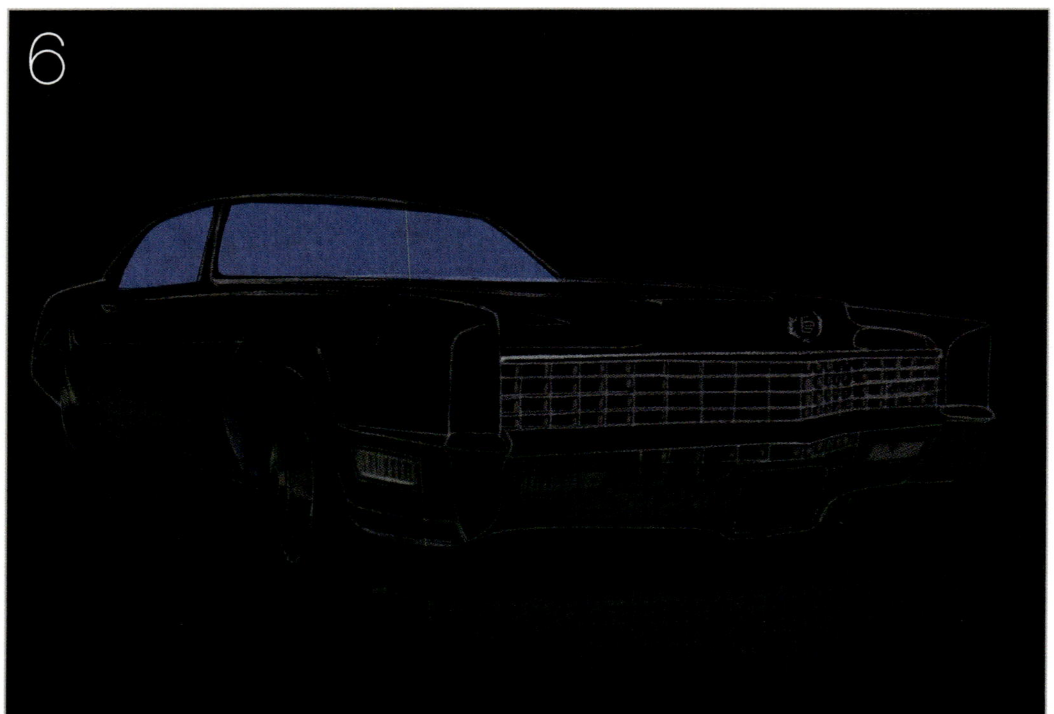

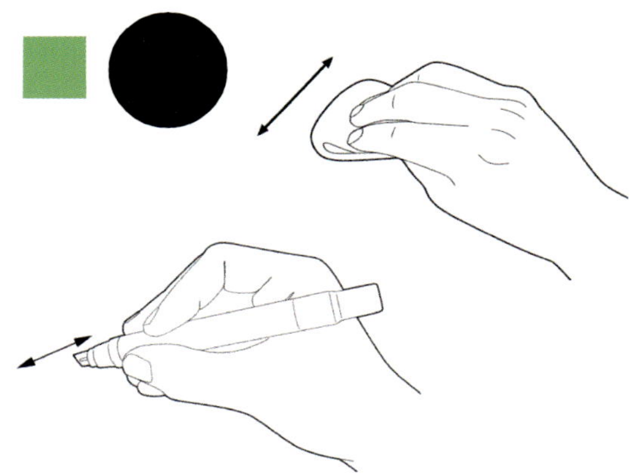

Now ghost in some green pastel on the surfaces that face the floor: The lower portion of the bodyside and the chrome bumper. Here, you begin to see how floor color is a critical component in describing the shape of car. Again, use a marker to clean up the pastel edge.

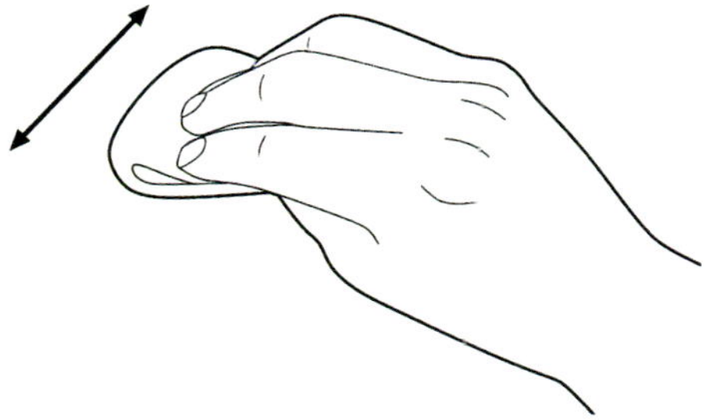

Now comes the light part. Load up a clean Webril pad with white pastel powder and ghost it in above the car's horizon line. At the top of the fender line, sneak in a small amount of blue pastel; this will represent the sky tone.

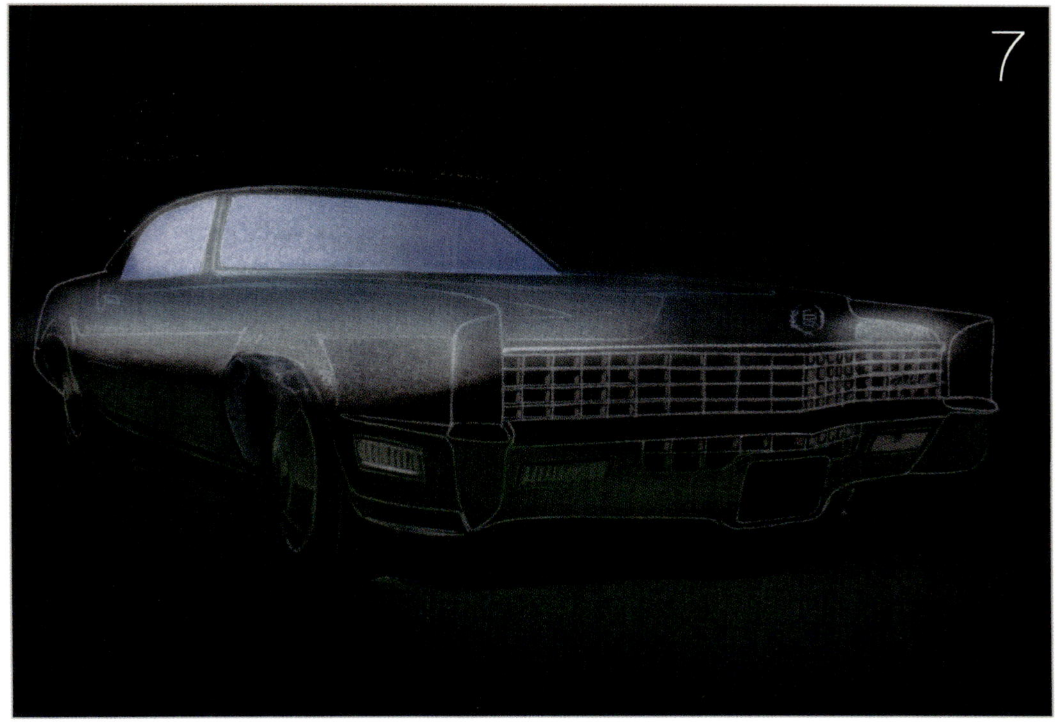

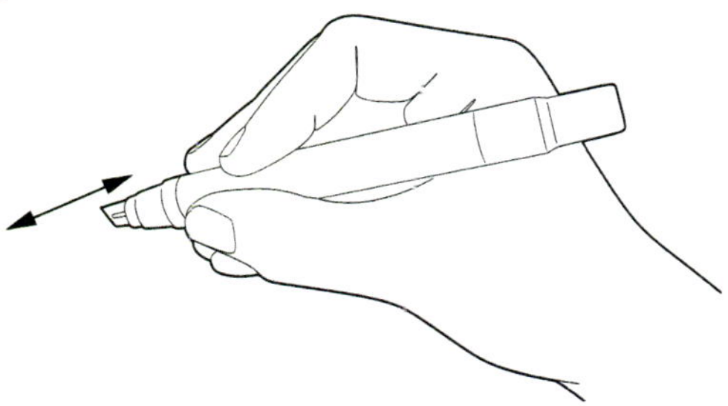

As you clean up the edge with a marker, you can see the full scope of the rendering taking shape.

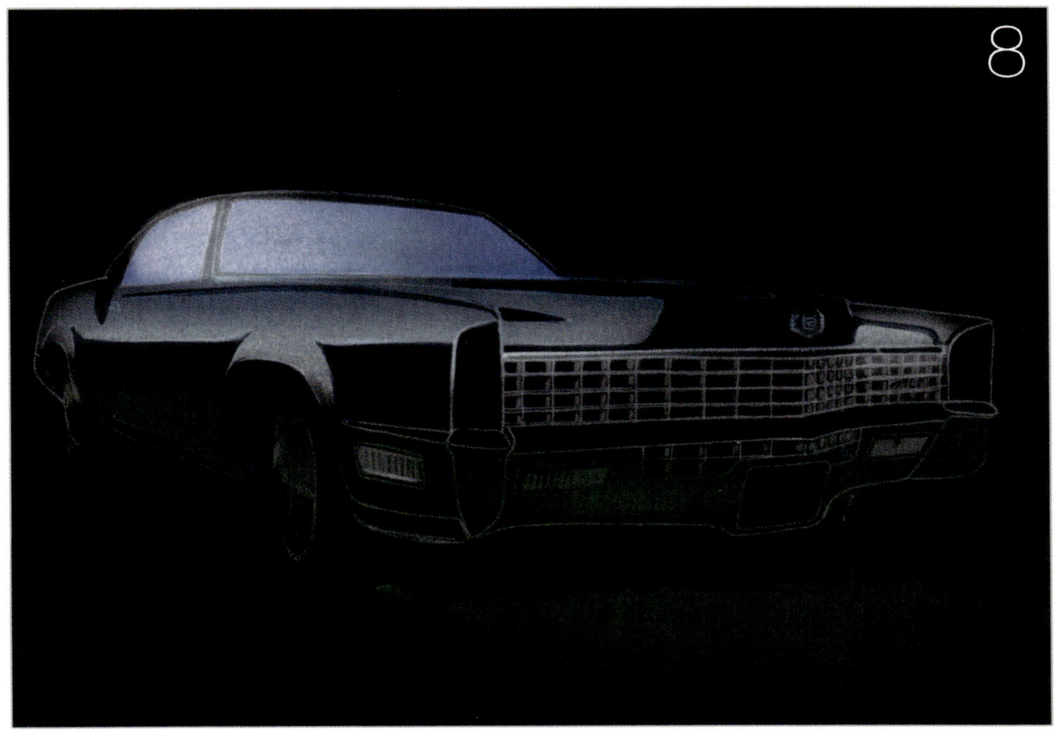

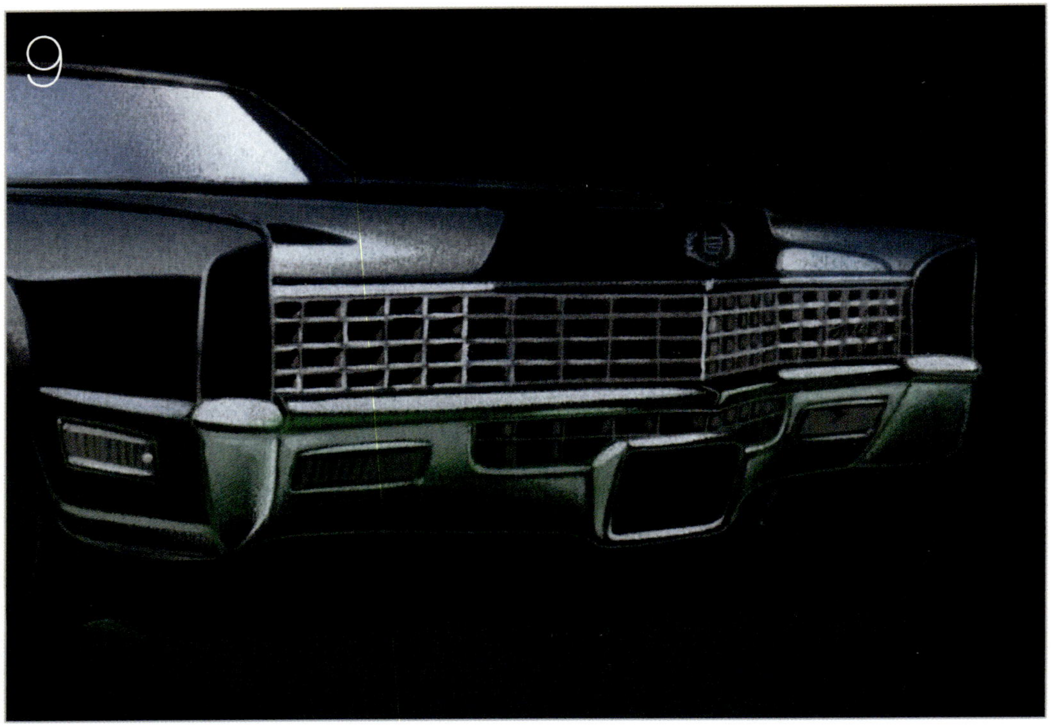

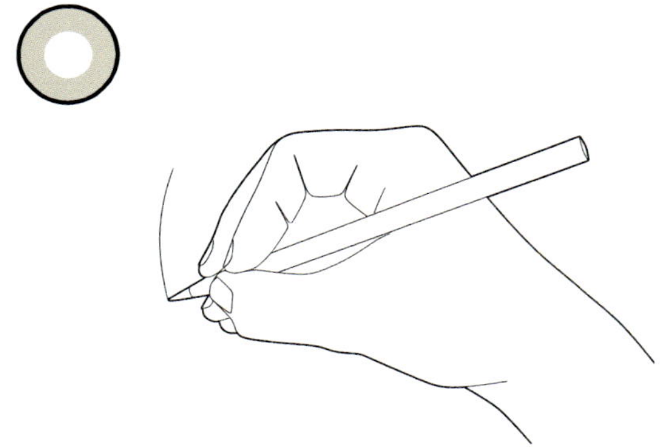

Now you can use a sharp, white pencil to punctuate the light parts of the car like the edges of the grille, the top surface of the bumper and the horizon on the far side of the glass.

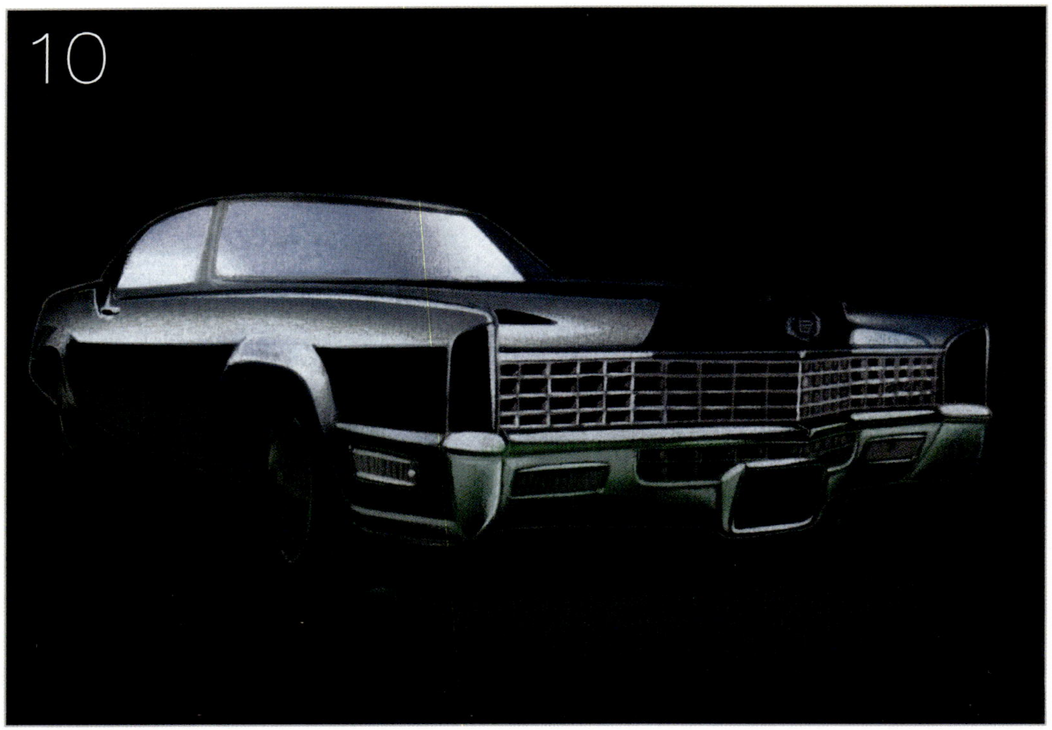

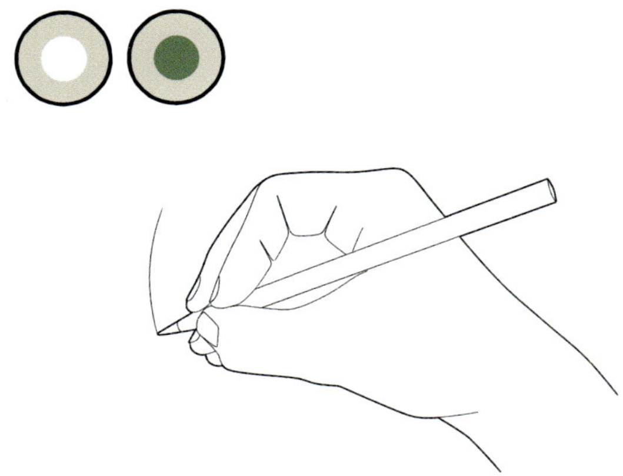

Continuing with the white pencil, add a bold highlight along the length of the bodyside horizon. You can also add punch to the underside of the bumper with a bit of Celadon Green pencil.

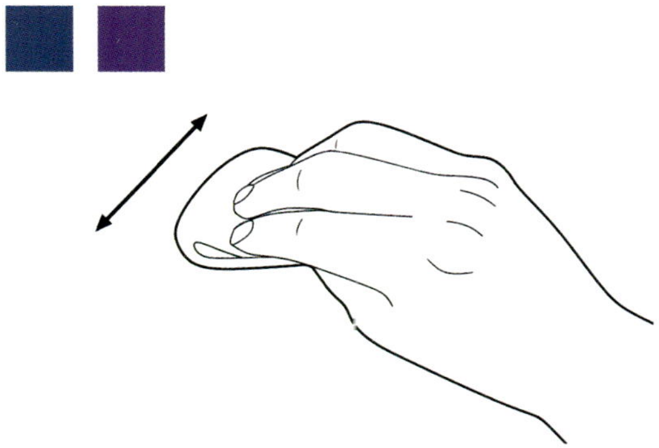

With a mixture of blue and violet pastel, add some punch to the sky tone in the windshield. Be mindful of how much pastel you are applying; try not to overload the paper.

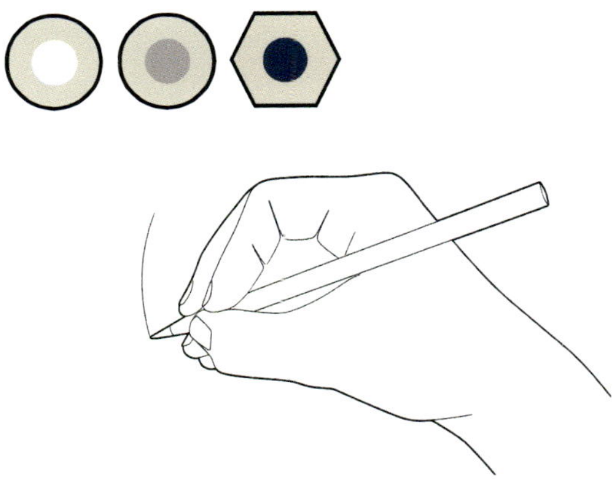

For the wheels and tires, try alternating between white and Cool Grey Prismacolor. For added depth, you may wish to add some blue Verithin in the areas shaded by the wheel wells.

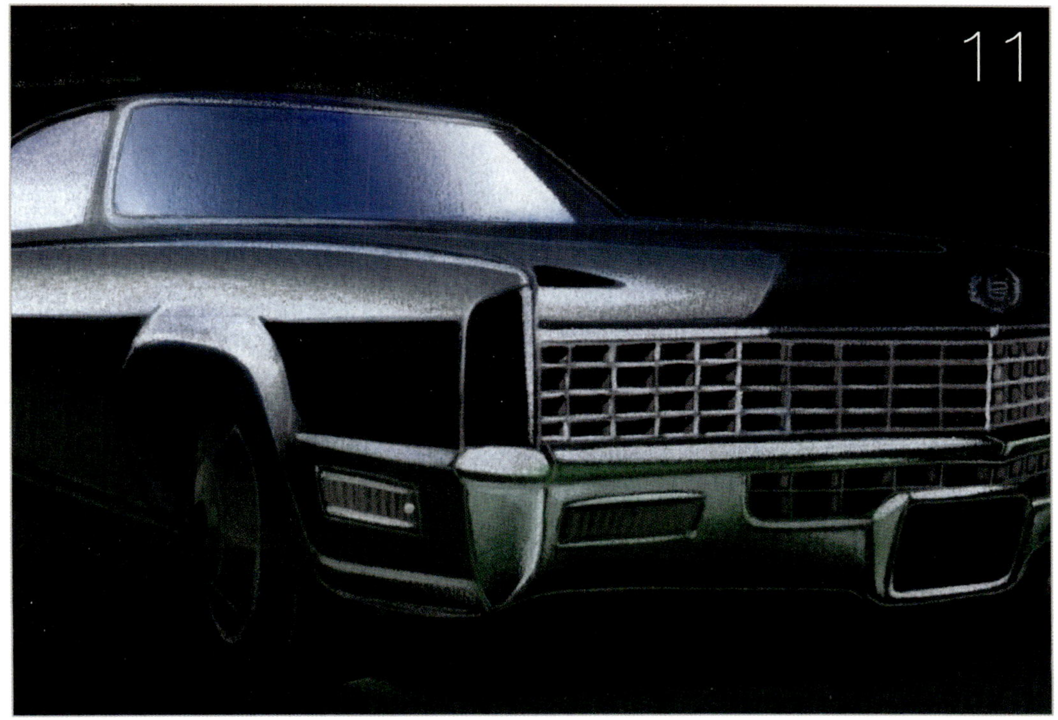

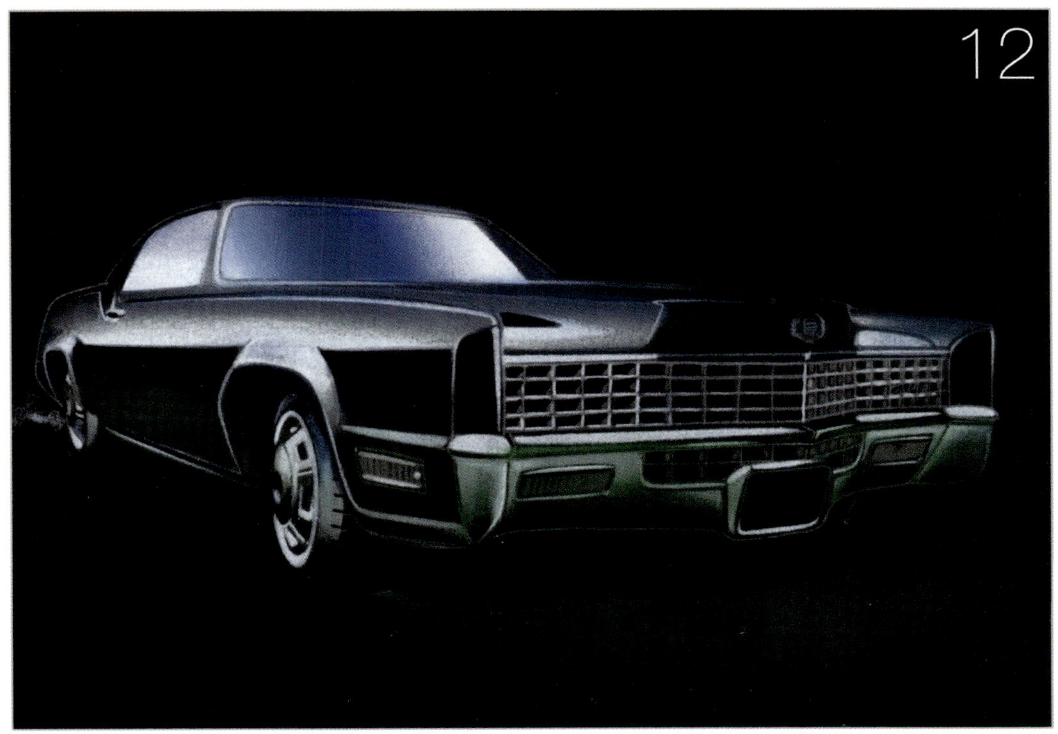

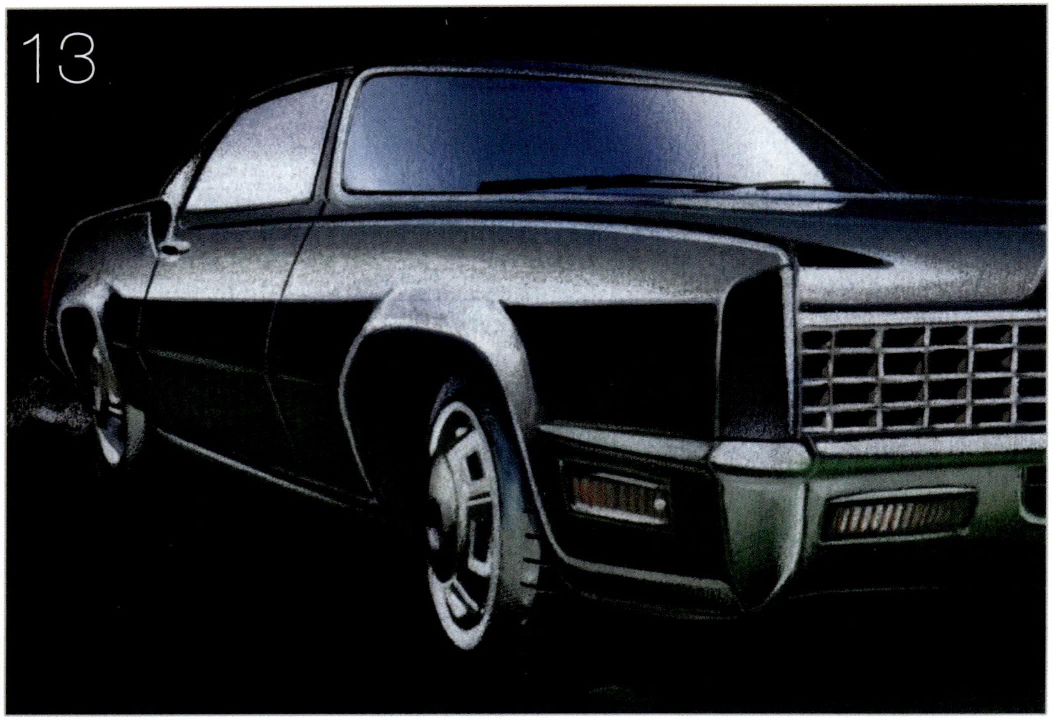

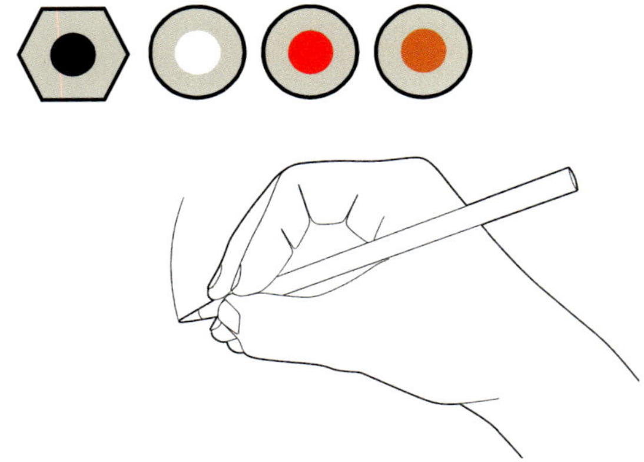

Now you can use a sharp, black Verithin to carve out the finer details like the door shuts and the windshield wipers. You can revisit the shut lines with some white Prismacolor, just make sure you highlight the edges that are facing you. Also, take a minute to add some red and orange for the lights.

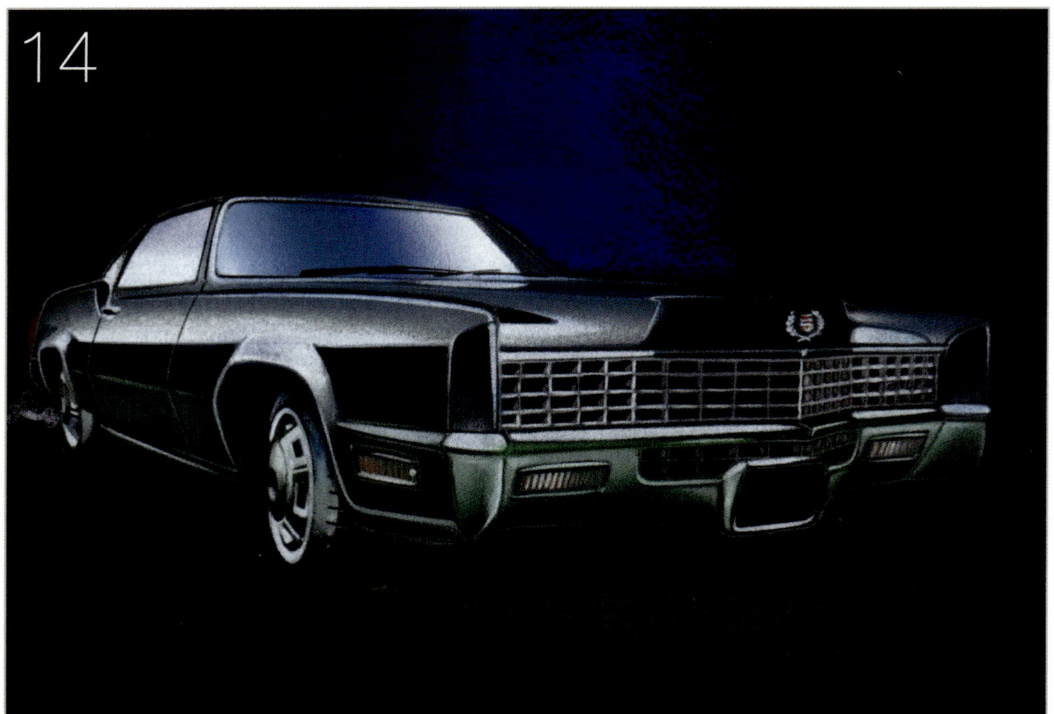

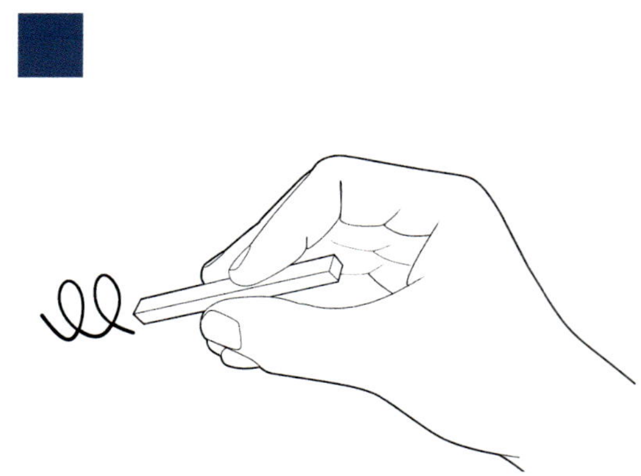

For the background color, apply a liberal amount of blue pastel directly to the paper. You can use different hues of blue across the width of the page if you like; this will add a bit more depth.

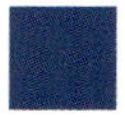

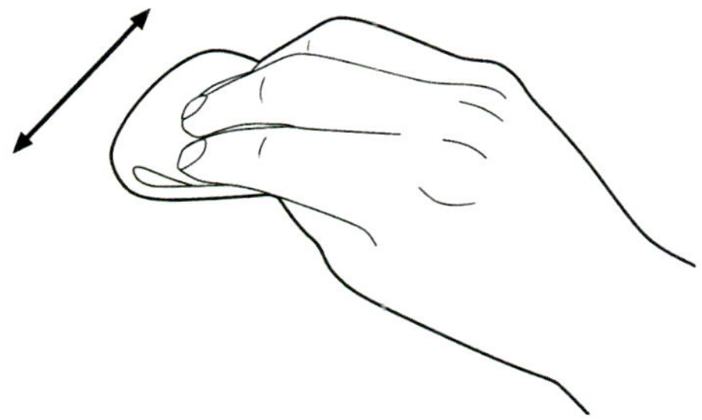

Blend out the background with a Webril pad. If the color loses some of its punch, apply a bit more pastel directly to the page and blend it out again.

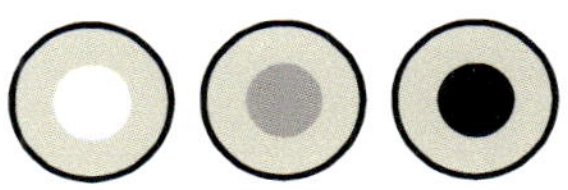

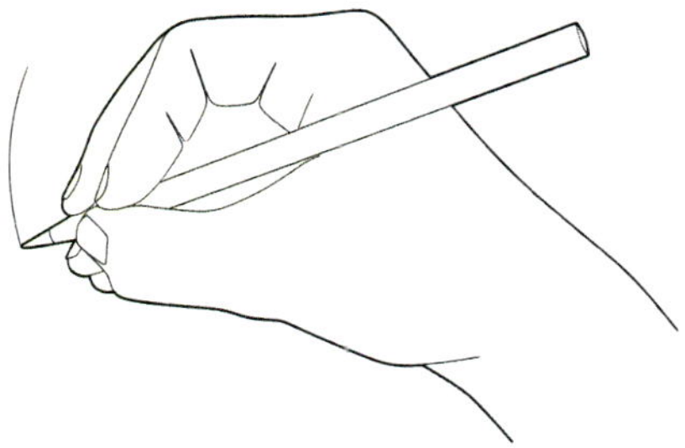

Add another layer of details using an array of pencils. For example: Rim and bumper highlights reflecting into the floor; the side mirrors and their shadows.

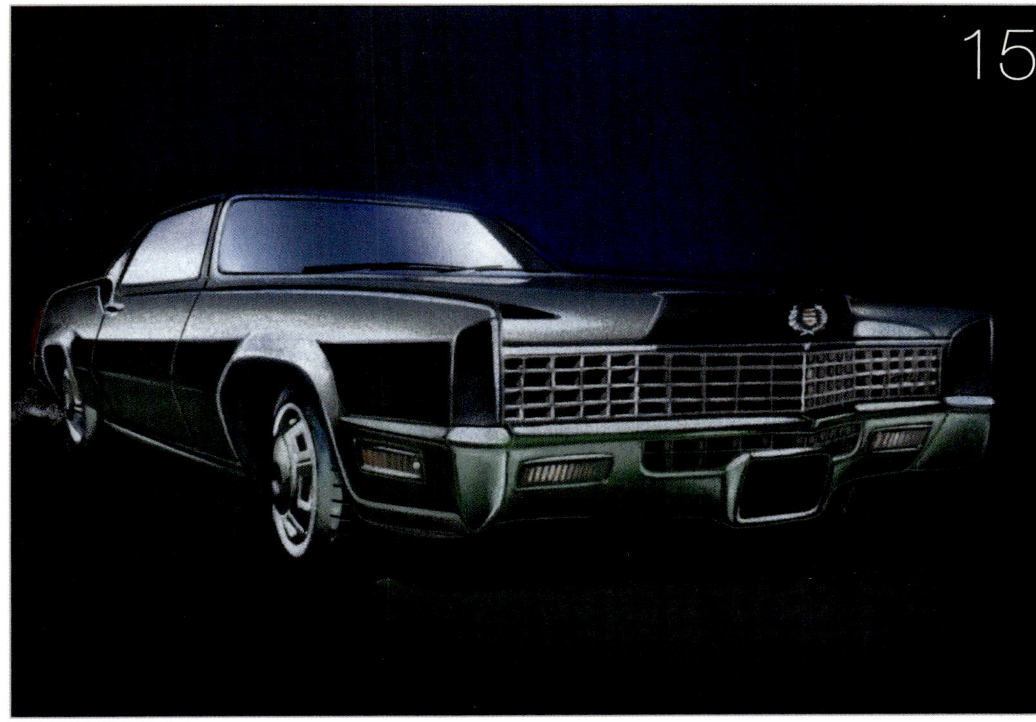

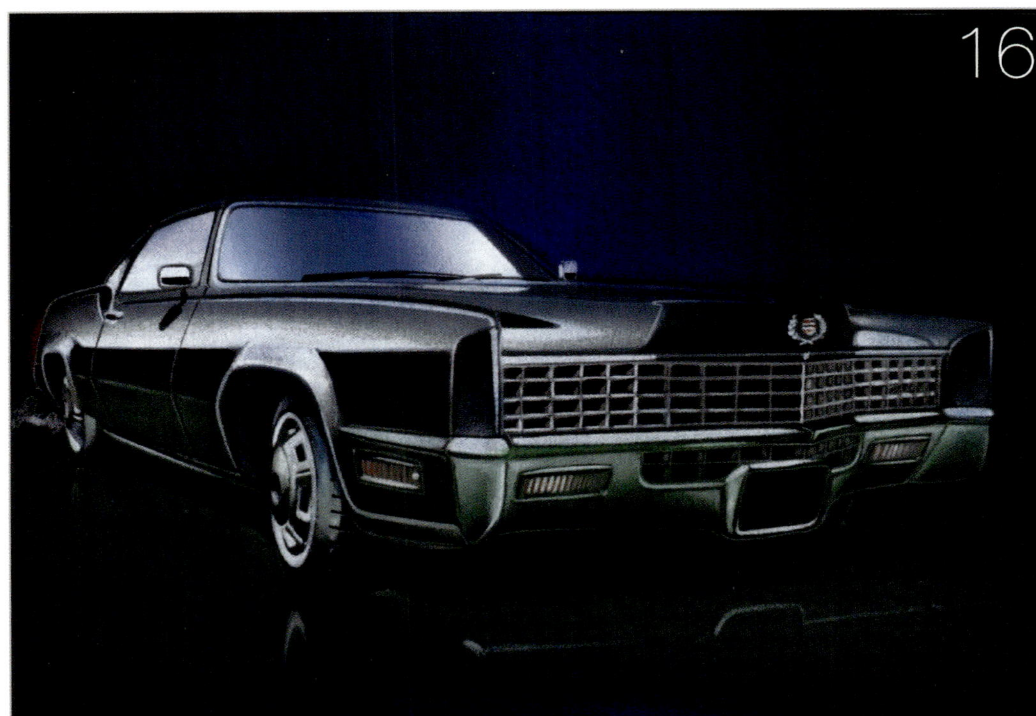

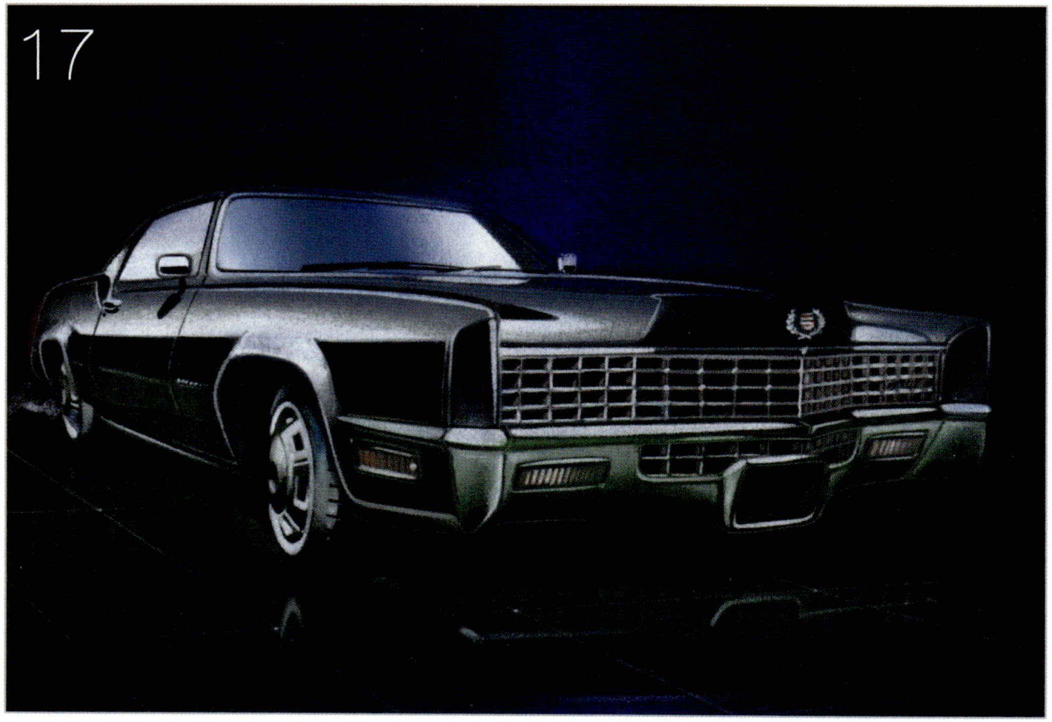

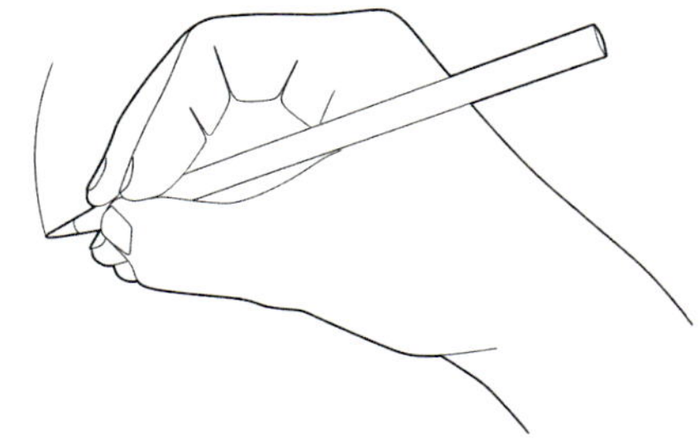

To give the floor a glossy, tiled appearance, strike in some black and white lines with a straight edge. Make sure the perspective of these lines matches the perspective of the car. Add some highlights on select edges to complete the effect.

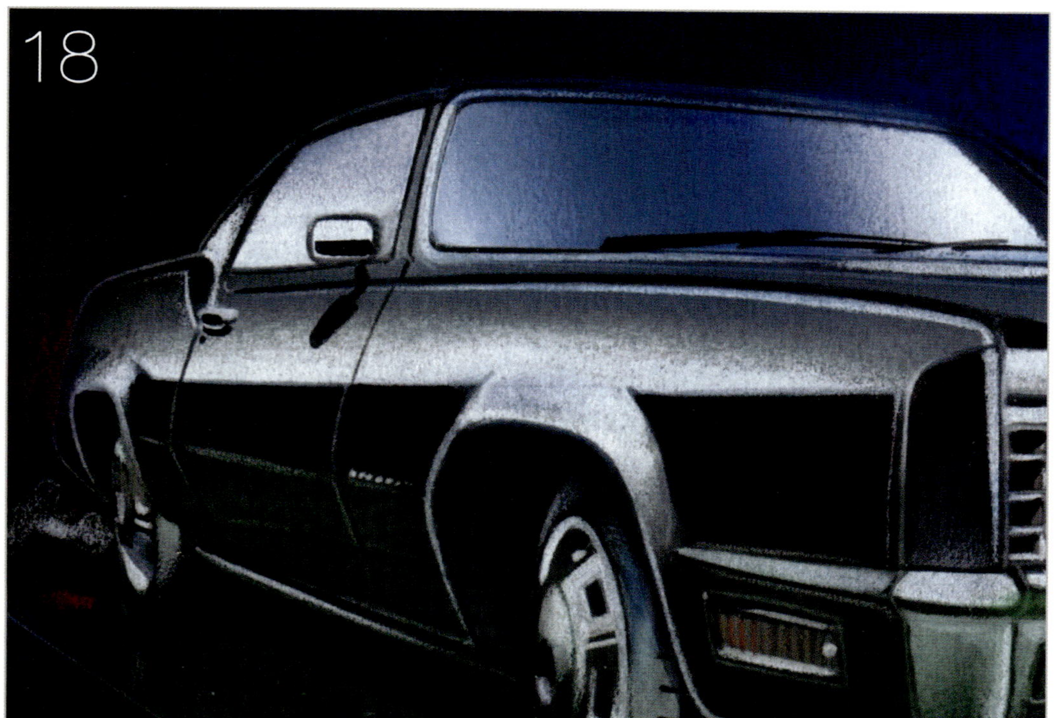

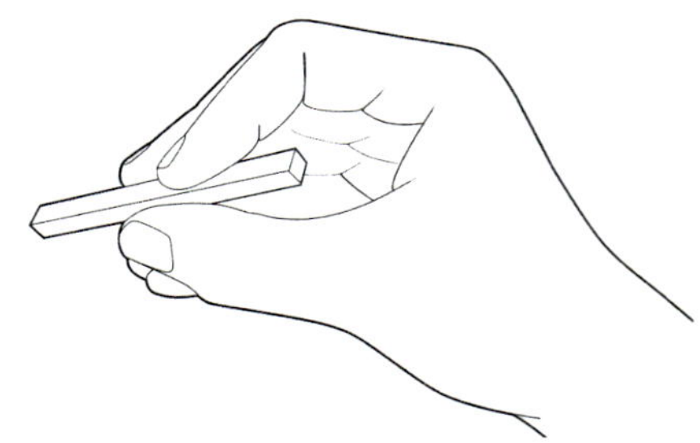

Apply some red pastel directly to the page near the taillight and blend it out with your fingertip.

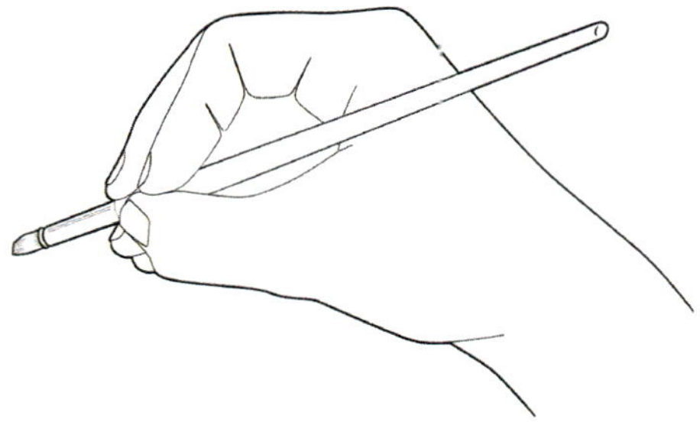

Carefully apply white gouache for the highlights. For the grille, apply small dots of gouache at the intersections.

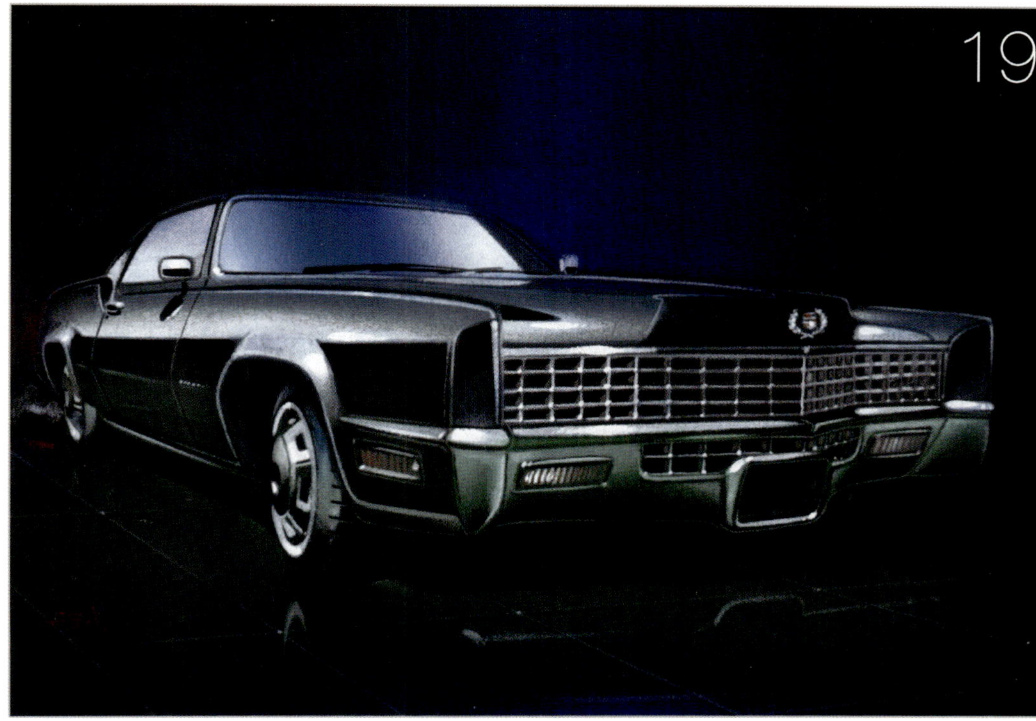

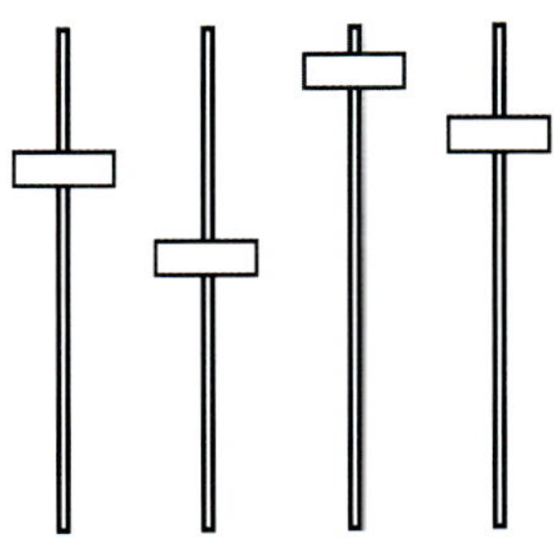

Scan in the rendering and turn up the color saturation in Photoshop. Use a white airbrush tool to refine the appearance of the white pastel.

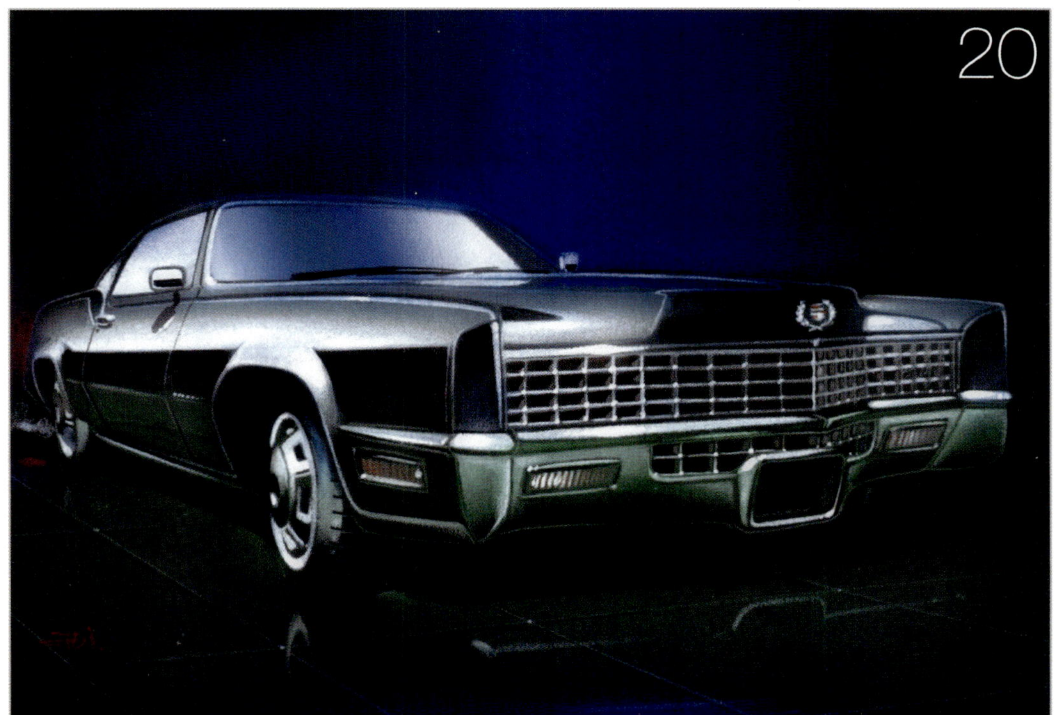

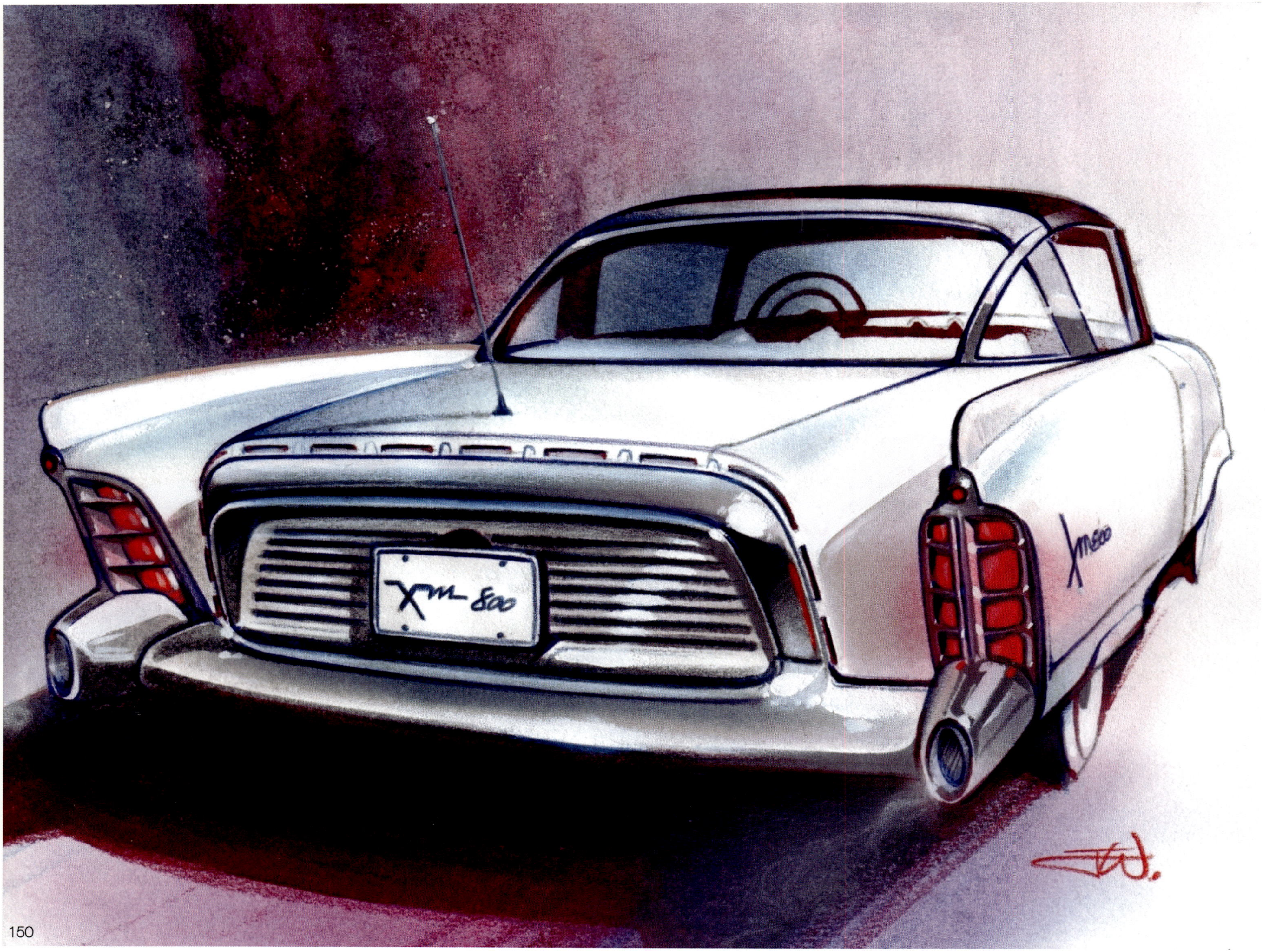
Xm-800
Xm800

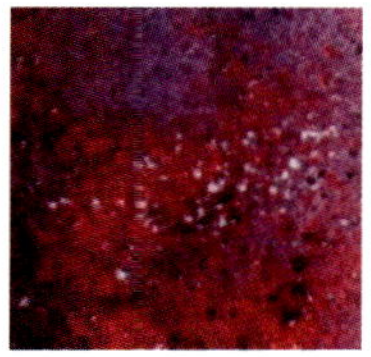

12. Welcome to the Space Age

Before Syd Mead made a name for himself in Hollywood, he produced an amazing body of promotional work for U.S. Steel. His painting style during this time had a spacey, almost otherworldly feel – thanks in part to some beautiful background compositions. What follows is a simple technique designed to mimic this mid-century look.

The Tools

Nupastels: White, Blue, Red, Violet.

Prismacolors: White, Cool Grey 20%, Deco Pink.

Verithins: Black, Indigo Blue.

Gouache: White, Red.

Prismacolor Markers: Indigo Blue, Mahogany Red, Cool Grey 20% & 70%.

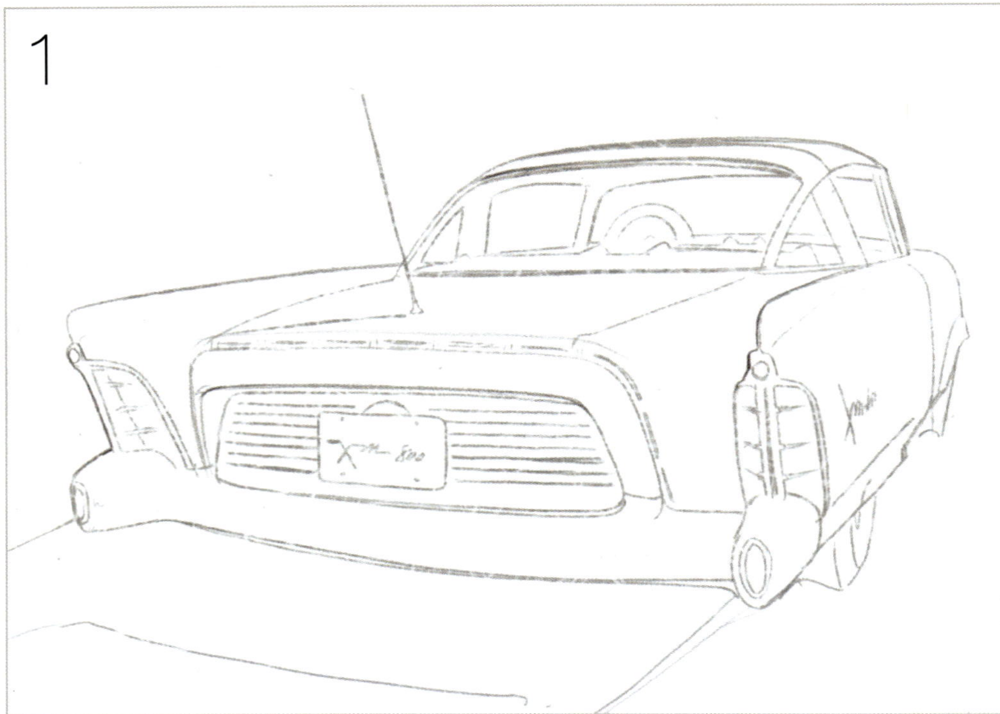

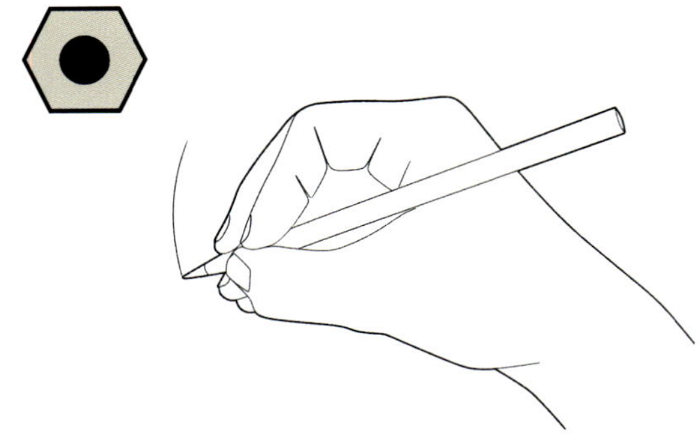

For this rendering, we'll be using white Canson, so try to keep your line-work clean and concise.

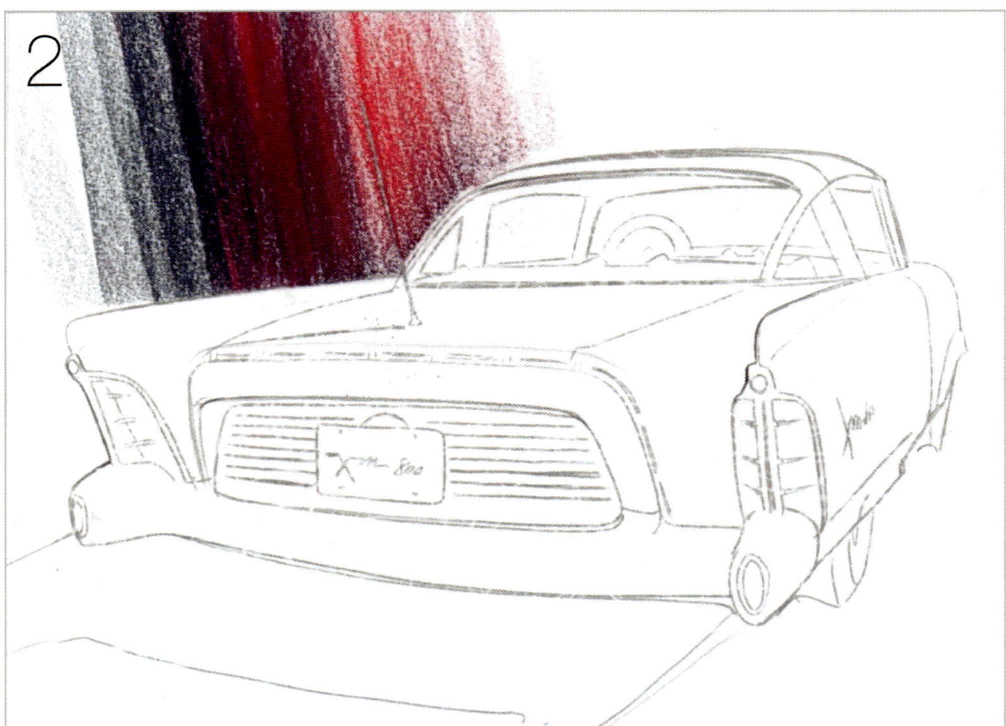

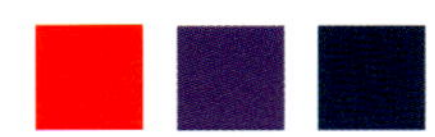

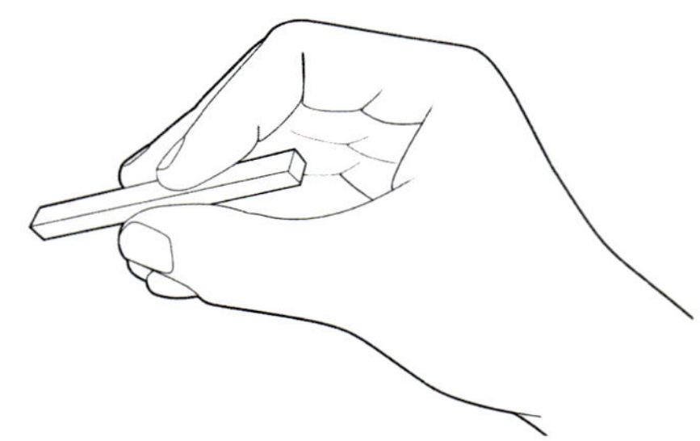

Apply red, violet and blue pastel directly to the paper in vertical strokes. You may want to mask off the car itself to be sure it stays clean.

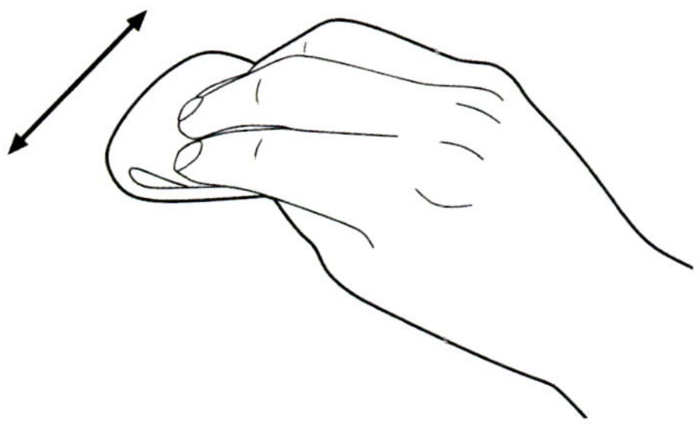

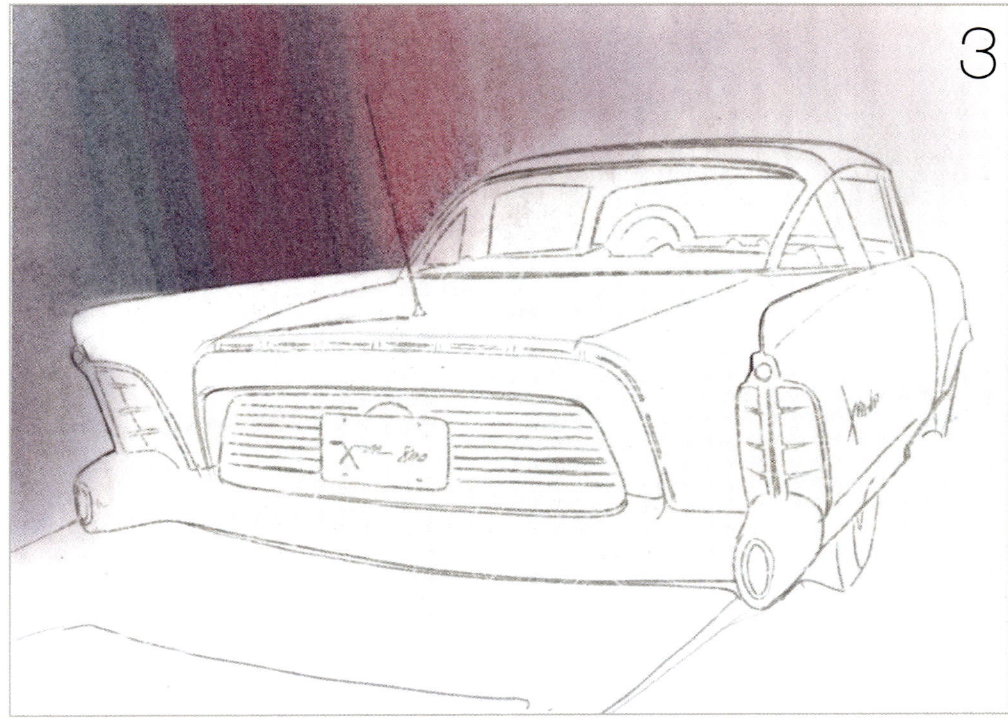

Blend out the background colors with a clean Webril pad, again, taking care not to involve the car itself.

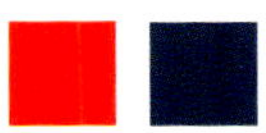

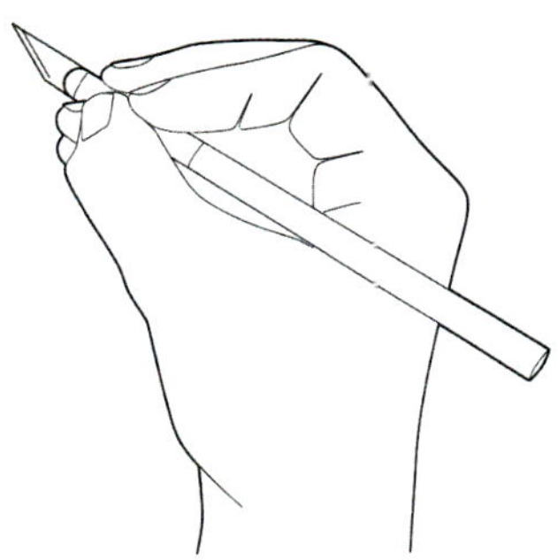

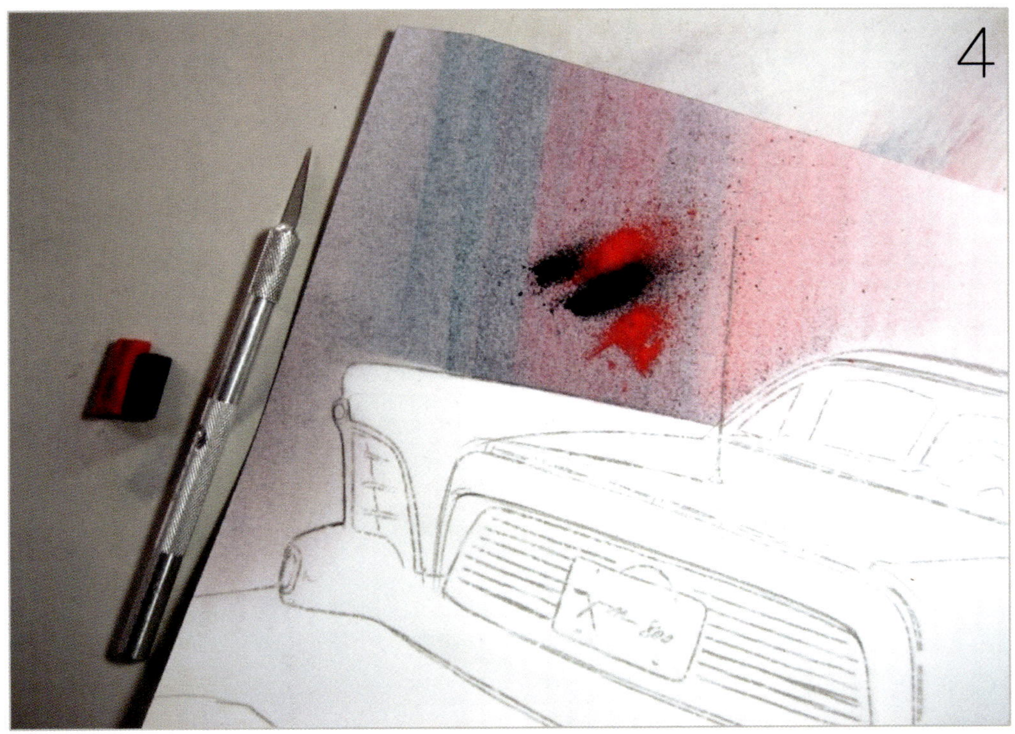

Using an X-acto knife, scrape some red and blue pastel directly onto the page in small piles.

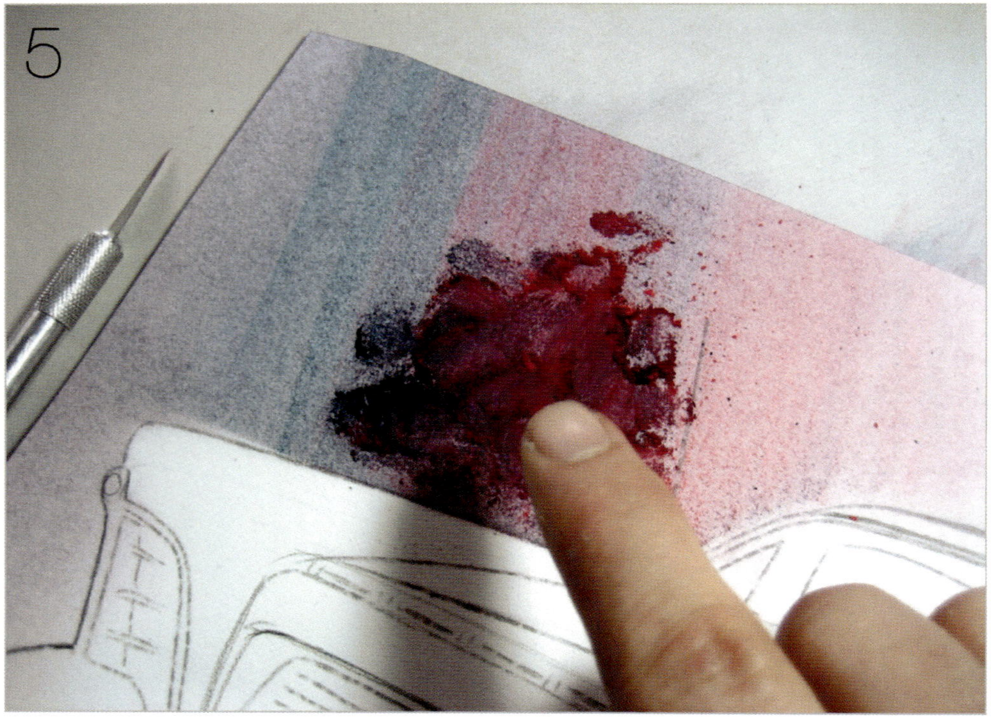

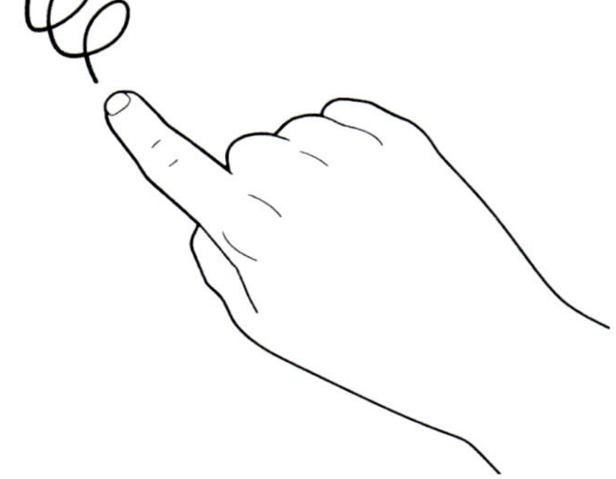

Using light circular motions, mix the colors together with your fingertip.

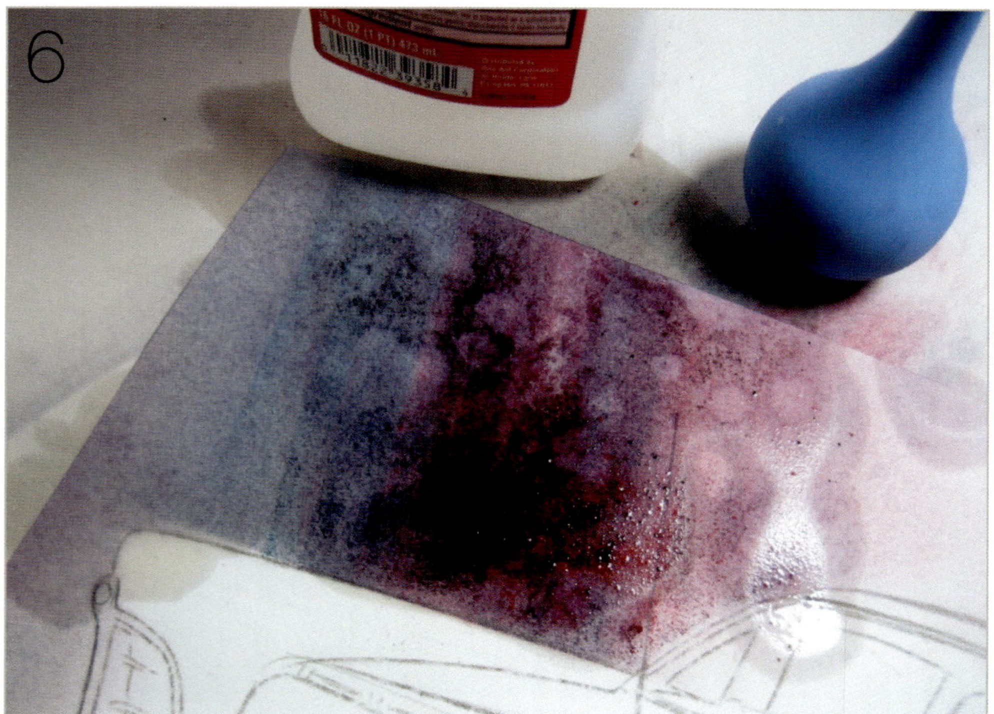

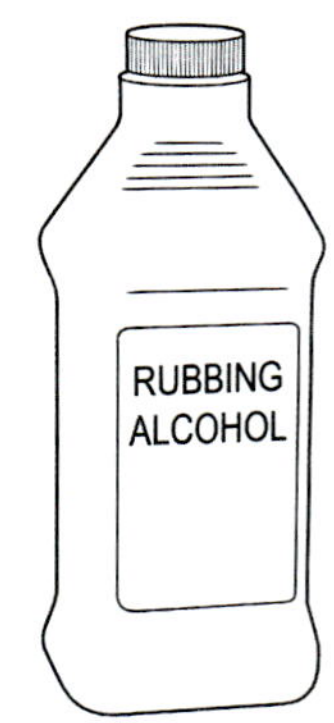

Using a dropper, apply rubbing alcohol in and around the loose pastels. This will create a rich, nebula-like effect in the background.

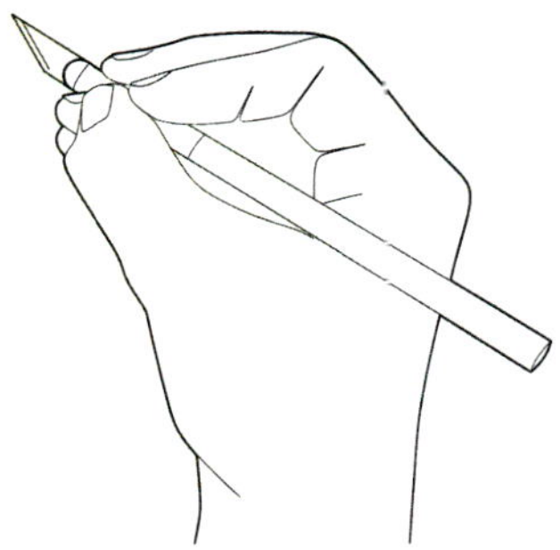

Before the alcohol evaporates, scrape a bit of white pastel into the mix; this will give the appearance of stardust.

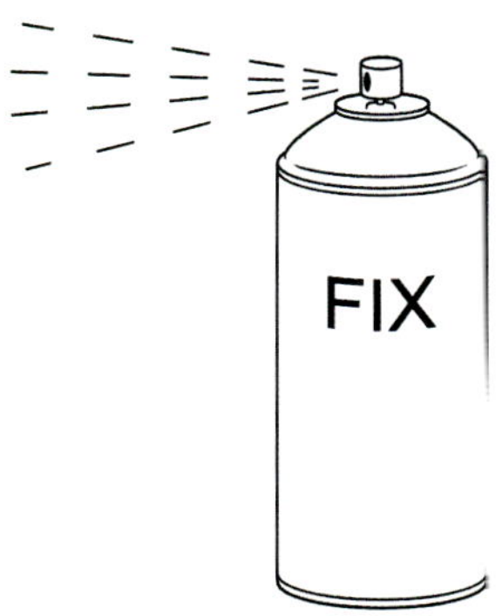

At this point, take a break and allow the alcohol to dry completely. You may wish to hit the background with some fixative to set the pastels, but be sure to read the warning labels first!

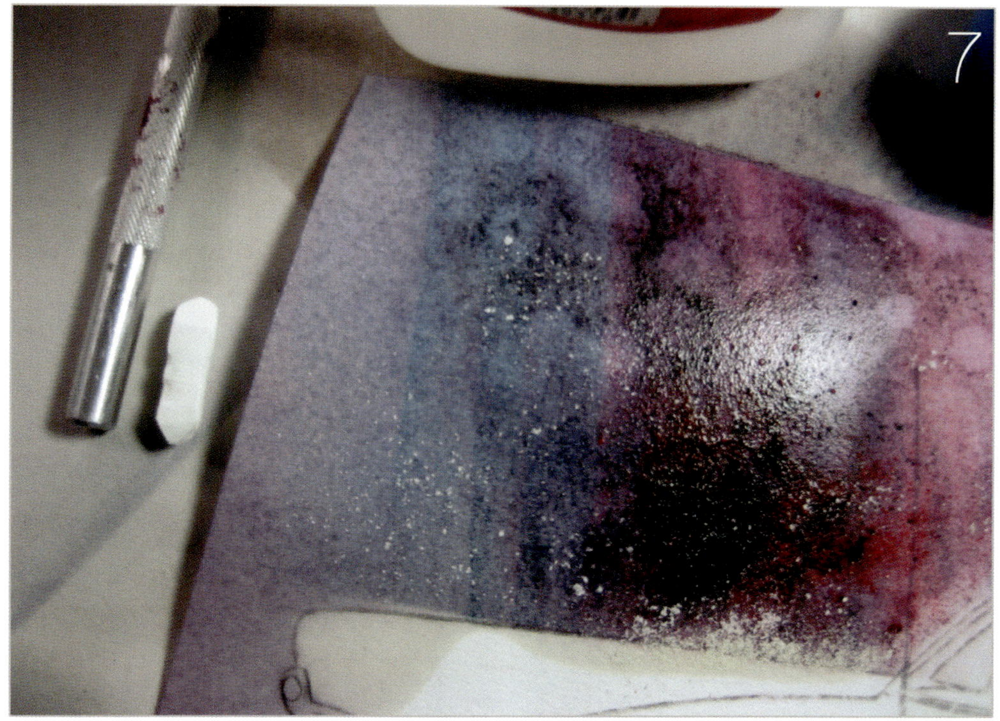
7

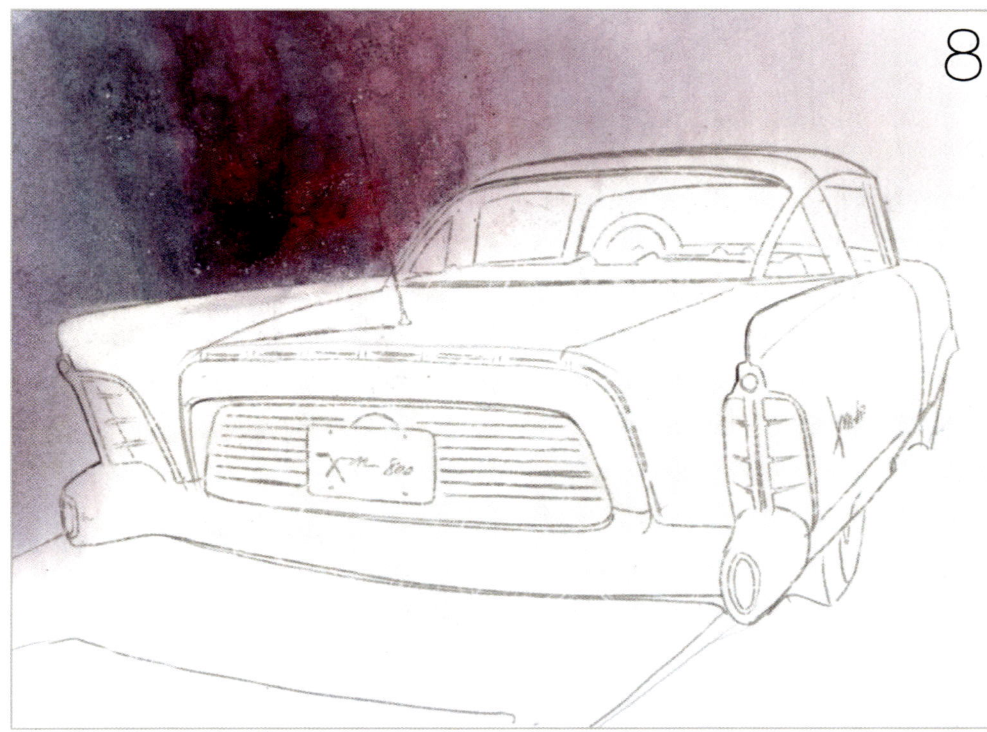
8

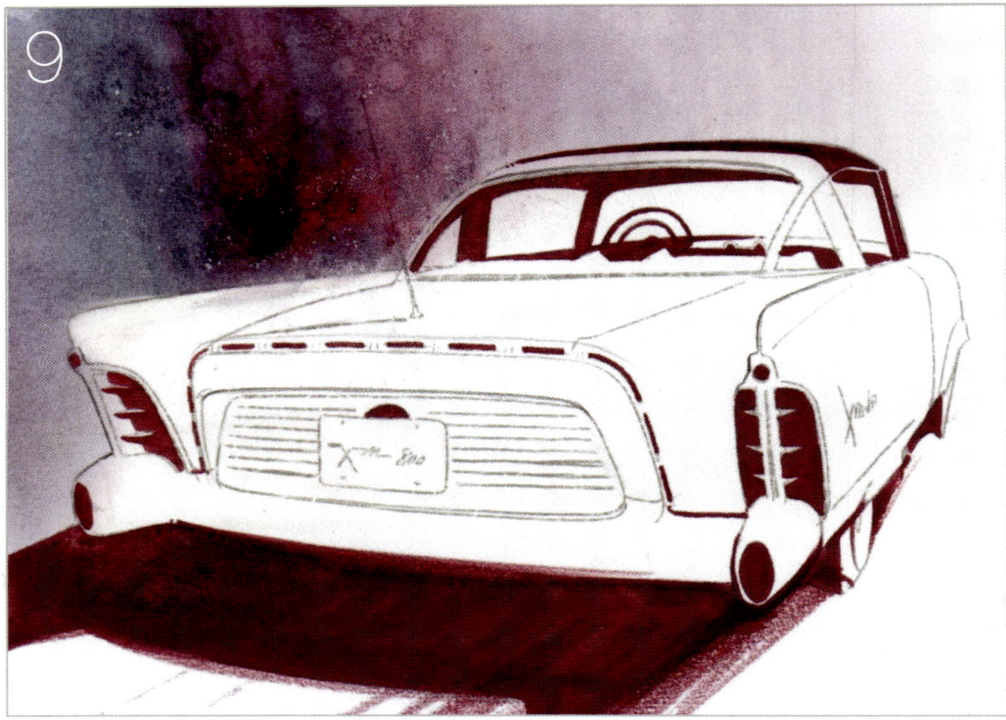

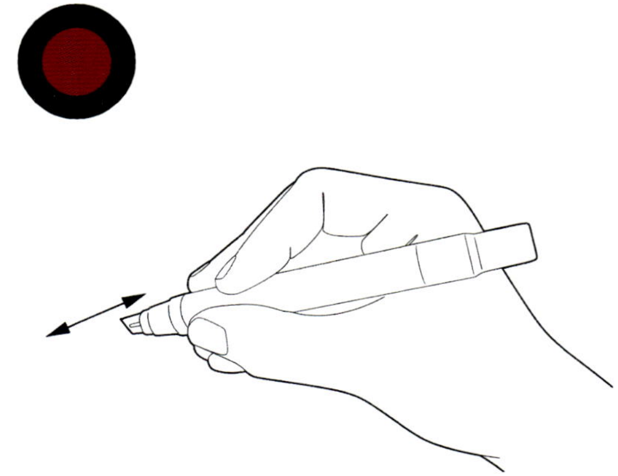

After the alcohol has dried, block in the roof, shadows, lamps and interior surfaces with a Mahogany Red marker.

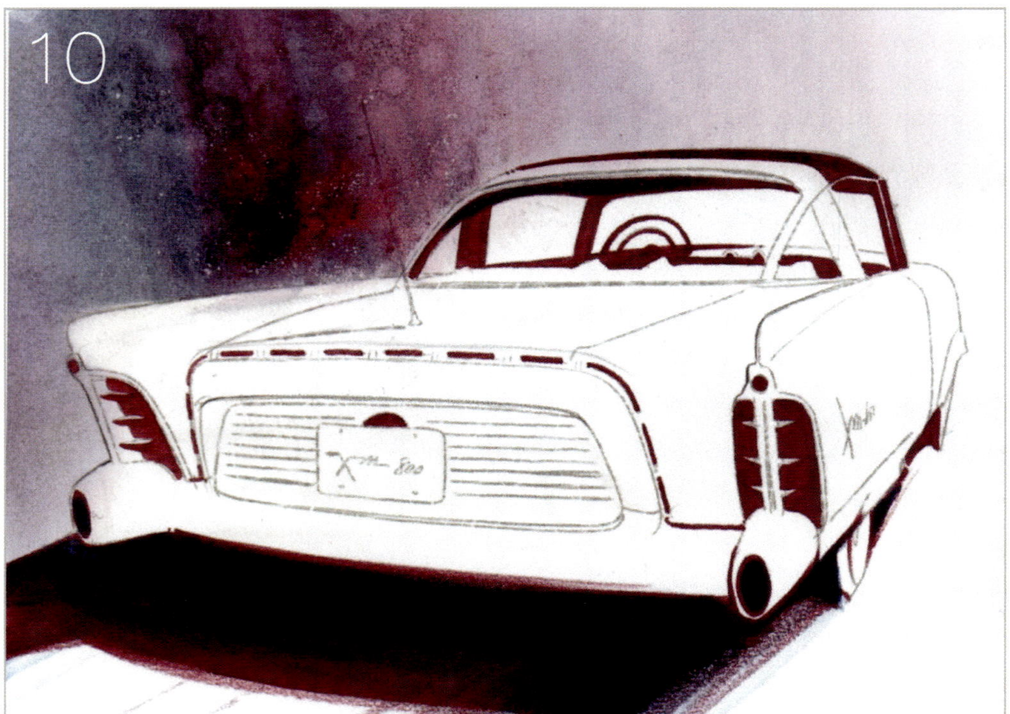

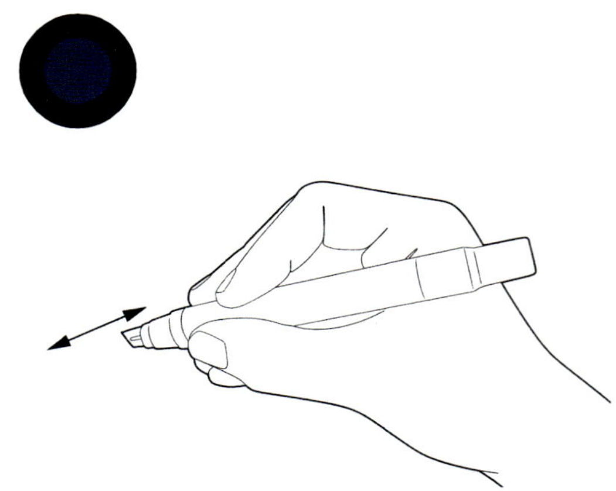

Next, add some depth to the shadow with an Indigo Blue marker.

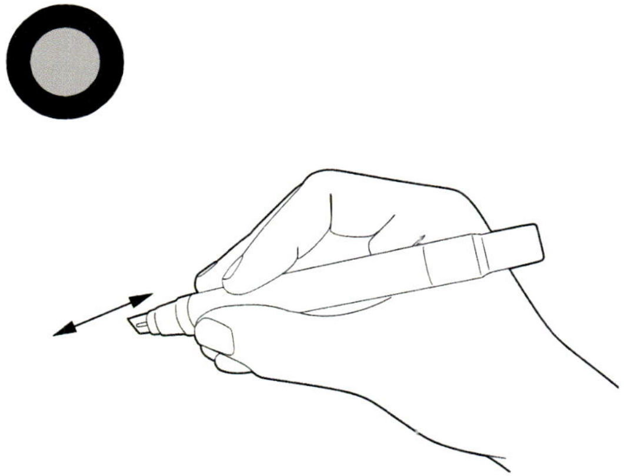

Now block in the chrome sky tones and the tailfin shadows with Cool Grey 20%. If the marker picks up a bit of red or blue, that's okay. It will help give the rendering a more cohesive feel.

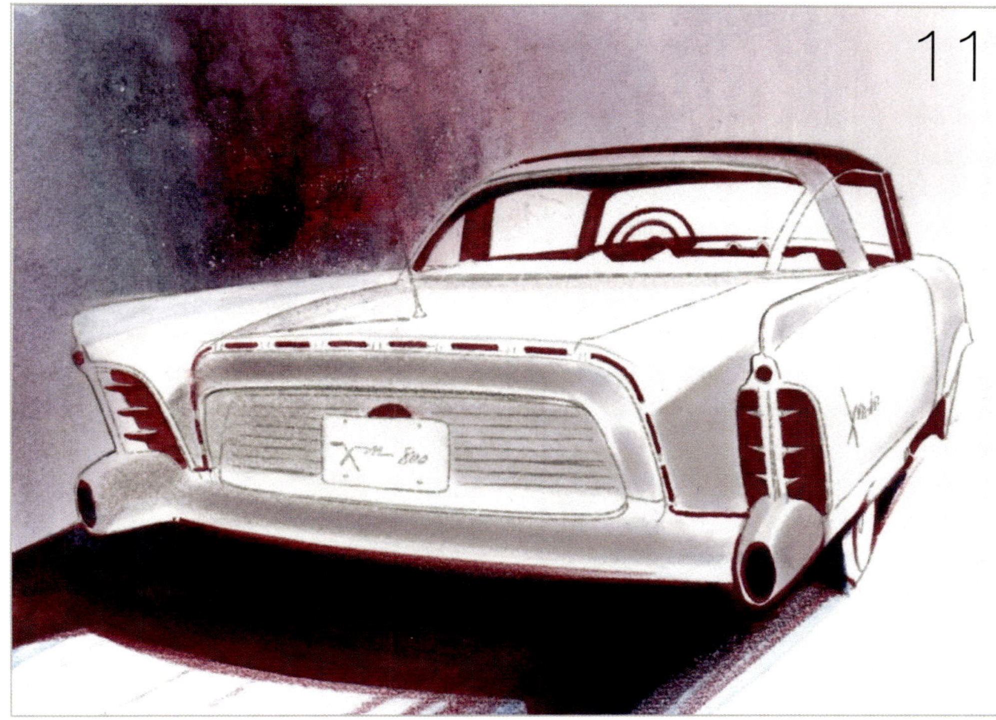

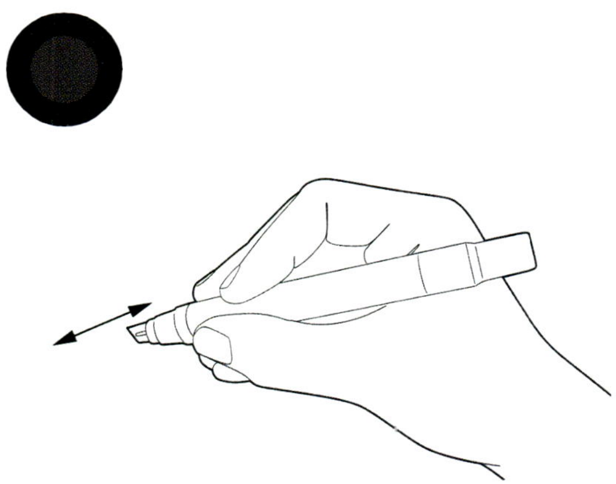

For the horizon lines on the chrome parts, use Cool Grey 70%. The exhaust pipes are essentially cylinders, so make sure your reflections follow suit.

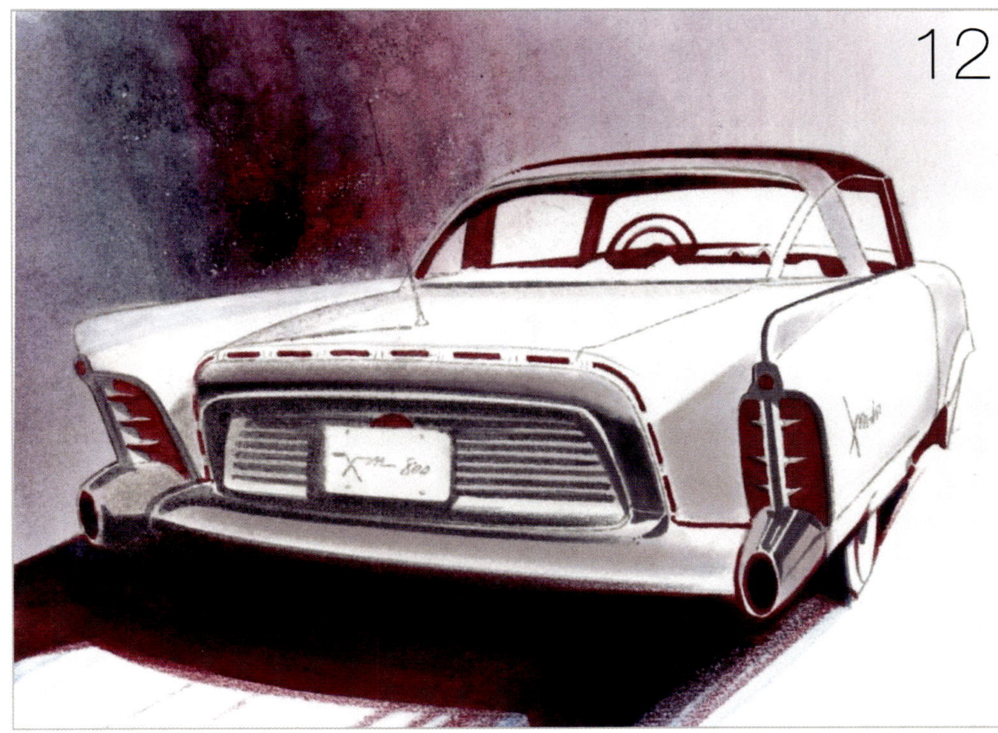

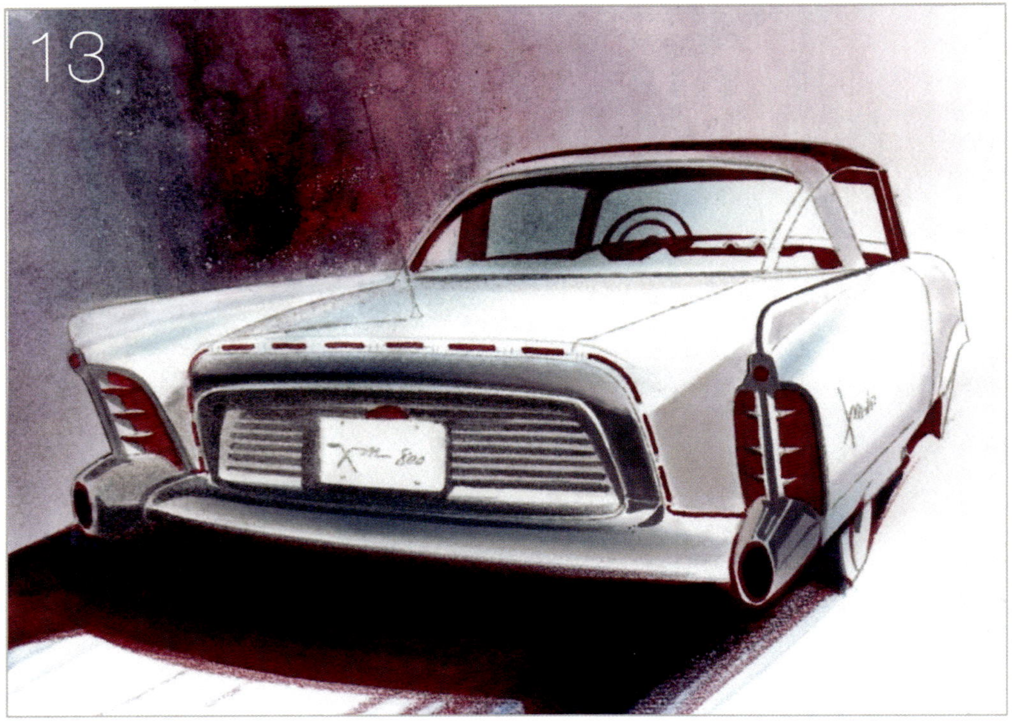

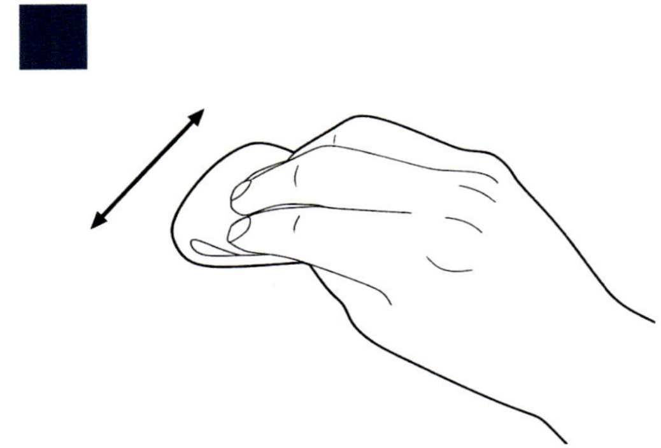

For the overall sky tone, ghost in a discreet amount of blue pastel powder.

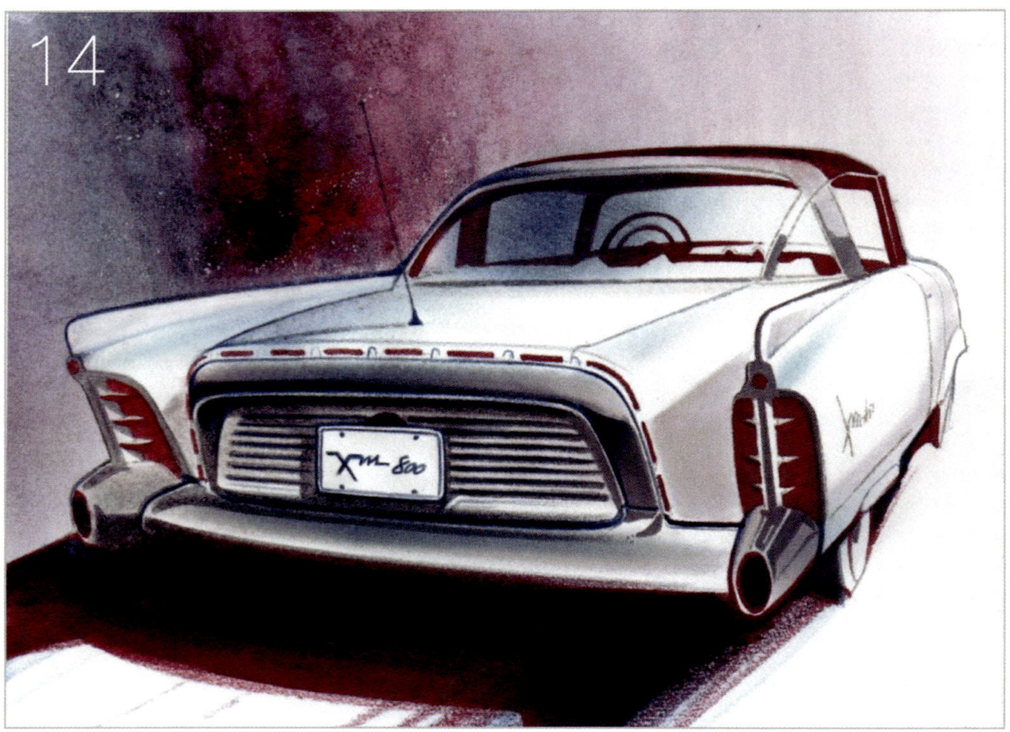

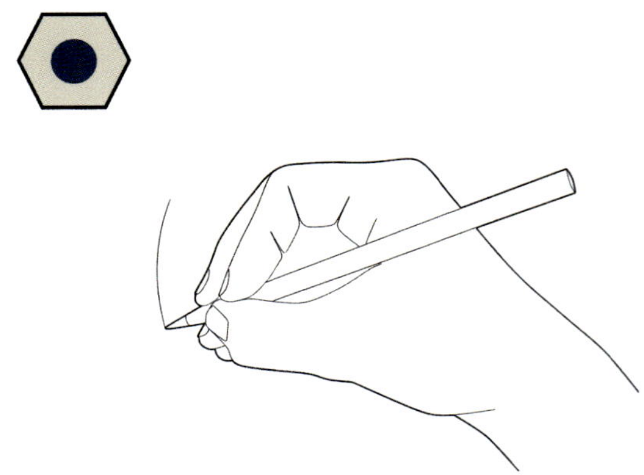

At this point, begin to define the details of the car with a blue pencil.

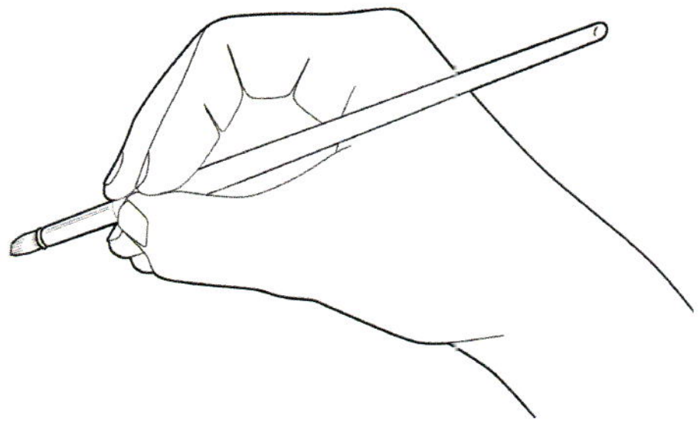

We usually leave the gouache highlights for the end, but try applying them now. This will give you a good idea of the overall contrast of the rendering; a sort of preventative measure to help you avoid overworking the piece.

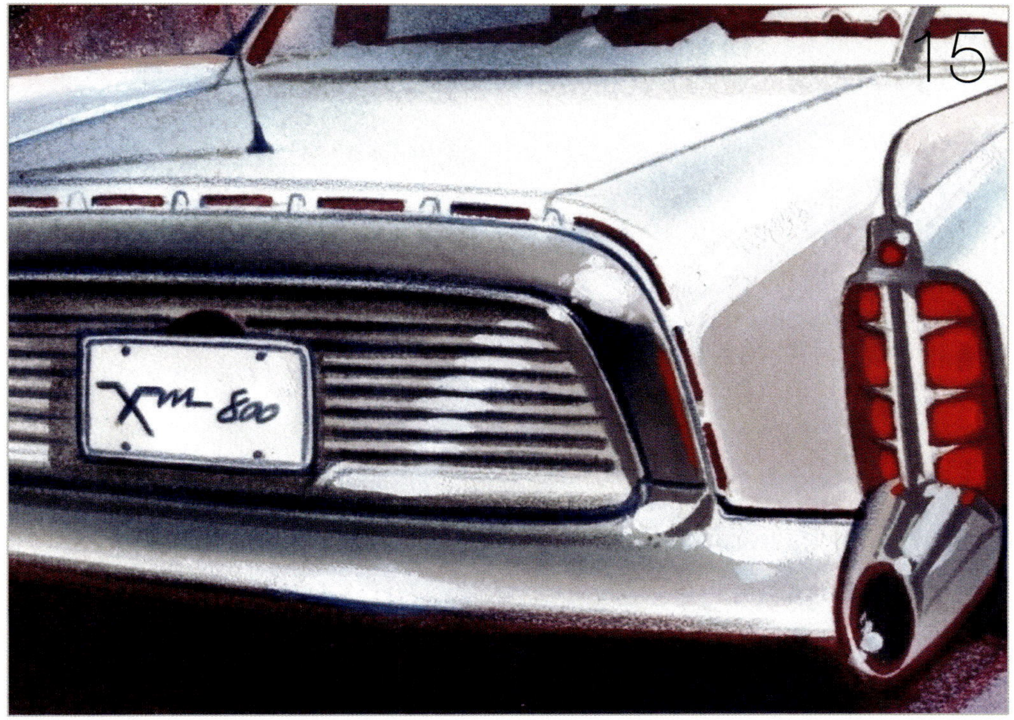

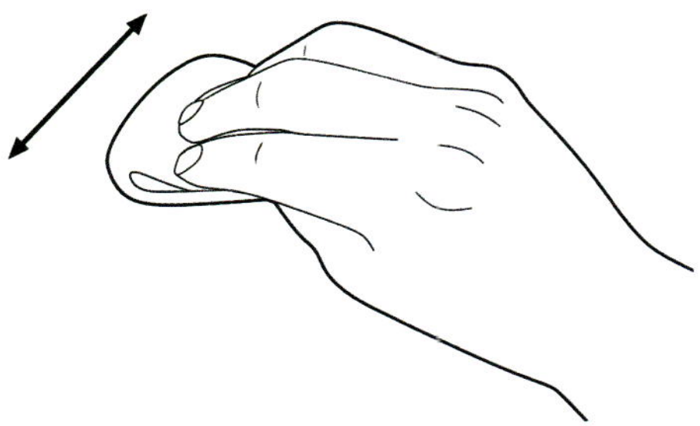

Using the same Webril pad you used to blend out the background, add some violet in the foreground under the car. This will help tie the piece together.

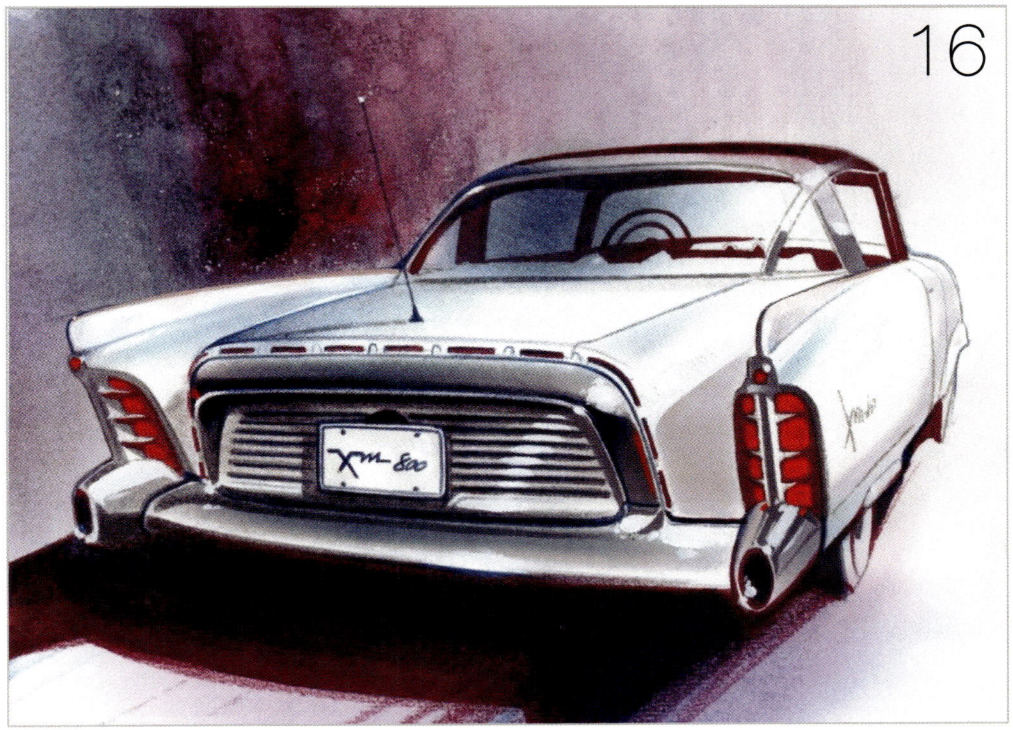

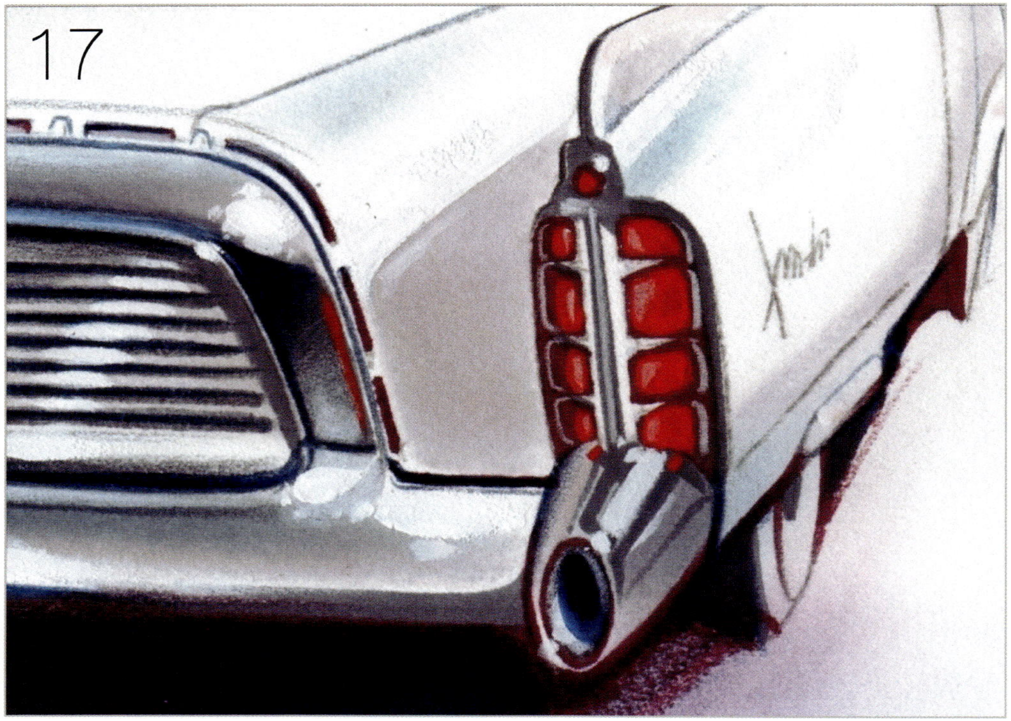

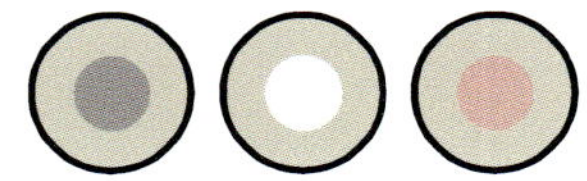

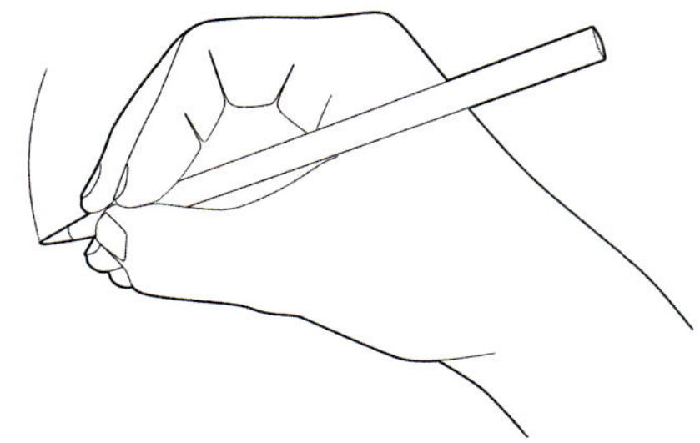

Once the red gouache is dry, add some detail to the lenses with an array of Prismacolor pencils.

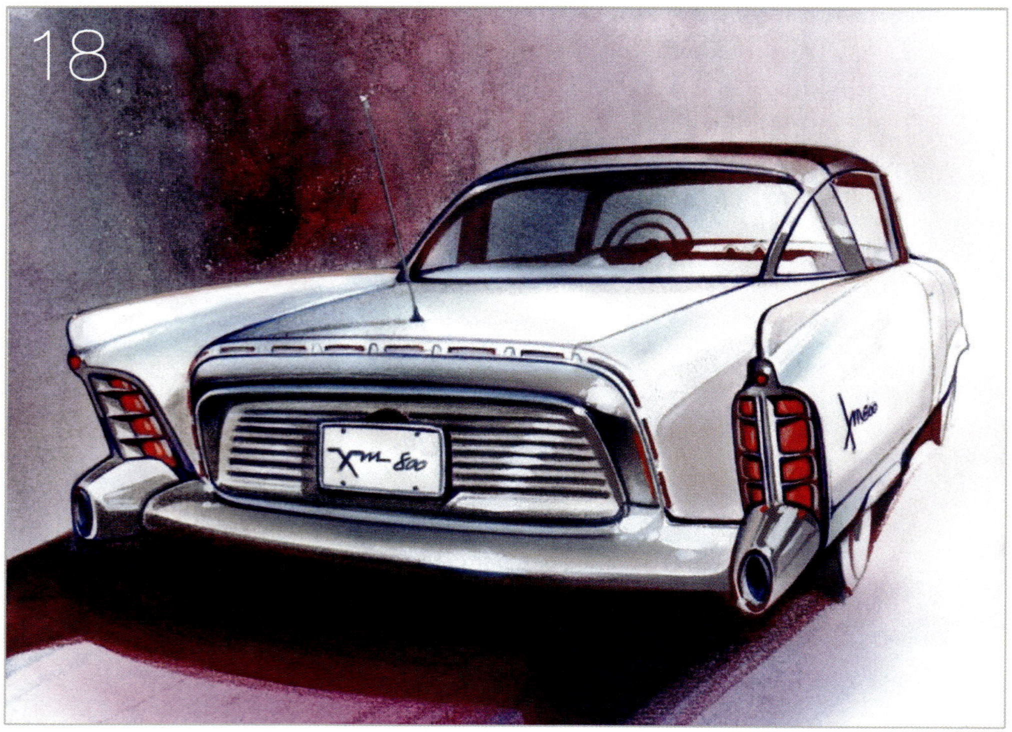

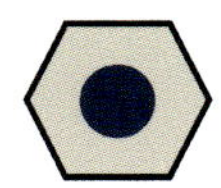

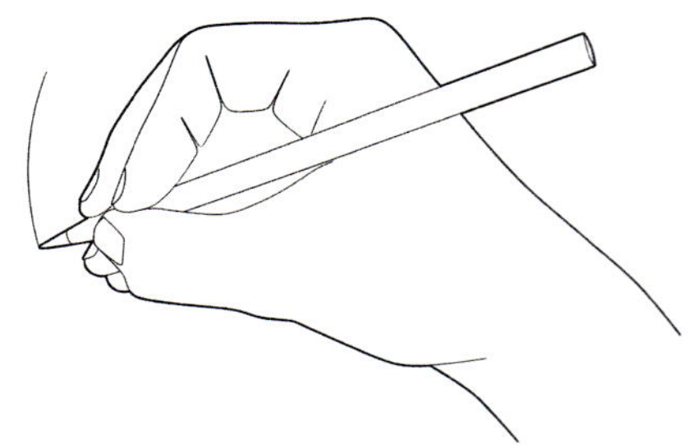

Continue adding line details with a blue Verithin. Be sure to sharpen often.

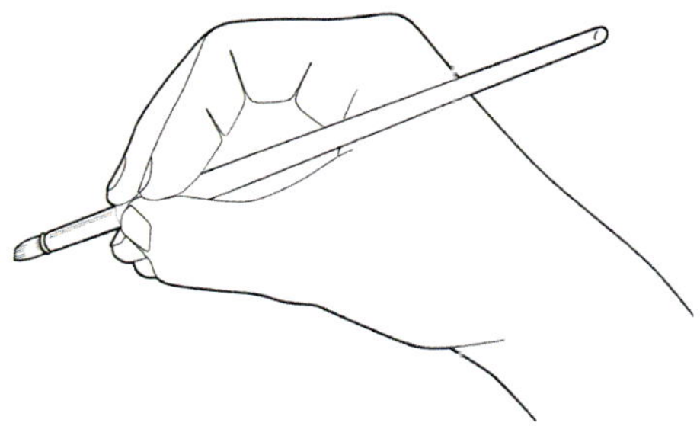

Apply one more round of white highlights, this time focusing on small areas like the corners of the chrome bezels and the point of the antenna.

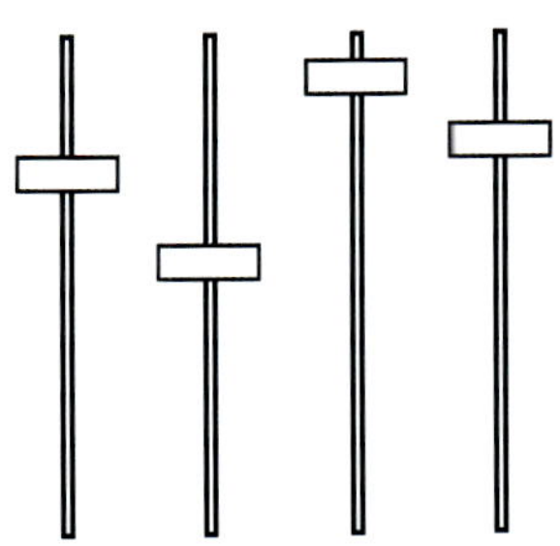

Scan in the rendering and turn up the color saturation in Photoshop. You may wish to add in some exhaust using an airbrush tool.

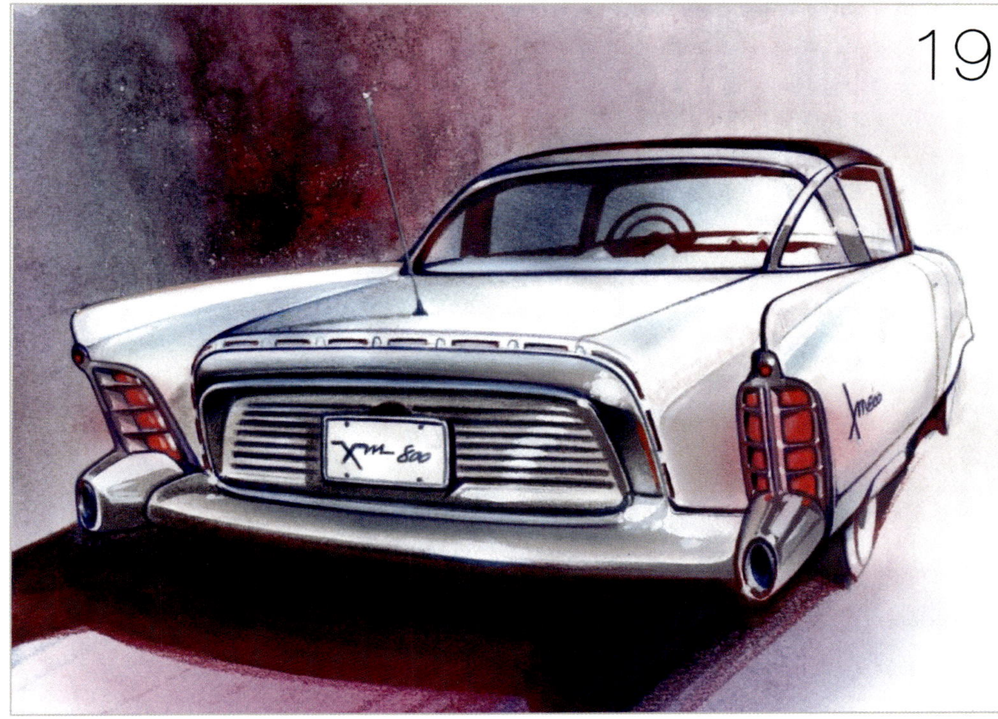

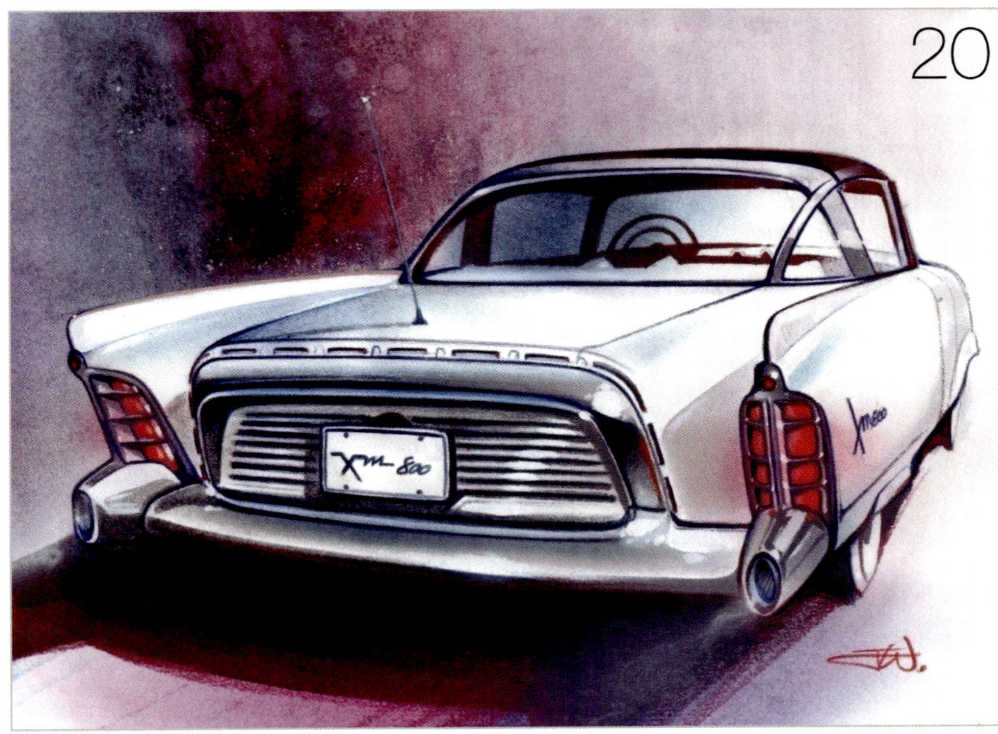

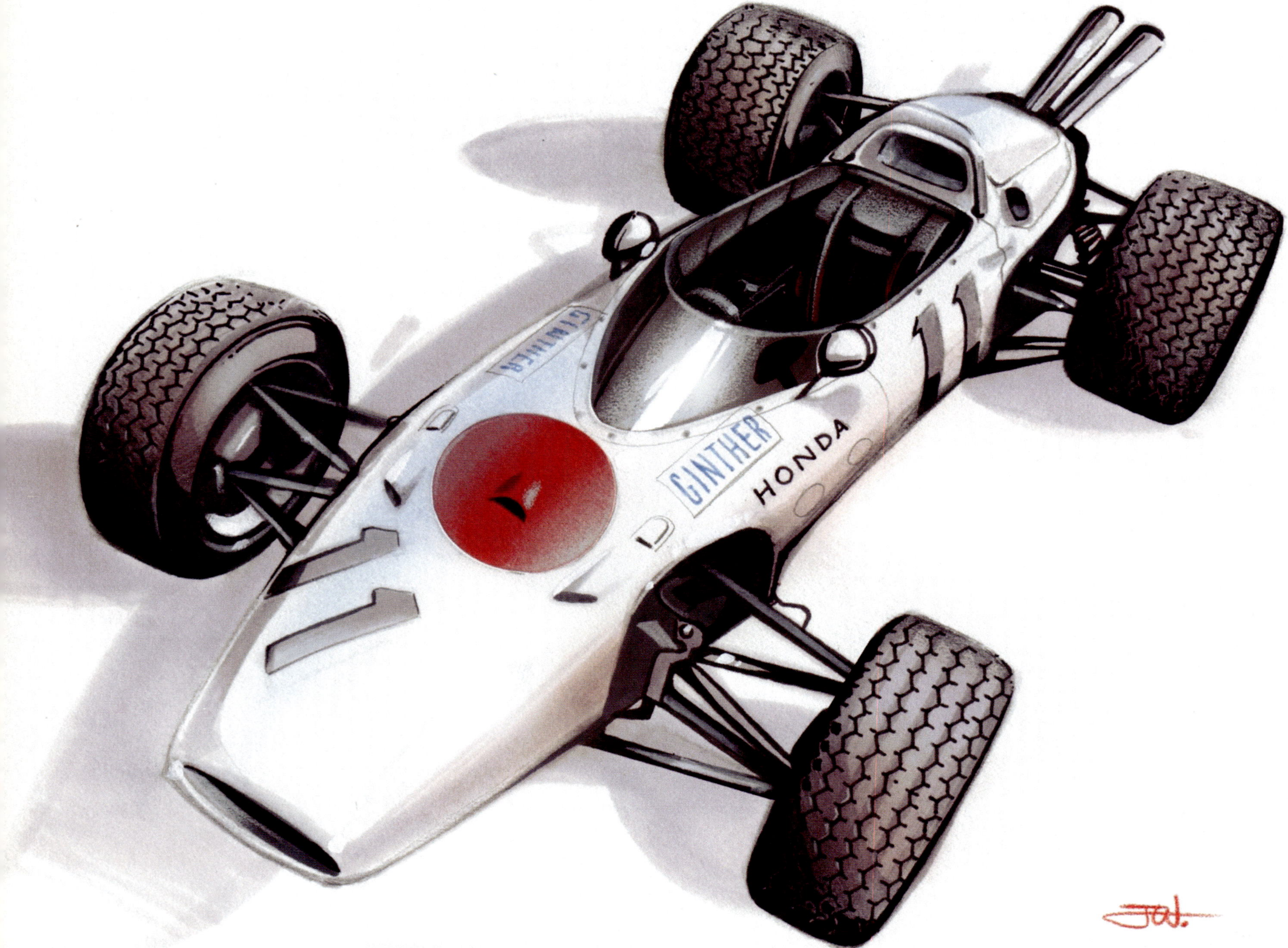
GINTHER
GINTHER
HONDA

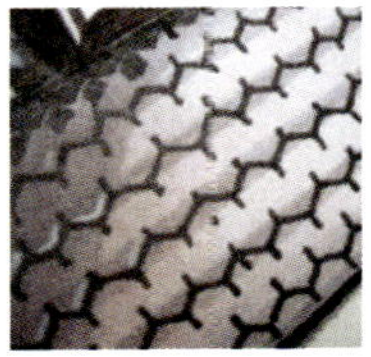

13. The Importance of Detail

As we saw in the previous rendering, a white car doesn't require very much in the way of sky tone or reflections. In cases like these, the details can often make or break the overall piece. Here is an example of a white Canson rendering that makes the most of this particular condition.

The Tools

Nupastels: White, Blue, Orange.

Prismacolors:
White, Cloud Blue,
Crimson Red, Cool Grey 20%.

Copic Pen:
Black.

Verithins:
Black.

Gouache:
White.

Prismacolor Markers: Black, Cerulean Blue, Cool Grey 20% and 60%.

1

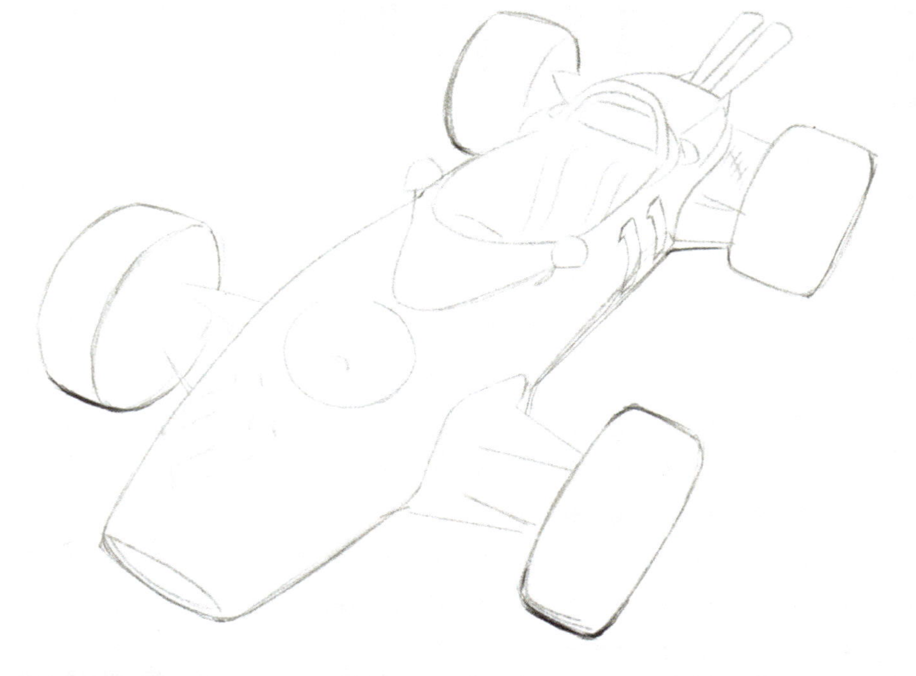

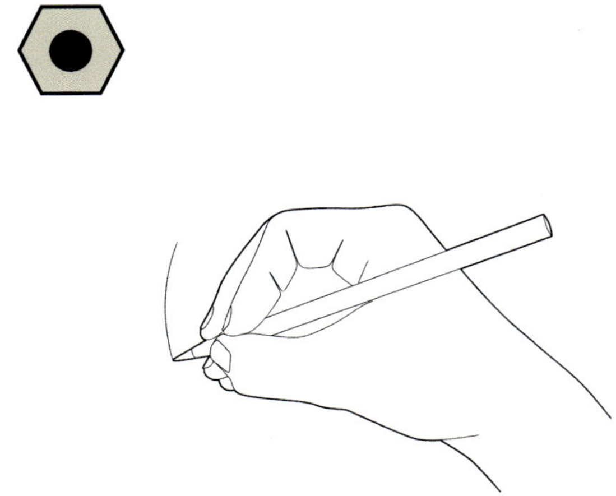

As with the previous rendering, try to keep your lines clean and concise.

2

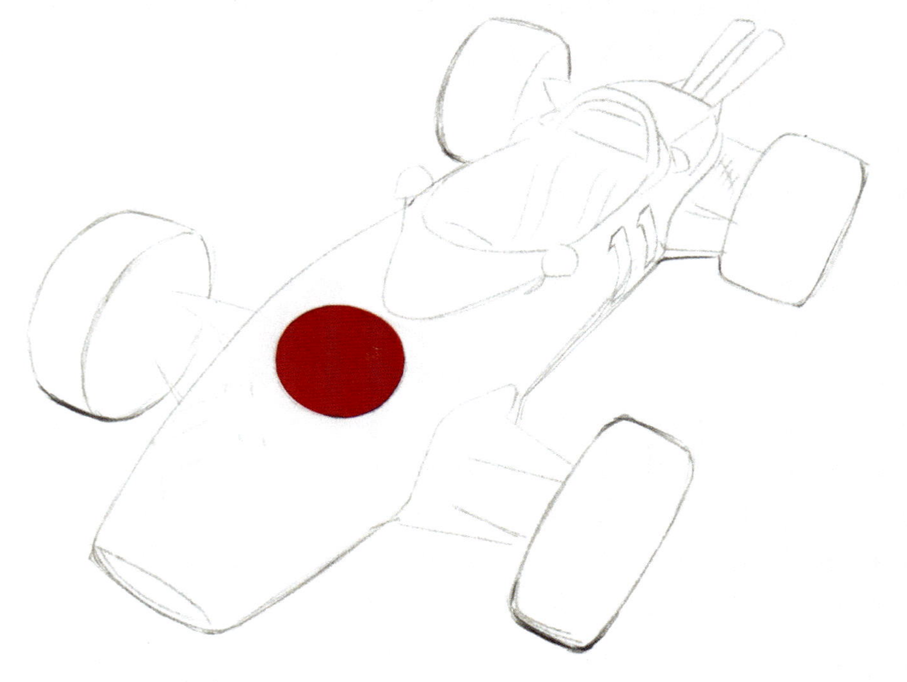

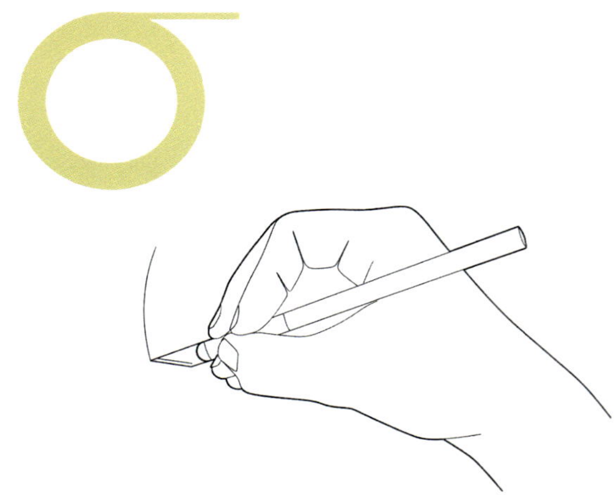

Splice in a piece of red Canson for the "rising sun" graphic. You may want to use an ellipse guide to ensure a clean shape with no kinks. Secure the red piece on the back with some tape.

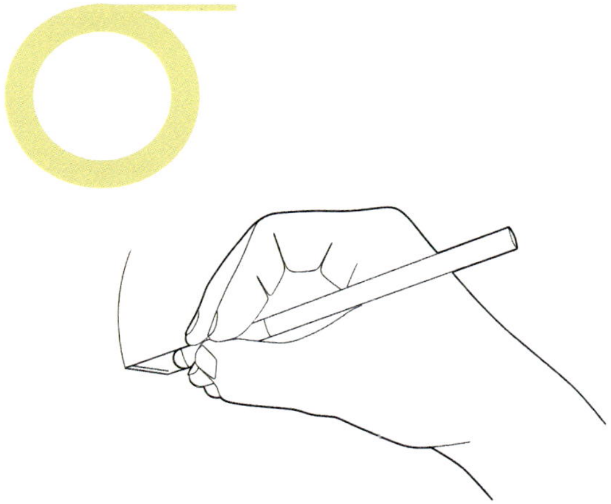

Now execute the same process for the cockpit with Warm Steel Canson.

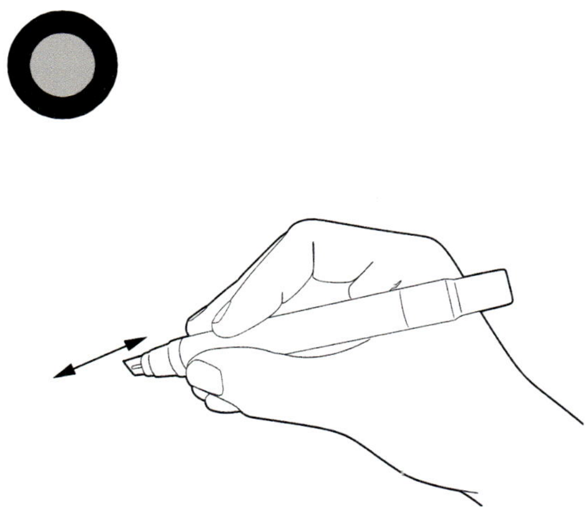

With the main fields of color in place, you can begin laying in the tires, shadows, numbers and other items with Cool Grey 20%.

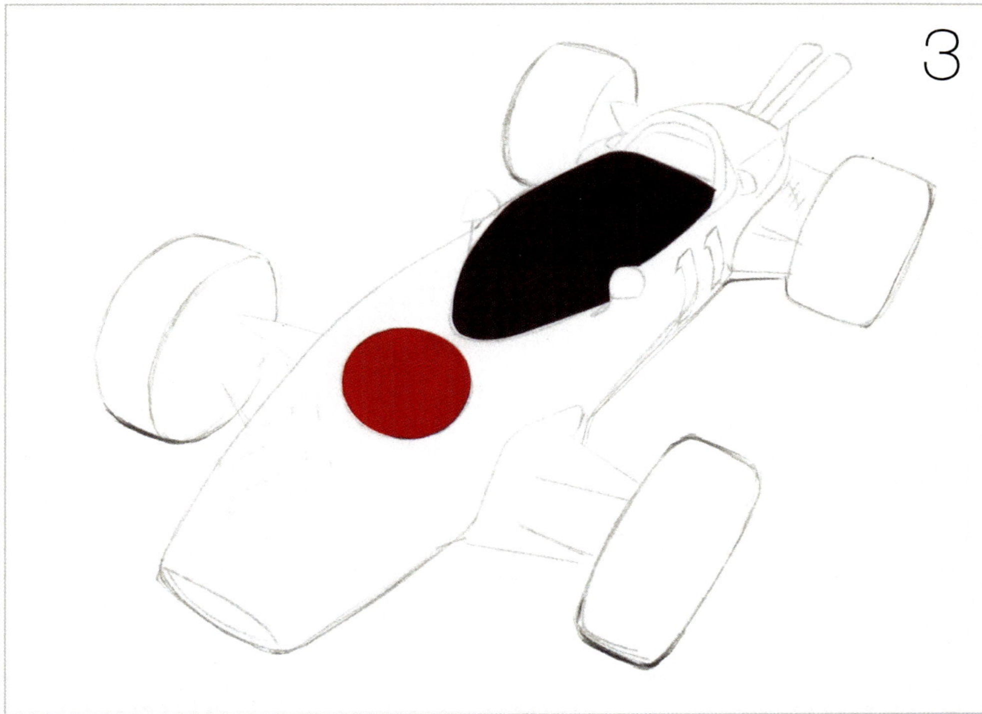

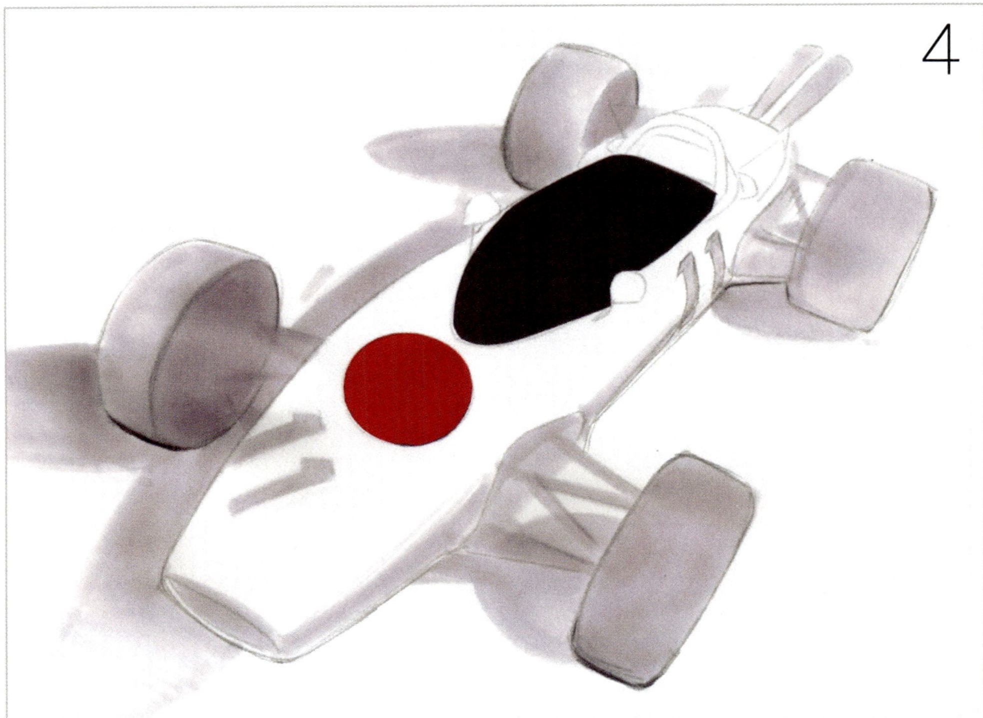

5

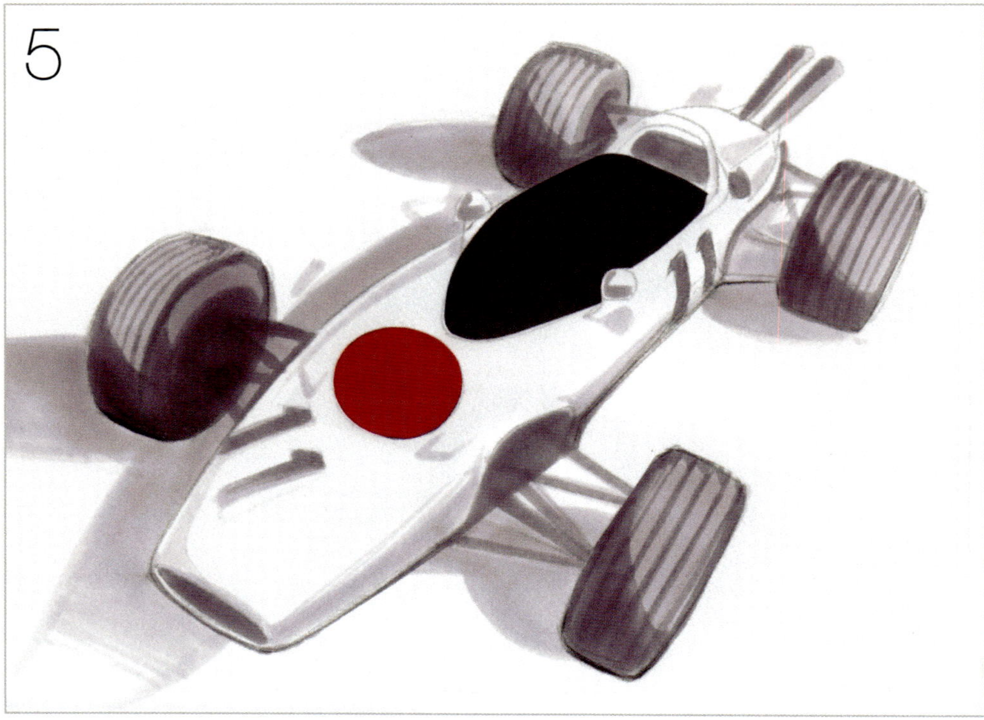

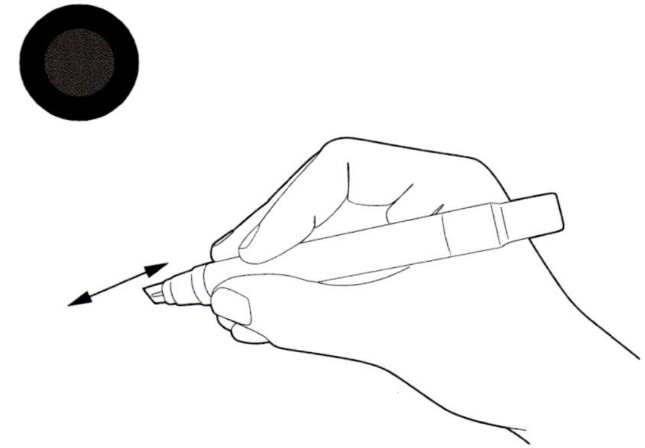

Now add tread detailing and tire shadows with Cool Grey 60%. The long strokes for the tread pattern may seem out of scale, but don't worry. These strokes will provide the base for the final, more detailed tread pattern.

6

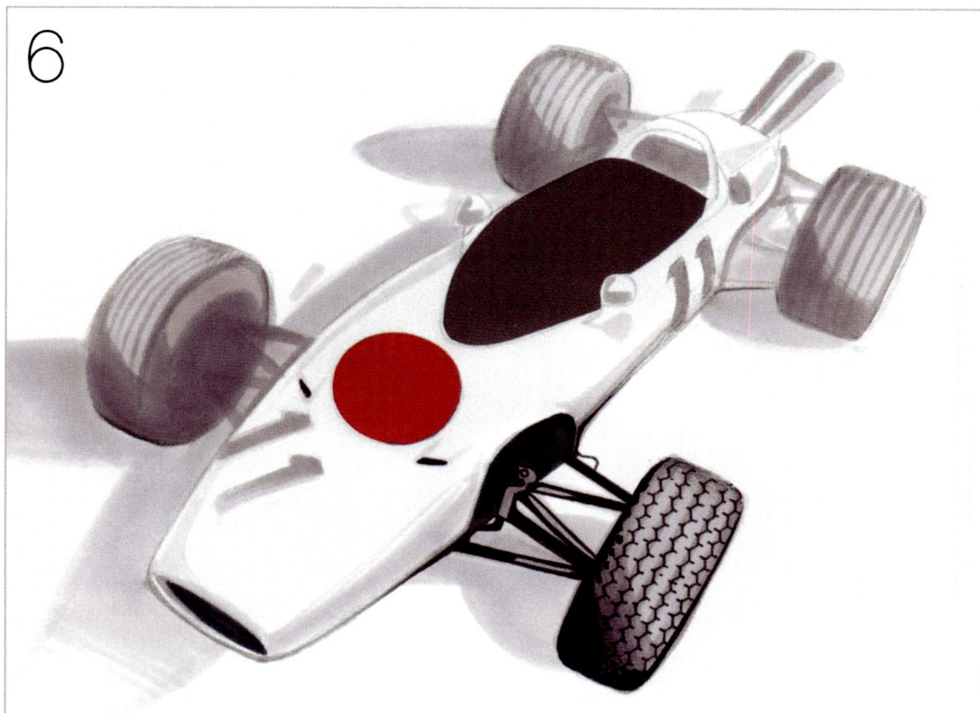

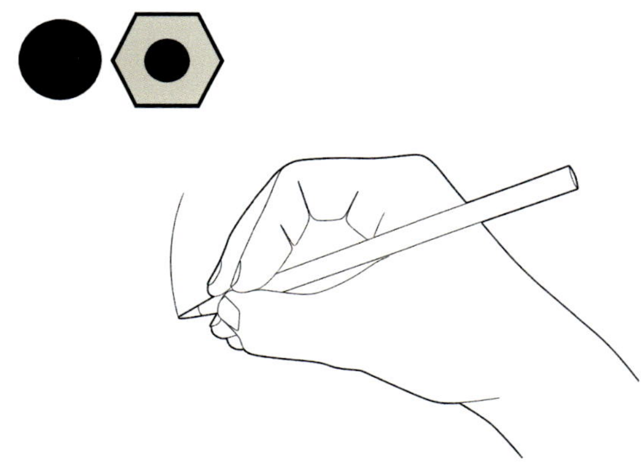

Using a fine-tipped black pen and a black Verithin, begin detailing one of the tires. Whatever you do, do not rush the final tread pattern. As tedious as it is, the end result will be worth it.

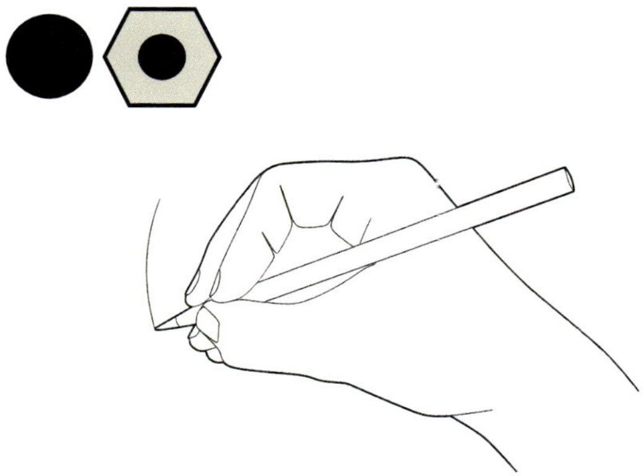

Hopefully, by the time you've moved to the next wheel, you'll have developed a sort of rhythm in your head. This will help you make sure that the tread pattern is consistent on all the tires.

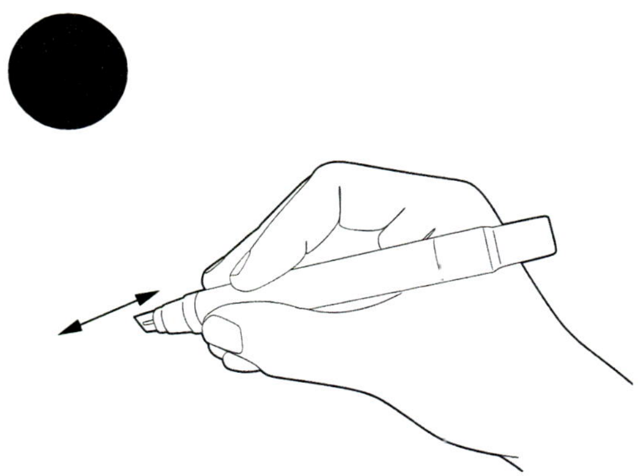

To give the suggestion of shiny decals, continue the grey horizon line through the first "1" with a black marker. Outlining the far edge will give the decals a slightly raised look.

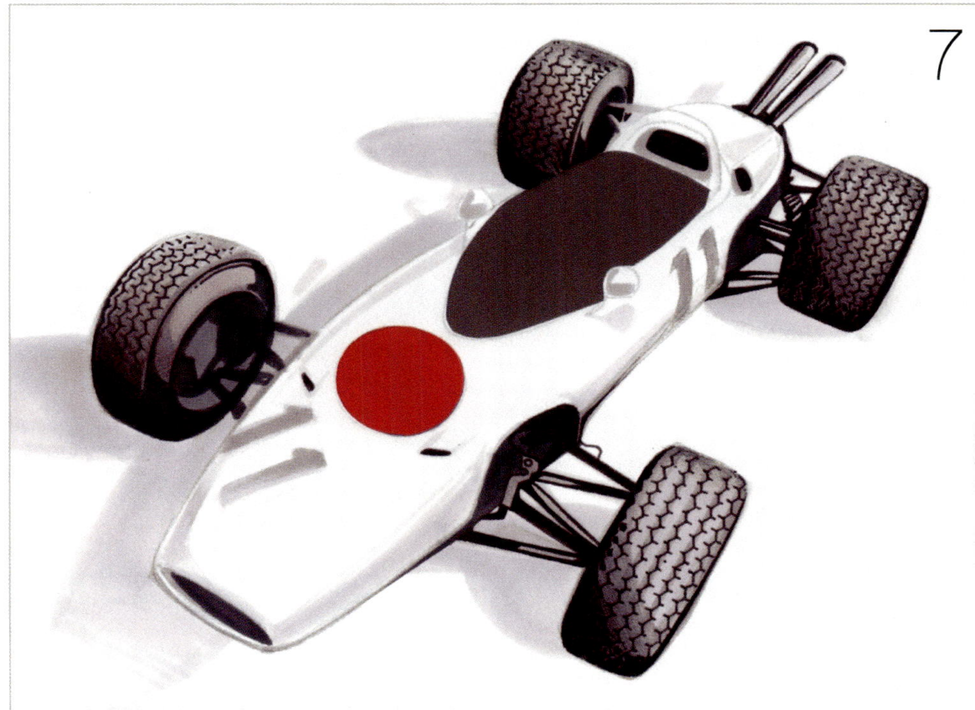

7

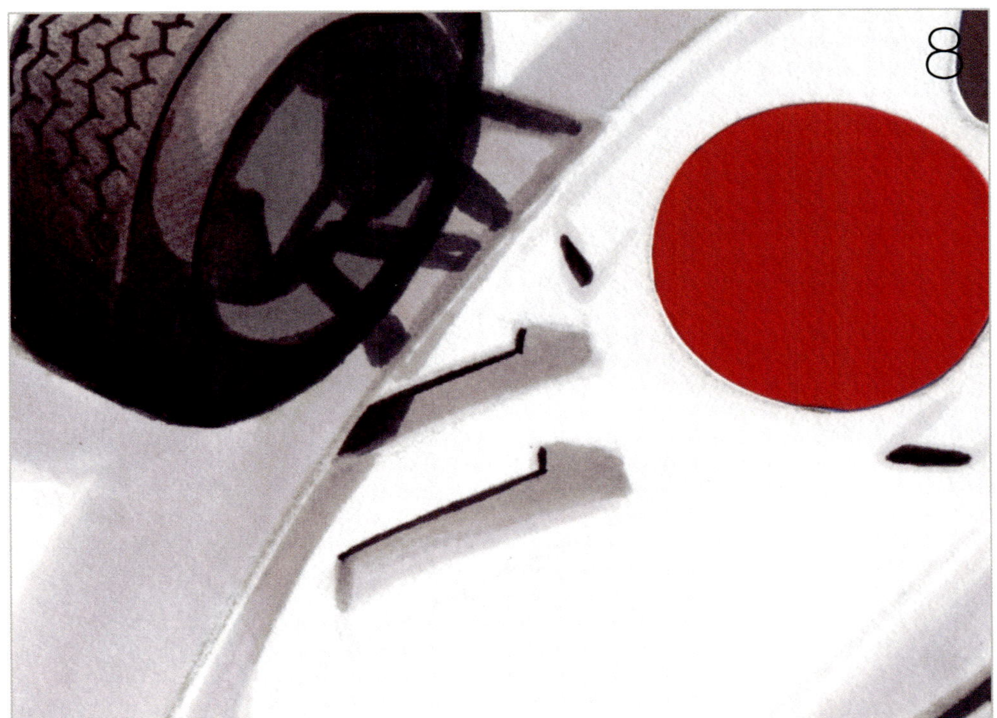

8

9

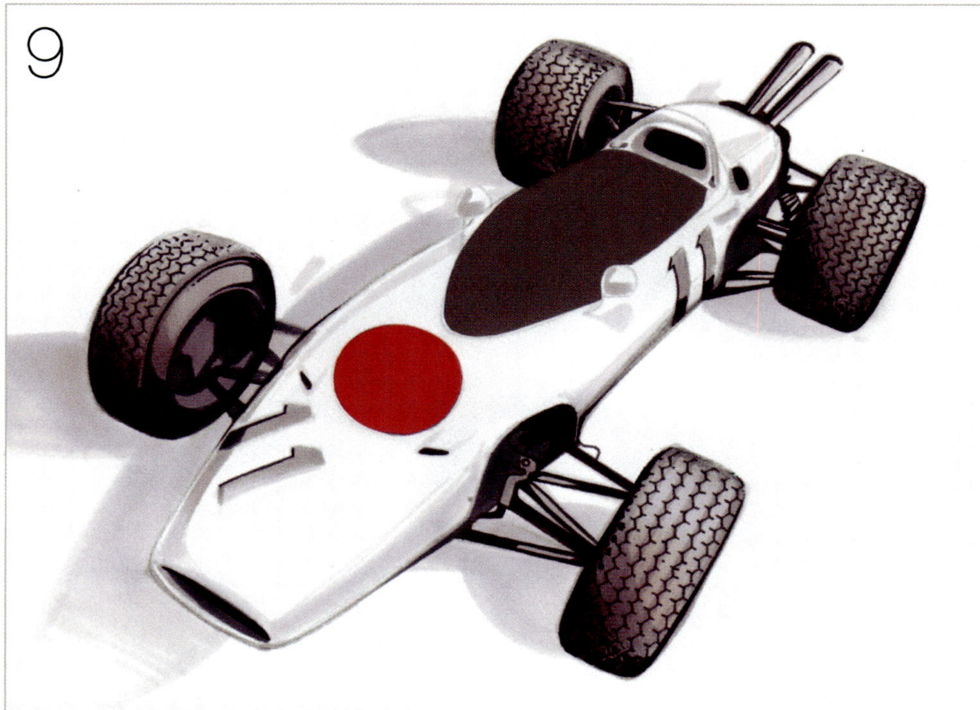

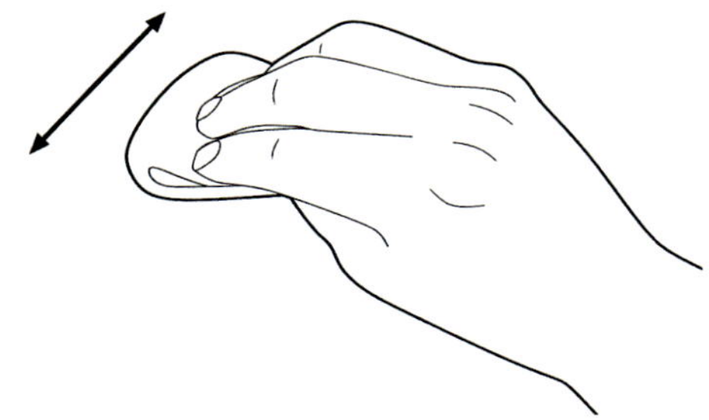

Lightly ghost in a bit of orange pastel powder at the base of the tires; this will serve to set them apart from the cool floor shadows.

10

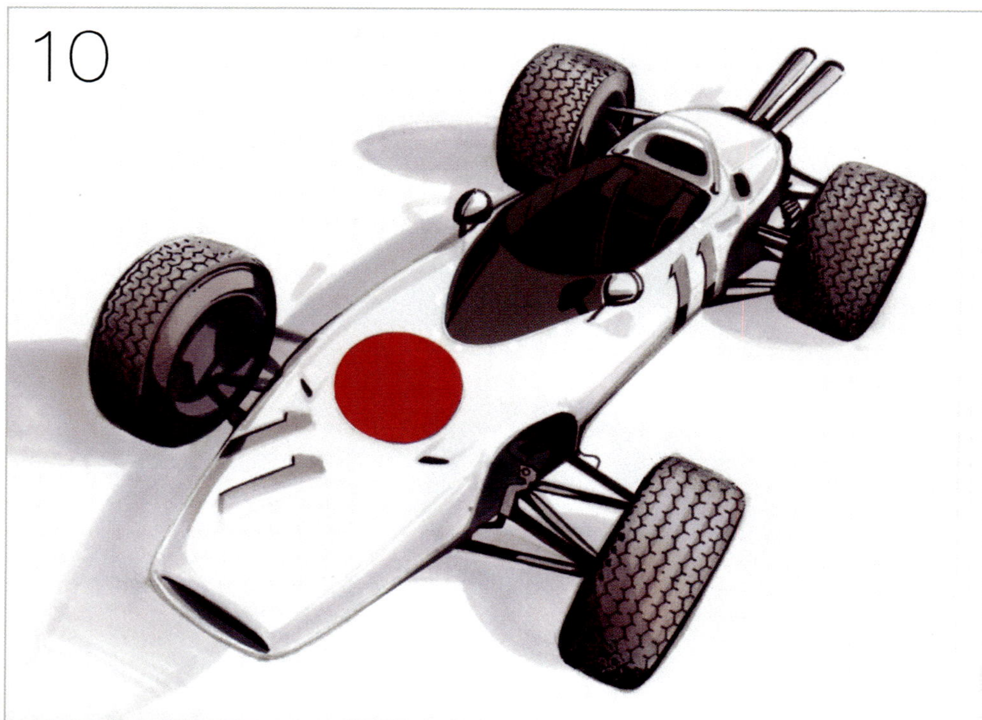

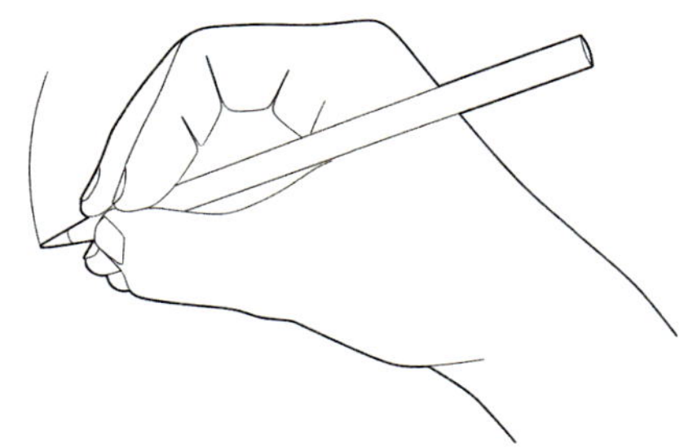

Returning now to the fine black pen, begin detailing the cockpit with shoulder belts, shadows, chrome side mirrors and windscreen reflections.

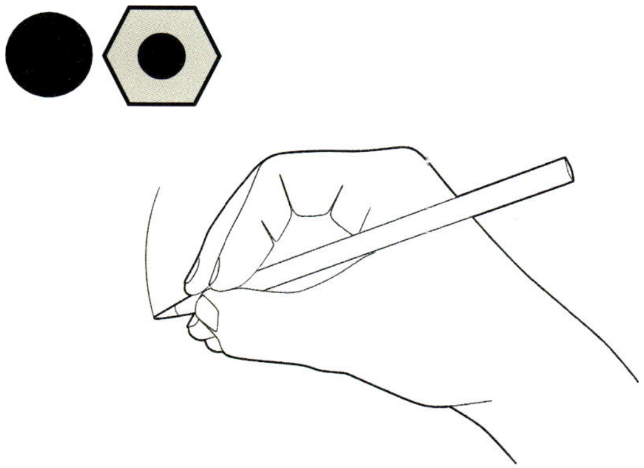

Similar to the cockpit, begin adding some smaller details along the fuselage of the car: The brand name, various rivets, etc.

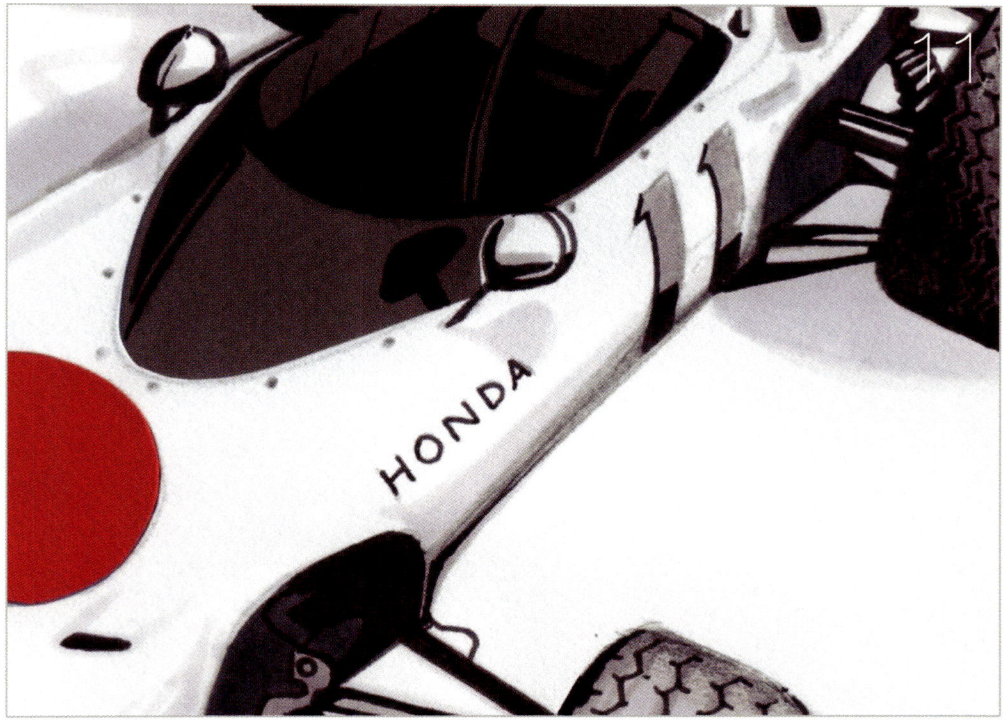

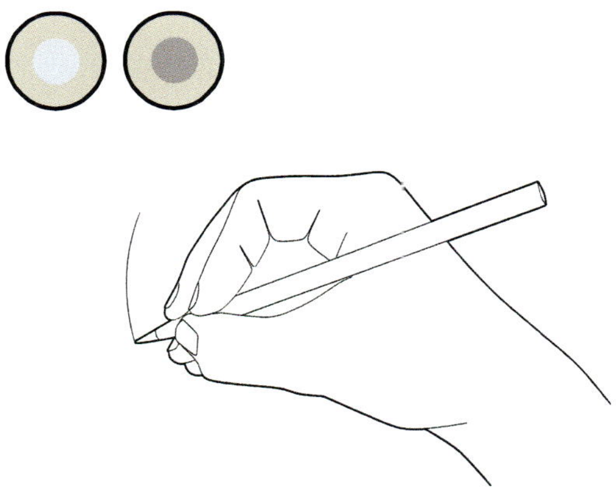

Instead of going straight to white pencil for the highlights, try Cloud Blue, or perhaps a Cool Grey. This will create an interesting color contrast on the tires.

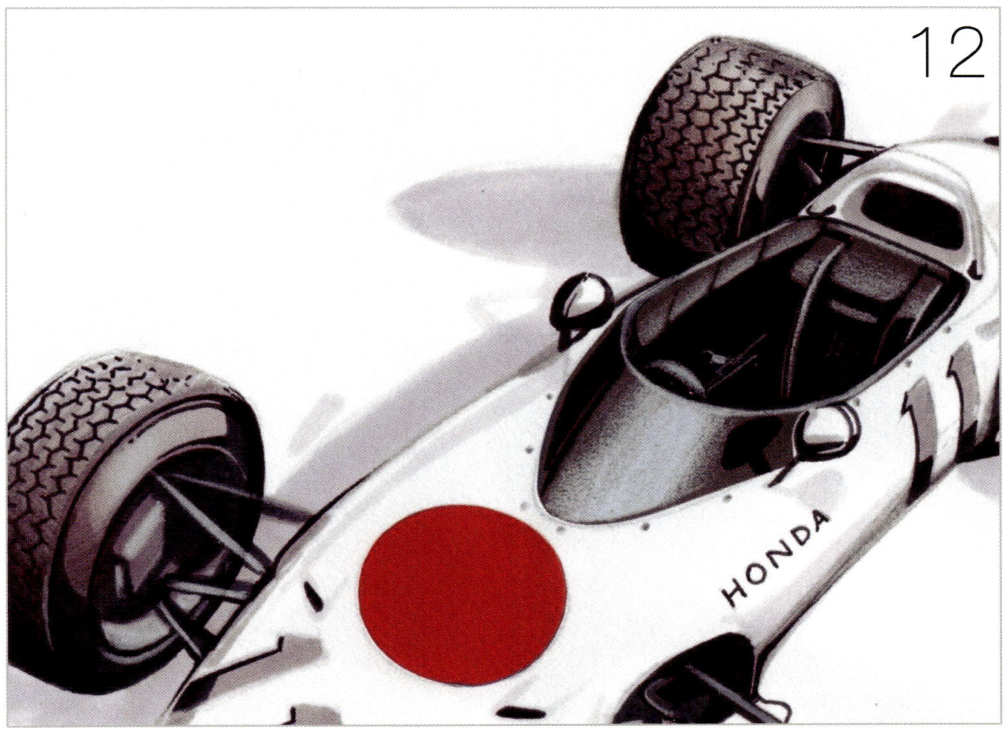

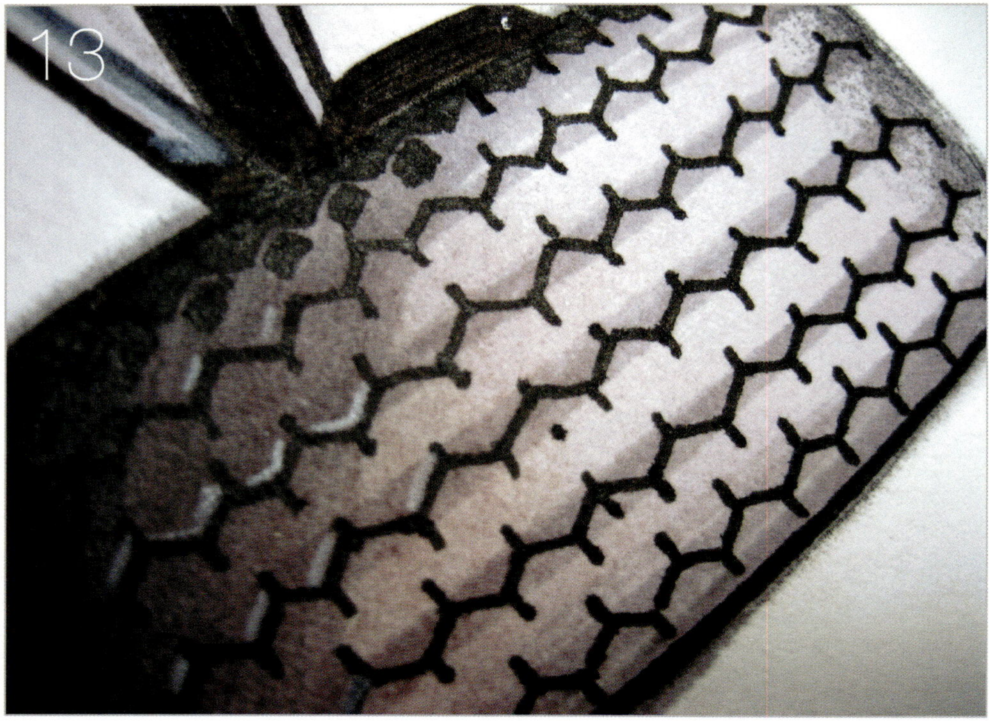

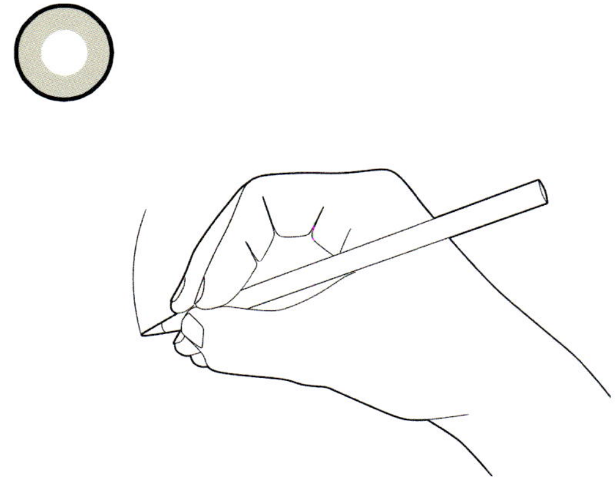

Use a sharp, white Prismacolor pencil to highlight the edges of the tread.

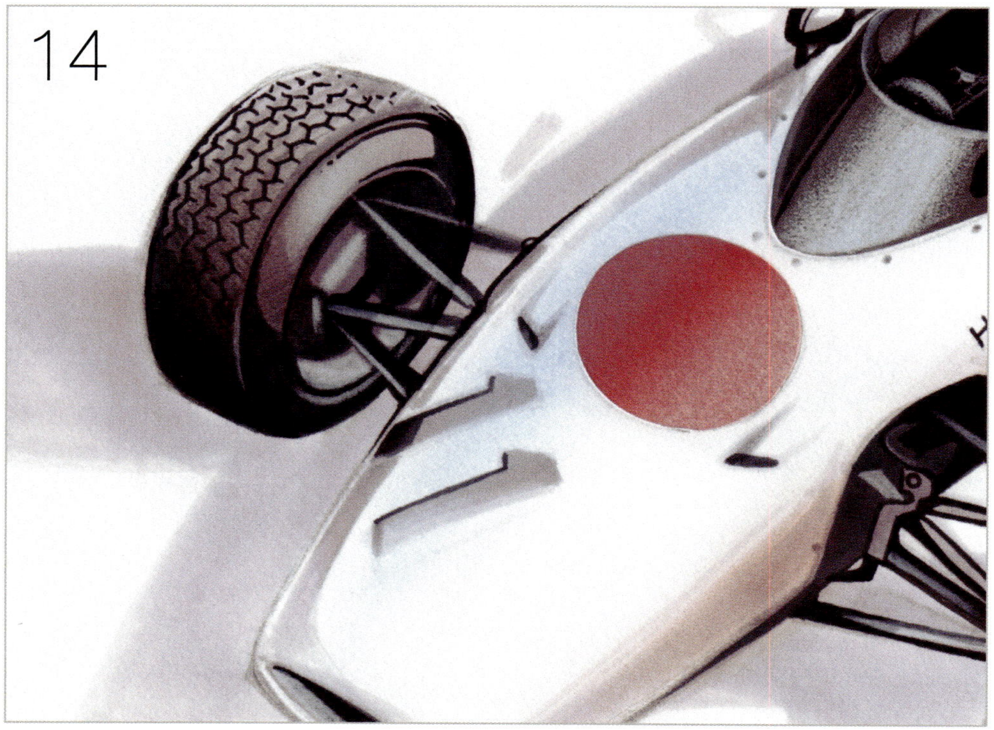

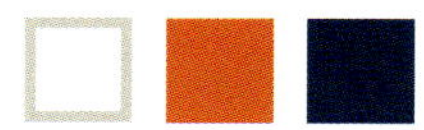

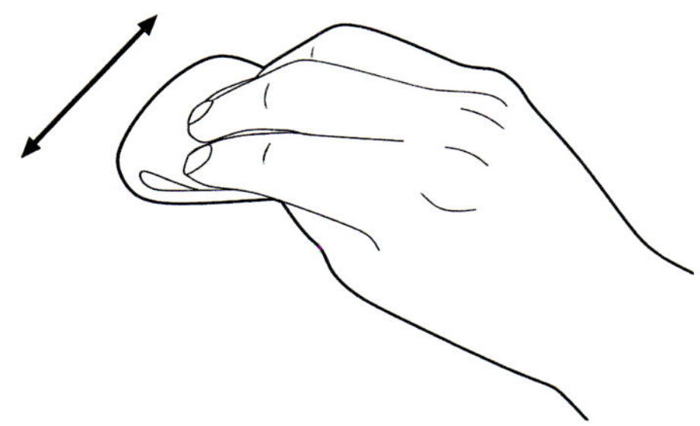

Similar to the previous rendering, ghost in blue pastel for the sky tone. Then add some orange down below for the earth tone. Use white in the middle to give the rising sun a bit of shape.

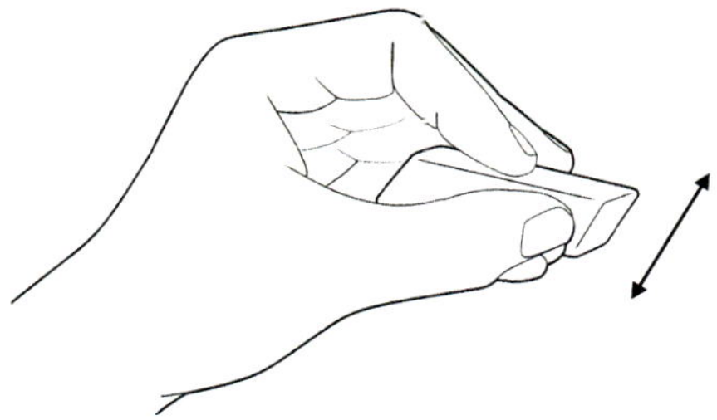

Take a clean eraser and bring back a corridor of the red Canson. Now you have a good balance between graphic punch and surface description.

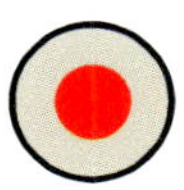

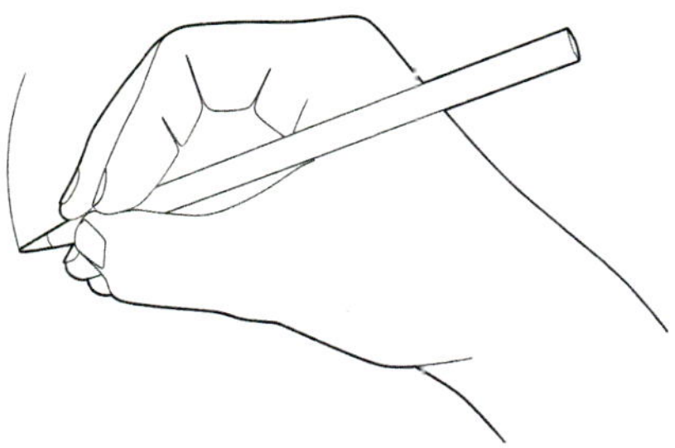

Add some minimal red details to the cockpit. These red accents will help tie this area to the rest of the car and keep the rising sun from looking isolated.

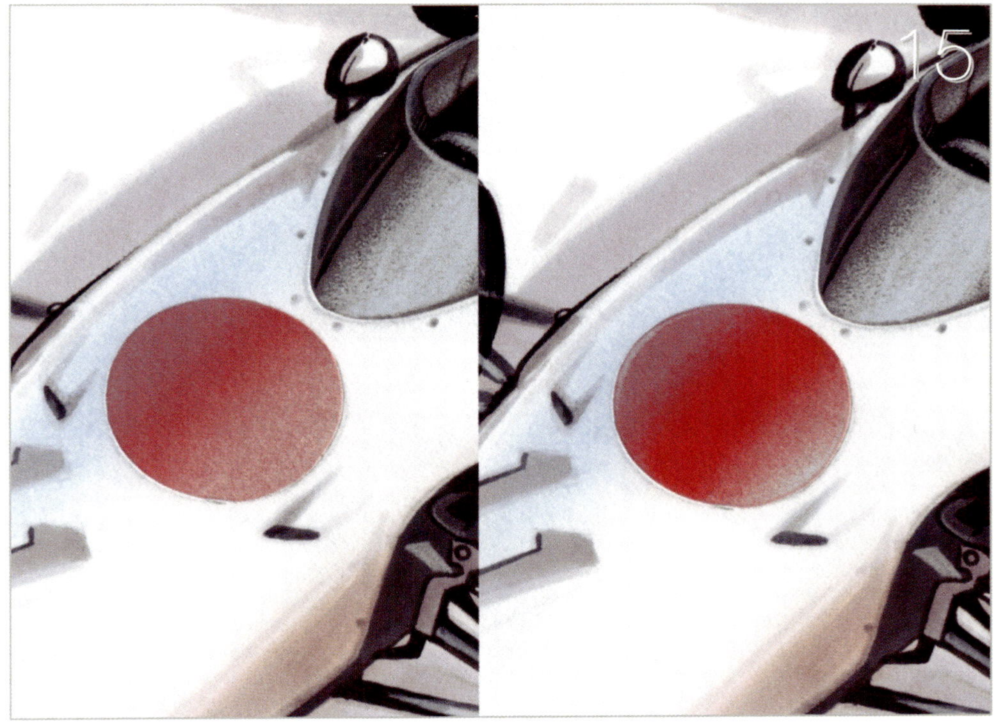

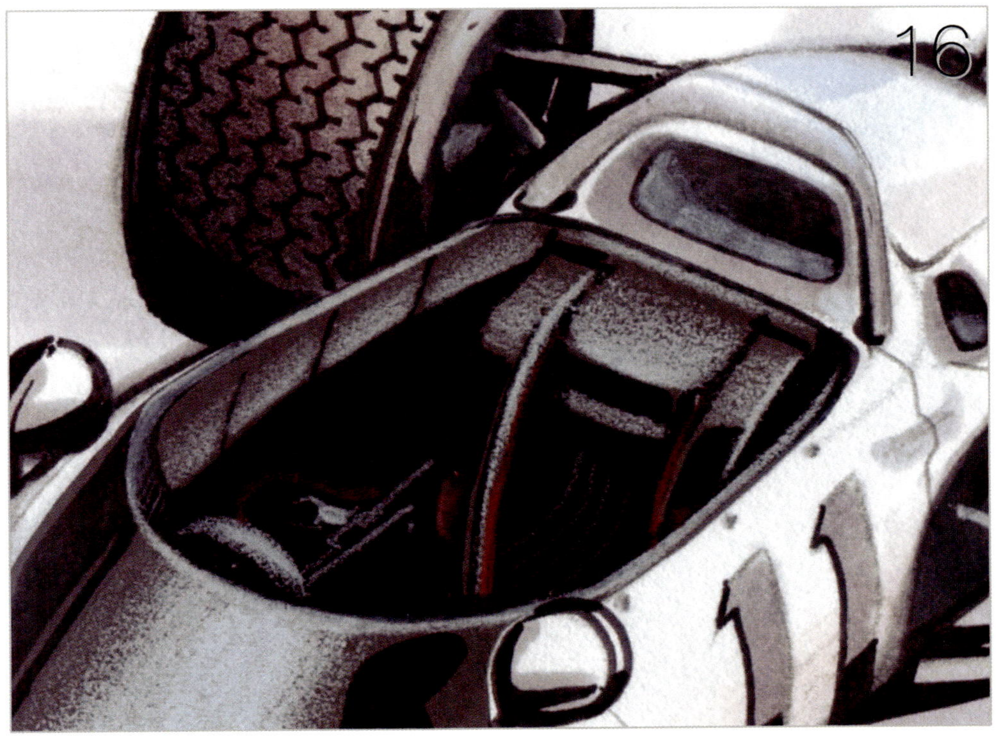

17

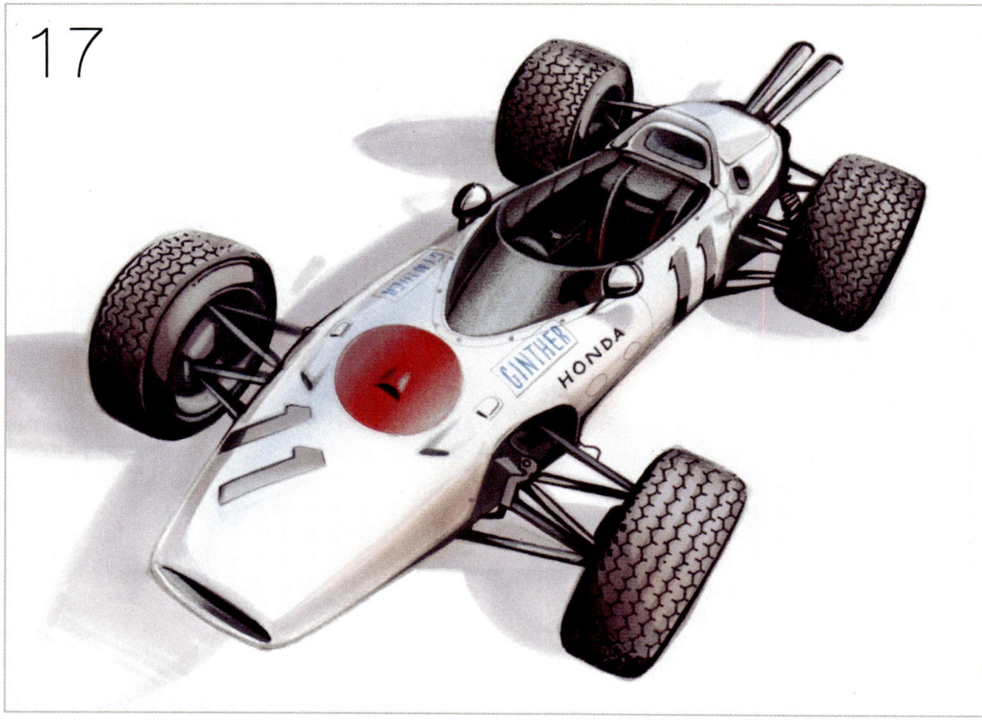

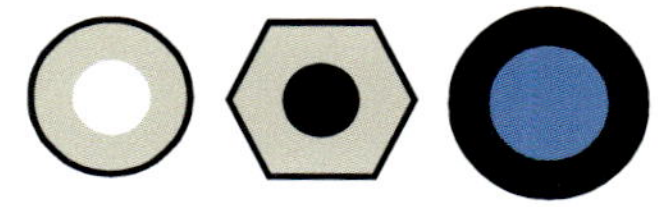

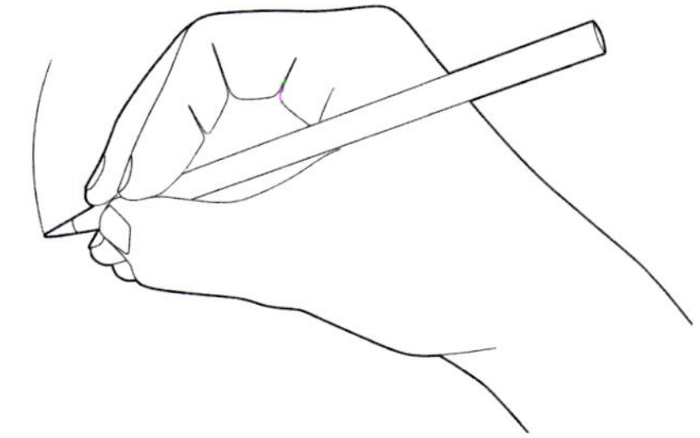

Add one more layer of detailing: Small vents, the driver's name and a few parting lines along the body of the car.

18

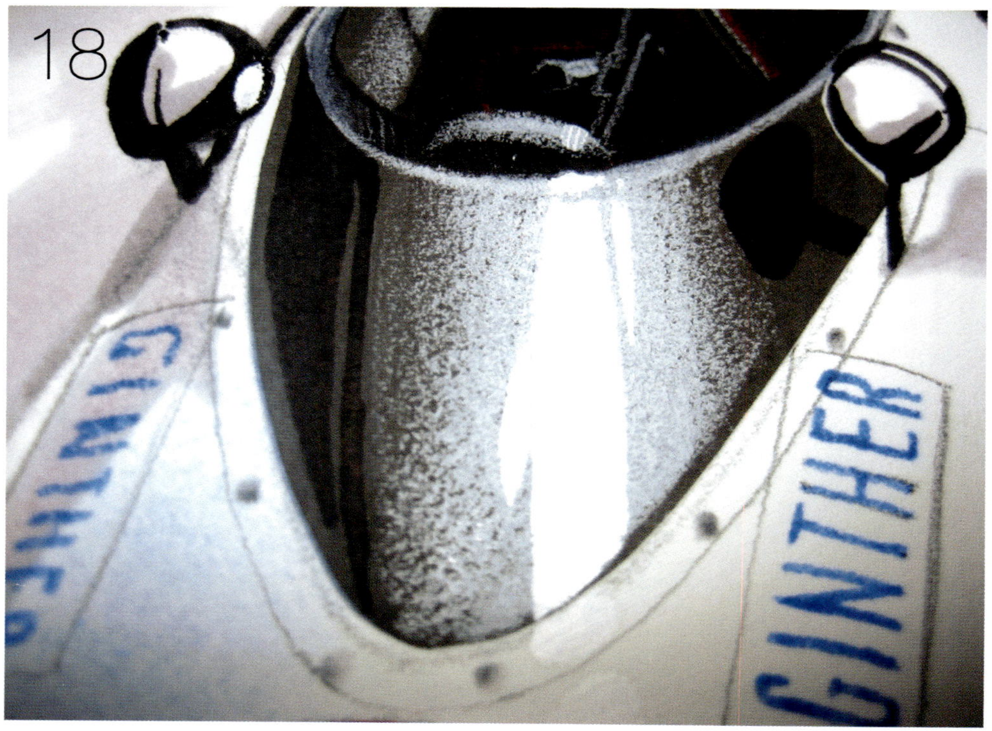

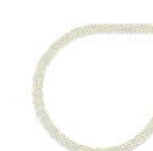

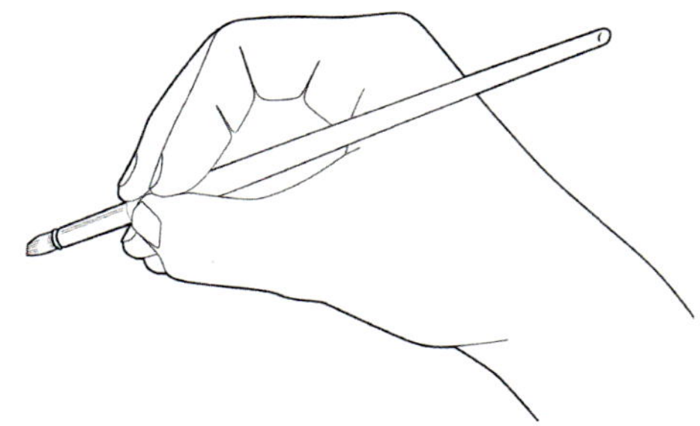

For the white gouache highlights, concentrate on the larger hot spots first, like those on the windscreen. This item is essentially a cropped cylinder, so the highlights should follow the shape closely.

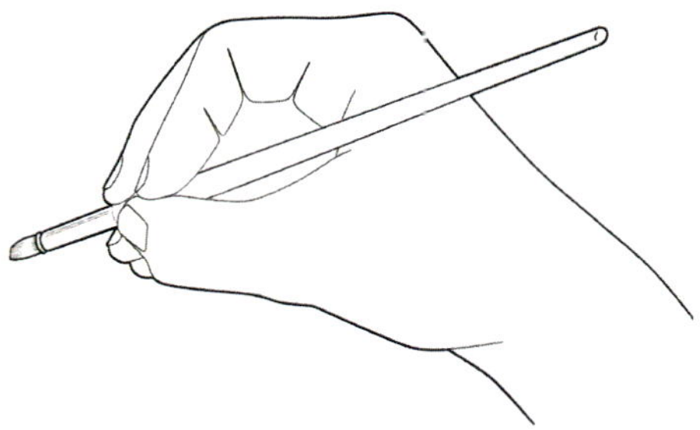

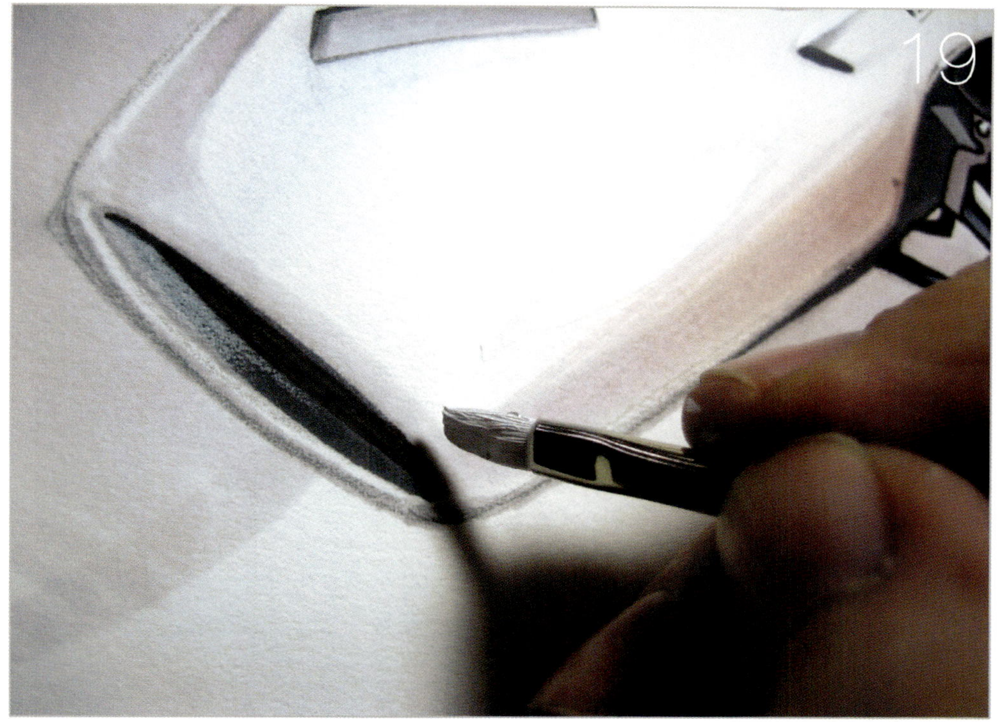

Even though the paper itself is white, highlights will still be visible along the car's horizon line.

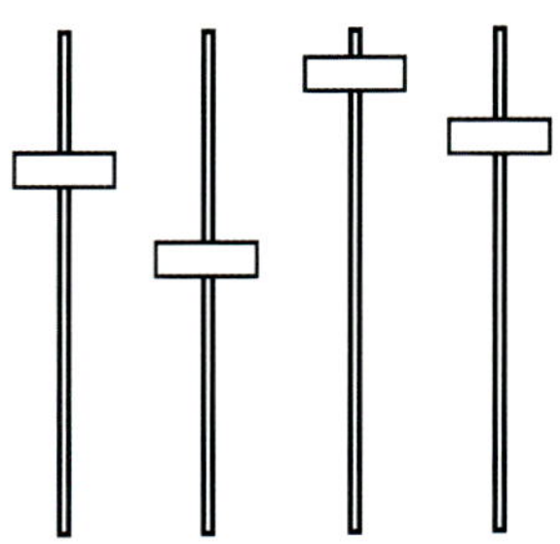

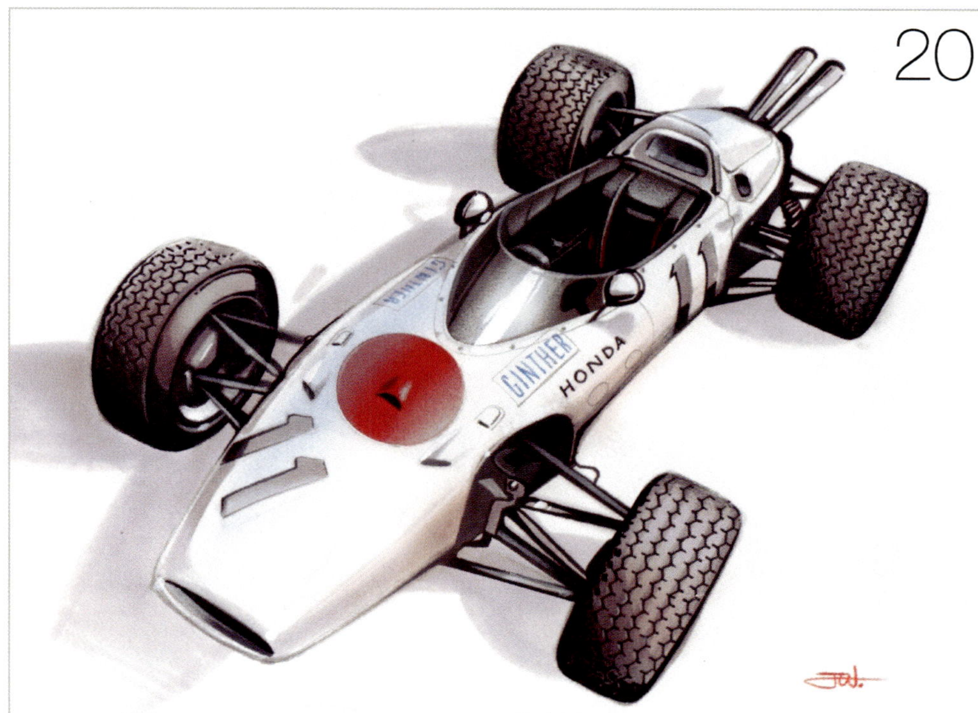

Scan in the rendering and turn up the color saturation in Photoshop. You may wish to clean up some of the highlights with an airbrush tool.

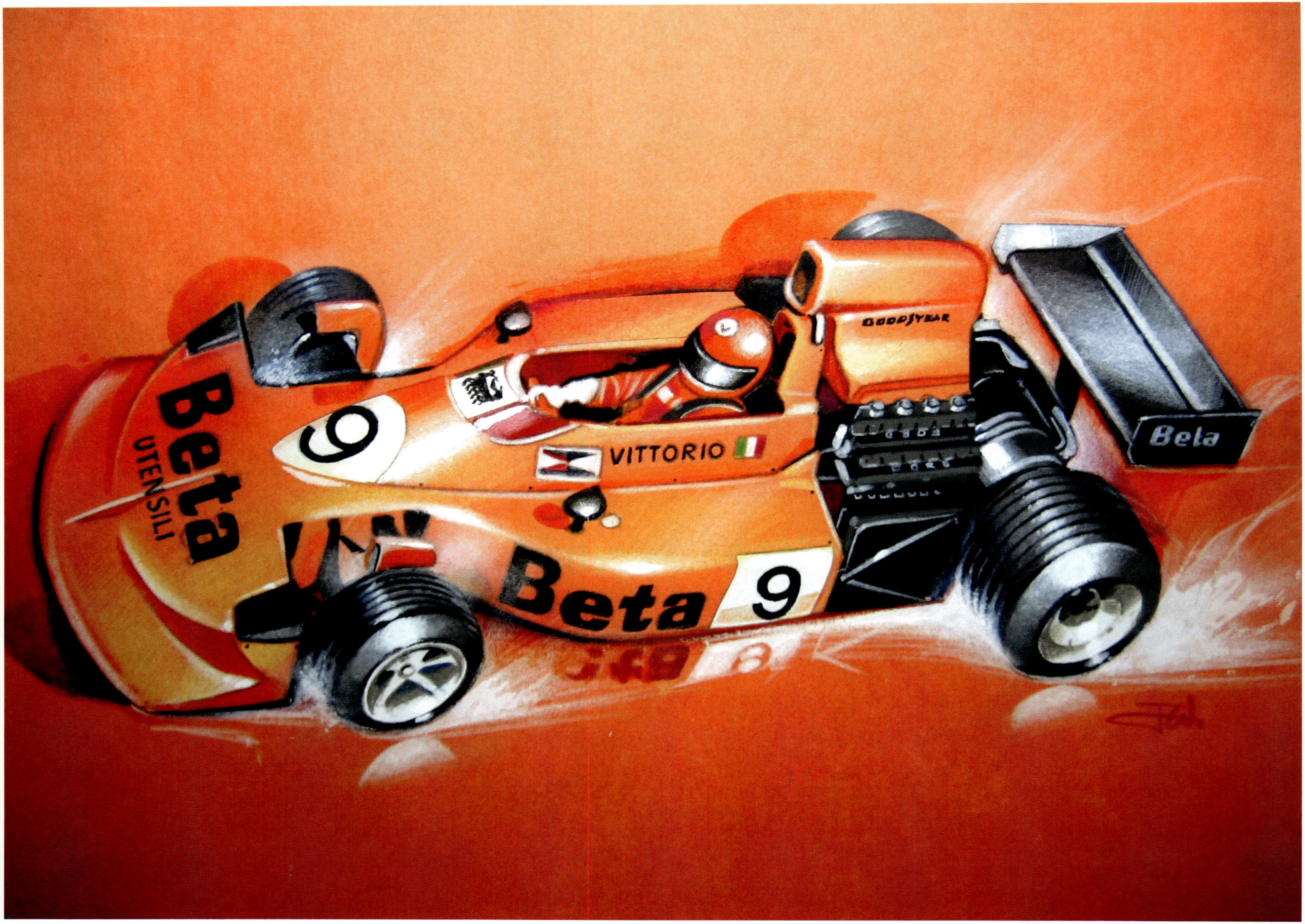
Beta
UTENSILI
9
VITTORIO
GOODYEAR
Beta
9
Beta

14. Orange Velvet

One of the drawbacks of Canson is the deep grain. Unless you sharpen your pencils frequently, your linework will end up looking like crayons. So, here's an alternative: construction paper. This soft, velvety surface yields nice results – provided you are patient and use a light touch.

The Tools

Nupastels: White, Yellow, Red.

Prismacolors:
White, Cool Grey 20%, Dark Green, Black, Yellow, Crimson Red, Mulberry, French Grey 20%.

Copic Pen:
Black.

Verithins:
Black.

Gouache:
White.

Markers: Crimson Red, Chrome Orange.

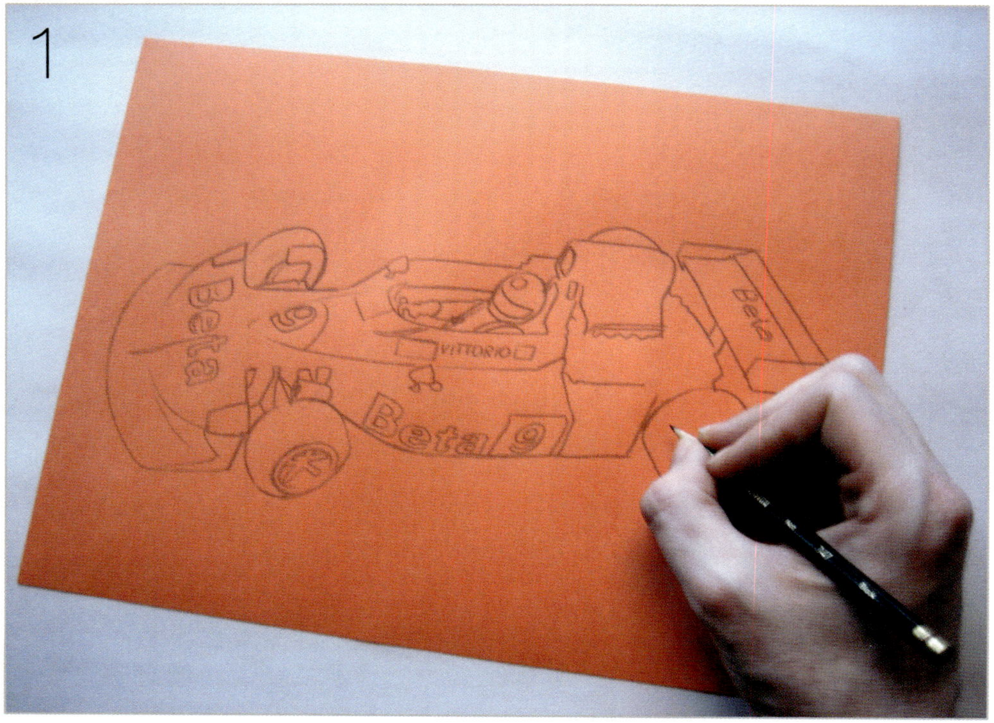

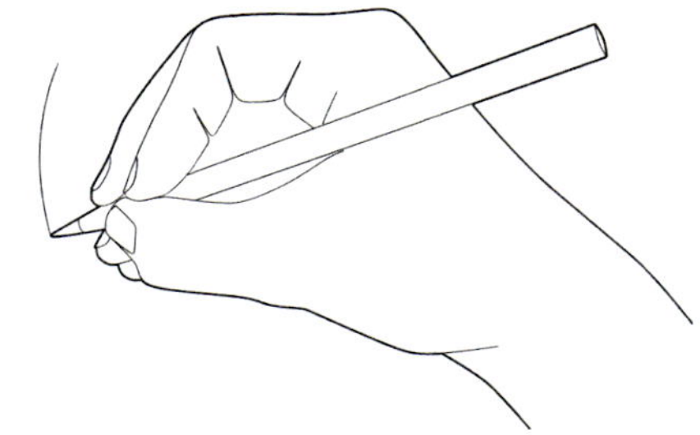

As with the previous rendering, try to keep your lines clean and concise.

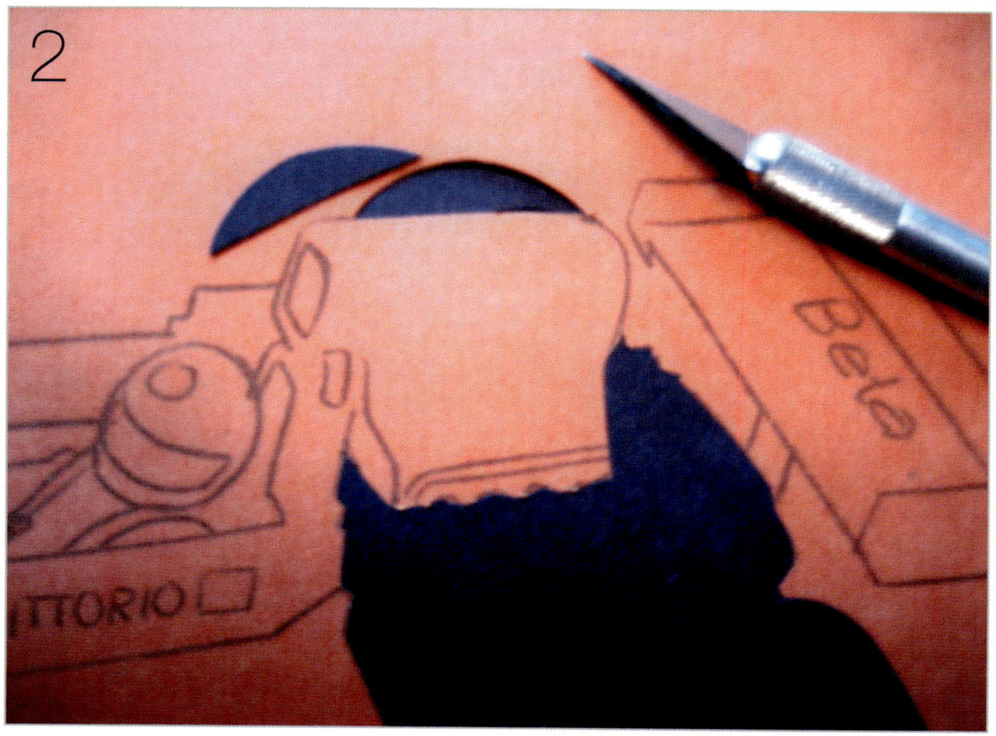

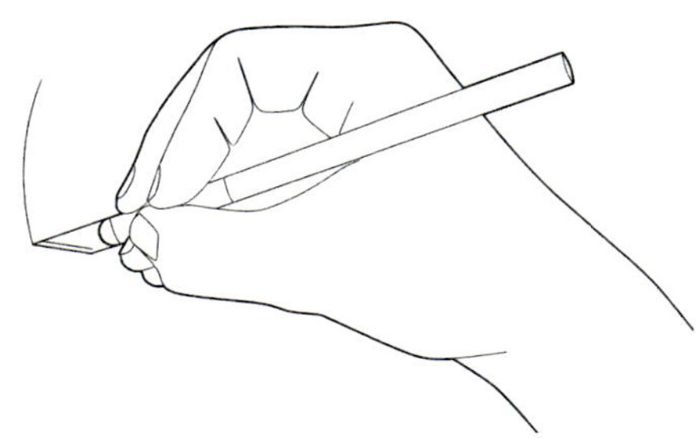

Construction paper is an ideal candidate for splicing in color. This is due in large part to its delicate nature, but also because it is particularly easy to slice through multiple layers.

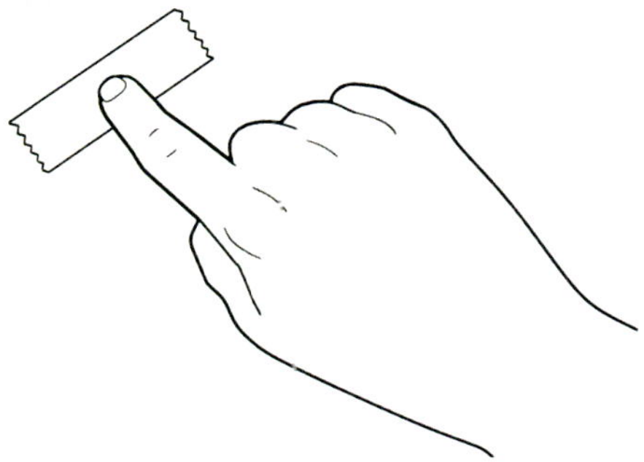

Since construction paper is thinner than Canson, use clear tape to secure the pieces instead of masking or painter's tape. This way, you'll avoid unwanted "witness lines" on the front.

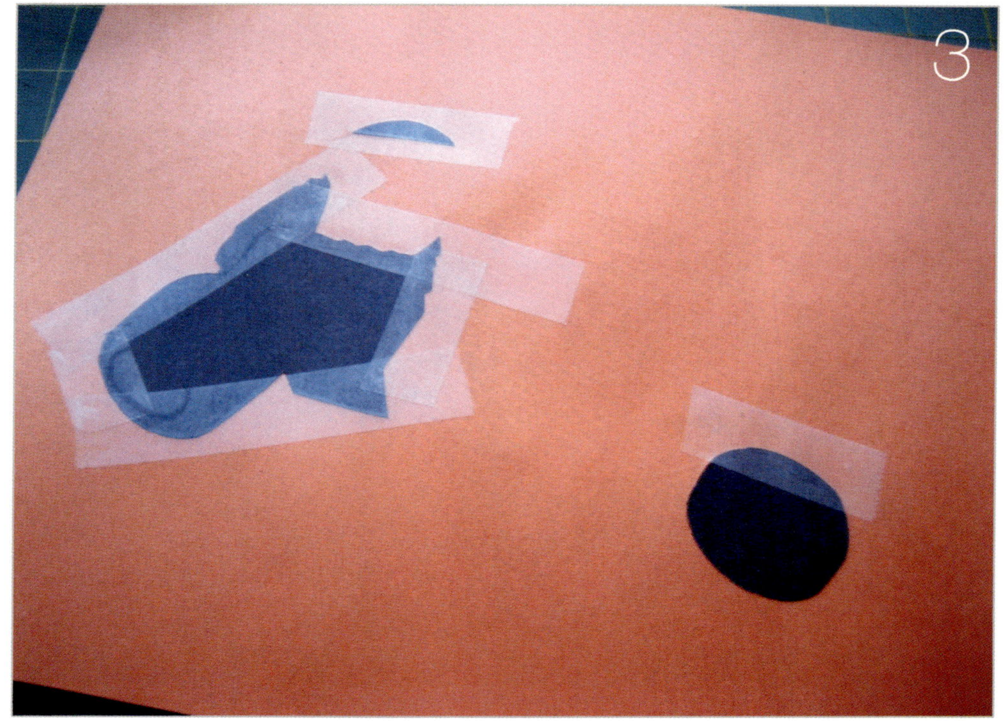

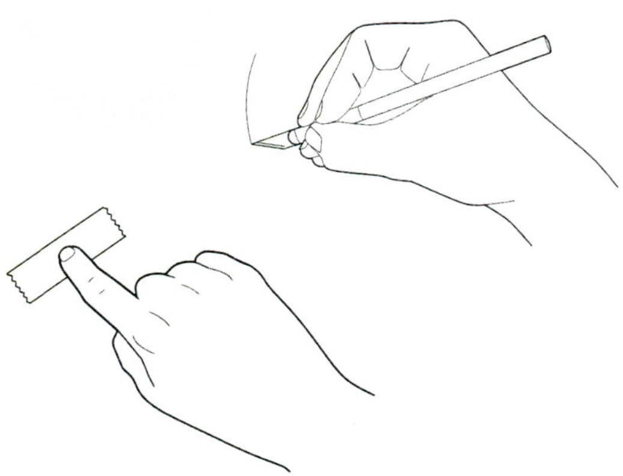

Once the darkest areas are all spliced in and secured, execute the same operation for the white graphics.

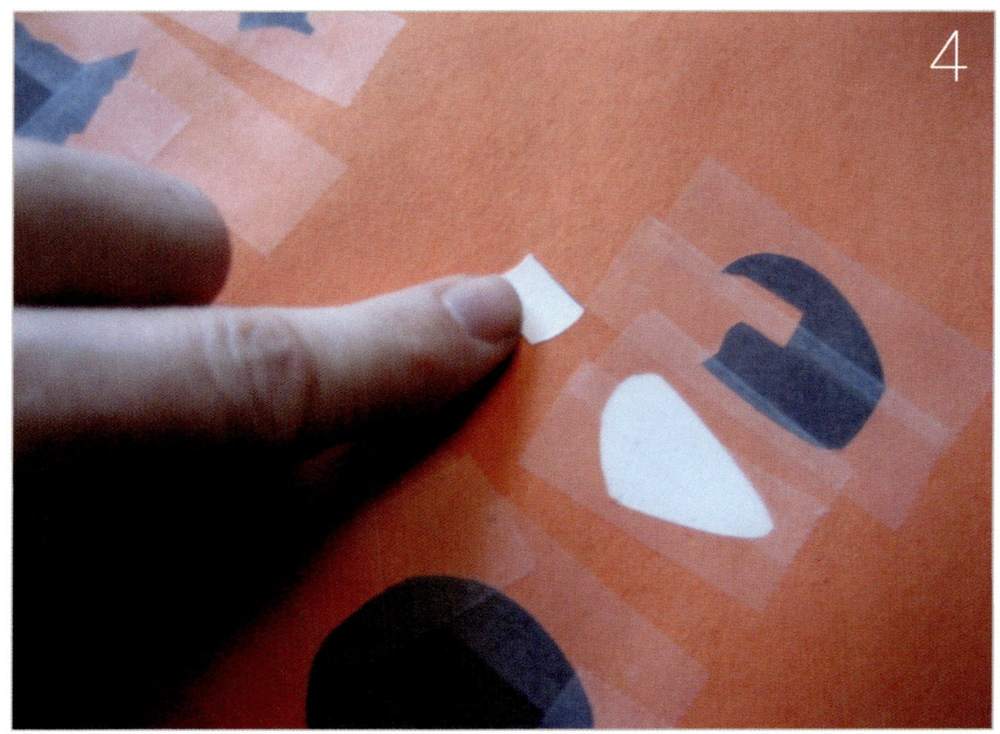

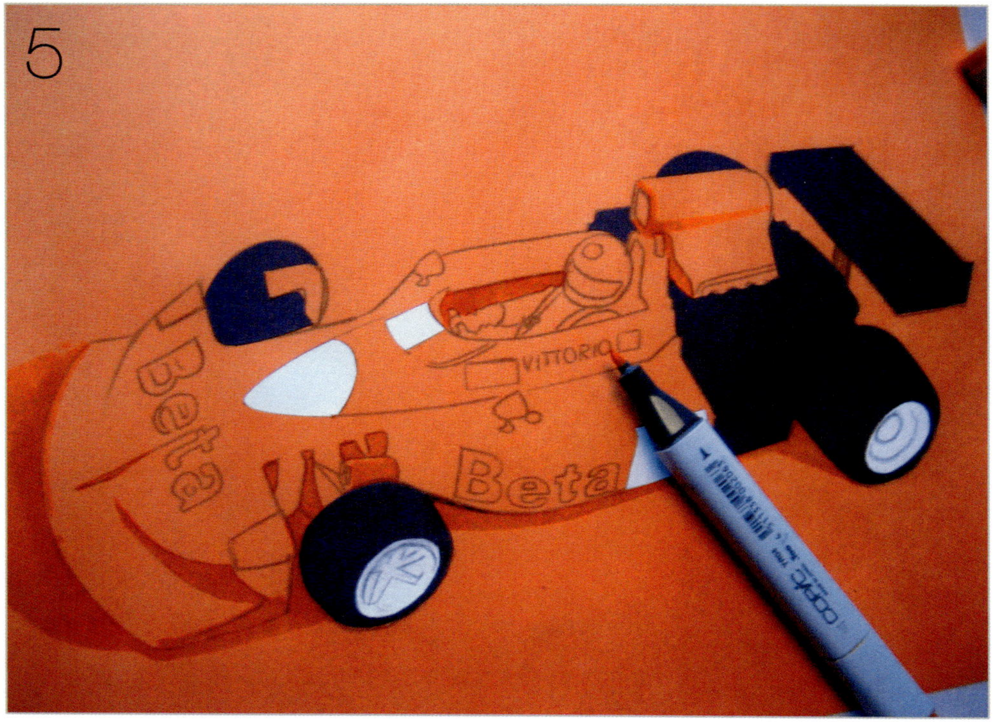

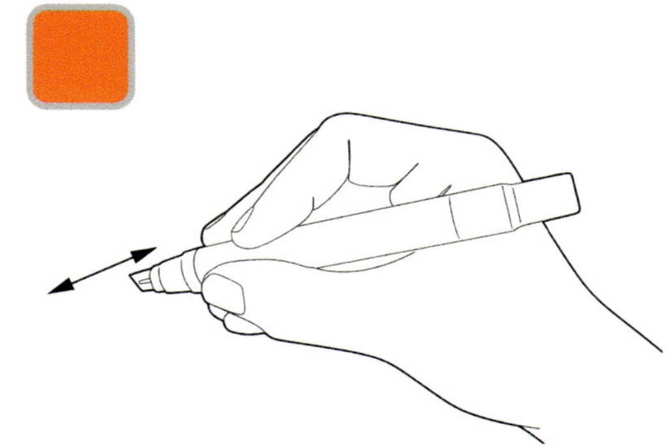

Begin blocking in the shadows and reflections with an orange marker.

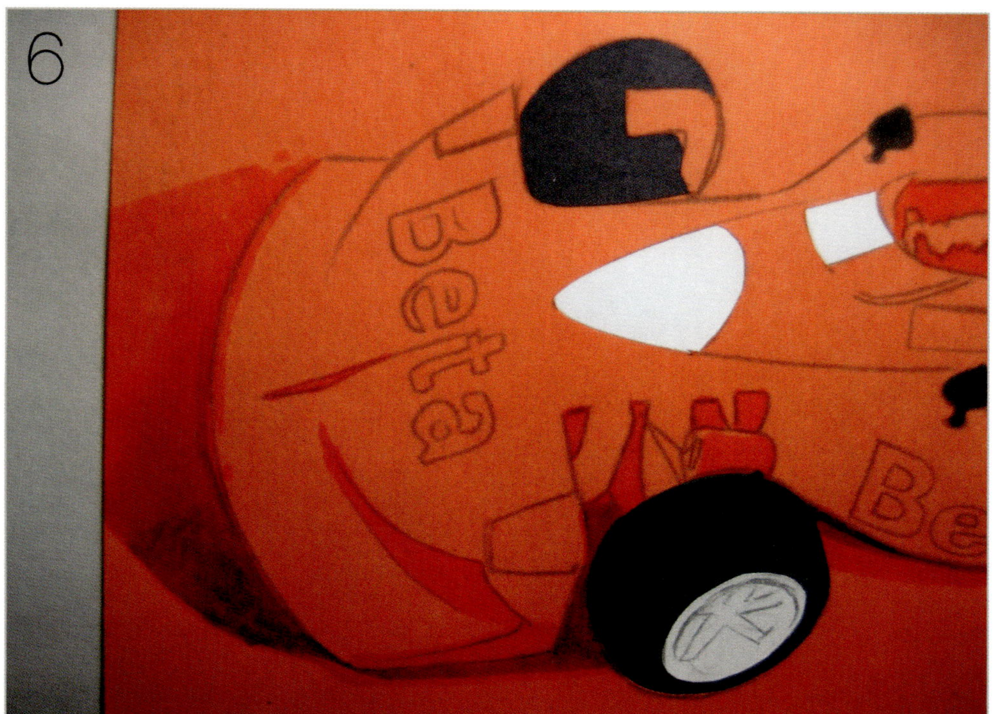

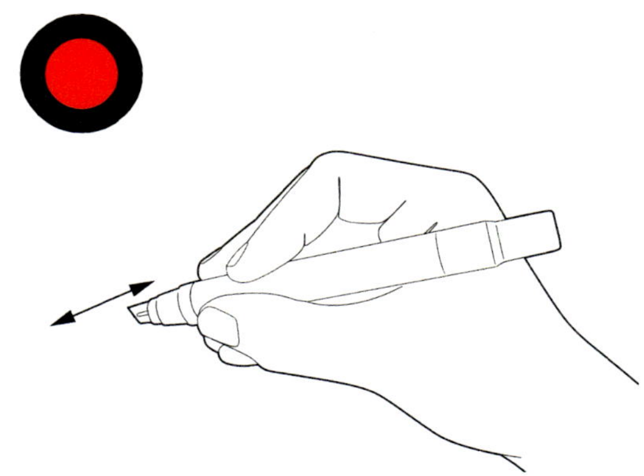

Use a red marker to add depth to the shadows. If the marker picks up some of the black pencil, this will only add to the depth.

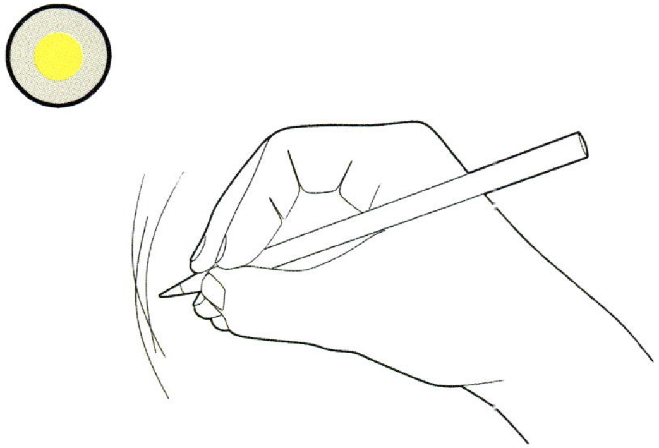

Using yellow pencil for the base highlights will add another dimension to the car's color.

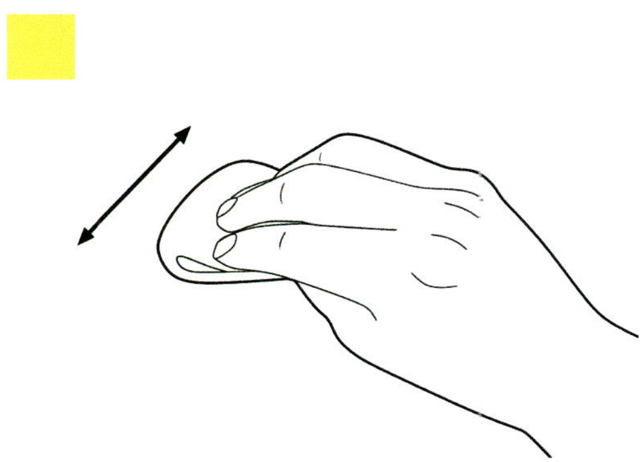

Make the yellow penc l feel more at home by ghosting in scme yellow pastel powder around the highlights.

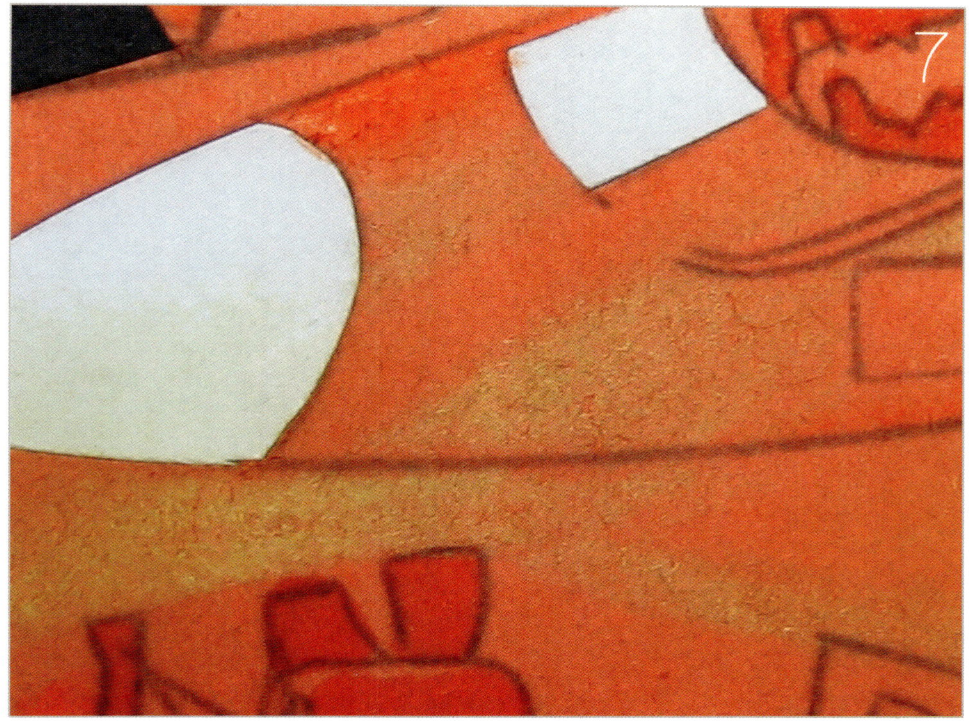

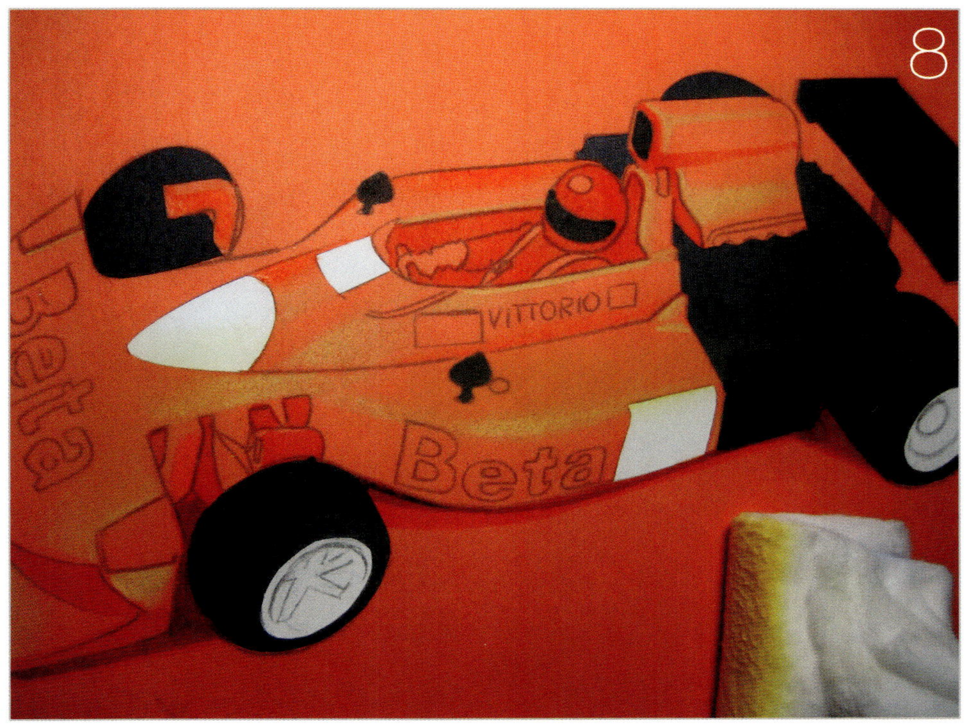

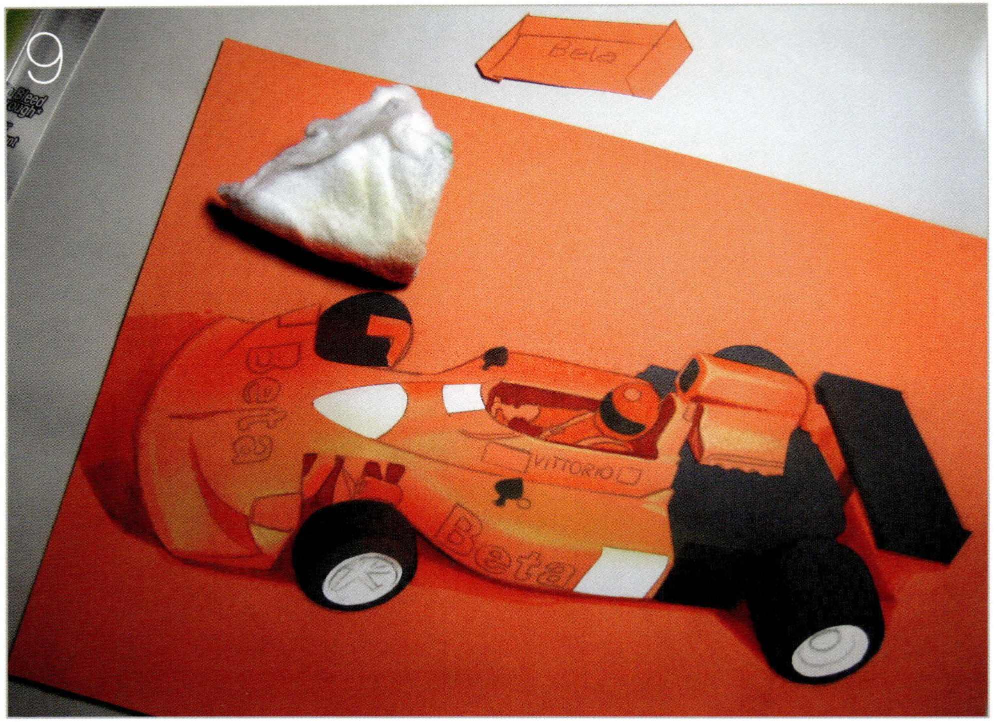

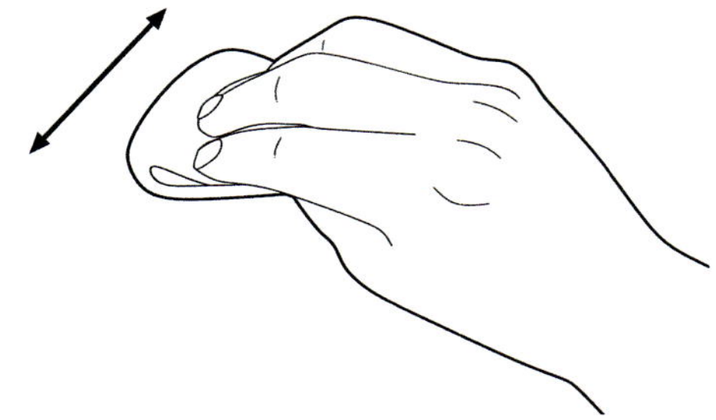

On the far side of the car, add a bit of red pastel. This, combined with the yellow highlights, yields a far better result than you would get with just white pencil and black pastel.

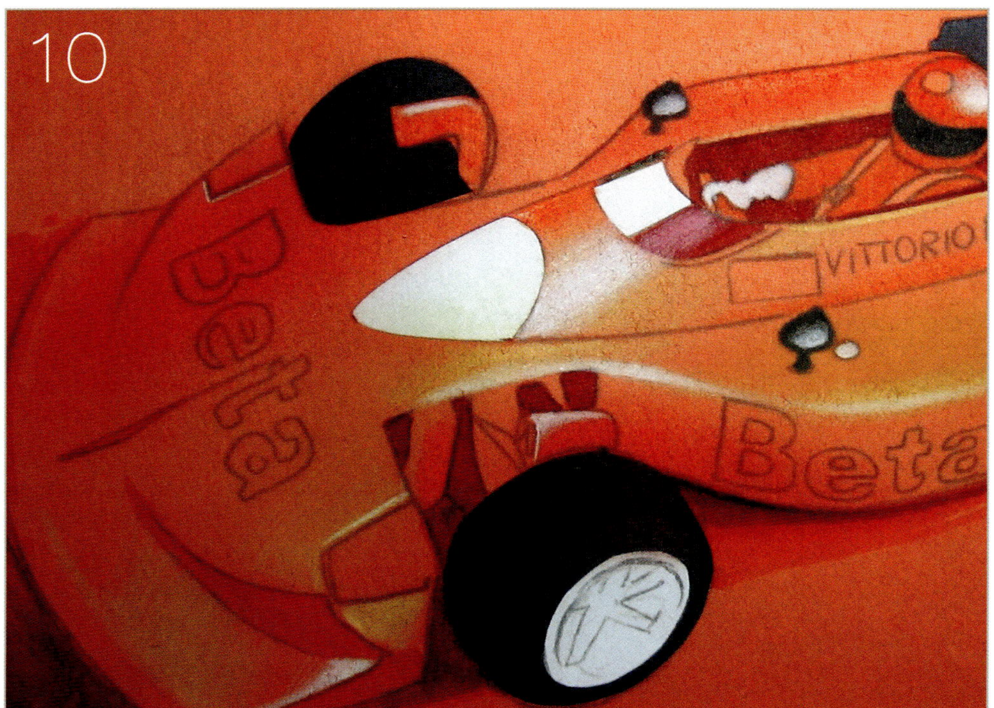

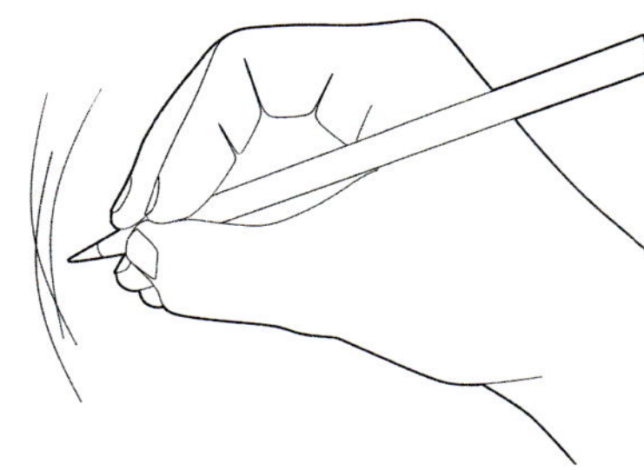

Now it's time for the white pencil. Concentrate most of the highlights on hard edges and part lines. Use a softer, broader highlight on the cockpit.

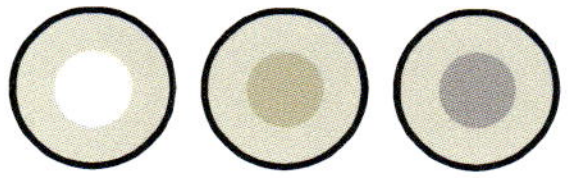

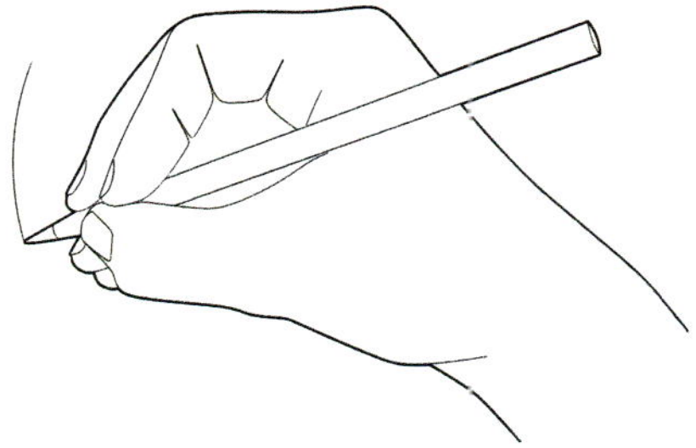

For the tires, use a combination of white and Cool Grey Prismacolor pencils. For the wheels, try a bit of French Grey. This will create a subtle color contrast.

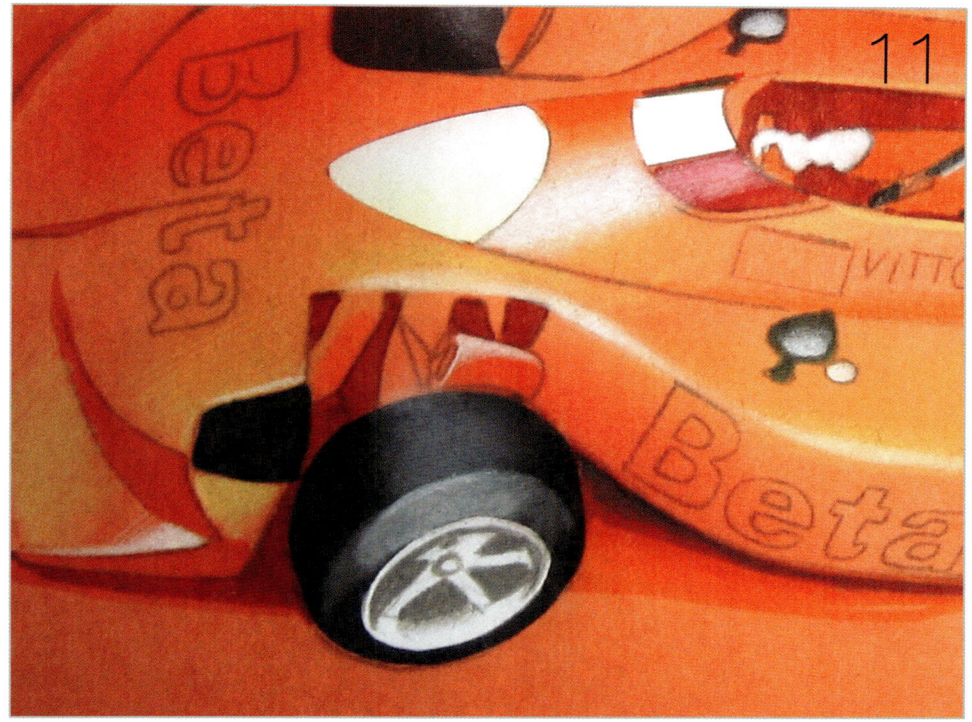

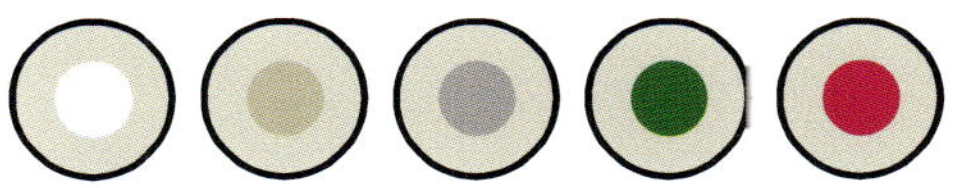

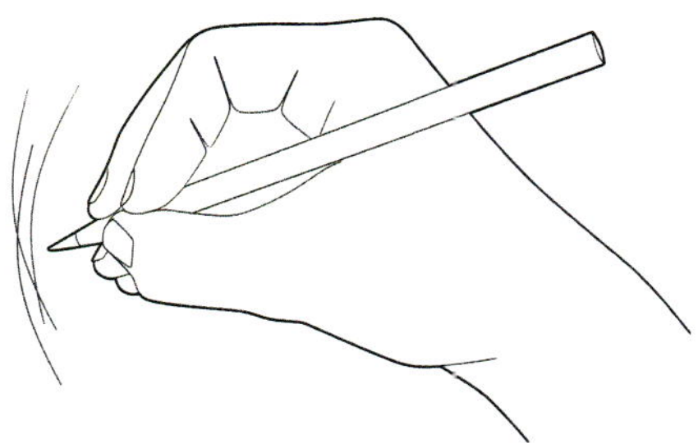

Using an array of pencils, add shading and detail to the back half of the car. Smooth out the highlights with a bit of white pastel if you like.

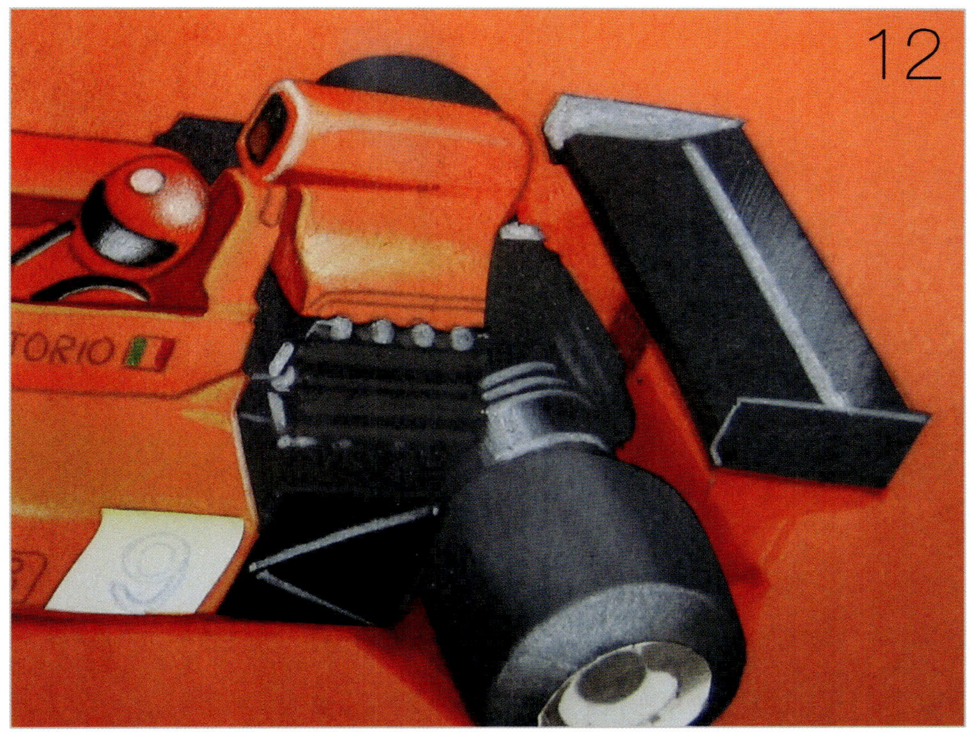

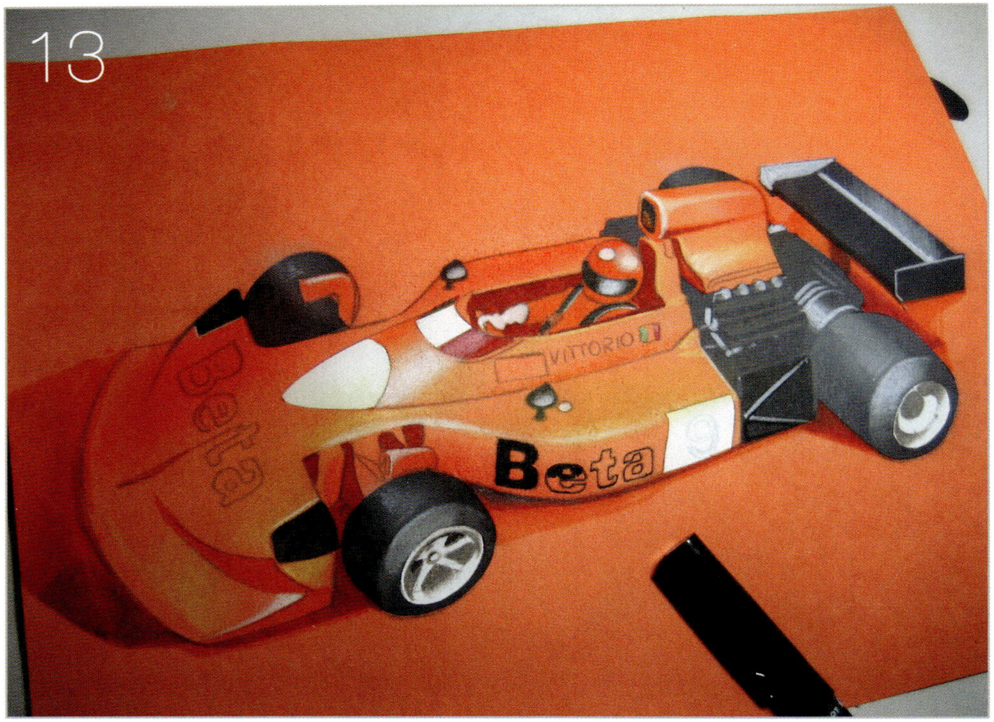

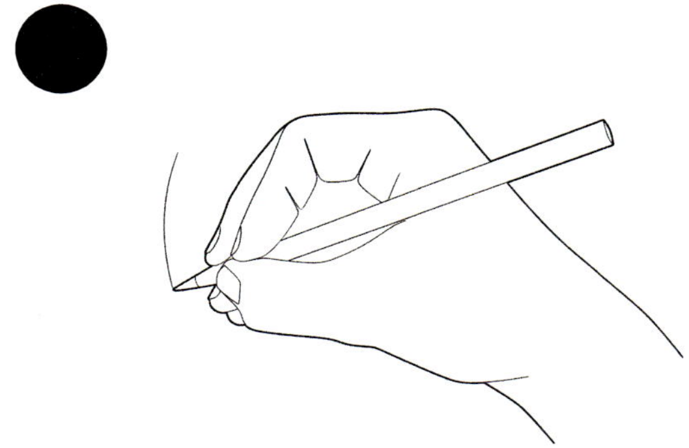

Use a fine-tipped black pen to block in the lettering and graphics.

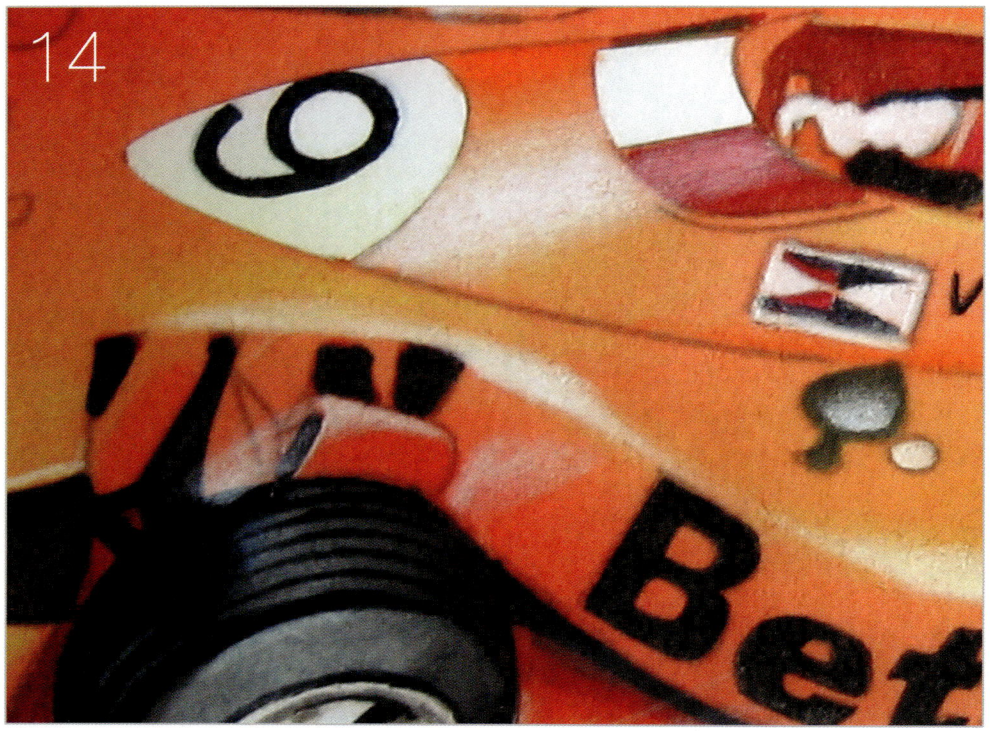

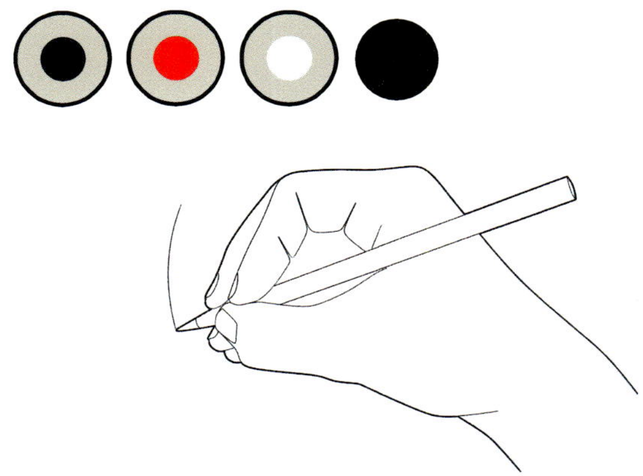

Continue adding details with a variety of tools. The black pen can also be used to lay in the tire treads. Since the wheels here are in motion, you'll only need longitudinal lines.

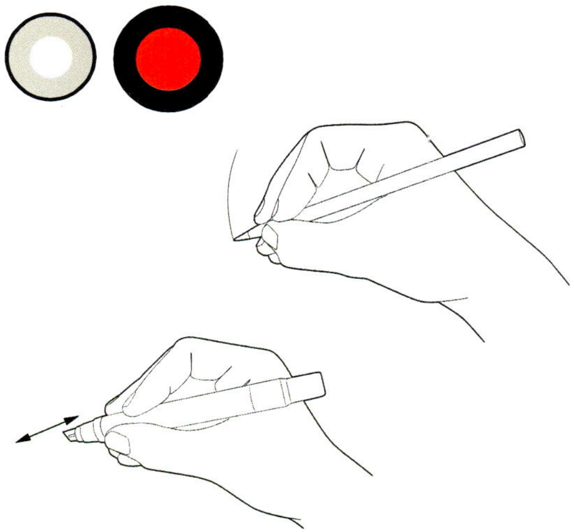

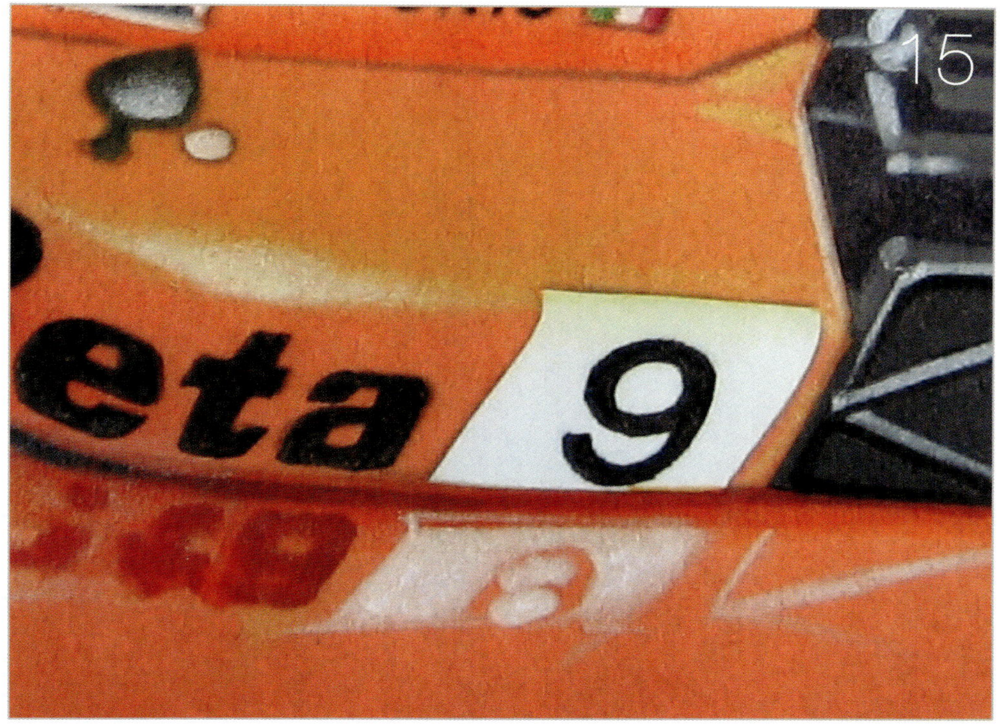

Take a short break from the graphics to consider the road surface. For the illusion of a wet track, use a white pencil and a red marker to mirror the side graphics onto the surface.

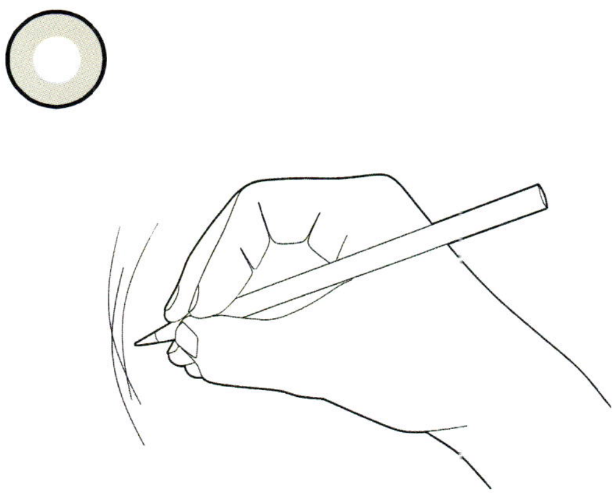

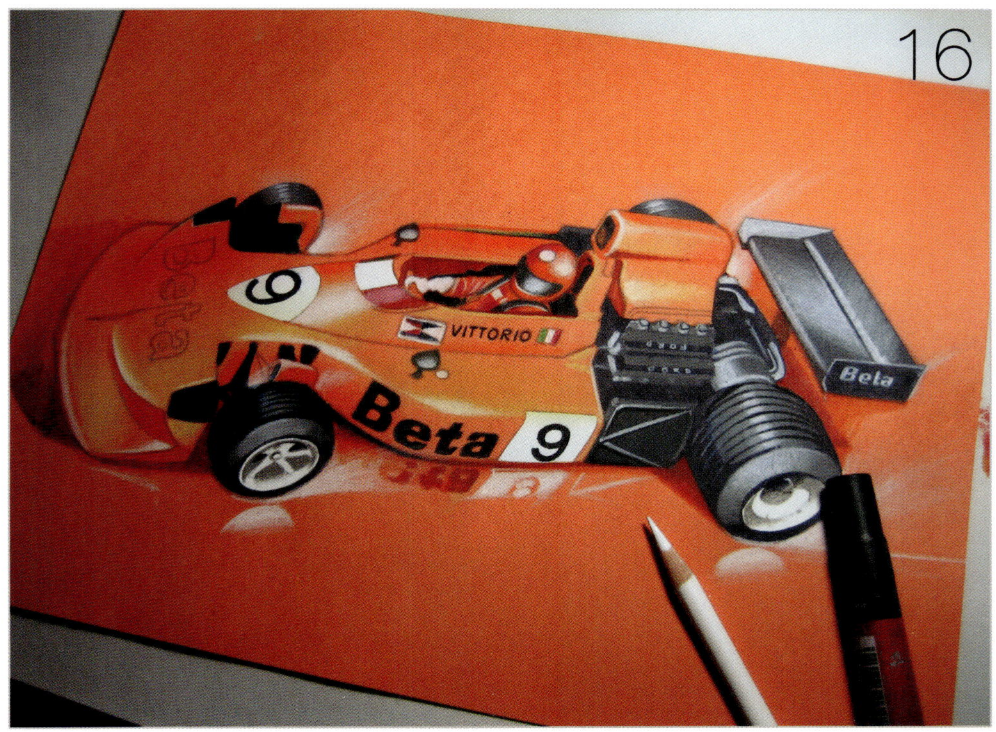

For the water spray, loosely lay in some white pencil behind the tires.

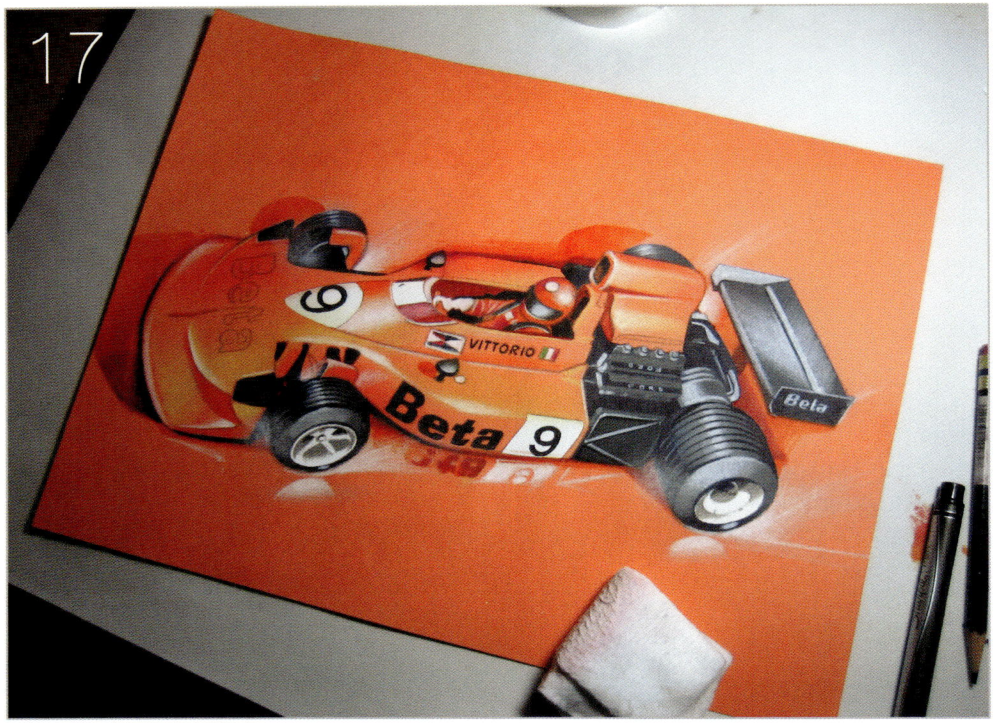

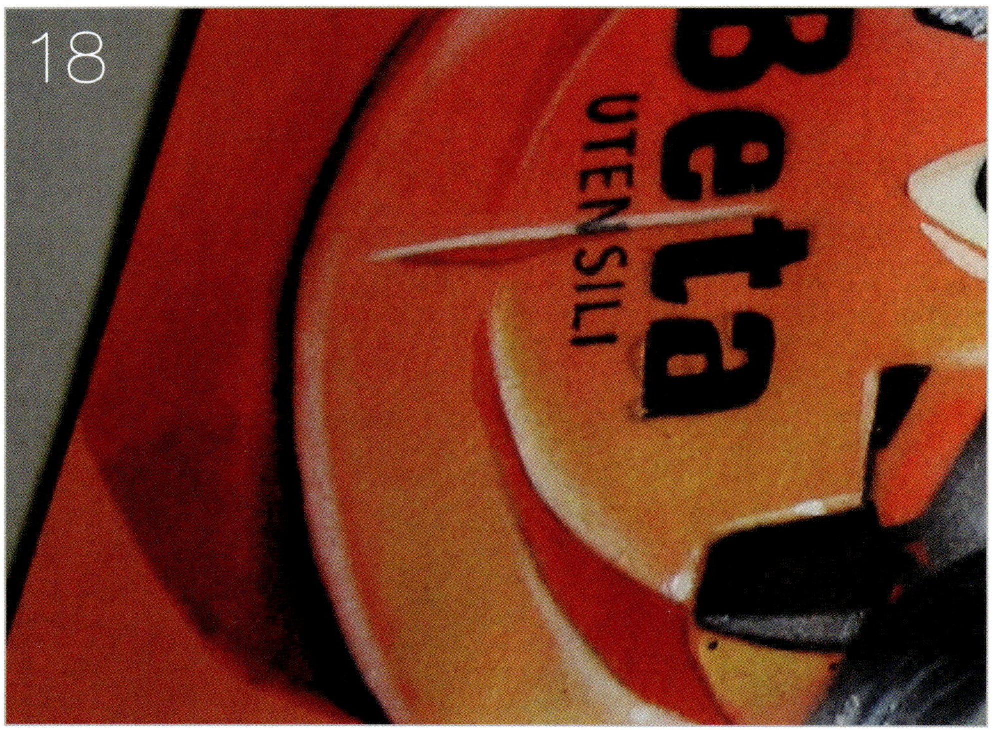

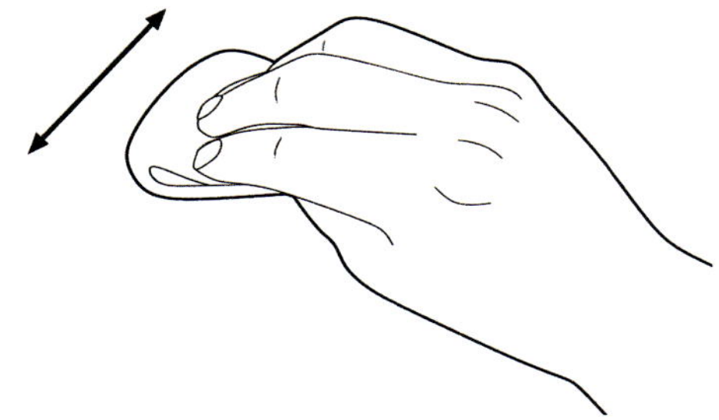

Ghost in some white pastel in front of and behind the tires to give the illusion of mist.

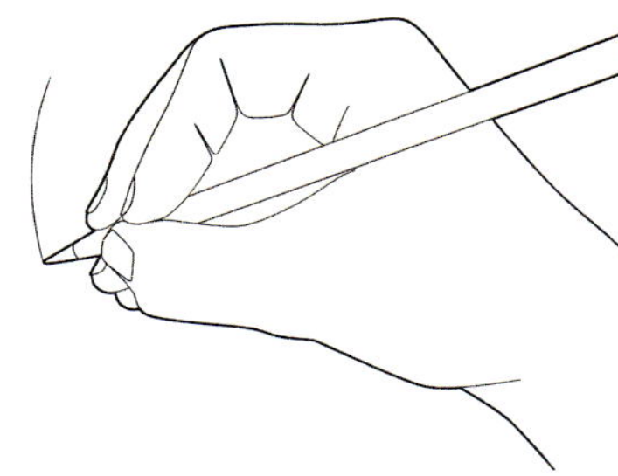

Before you break out the white gouache, finish up the front graphics. Indicate any highlights and part edges that may have slipped by.

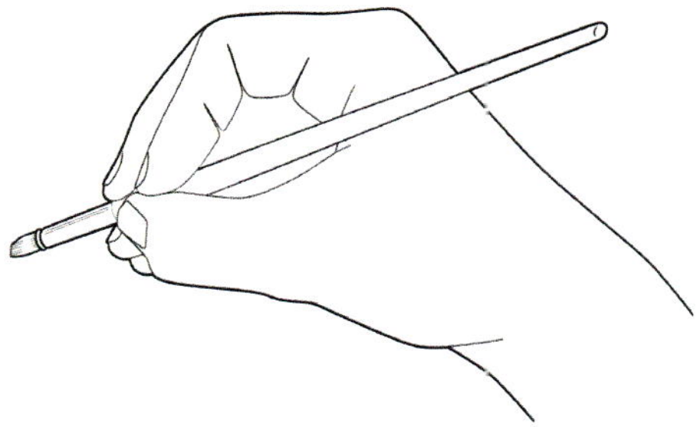

Exercise restraint with the body highlights, but feel free to be expressive with the water spray highlights.

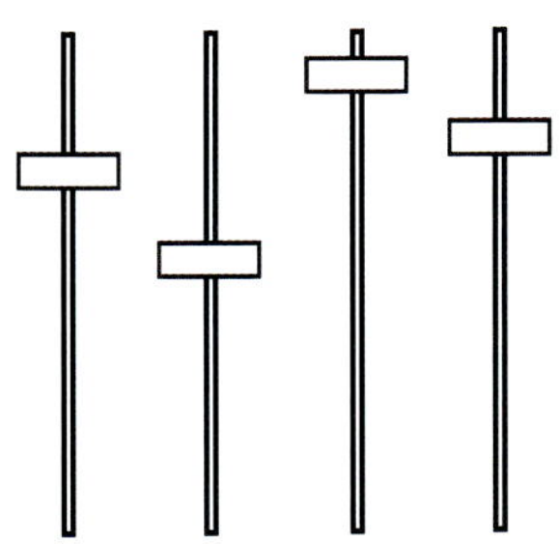

Scan in the rendering; turn up the color levels in Photoshop, if desired.

19

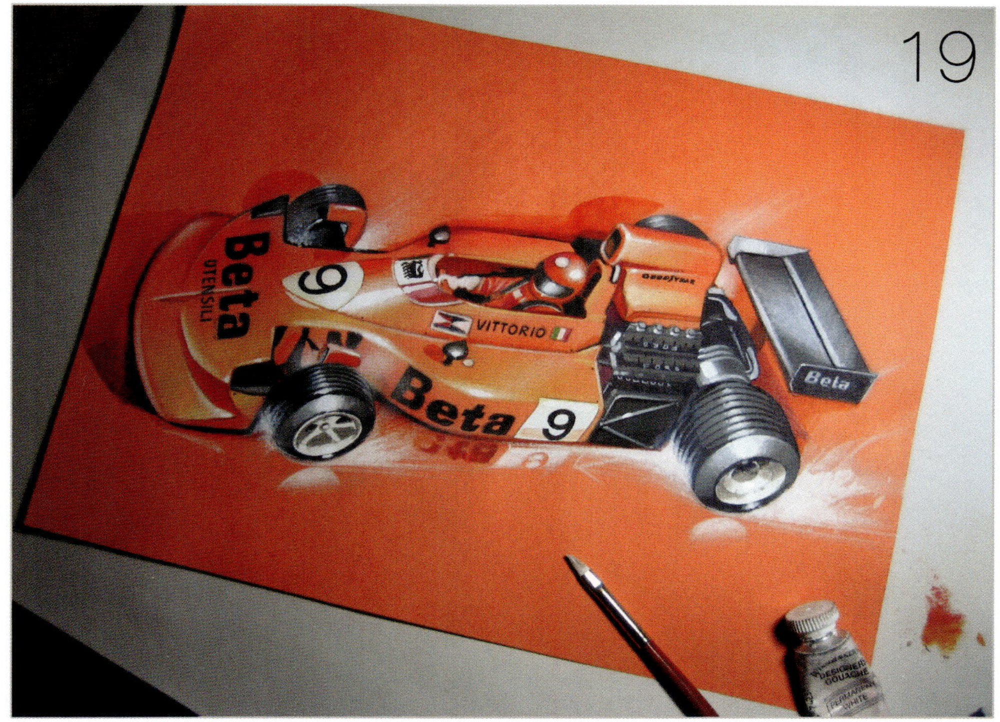

20

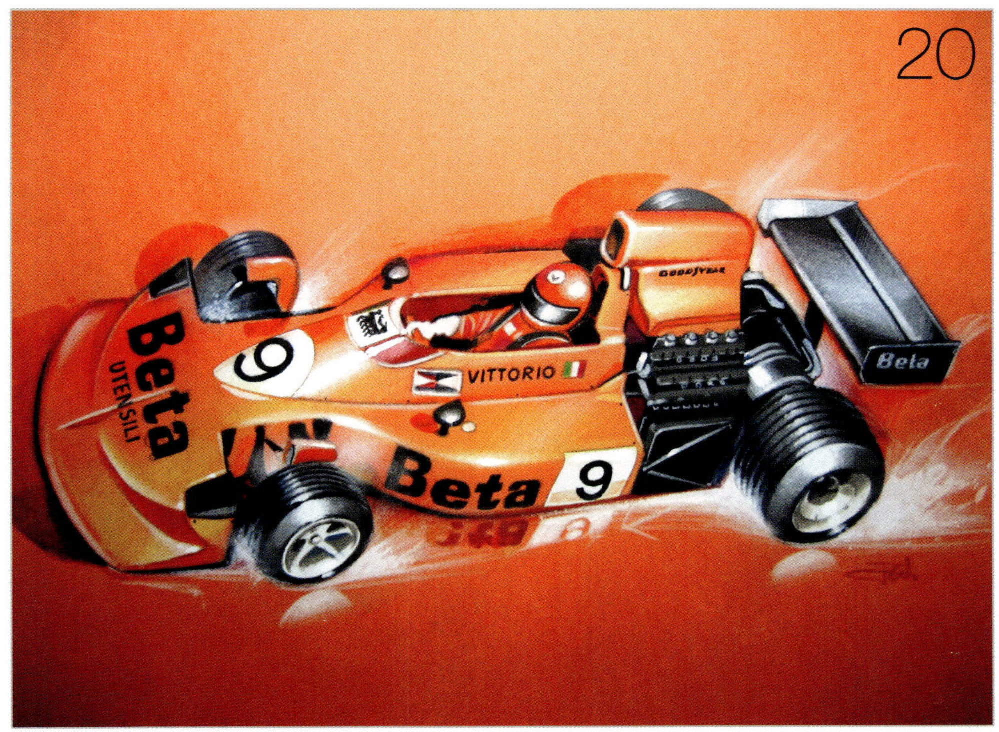

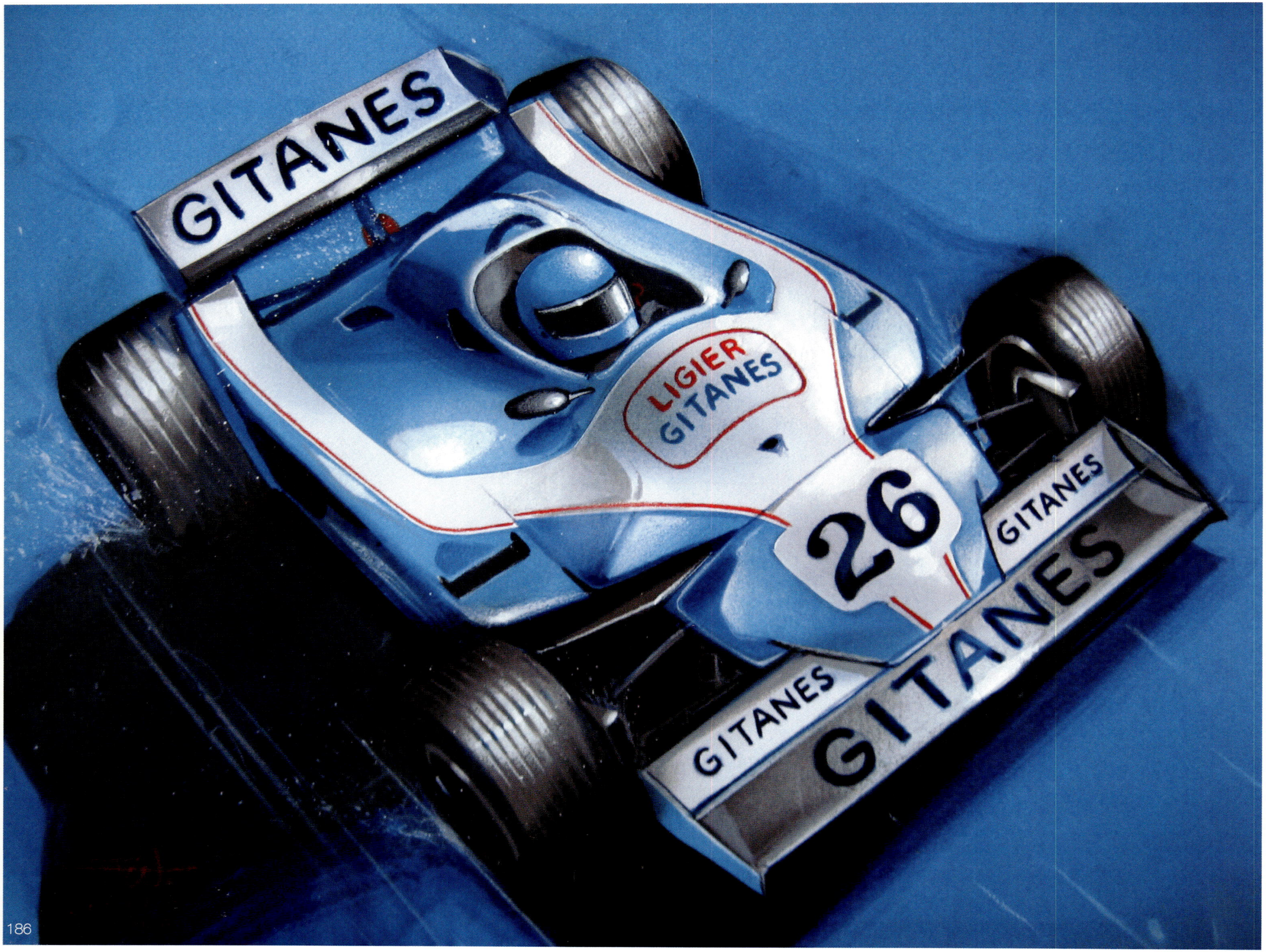
GITANES
LIGIER
GITANES
26
GITANES
GITANES
GITANES

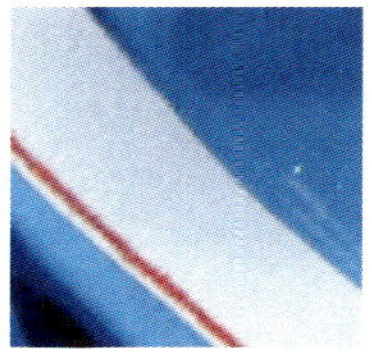

15. Bold Graphics Made Simple

As you've seen in the previous renderings, splicing in color can be a great alternative to filling in large areas with marker or pencil. If your subject has bold graphics that stretch over a wide array of surfaces, then splicing can be even more useful. This tutorial features blue Canson -- with Warm Steel and Light Grey Canson in supporting roles.

The Tools

Nupastels: White, Indigo Blue.

Prismacolors:
White, Cool Grey 20%, Indigo Blue, Crimson Red, True Blue.

Verithins:
Indigo Blue.

Gouache:
White.

Prismacolor Markers: Indigo Blue, Cerulean Blue, Navy Blue.

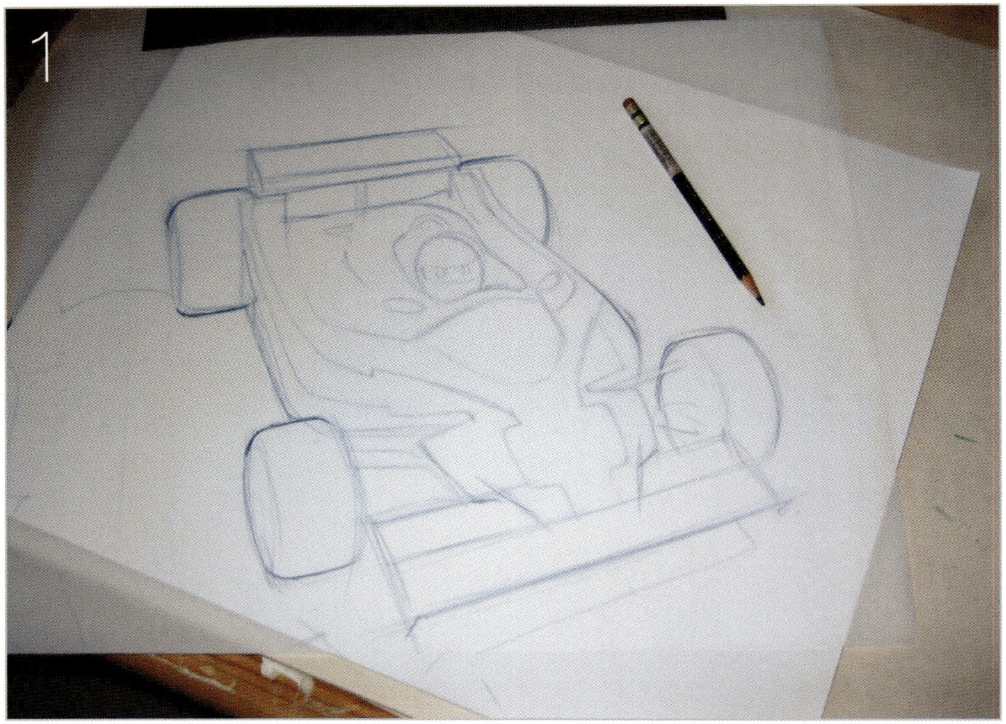

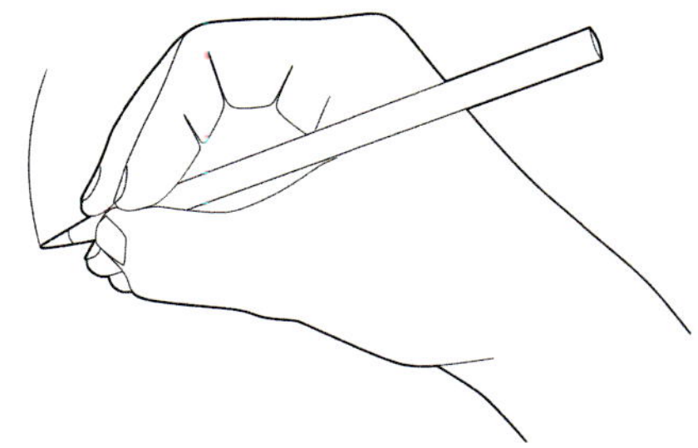

For this rendering, step up to a larger format like 18"x24"; this will make the intricate graphics appear much cleaner overall. Before you get out the Canson, hone in the perspective on a piece of vellum or tracing paper.

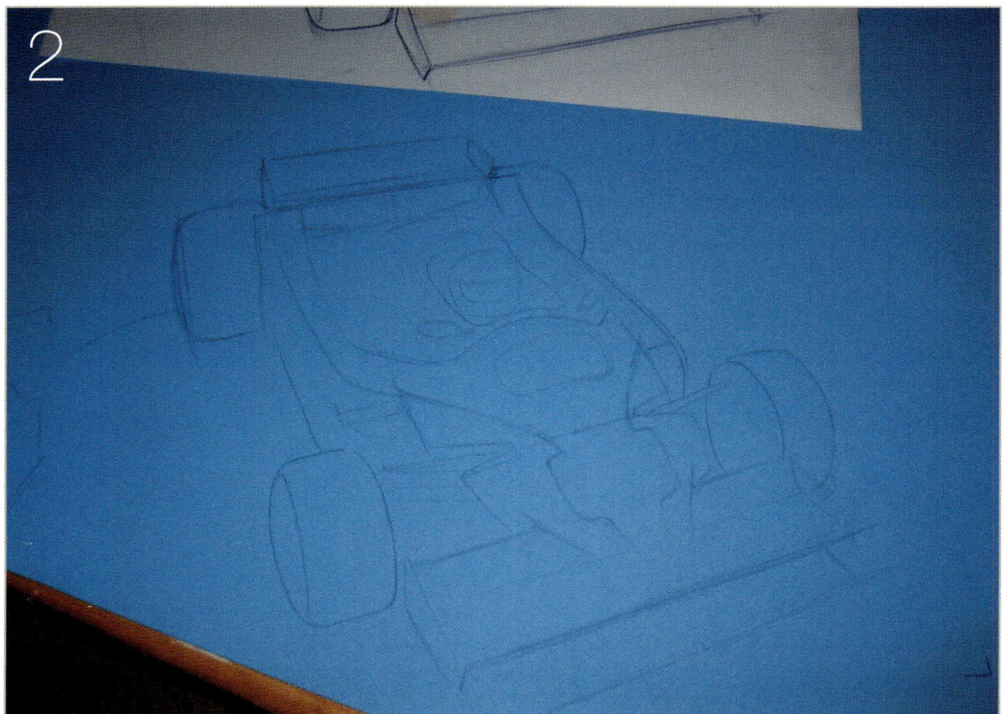

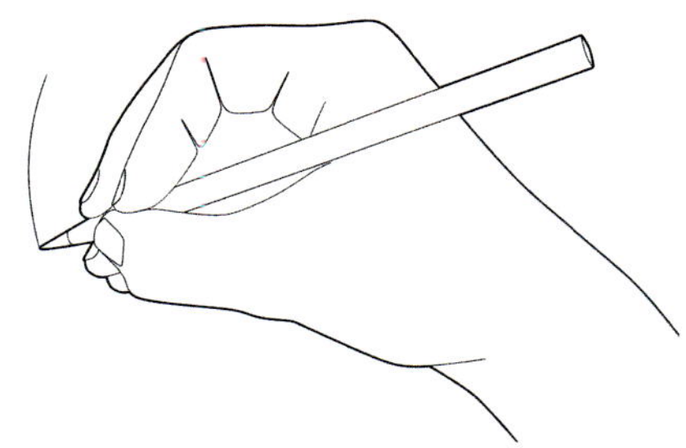

Transfer the drawing to the blue Canson using a piece of carbon paper.

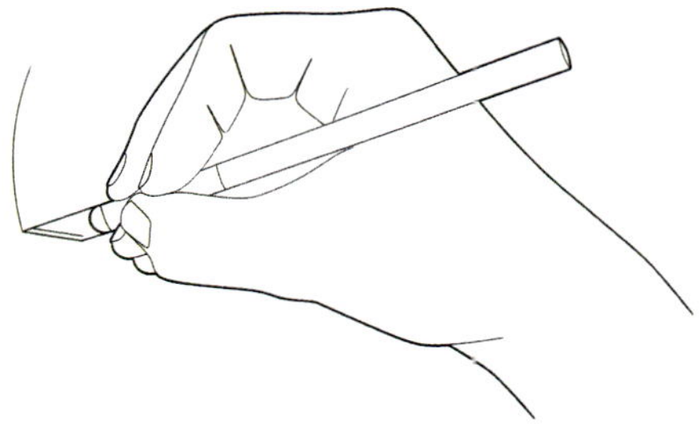

Put a piece of Light Grey Canson under the blue piece and begin cutting out the white graphics with an X-acto.

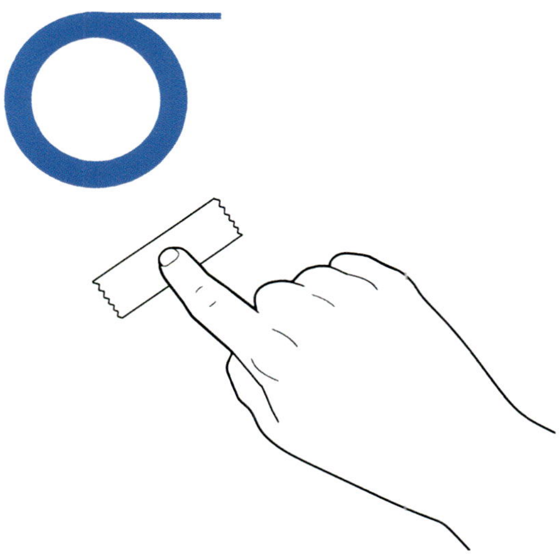

Discard the blue bits and insert the white ones. Secure the pieces on the back with some wide tape.

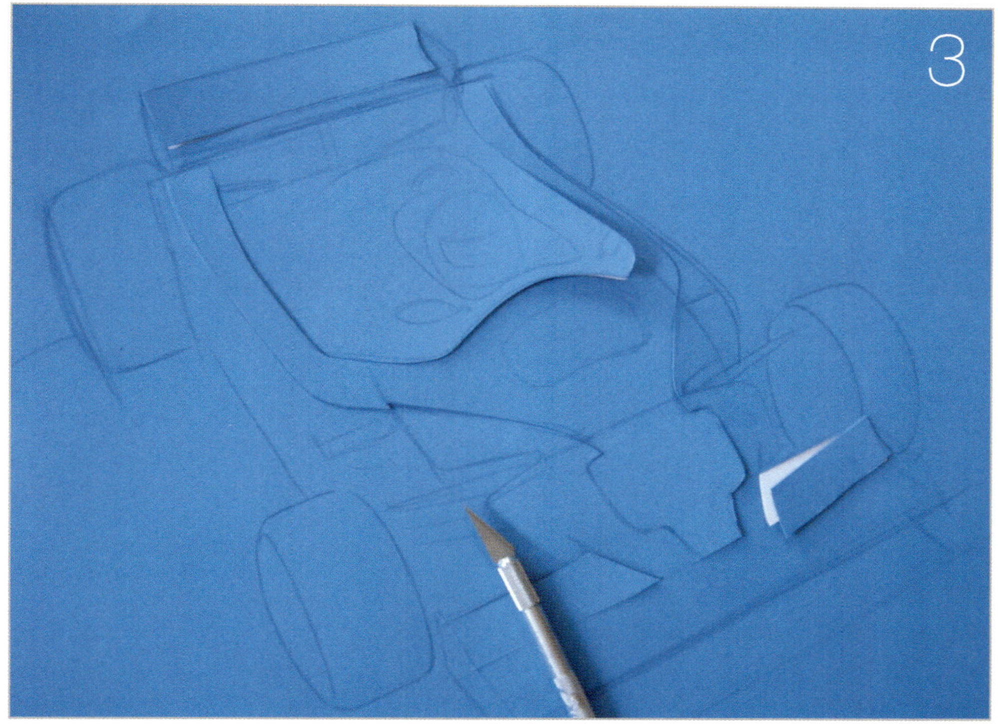

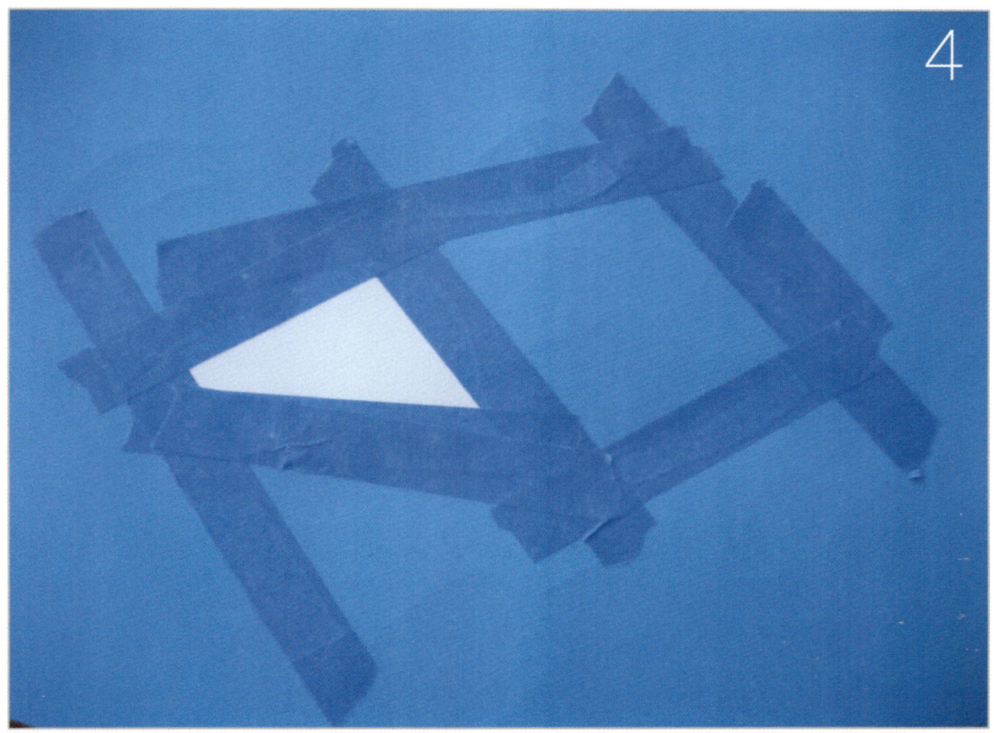

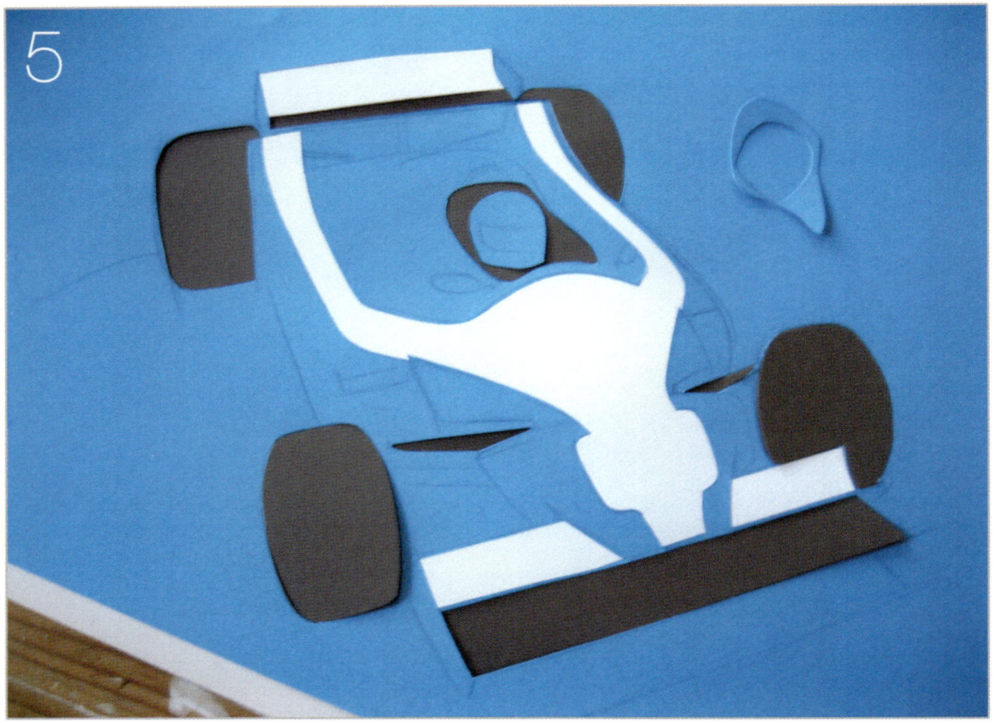

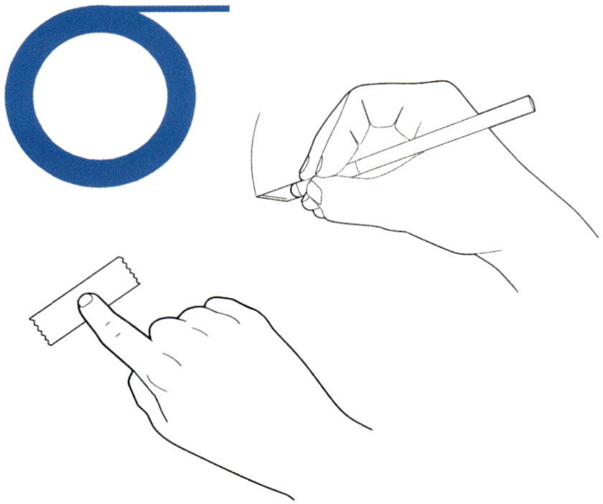

Perform the same process with the Warm Steel Canson. This shade of grey will provide an excellent base for the tires and the front wing. It also offers a nice color contrast next to the blue.

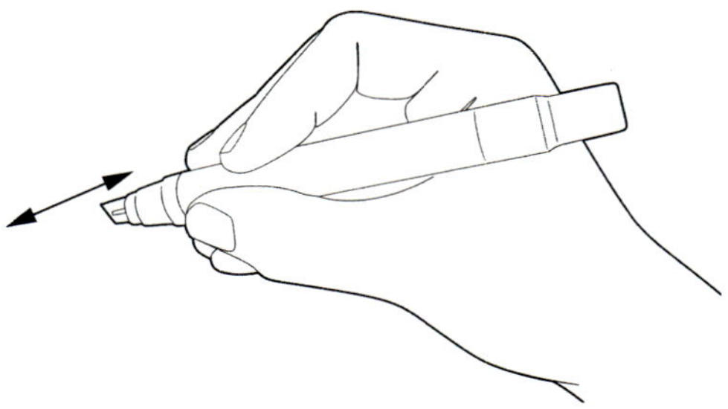

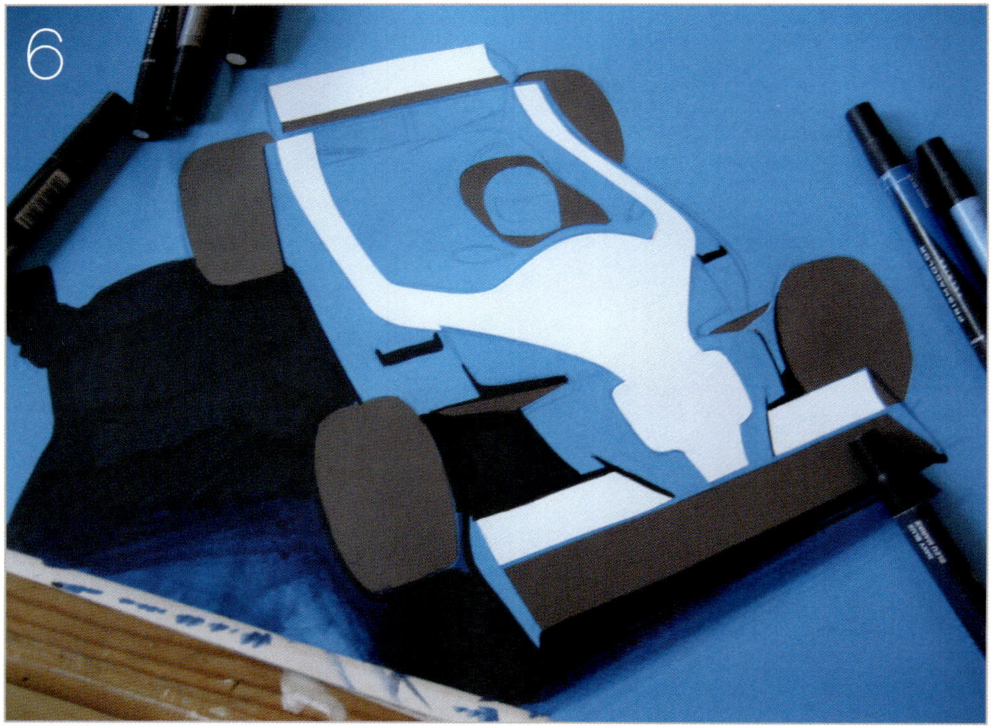

Using a series of blue markers, begin laying in the main shadow.

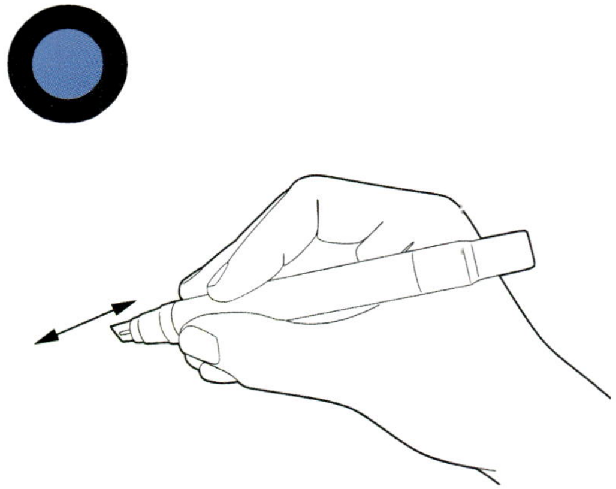

In this rendering, the light source is on the right, so some portions of the body will be in shadow. Block in these parts of the car with Cerulean Blue.

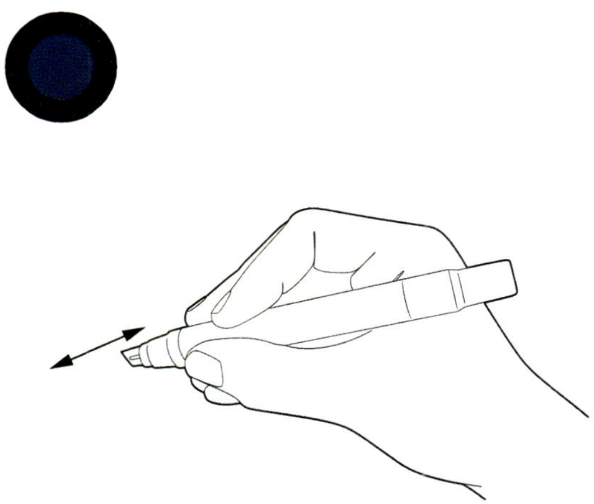

Add in the main reflections using a modest amount of Indigo B ue marker.

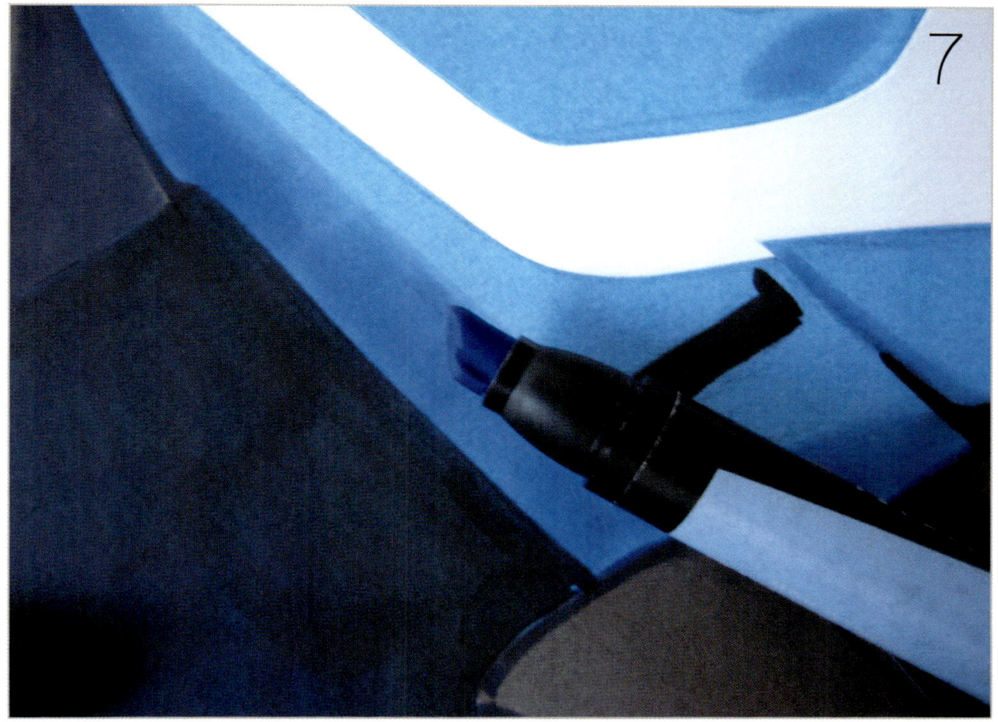

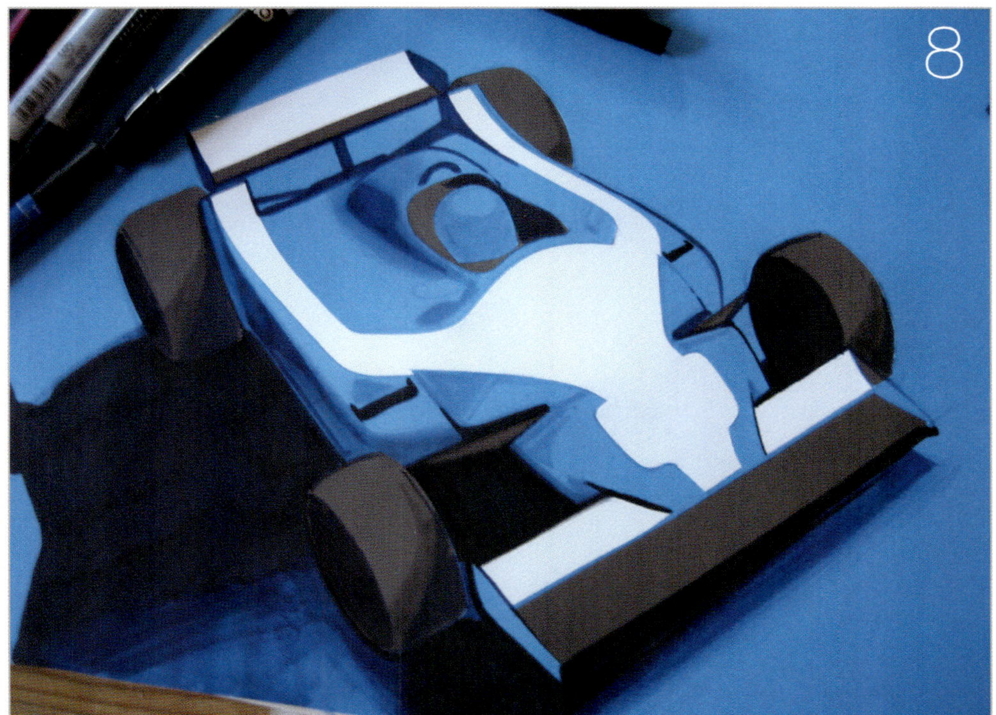

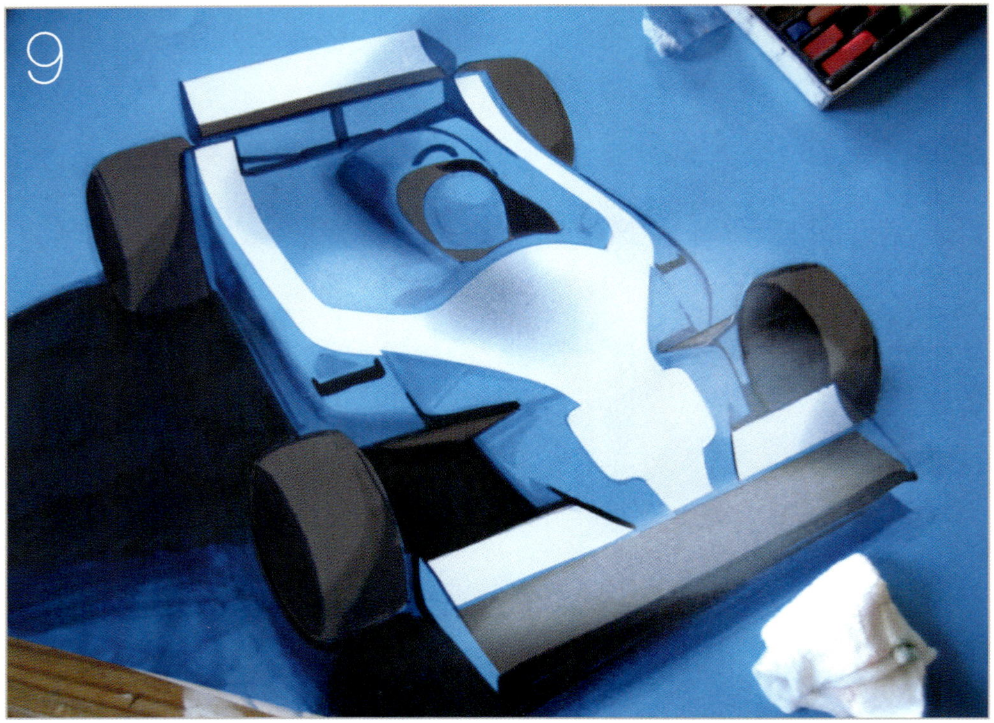

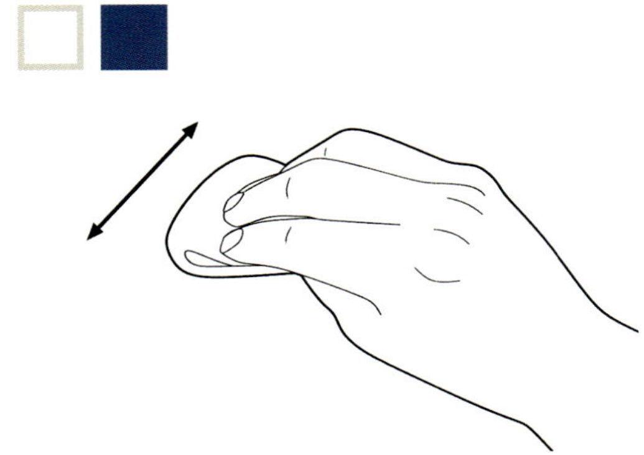

Describe the soft shapes of the bodywork by applying white and blue pastels with separate Webril pads.

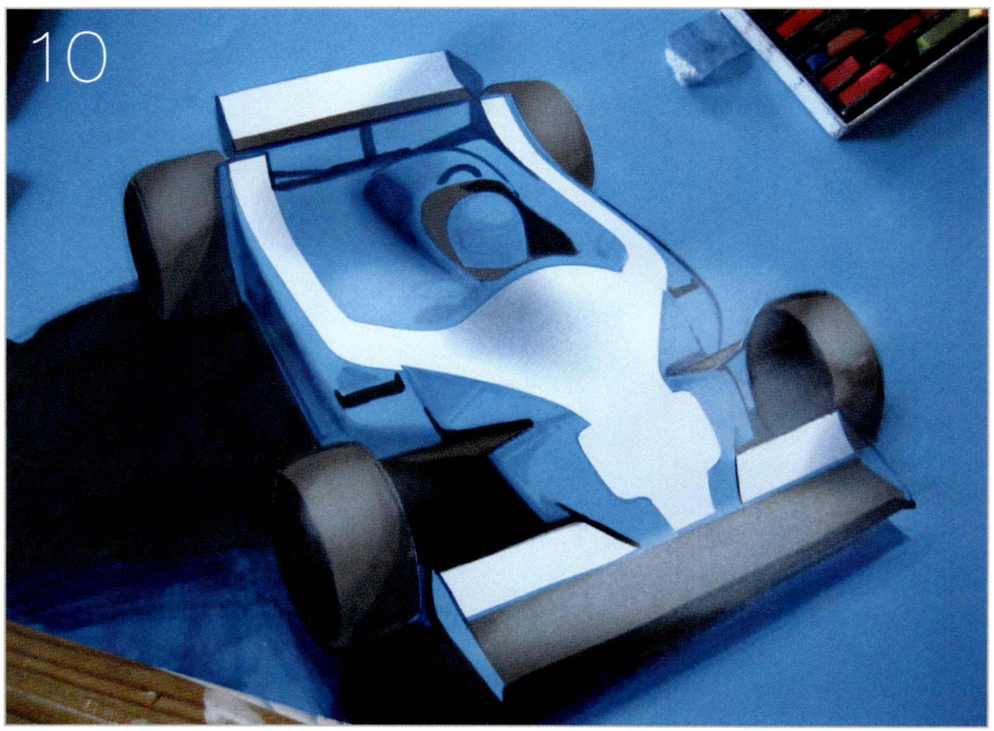

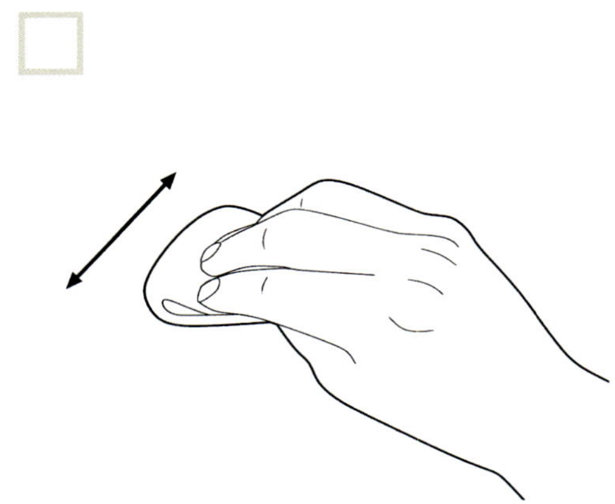

Next, ghost in soft highlights on the top of each tire with white pastel.

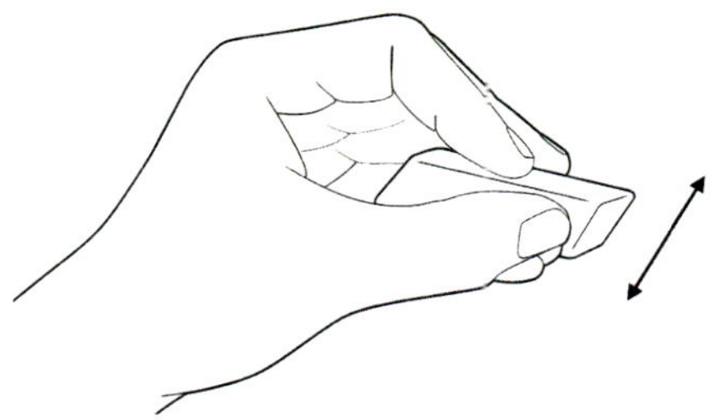

Clean up the white and blue pastel edges with an eraser.

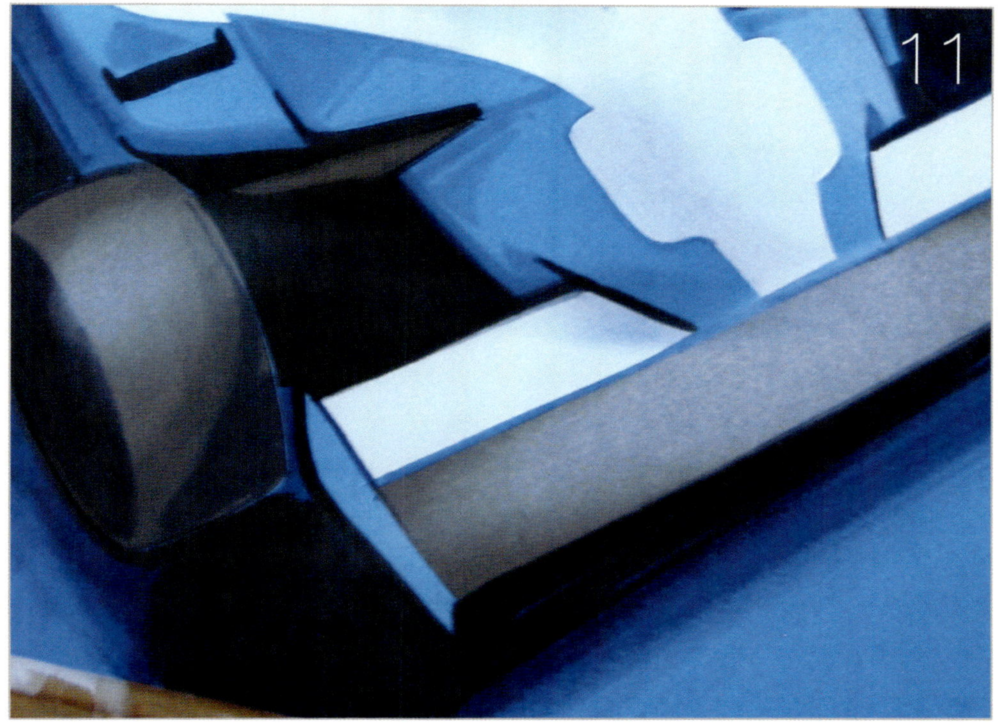

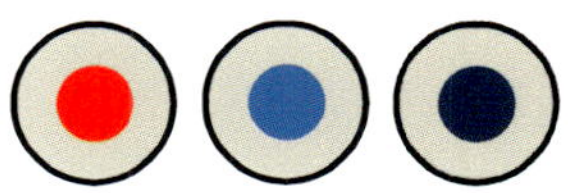

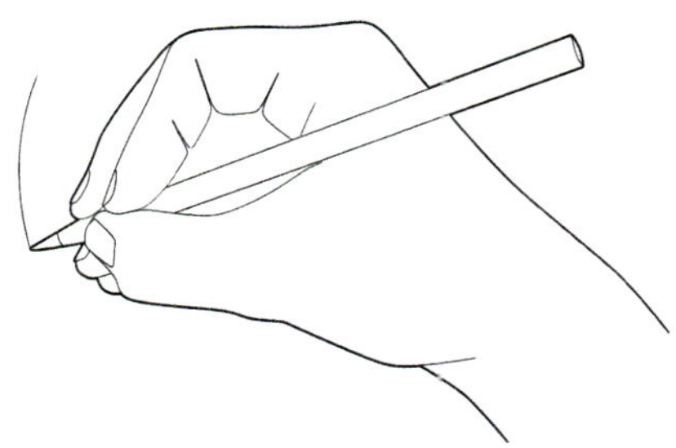

With the pastel shading in place, begin laying out the graphics in pencil.

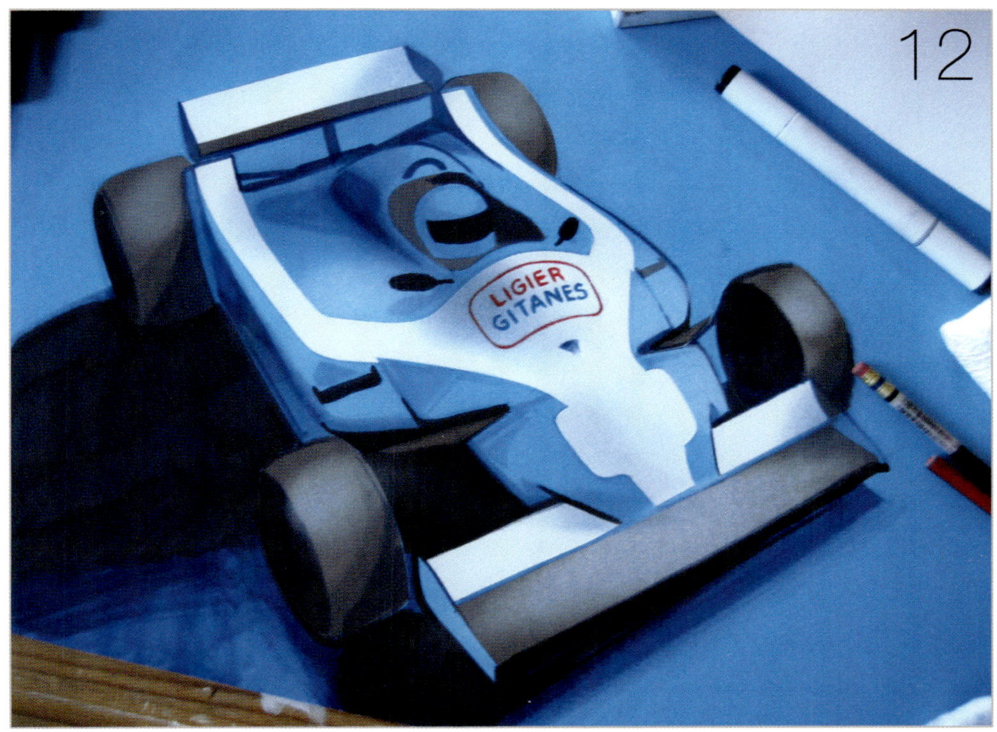

13

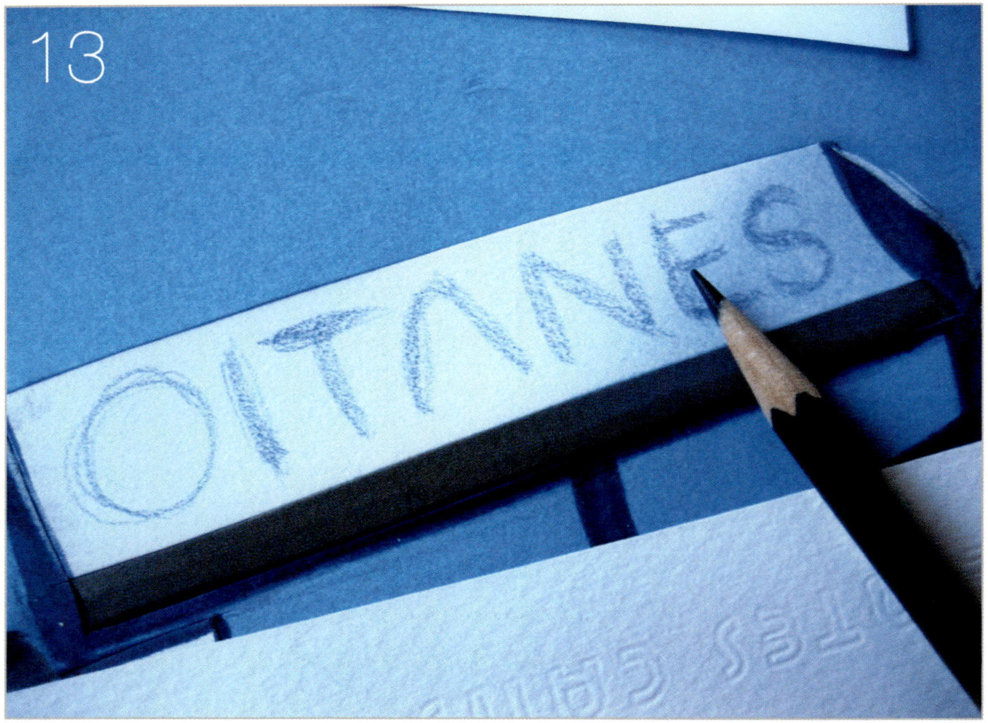

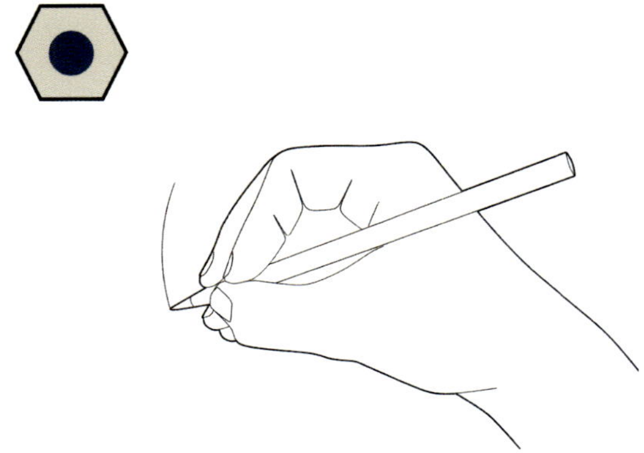

For the larger graphics, lay them out first in blue Verithin to avoid making a mistake with the Prismacolor pencils.

14

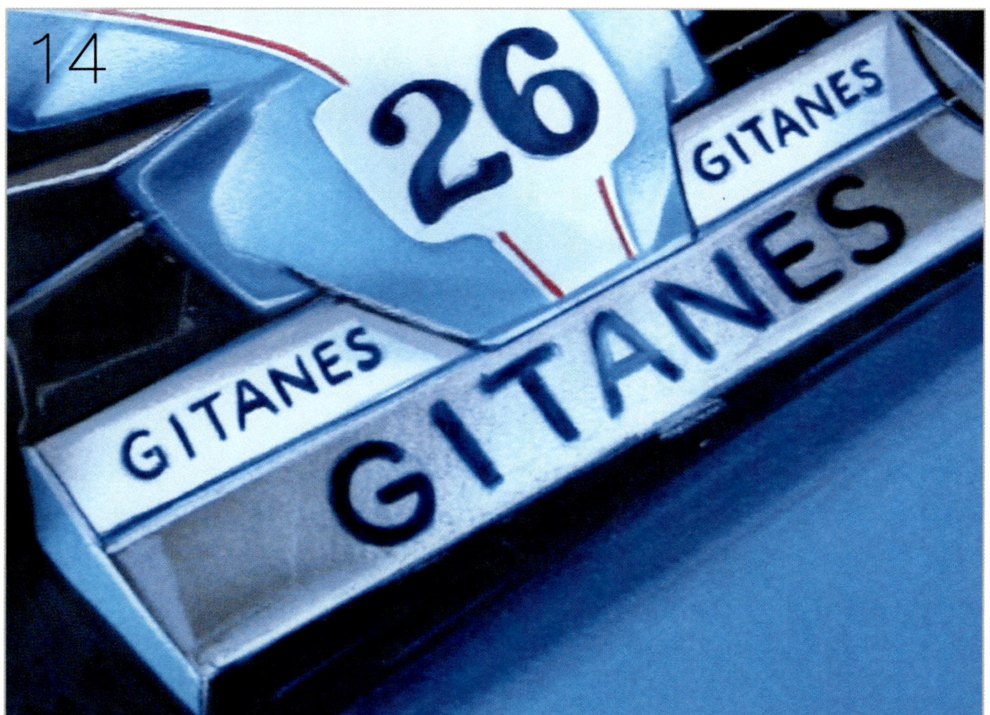

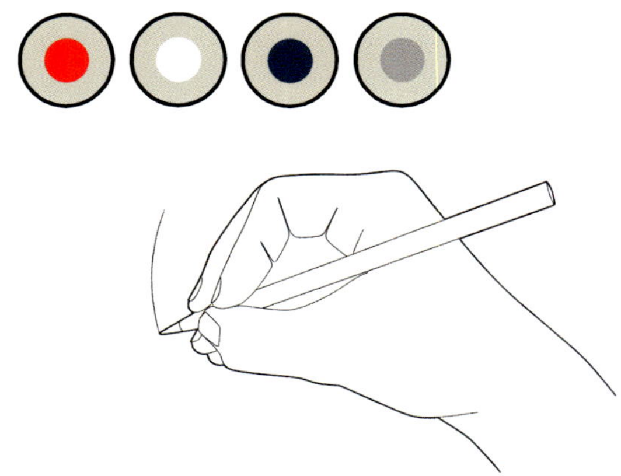

Before applying the front wing graphics, take a moment to lay in a white hot spot across the middle that fades on both ends. This will represent the sheen of the metal wing.

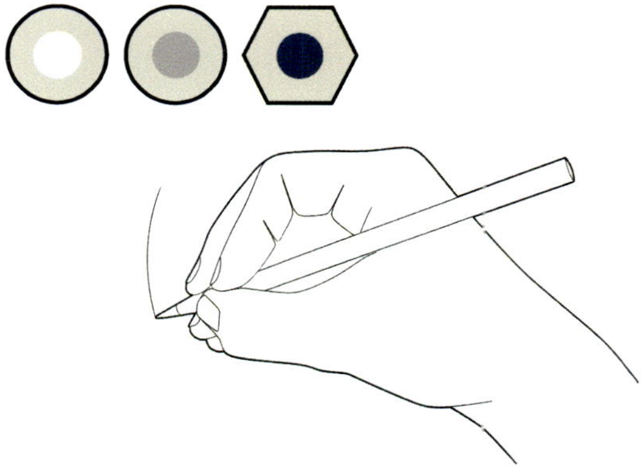

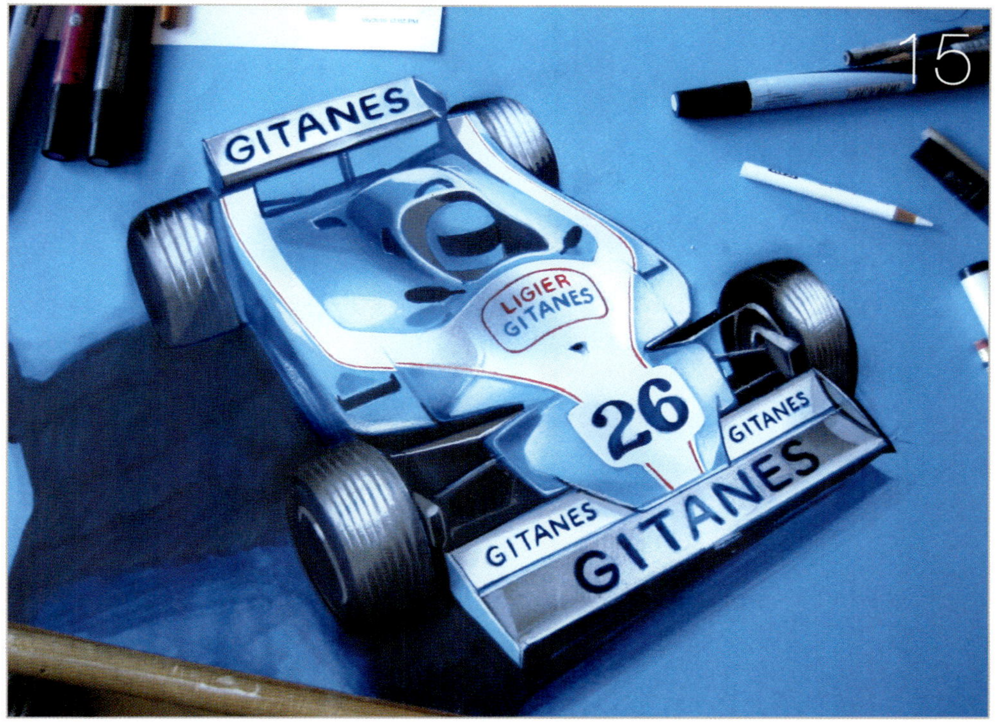

Use a combination of white and Cool Grey Prismacolor pencils to lay in the bodywork, tire and A-arm highlights. A blue Verithin pencil can be used to carve out the tread valleys.

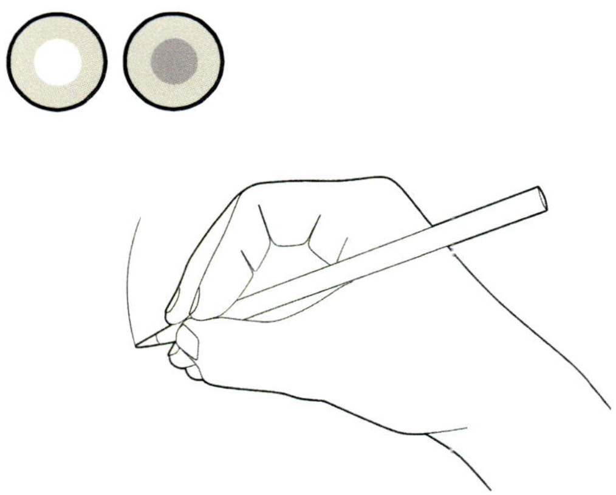

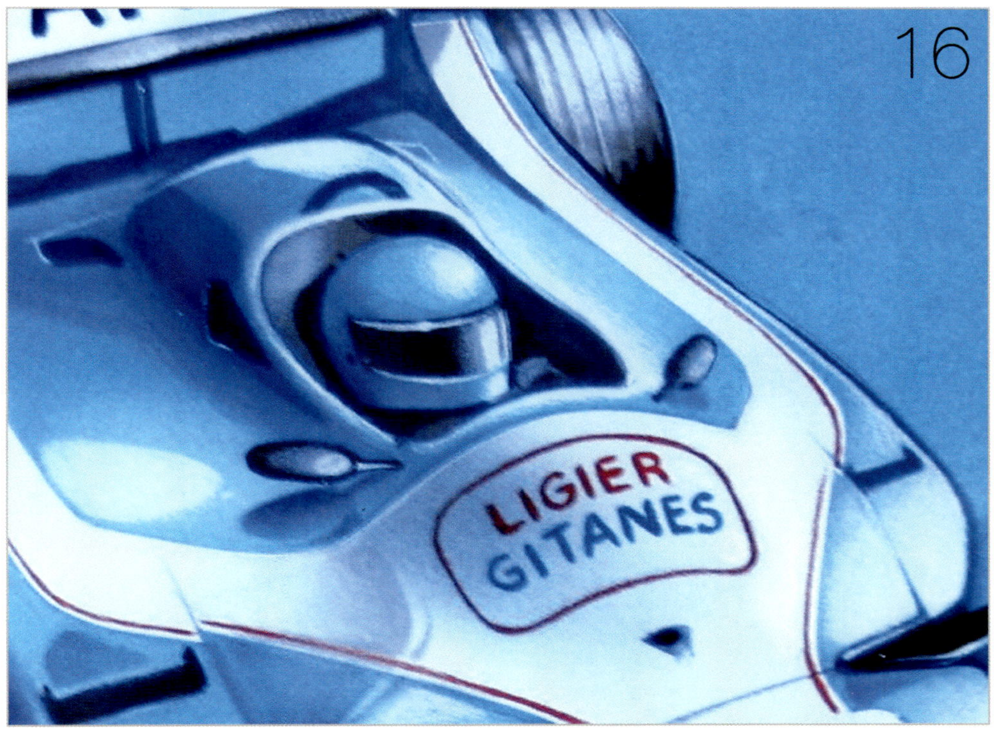

Similar to the first tutorial, light the visor as a cylinder and the helmet as a sphere. The dark side mirrors should have elliptical highlights.

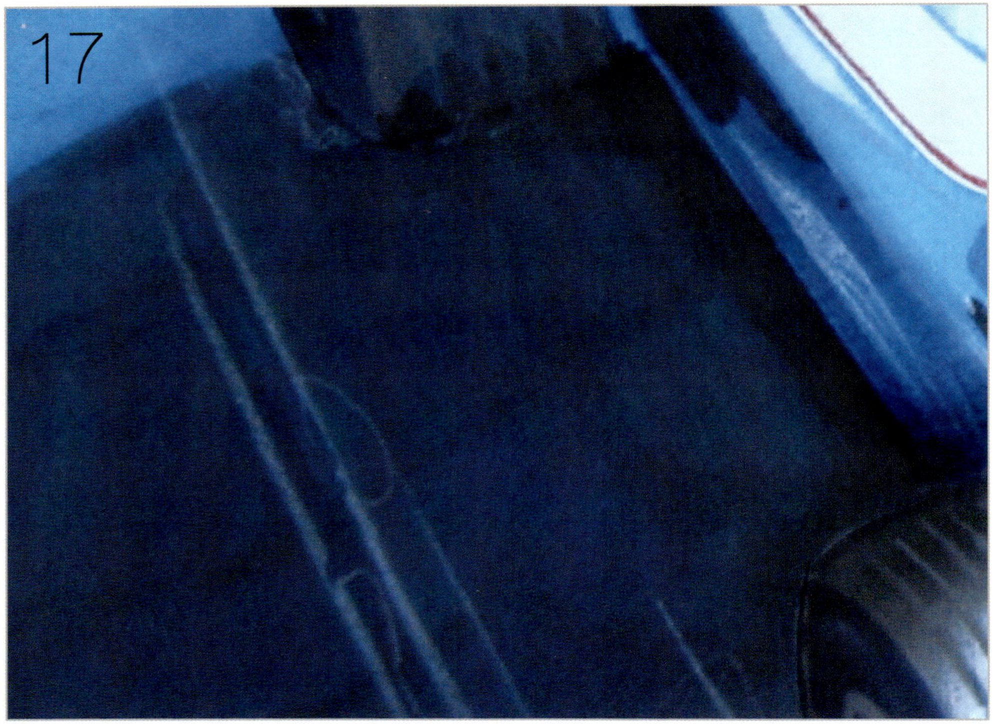

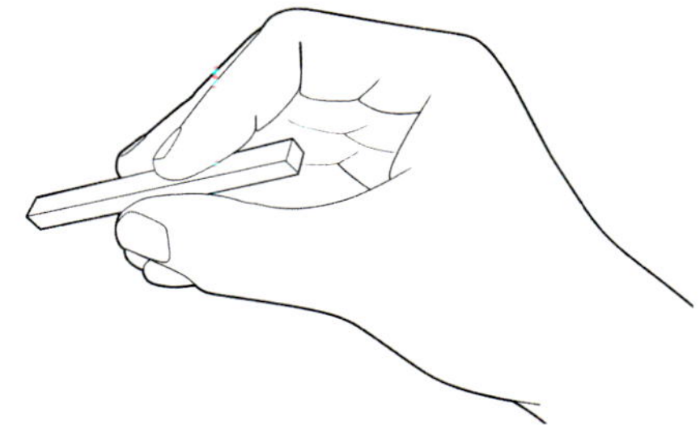

Directly apply a bit of white pastel next to and in front of the car in long, loose strokes. This will further enhance the suggestion of motion.

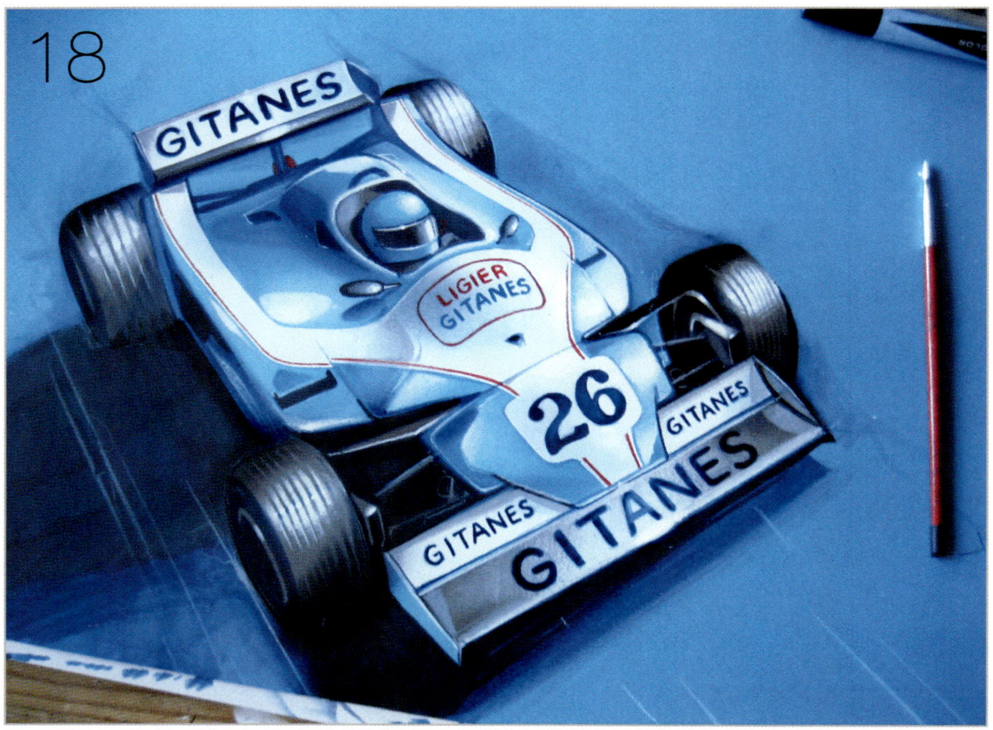

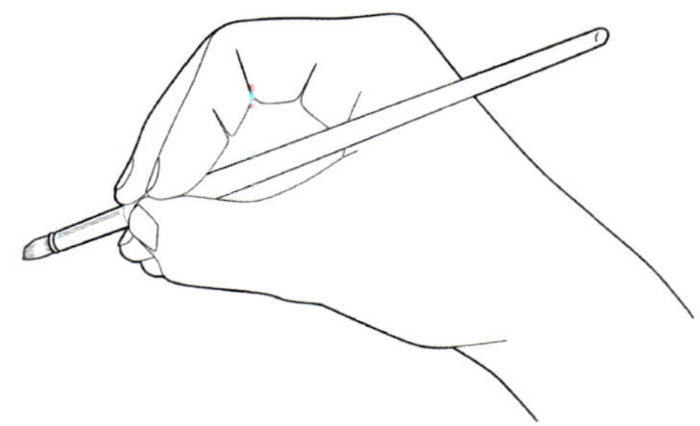

Concentrate the tread highlights on the top of each tire, in line with both axles. This will help give the appearance of wet rain tires. Keep all the bodywork highlights minimal and clean.

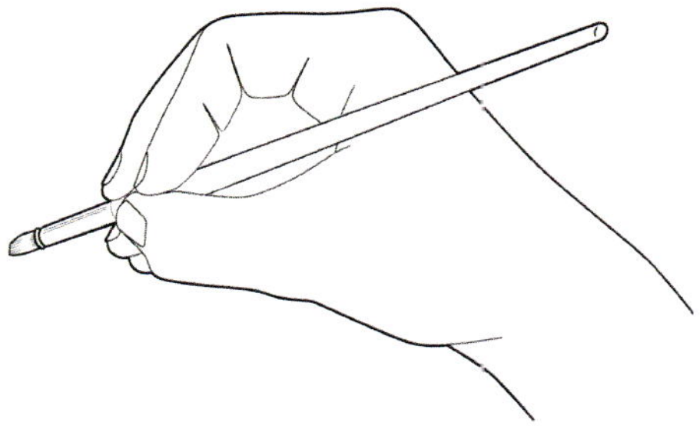

Use an old toothbrush to apply a spatter effect behind the tires. Mix a bit more water into the gouache than you would under normal circumstances. Use light brush strokes to enhance the spray.

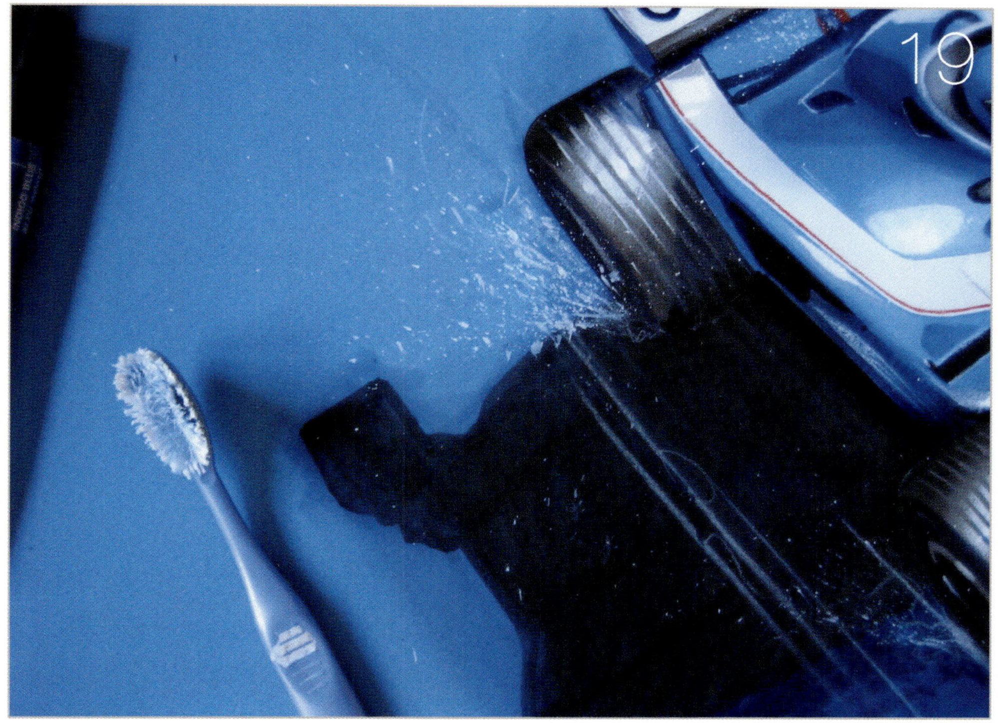

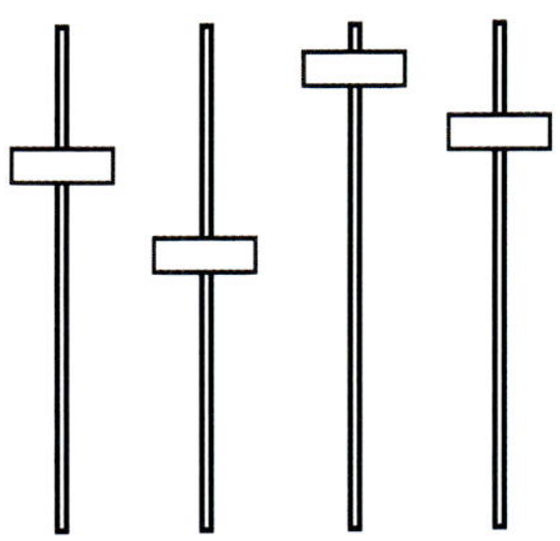

Scan in the rendering and adjust the levels to your liking in Photoshop.

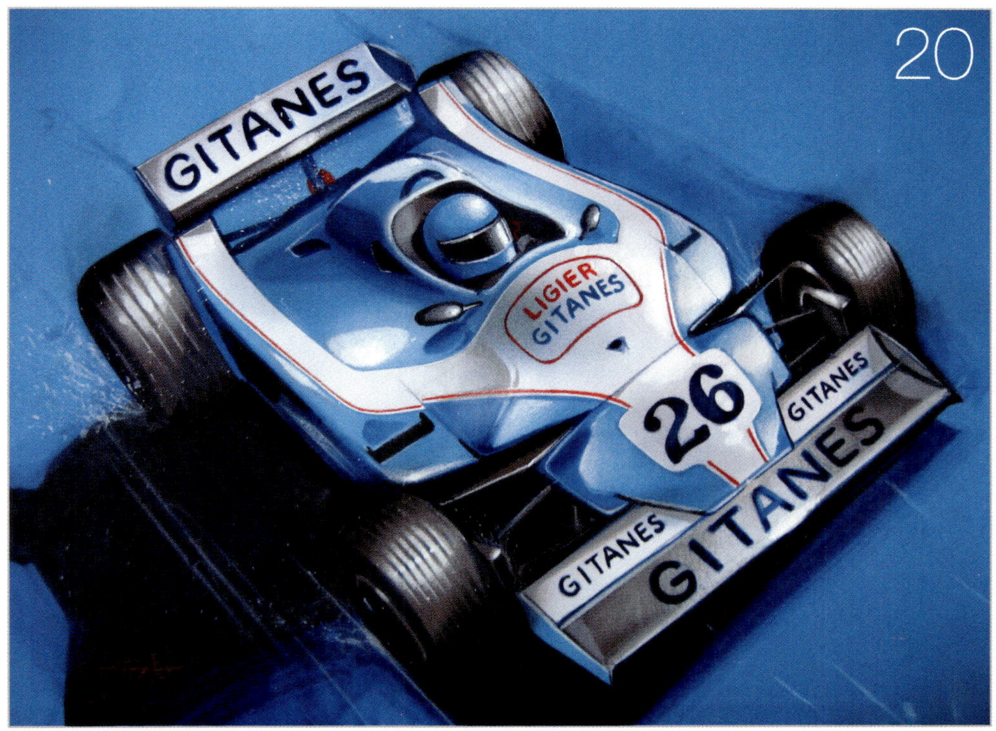

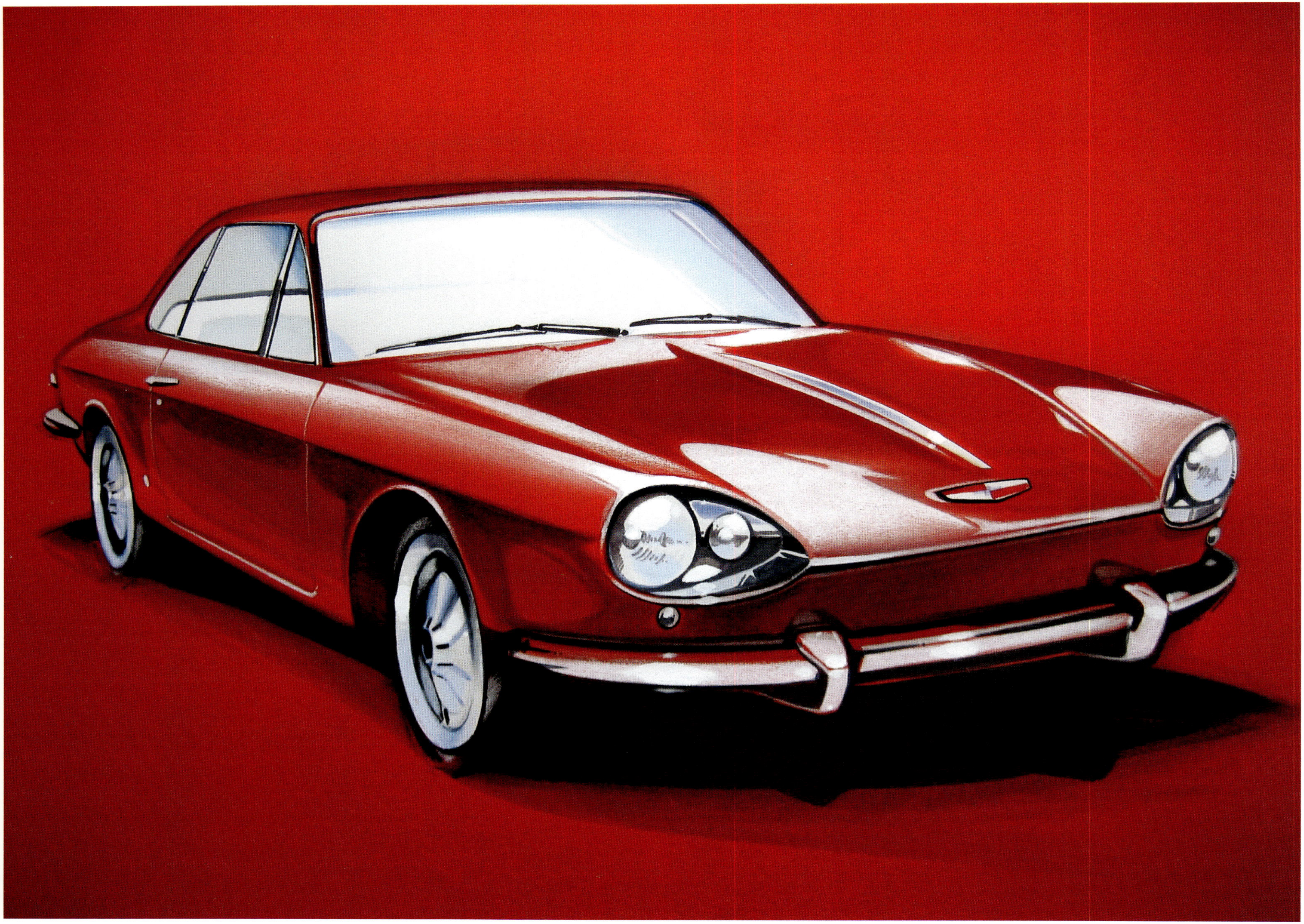

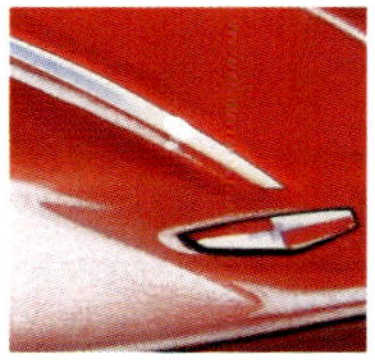

16. Seeing Red

I've seen red Canson renderings grab viewers from clear across a room. Granted, the medium is hard to ignore, but simply drawing a car on red paper won't make the car, or the drawing, hot by default. In this tutorial, you'll learn a simple red Canson technique that relies on contrast to draw an audience, not just raw color.

The Tools

Prismacolors:
Crimson Red, White,
Cloud Blue, Cool Grey 20%.

Copic Pen:
Black.

Nupastels:
White.

Verithins:
Black.

Gouache:
White.

Markers: Black, Cool Grey 40%, Crimson Red, Mahogany Red, Cool Grey 70%, Cerulean Blue, Lemon Yellow.

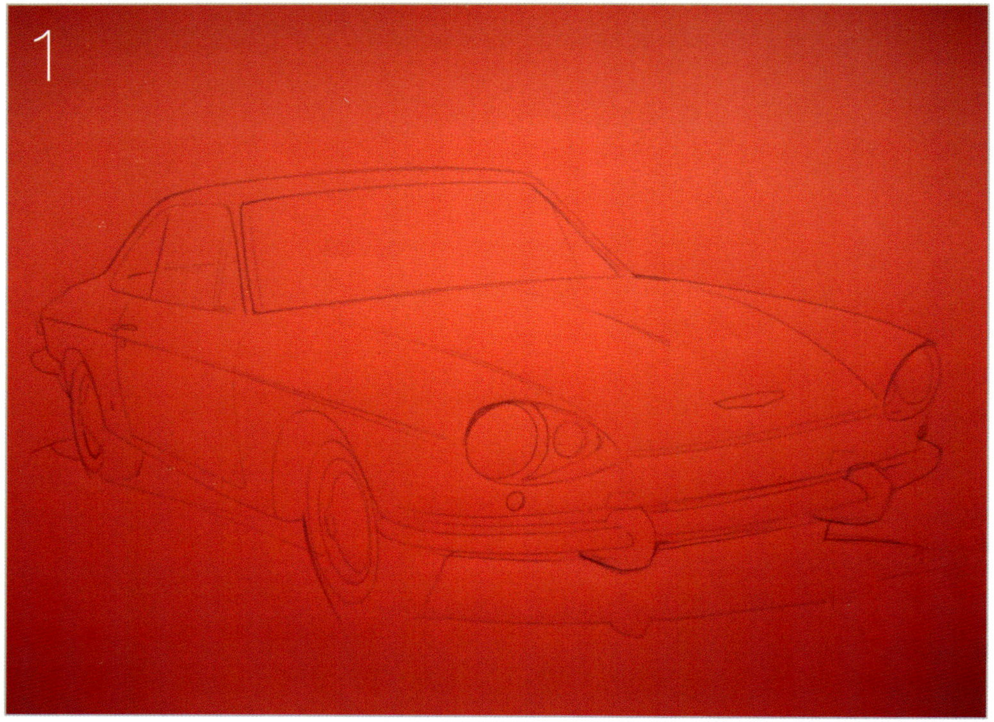

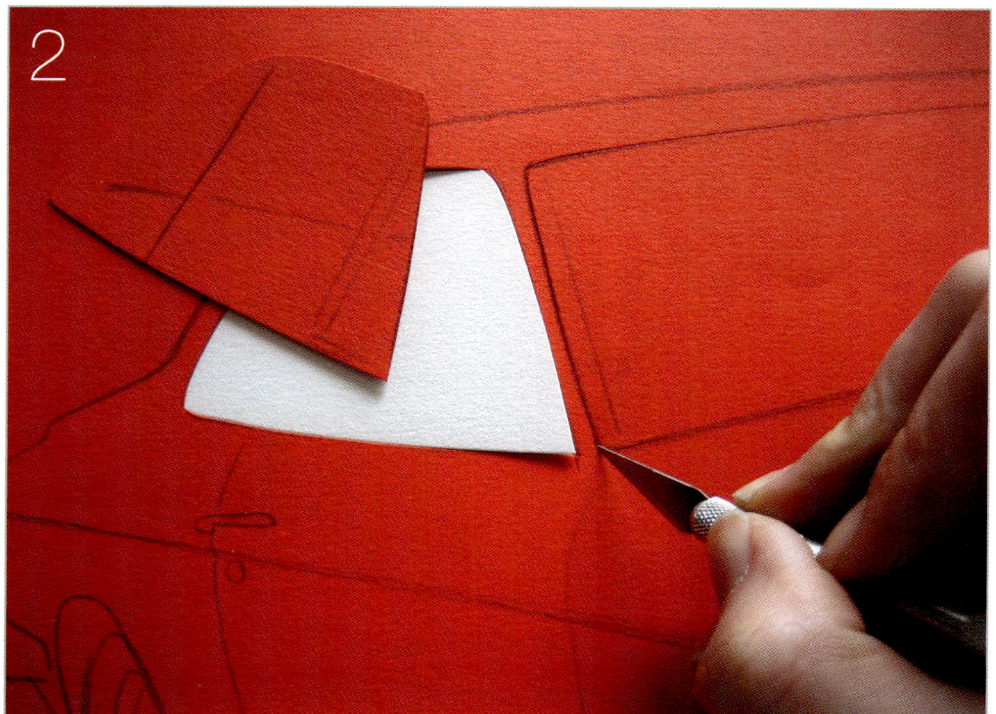

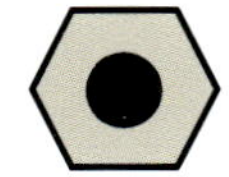

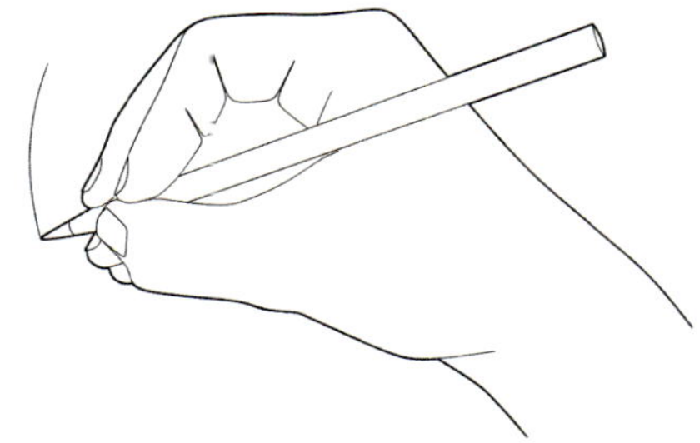

Start by laying out the basic lines of the car with a black Verithin pencil.

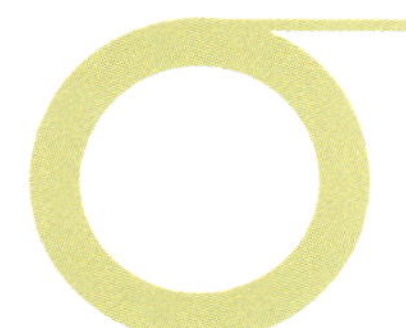

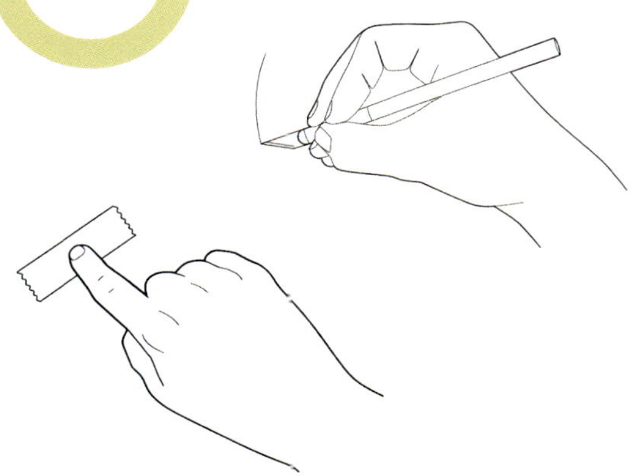

Place a piece of white Canson under the red piece and splice in the light parts of the car. Secure the loose bits on the back with some tape.

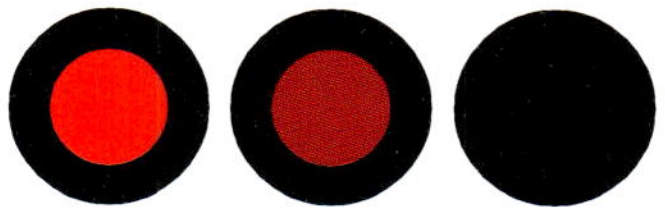

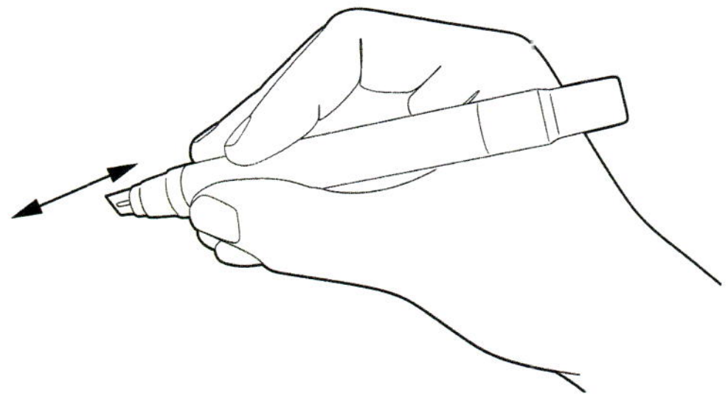

Block in the main reflections with Crimson and Mahogany Red markers. Like the Mercedes convertible rendering, this one features a cross-car reflection from an awning or a tree. Apply black marker for the wheel wells and shadow.

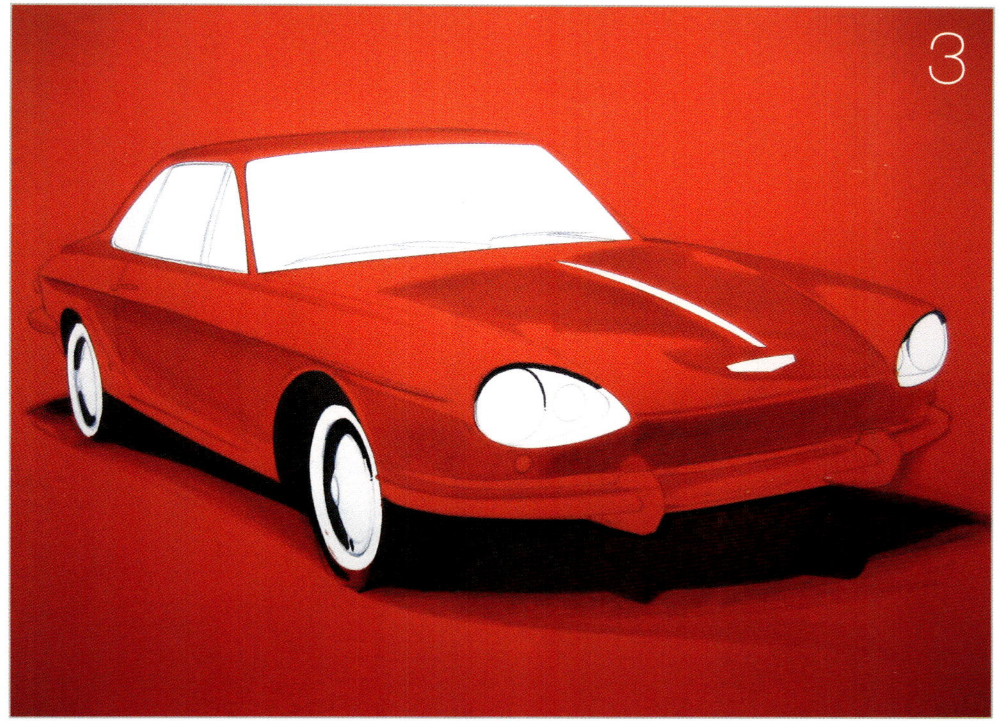

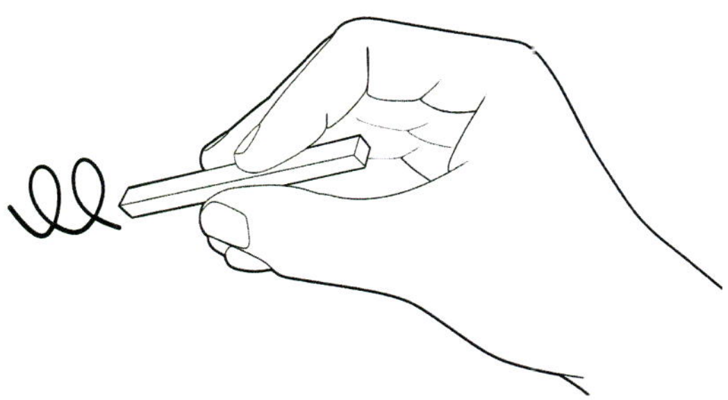

Locate the light parts of the body and directly apply some white pastel.

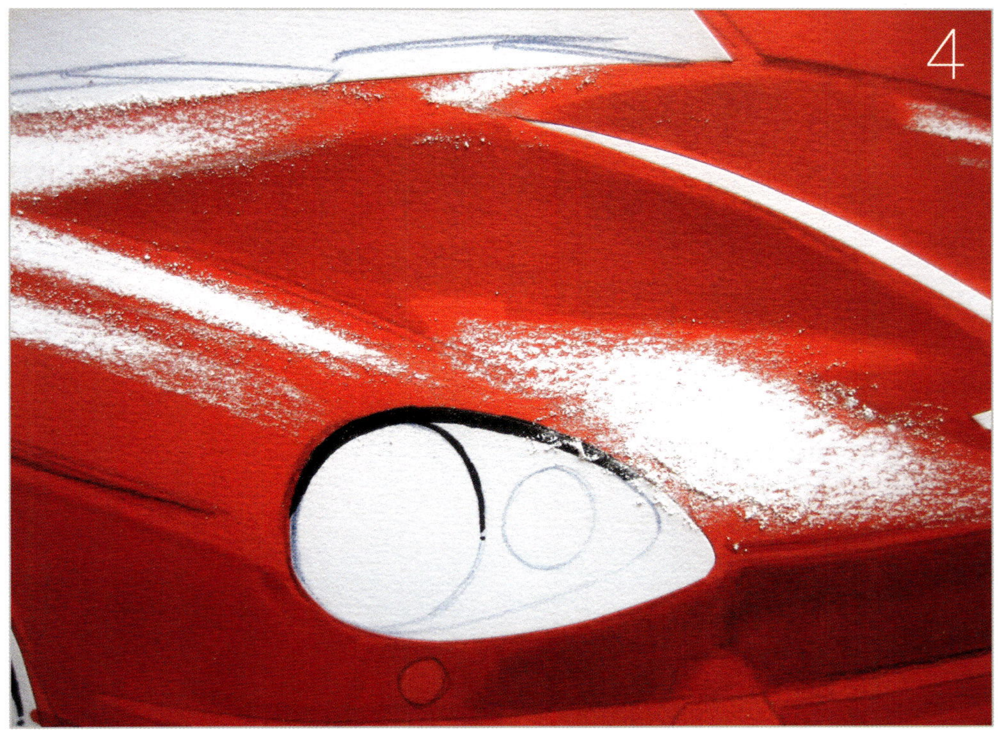

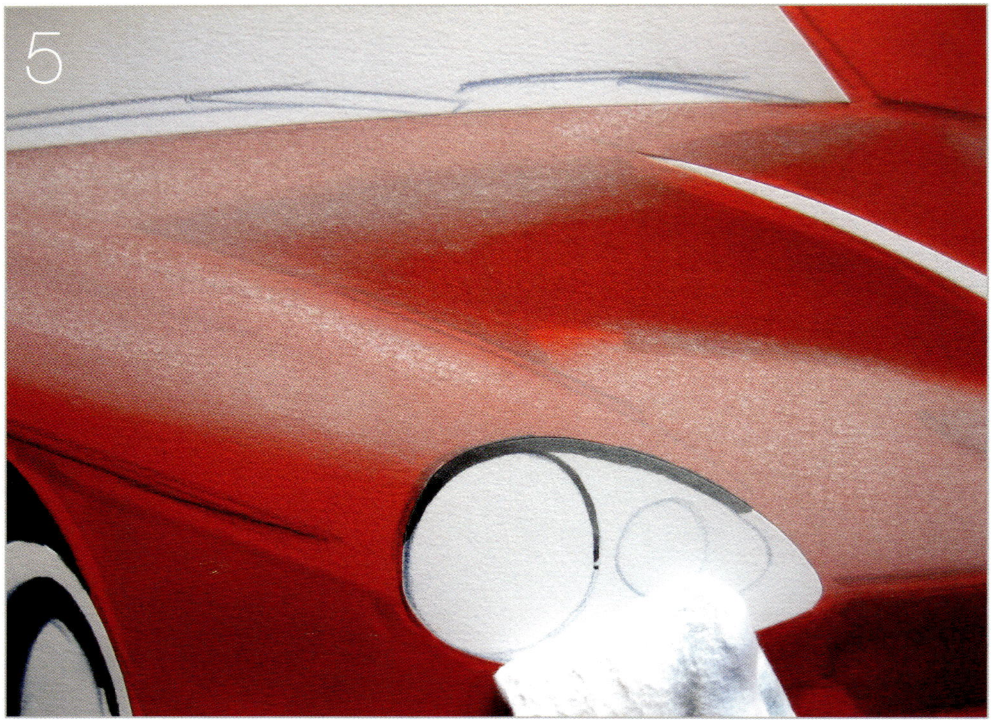

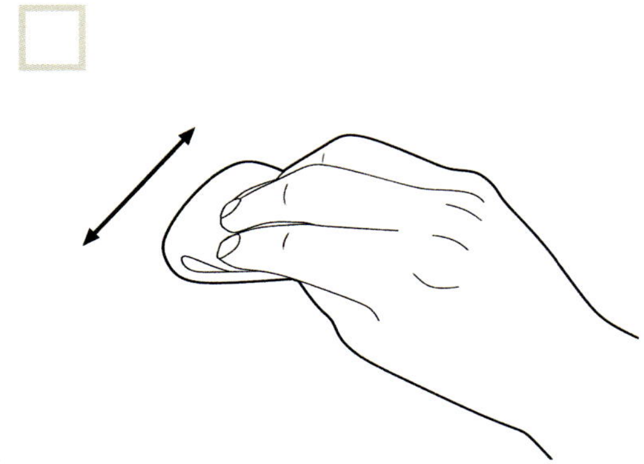

Next, use a clean Webril pad to spread and smooth out the white pastel.

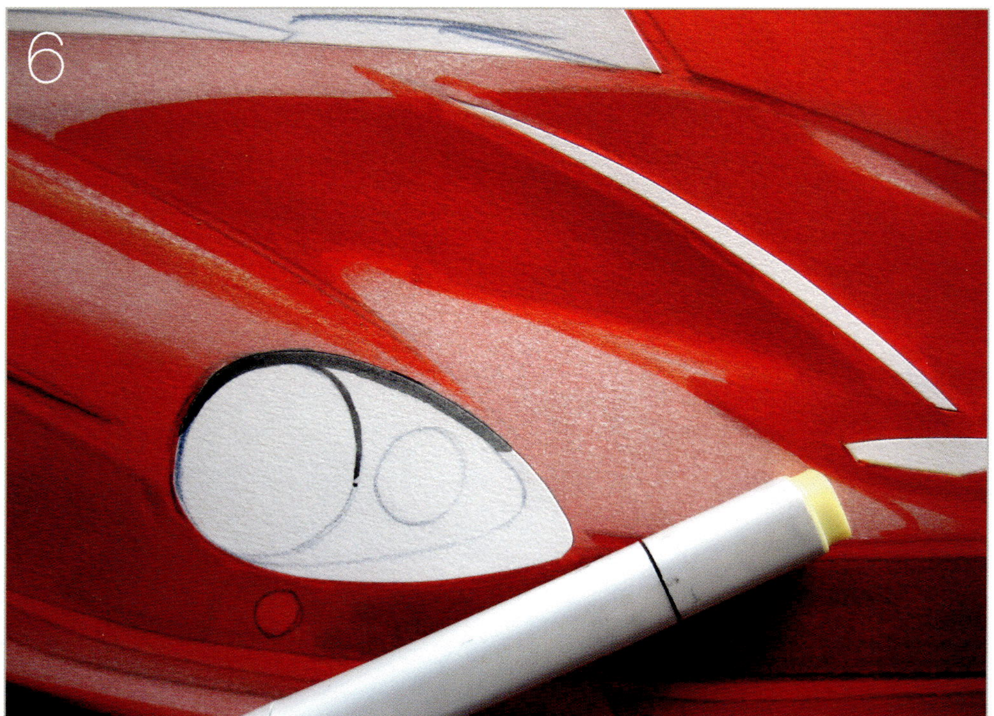

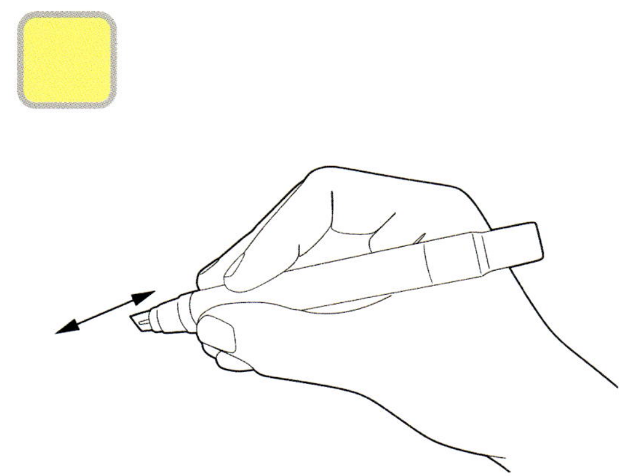

Instead of cleaning up the edge of the pastel with an eraser, try using a Lemon Yellow marker. This will give the reflections a rich, warm edge.

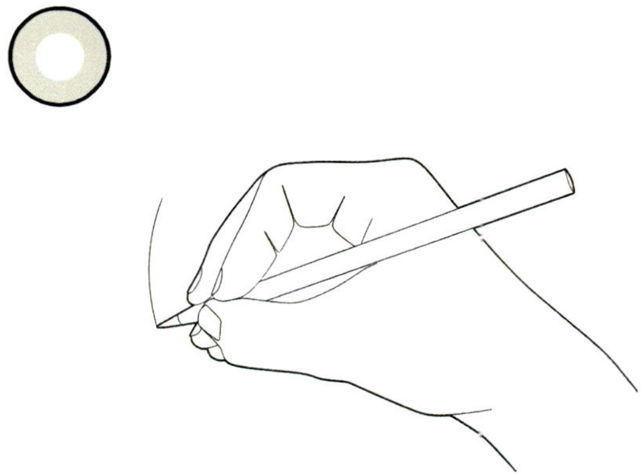

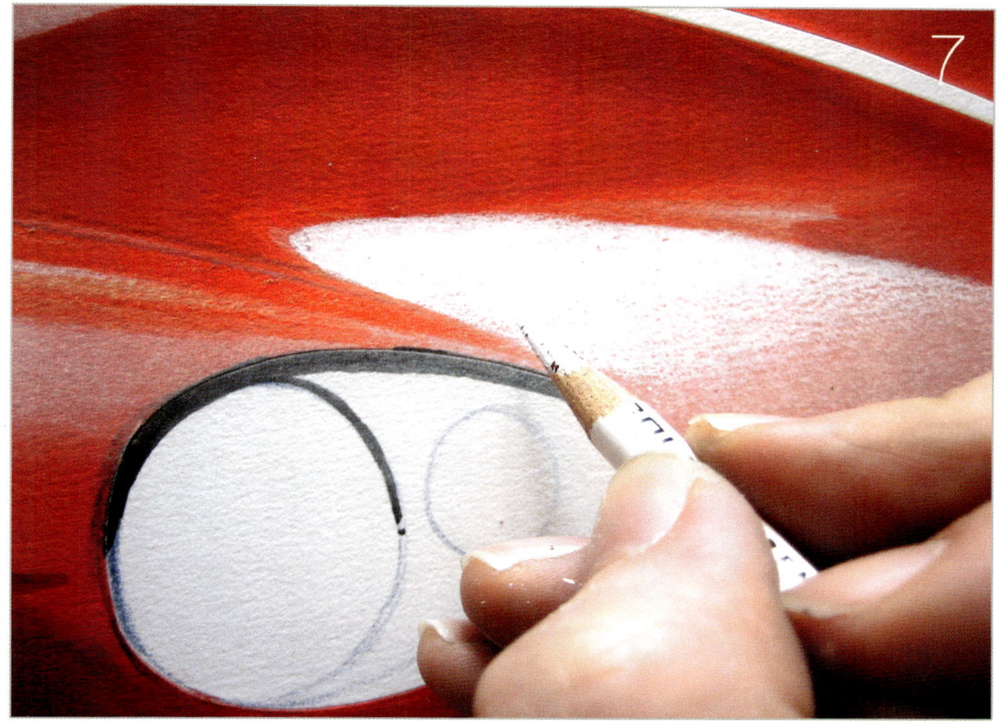

White pastel tends to lack punch on its own, so bolster the edge of the horizon using a white Prismacolor pencil.

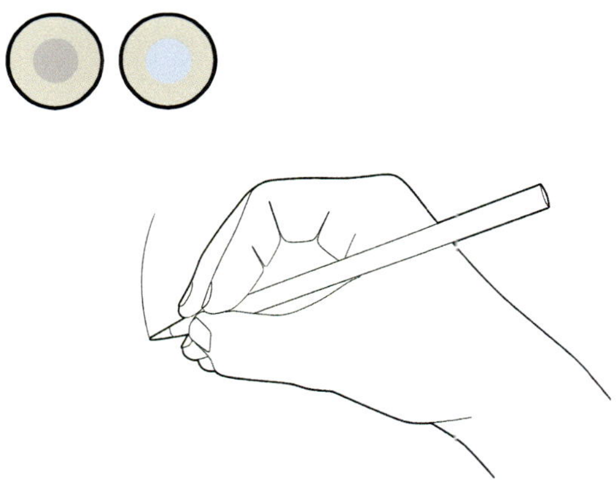

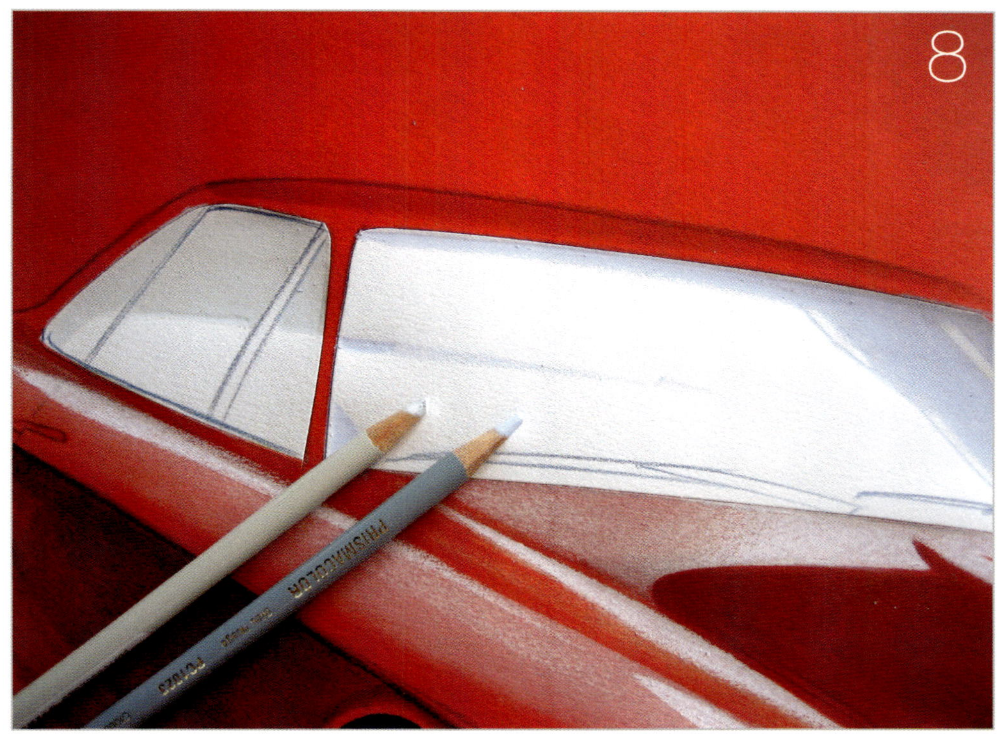

For that mid-60s Canson look, try keeping the windows relatively clean. A little Cool Grey and Cloud Blue pencil should provide adequate description.

9

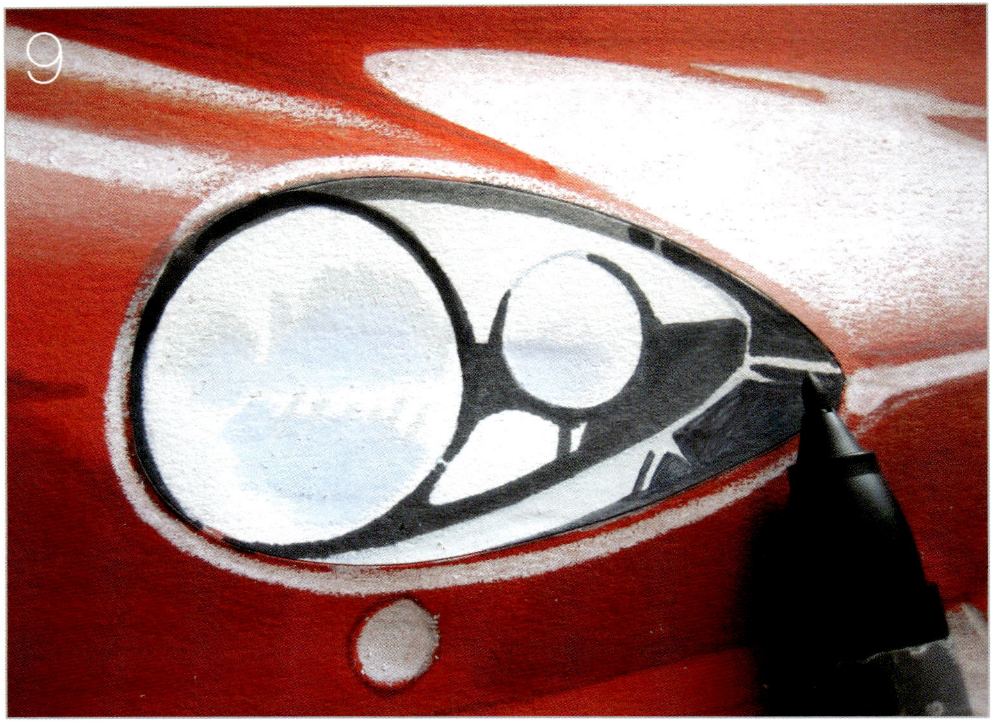

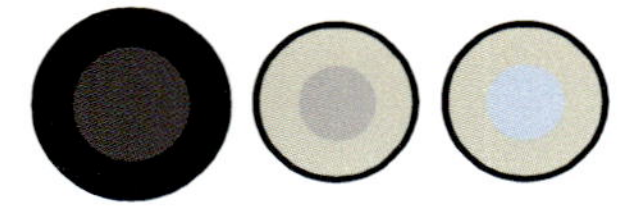

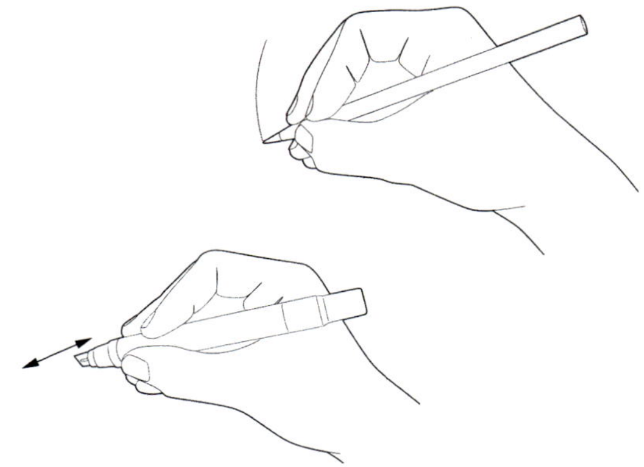

For the headlamps, begin by describing the horizon on the flat bezel with a Cool Grey 70% marker. Use Cool Grey and Cloud Blue pencils to add some definition to the lamps themselves.

10

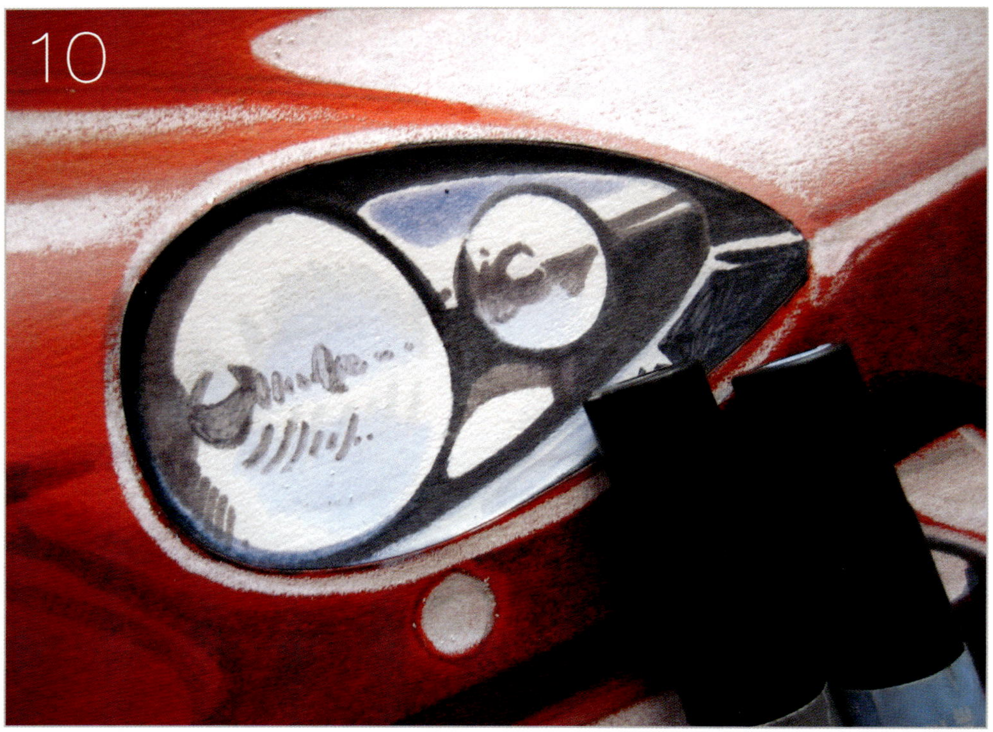

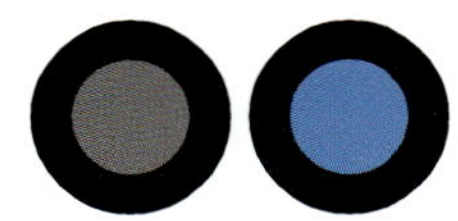

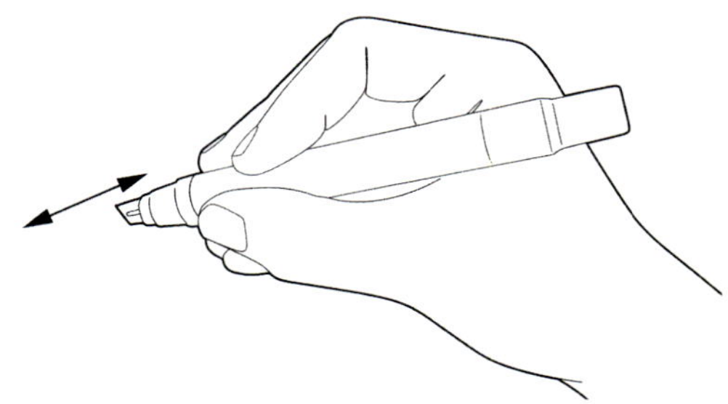

Now, using a Cool Grey 40% marker, define the bulb shields and give indications of the fluting. Use Cerulean Blue to block in sky tone.

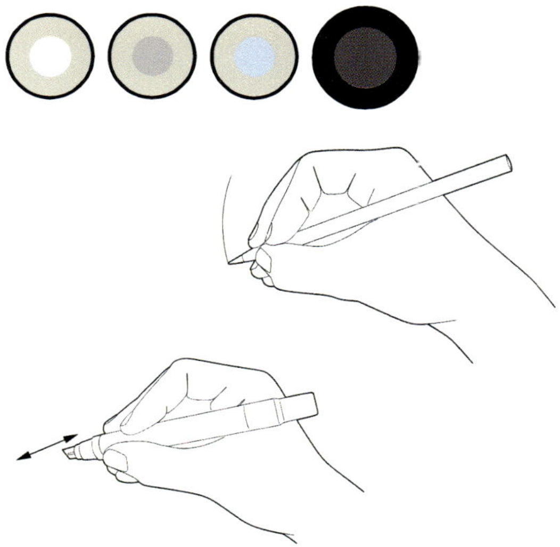

Once the lamps are complete, add detail to the front bumper with an array of light pencils. Try using a Cool Grey marker to indicate the bumper's reflection on the front fascia.

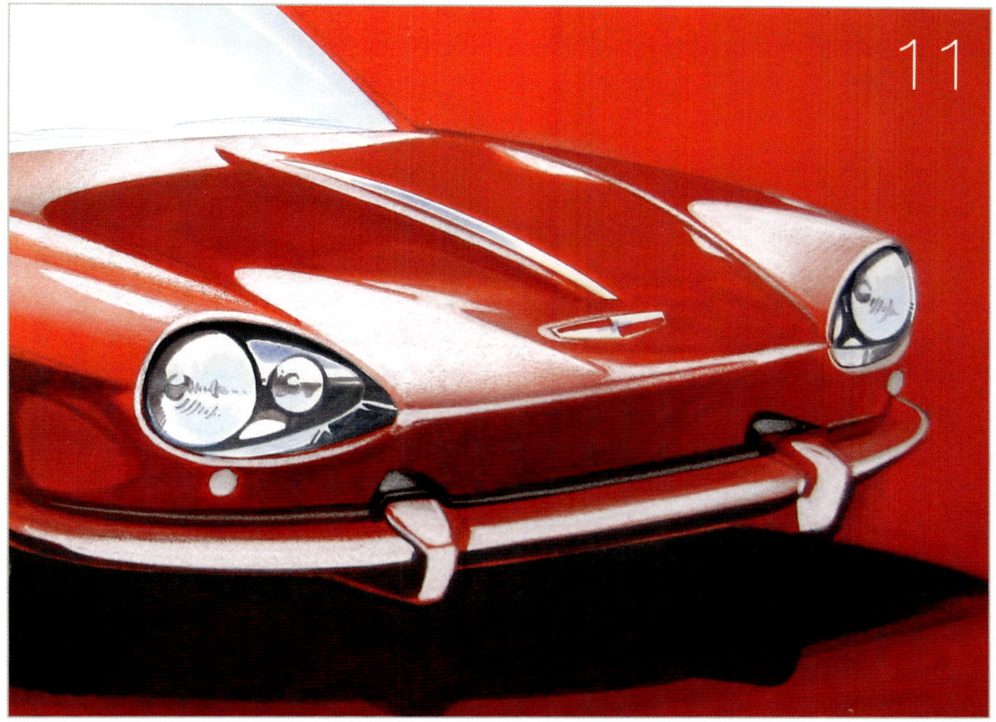

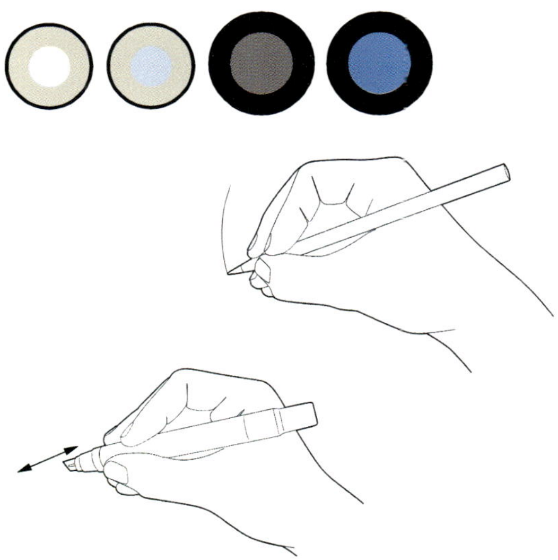

Use a white pencil to show the shut line edges and the wheel flare highlights. Indicate shadows on the tires with a Cool Grey 40% marker. Use blue markers and pencils to detail the wheels.

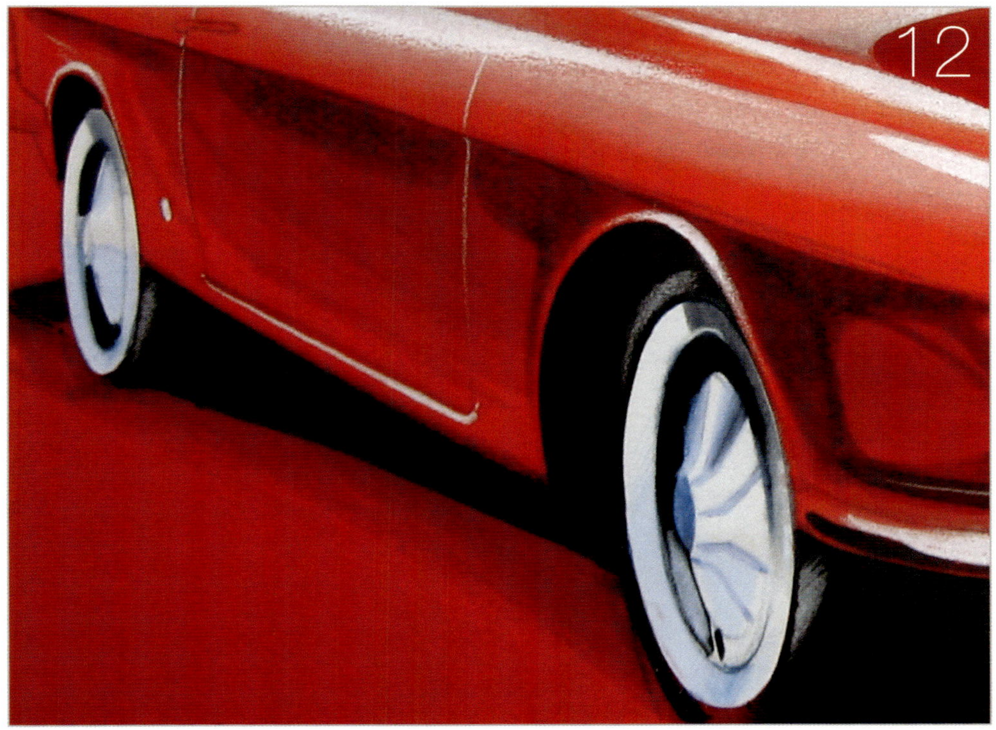

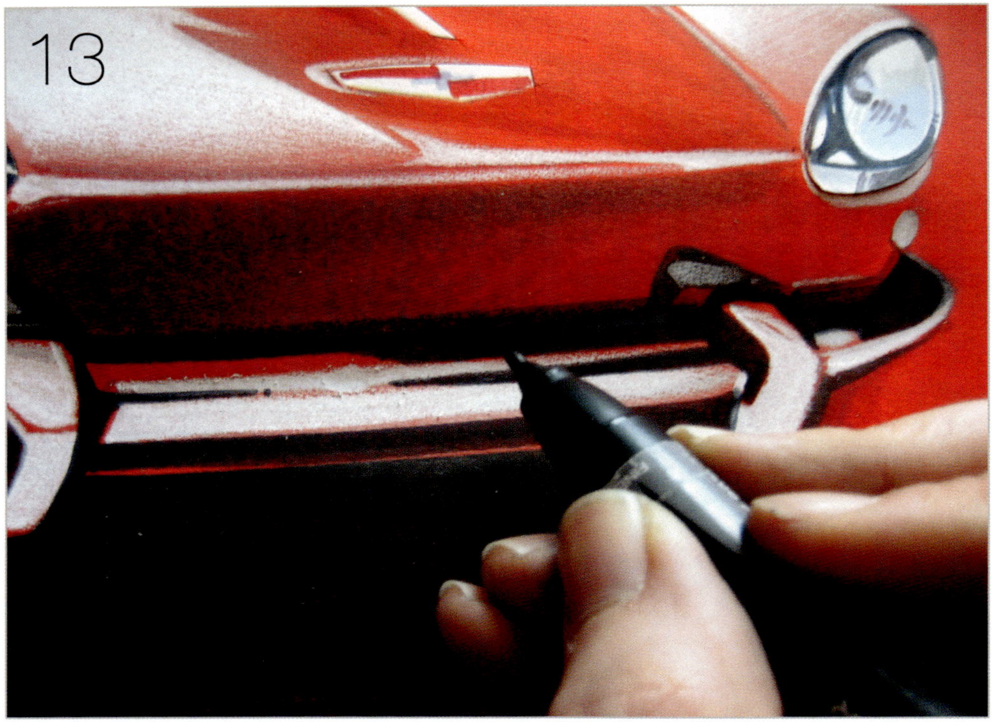

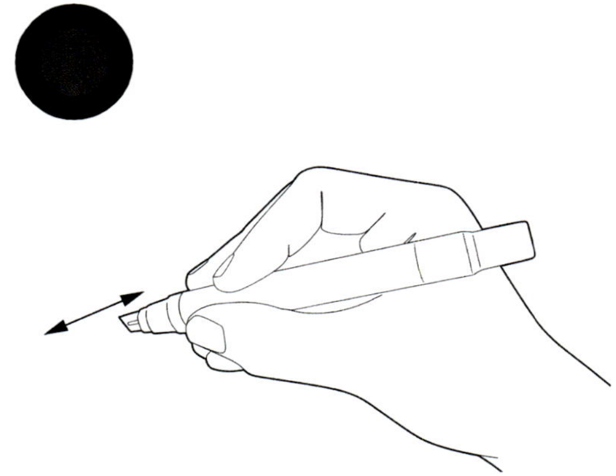

Similar to the headlamp bezels, lay in some Cool Grey 70% for the bumper's hard reflections. Leave some room for a highlight near the middle.

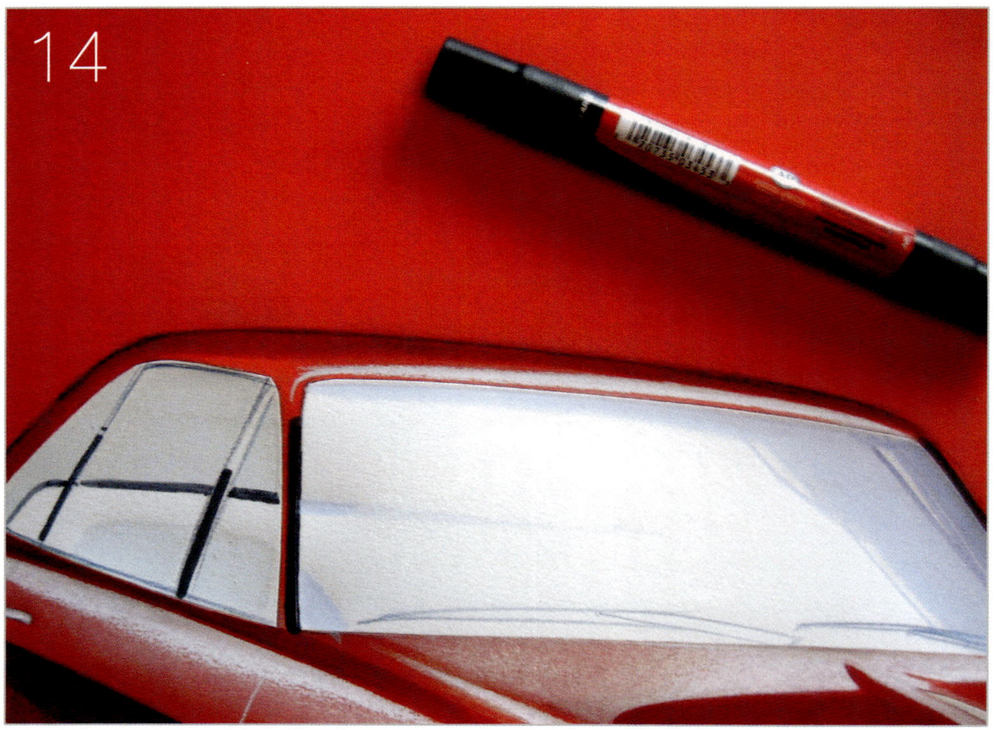

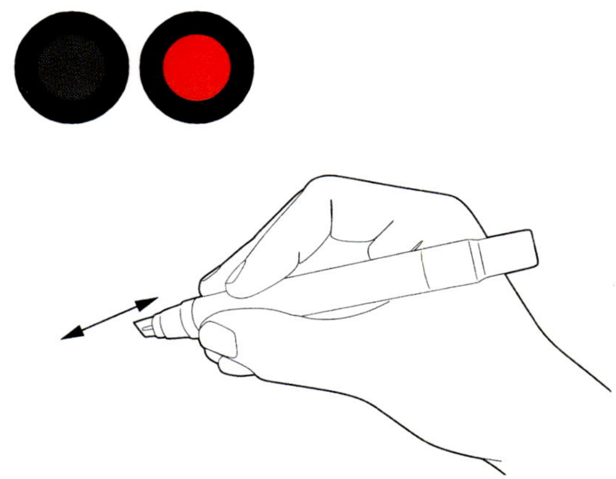

Continue with Cool Grey 70% to add horizon lines in the chrome window frames. Add some definition to the roof with a Crimson Red marker.

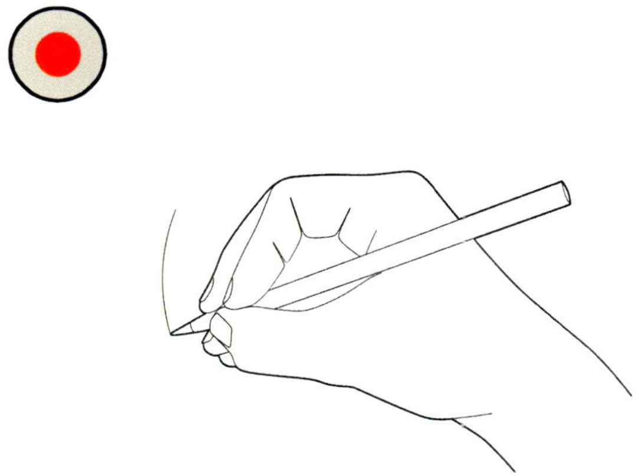

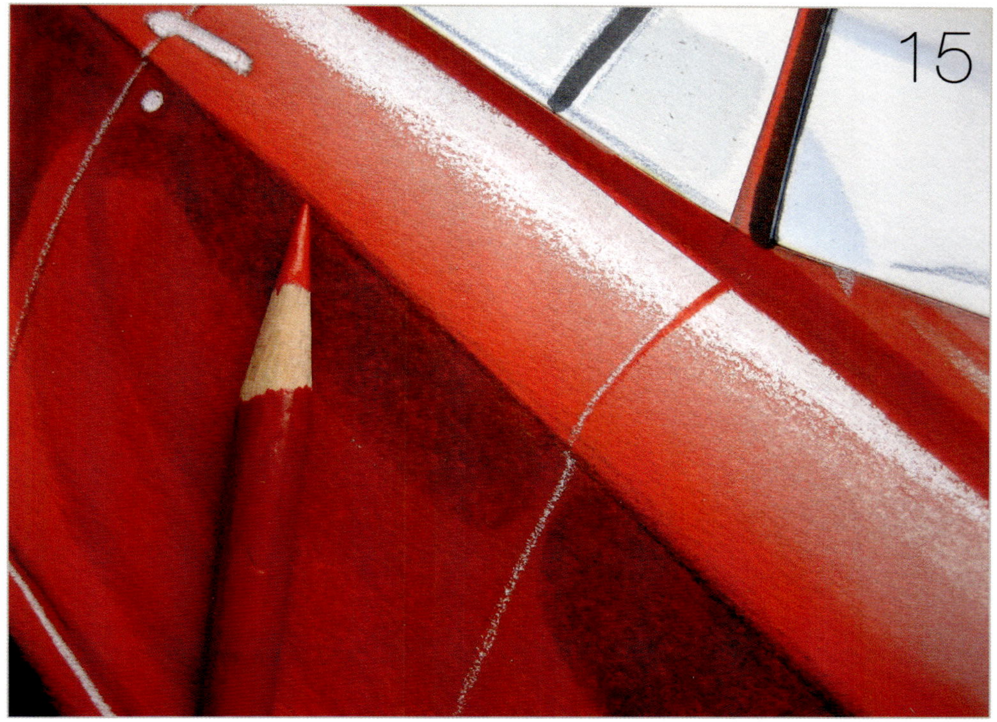

Instead of using black pencil to indicate the shut line itself, try a Crimson Red Prismacolor pencil.

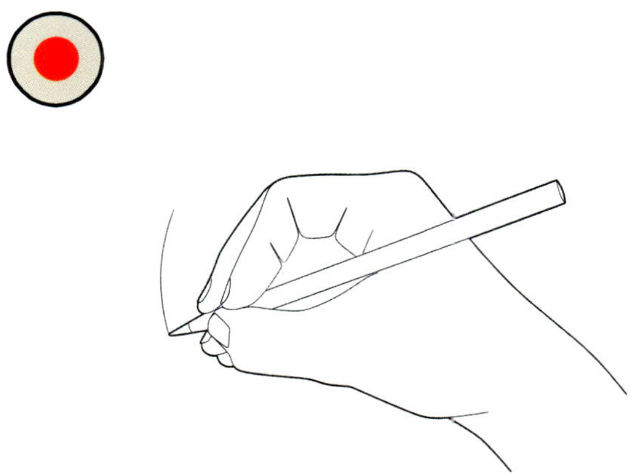

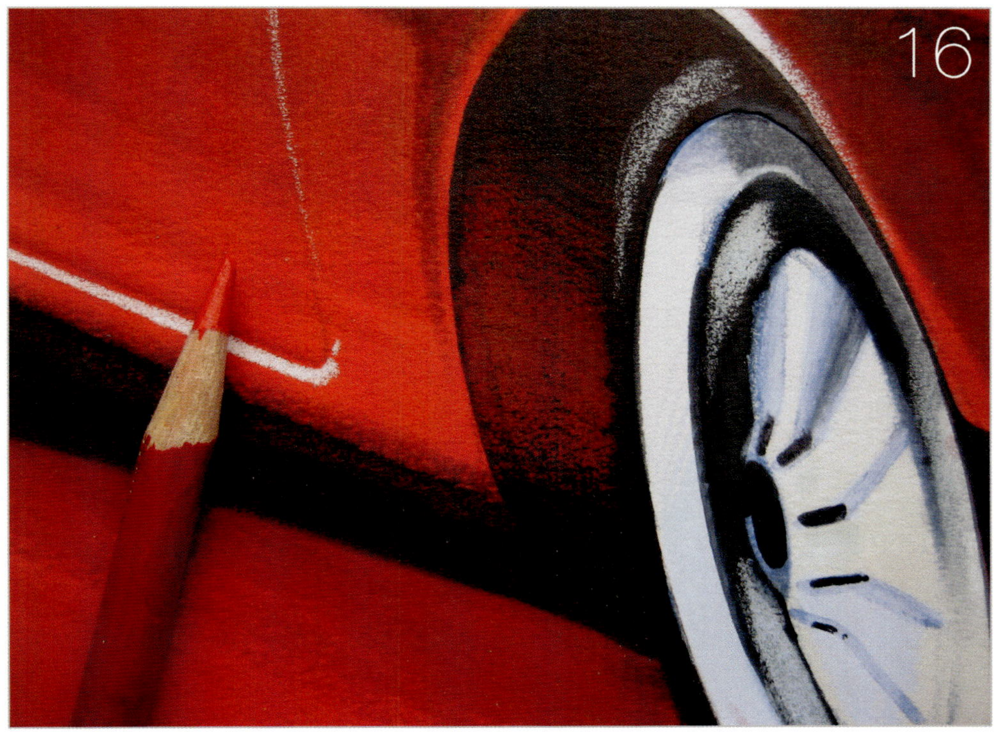

Staying with the same pencil, add a bit of color behind the front and rear tires; this will suggest that the red paint continues into the wheel wells.

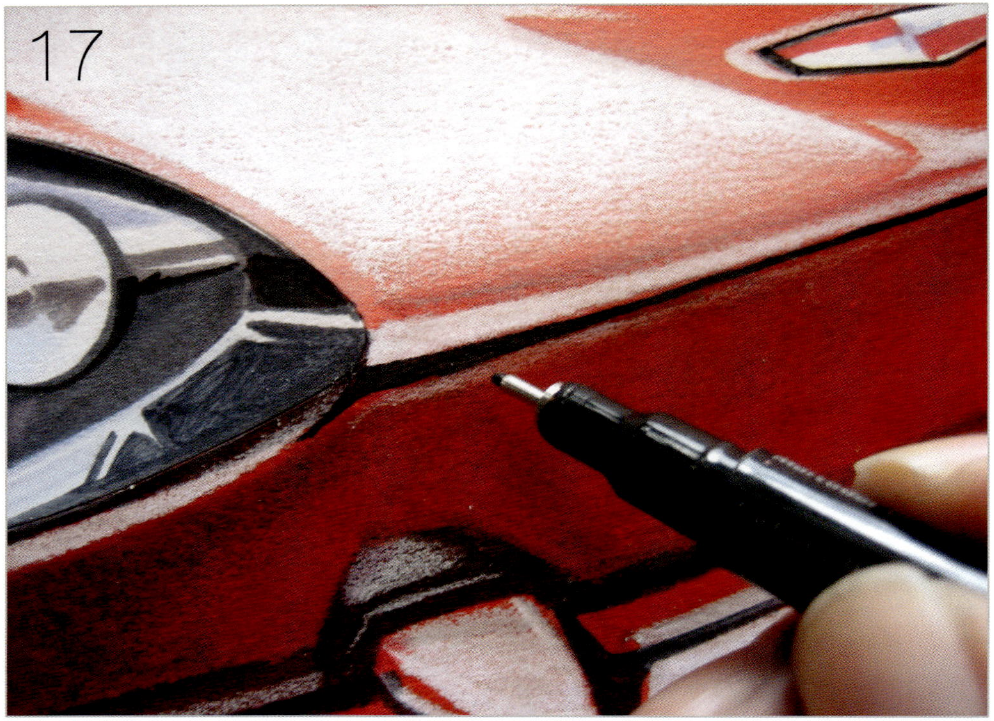

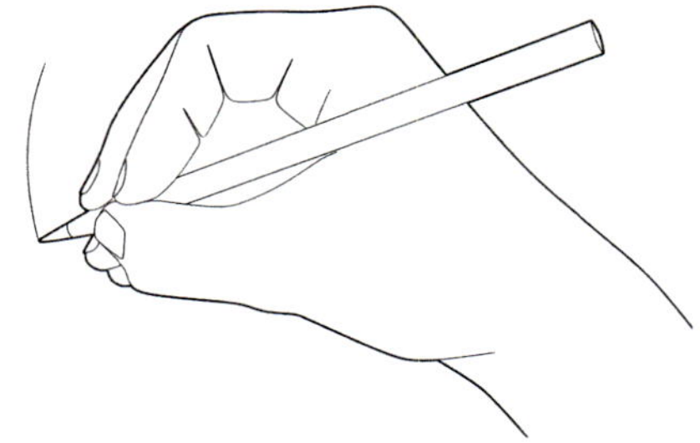

For shut lines that have larger gaps, like the hood shut, try using a fine-tipped black pen. While you're working in this area, add some punch to the horizon lines in the headlamp bezels.

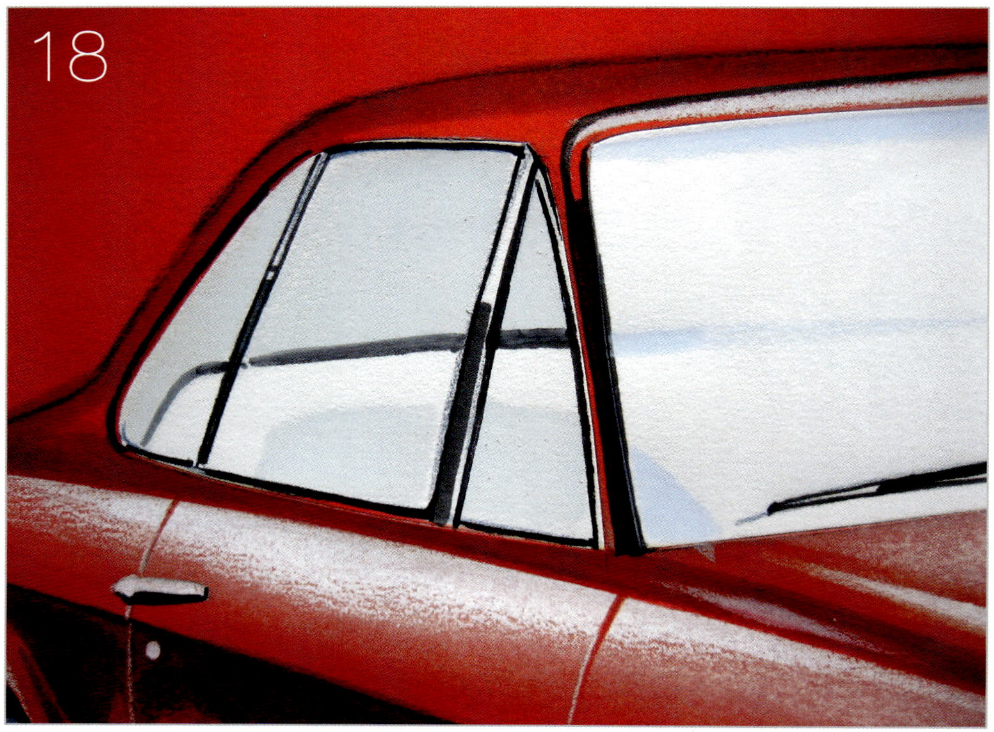

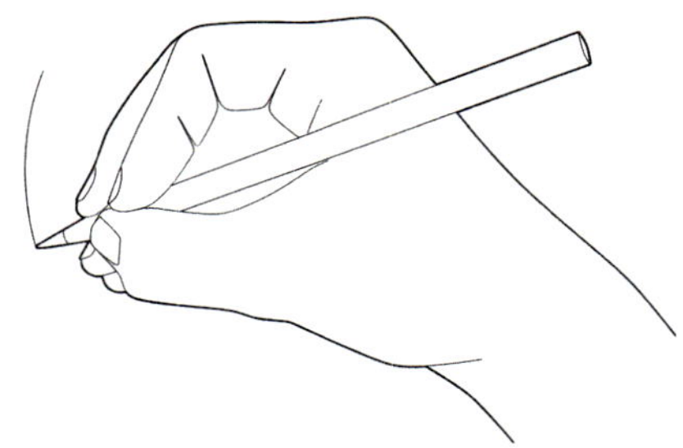

Define the edges of the chrome window frames with the same black pen. Hit the windshield wipers and door handles too, if you like.

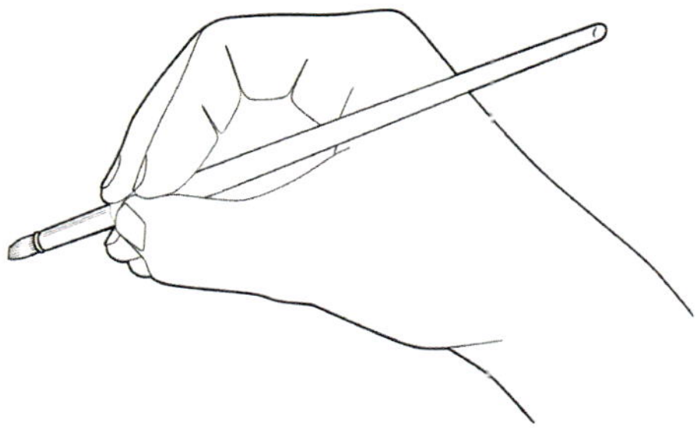

Remember that highlights like to line up with each other like moons in the same orbit. This also tends to enhance the illusion of shiny chrome.

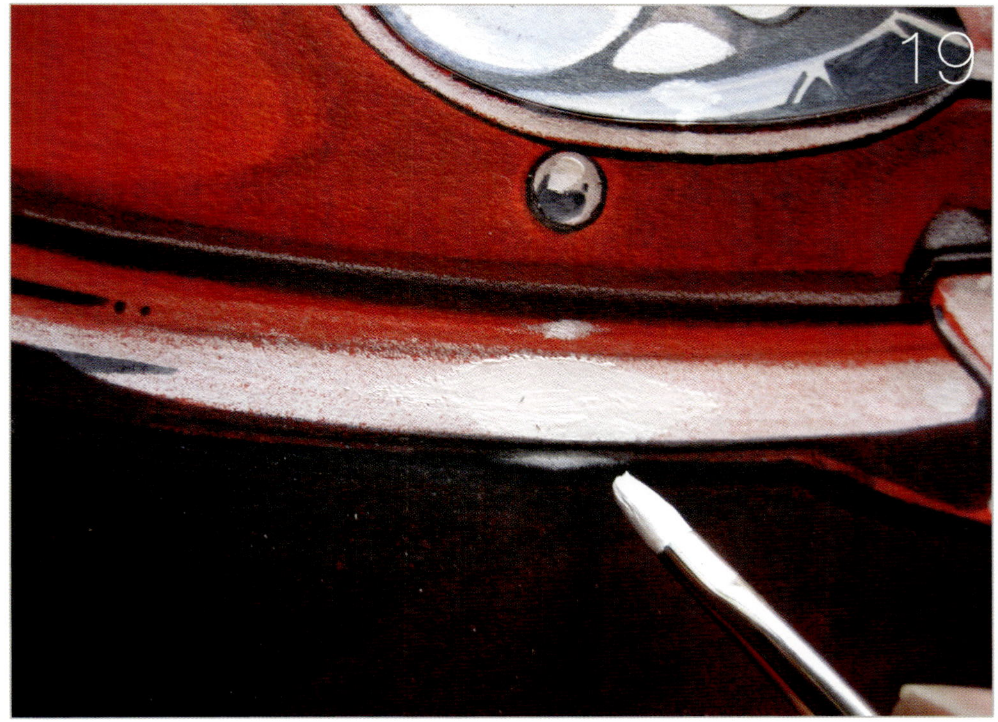

19

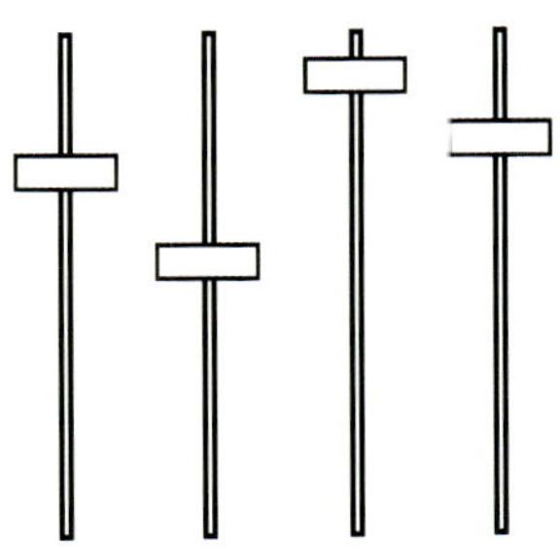

Scan in the rendering and adjust the levels to your liking in Photoshop. Use a white airbrush tool to soften the highlighted areas and add a bit of glow.

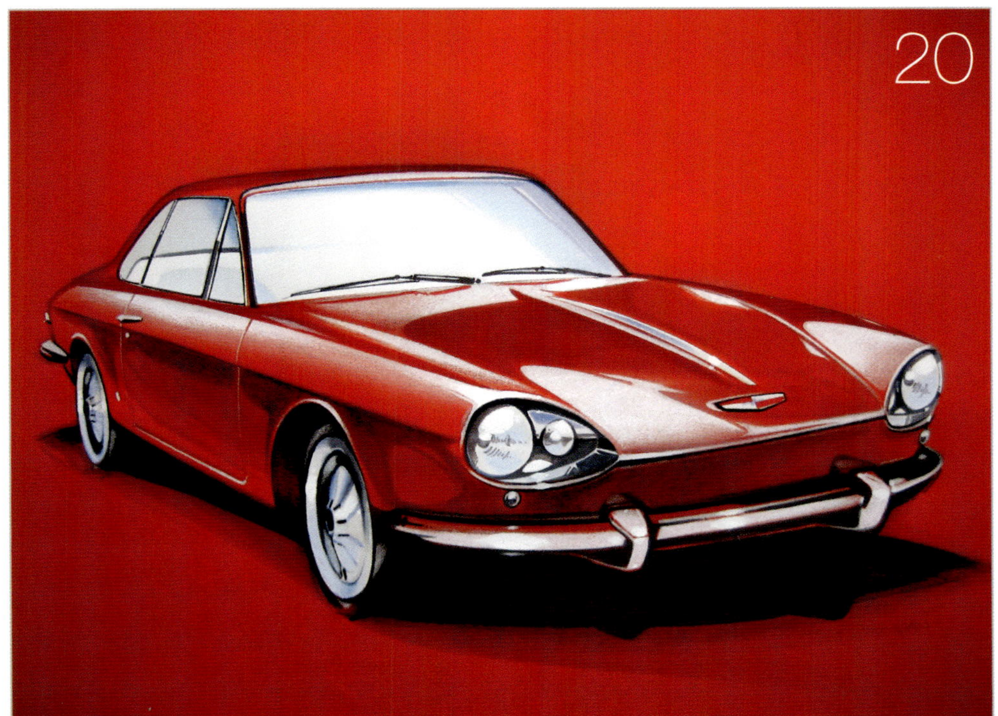

20

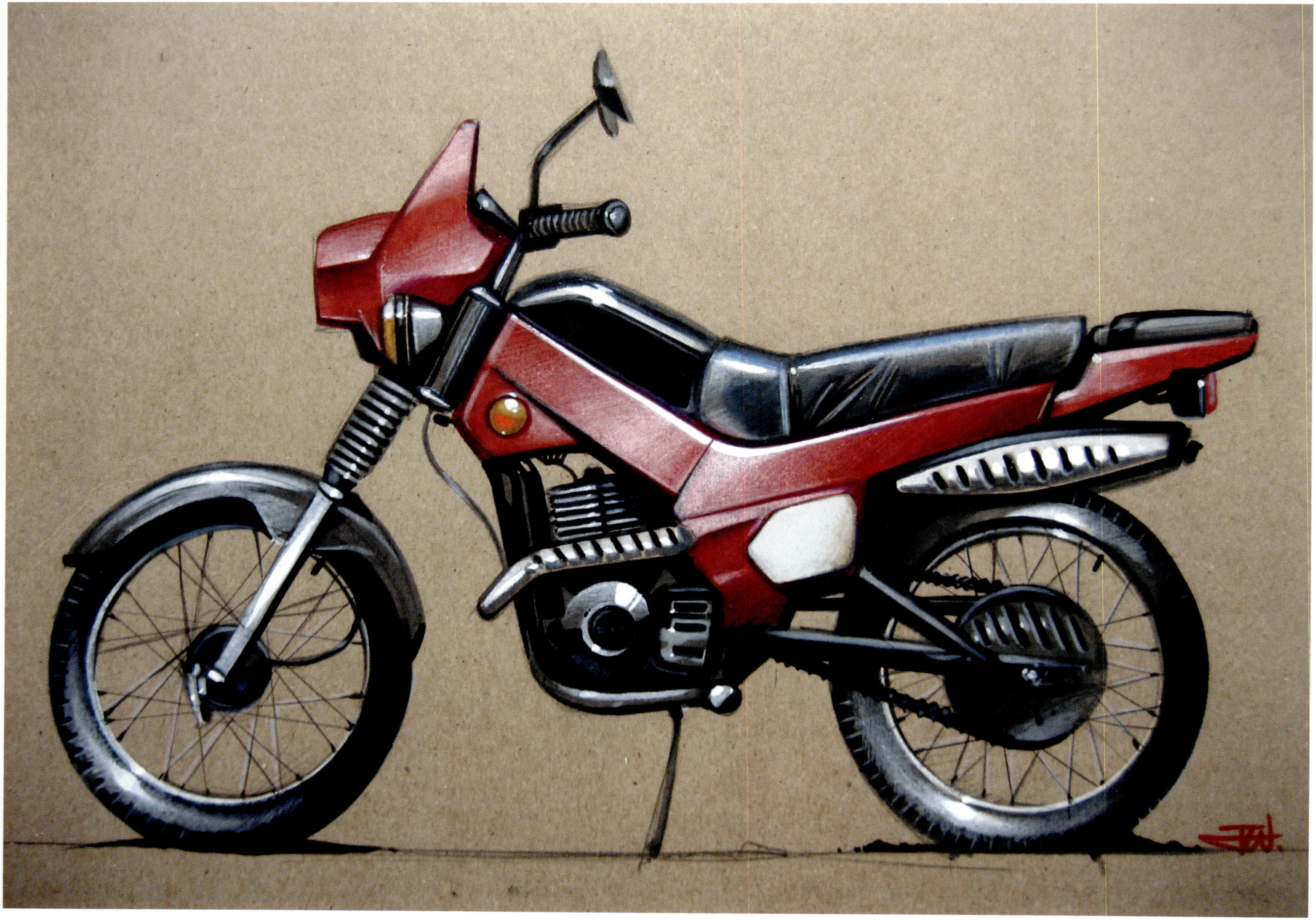

17. Recycling

Chipboard is another word for the backside of vellum and newsprint pads. I never throw these away. Not only is the surface of this material a pleasure to draw on, but it also works surprisingly well with markers. It's similar to Canson, in that it provides something of a rendering middle ground. But where Canson is textured and rough, chipboard is race track smooth – almost like Masonite. If you're patient, you can achieve a level of precision and realism here that you can't get on other drawing surfaces.

The Tools

Prismacolors:
White, Cool Grey 20%,
Mahogany Red, Deco Pink.

Verithins:
Black.

Felt Pen:
Black.

Gouache:
White.

Prismacolor Markers: Black, Cool Grey 40%, Crimson Red, Cool Grey 70%, Cerulean Blue, Orange.

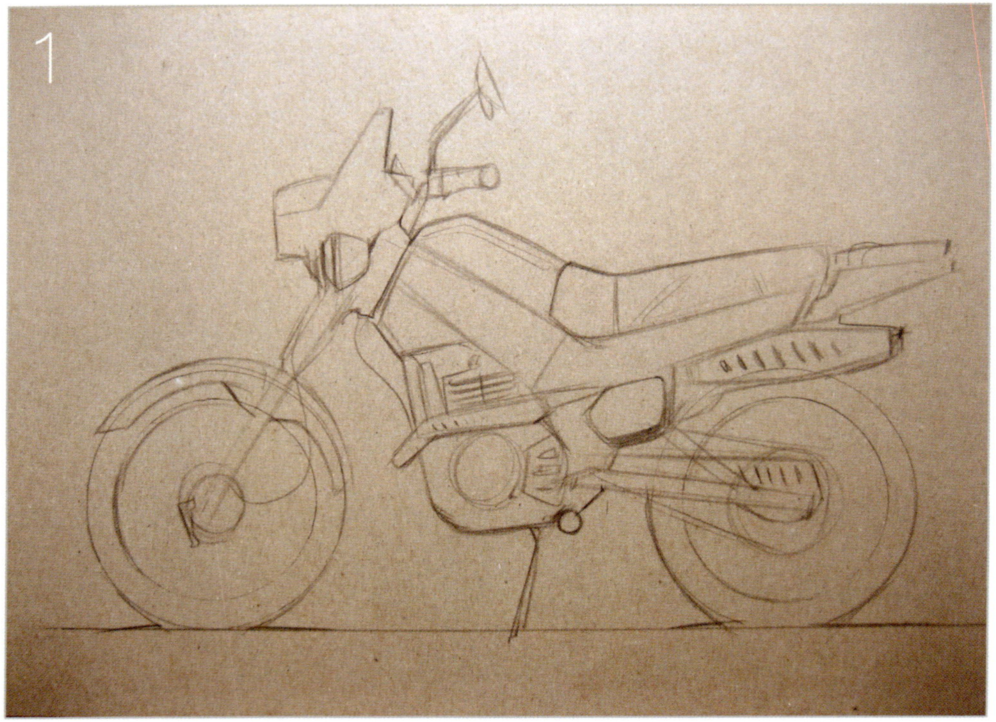

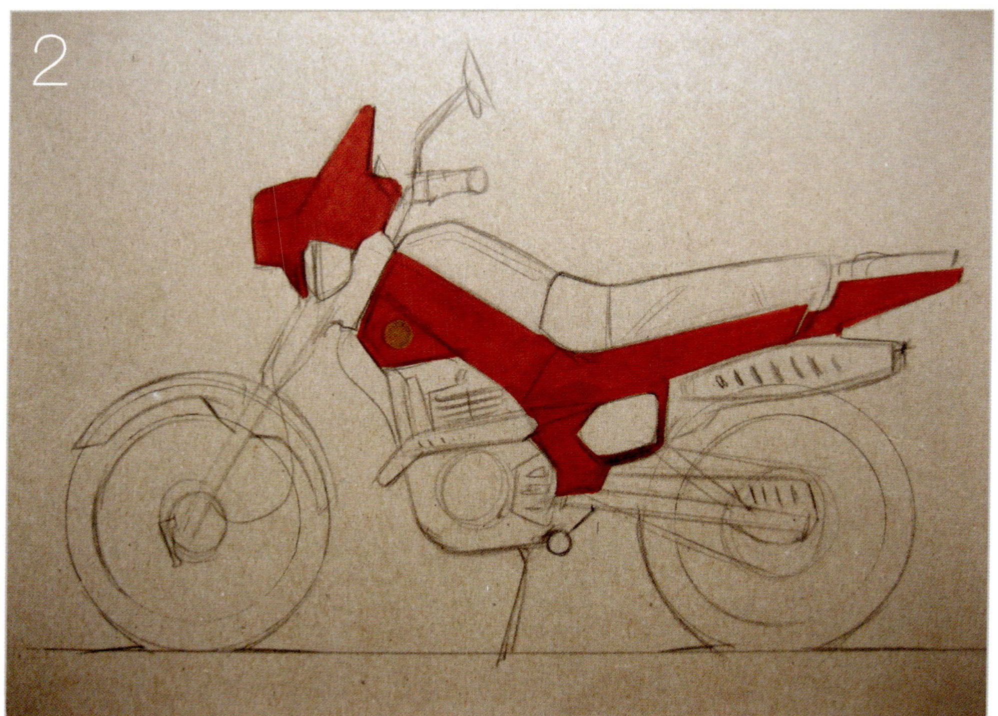

Start by laying out the basic lines of the bike with a black Verithin pencil.

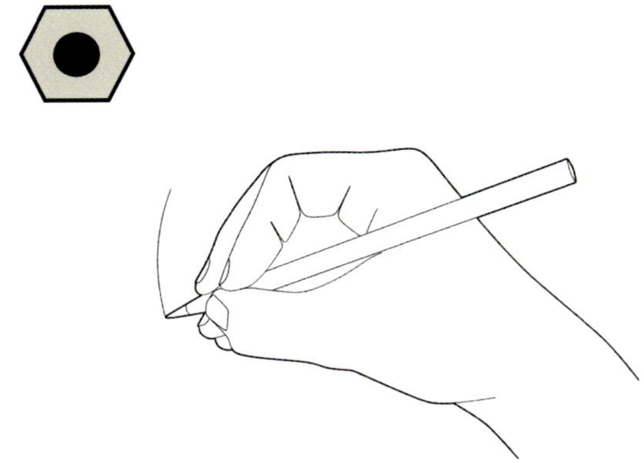

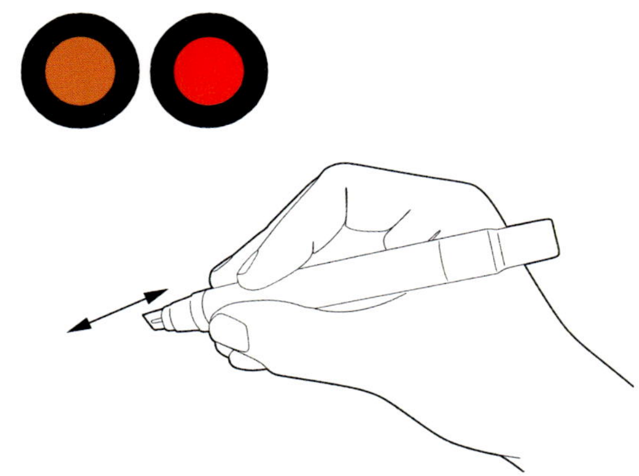

It is critical to apply the red marker *before* you apply the black marker (red ink can easily pick up the black ink and turn discolored). The red marker will provide the mid-tone color for the body of the bike, while light and dark tones will be added with pencil.

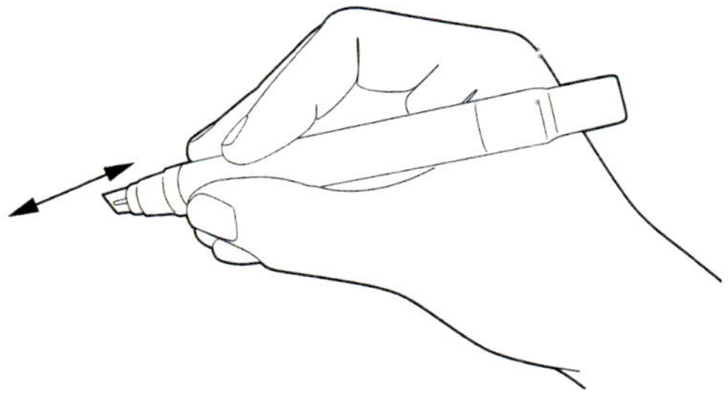

It may be tempting to reach for the black markers right out of the gate, but the smarter move is to start with Cool Grey. This way, you'll have a nice middle ground from which to add light and dark tones, similar to the way the red body will be described.

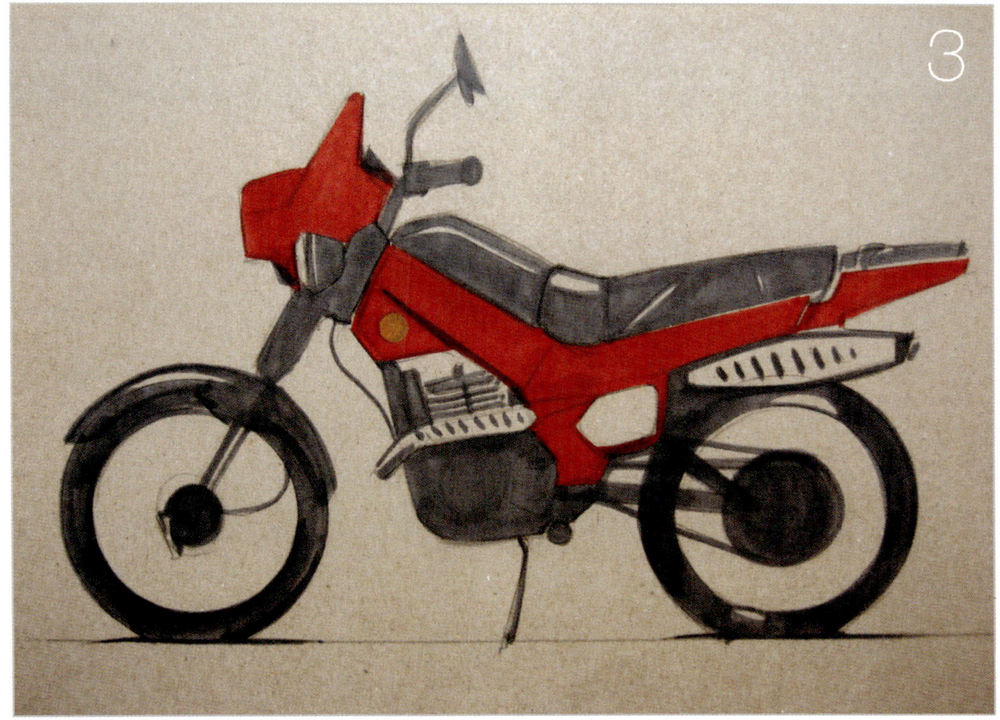

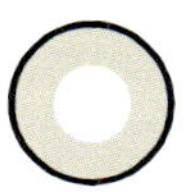

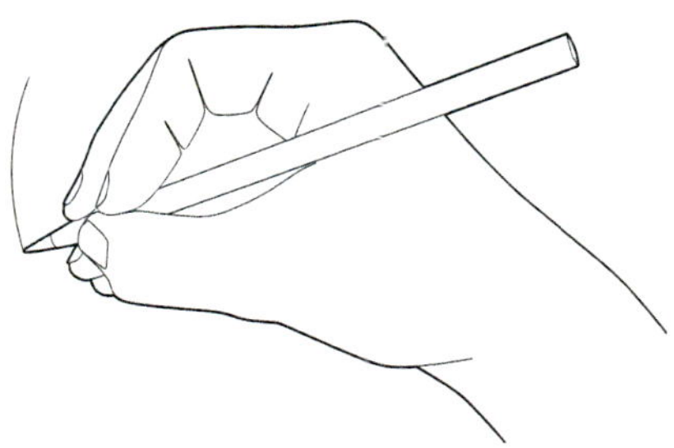

Before you add any more black linework, it's a good idea to lay down some of the bolder highlights with a sharp, white Prismacolor pencil. Use an improvised guide like this roll of tape in order to get a crisp, true highlight.

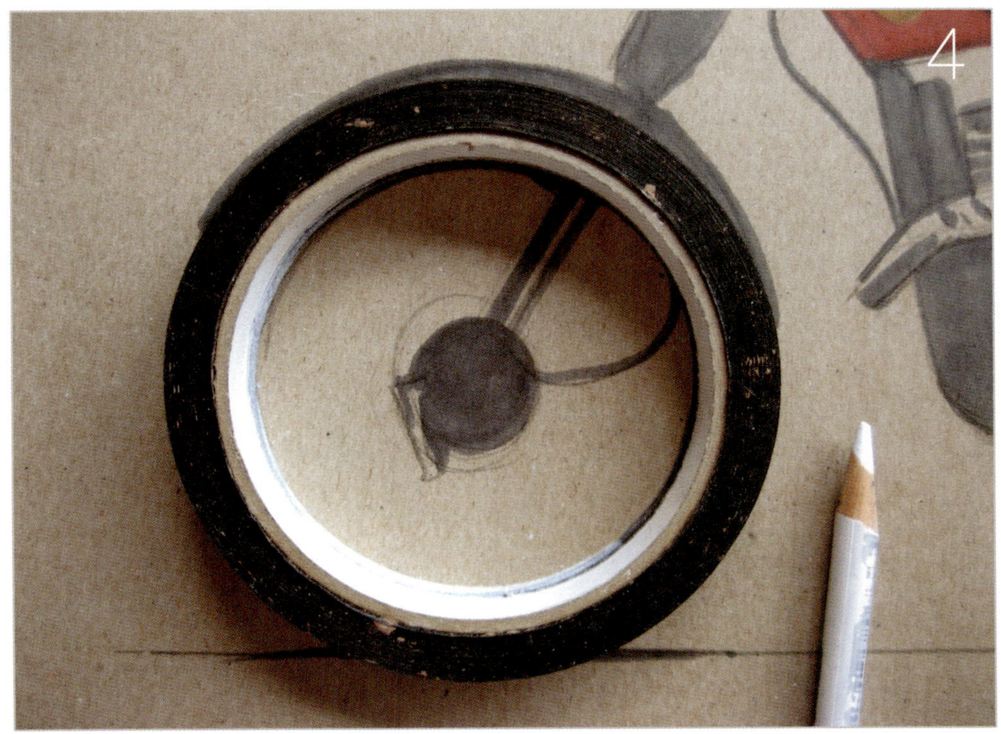

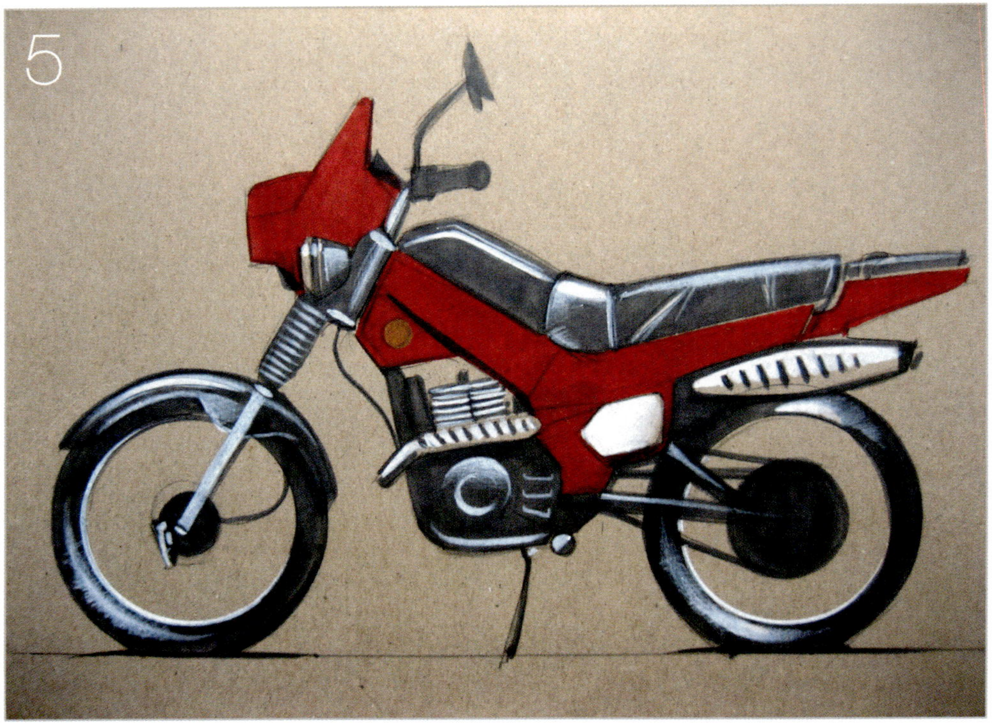

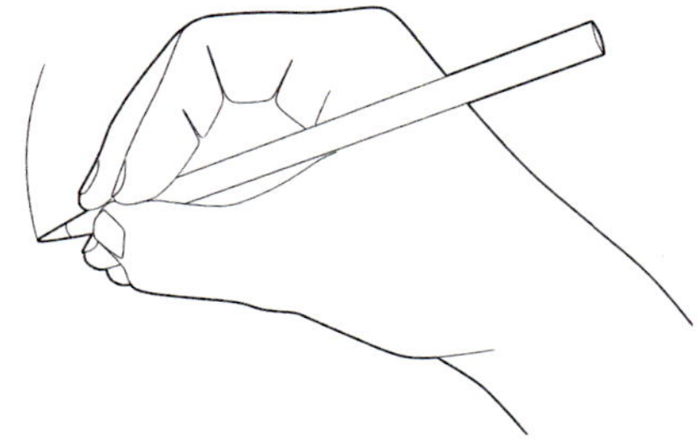

Continue by applying the bolder highlights, but don't confine yourself to white pencils. Use French Greys or Cool Greys in areas where the highlights are a bit broader, then punctuate them with a white hot spot (as seen on the rear tire).

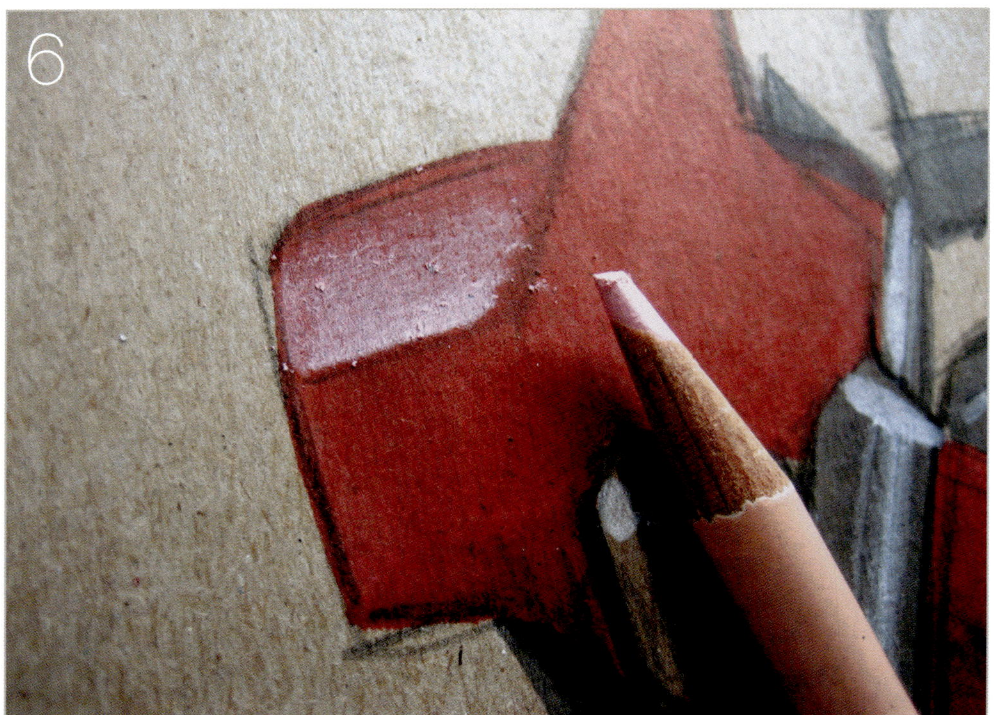

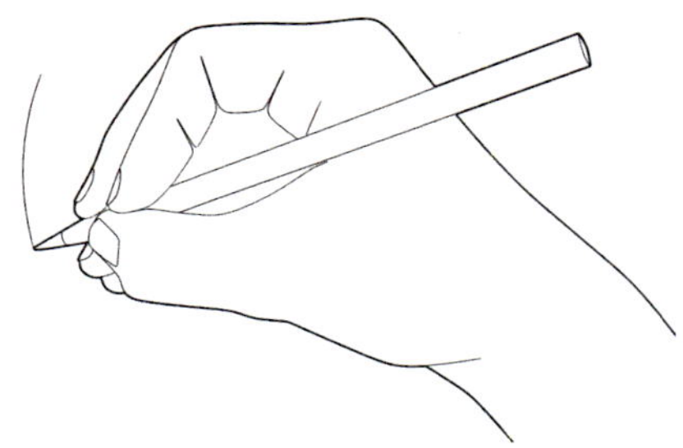

For the lamp housing highlight, start with a warm Prismacolor pencil, such as Deco Pink. You can use a white pencil to create some hot spots later on, but starting with a pencil keyed slightly to the mid-tone will give the body color a lot more interest.

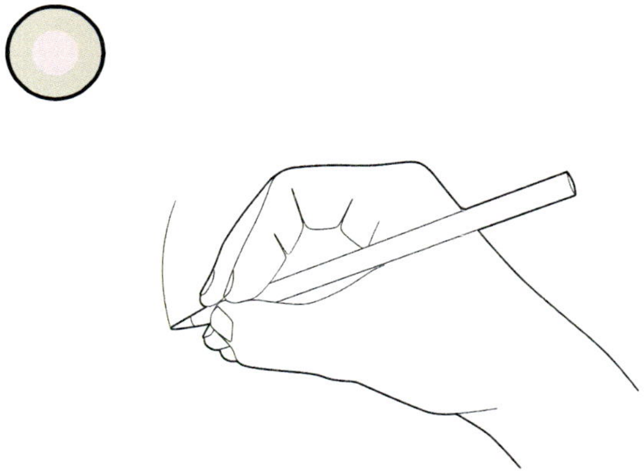

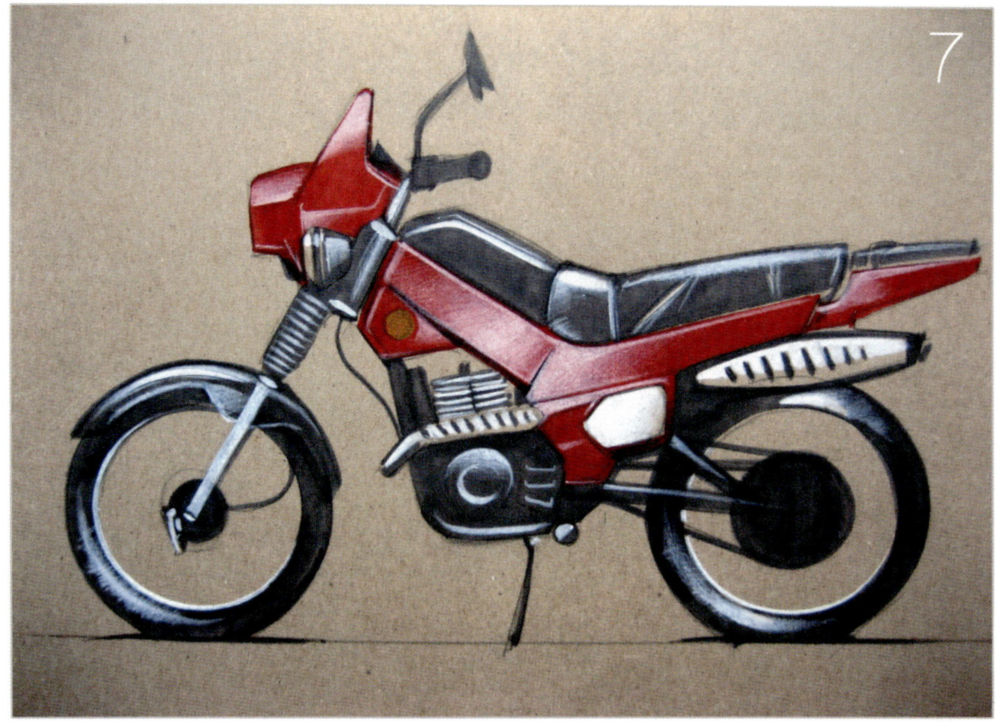

Continue applying Deco Pink on the frame of the bike, concentrating the light in the center to create a focal point.

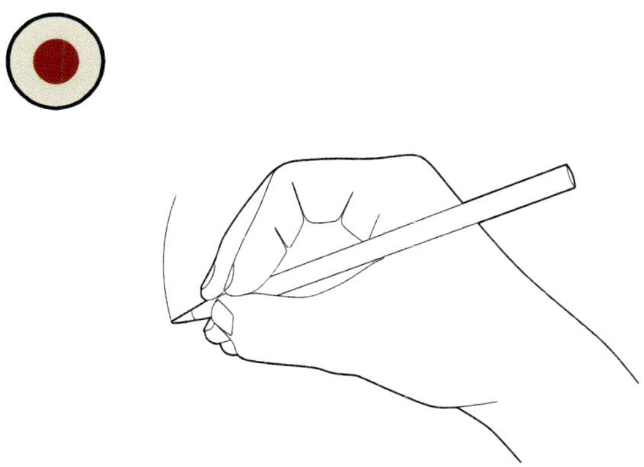

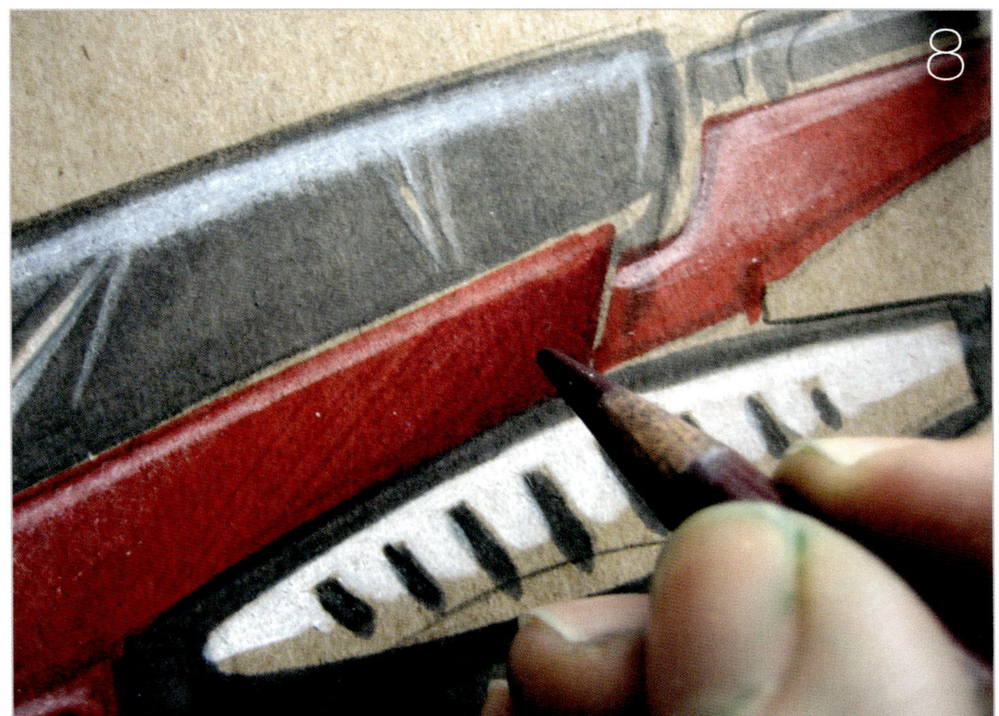

For the darker parts of the frame, try using a Mahogany Red pencil or maybe even a purple pencil. Similar to the Deco Pink, this will make the overall red far more interesting.

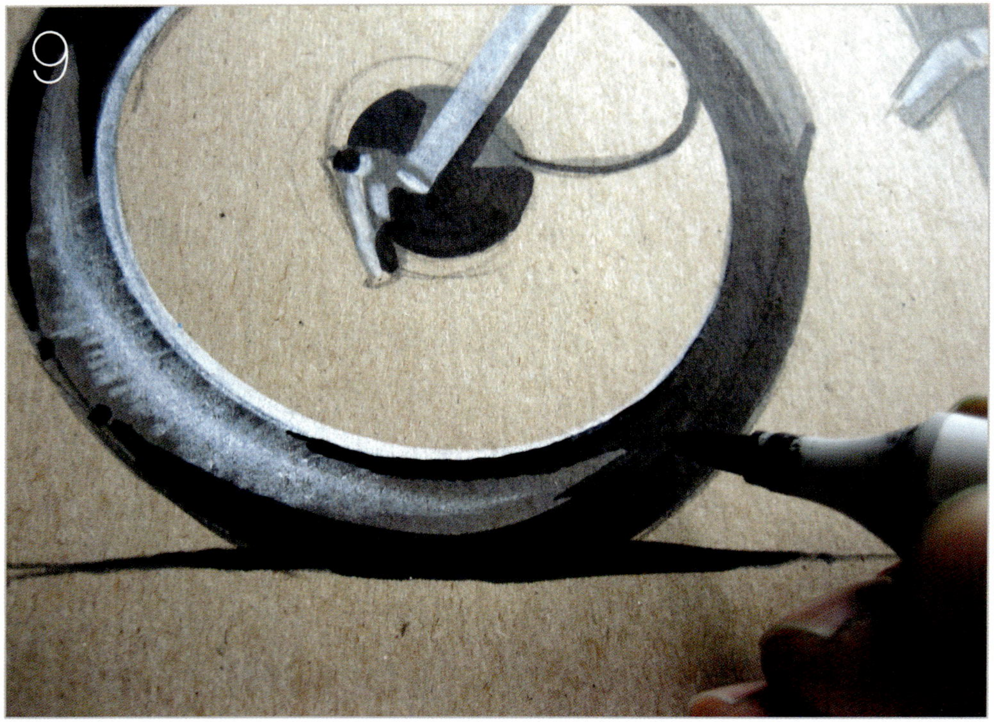

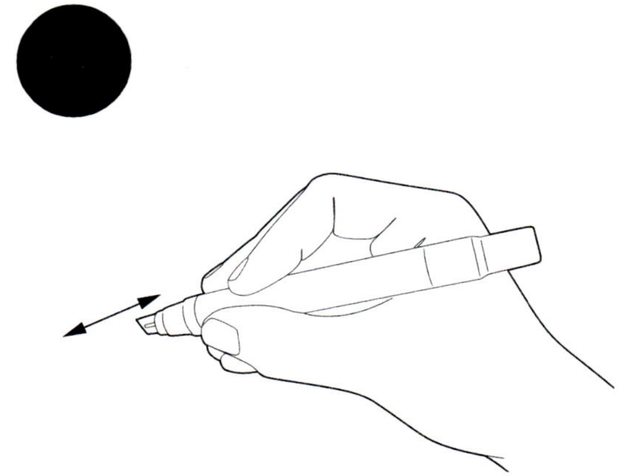

Turning to the tires, lay in black marker in areas that are not catching highlights. Be sure to leave a generous amount of Cool Grey marker exposed around the highlight itself; this will serve as the mid-tone for the tires.

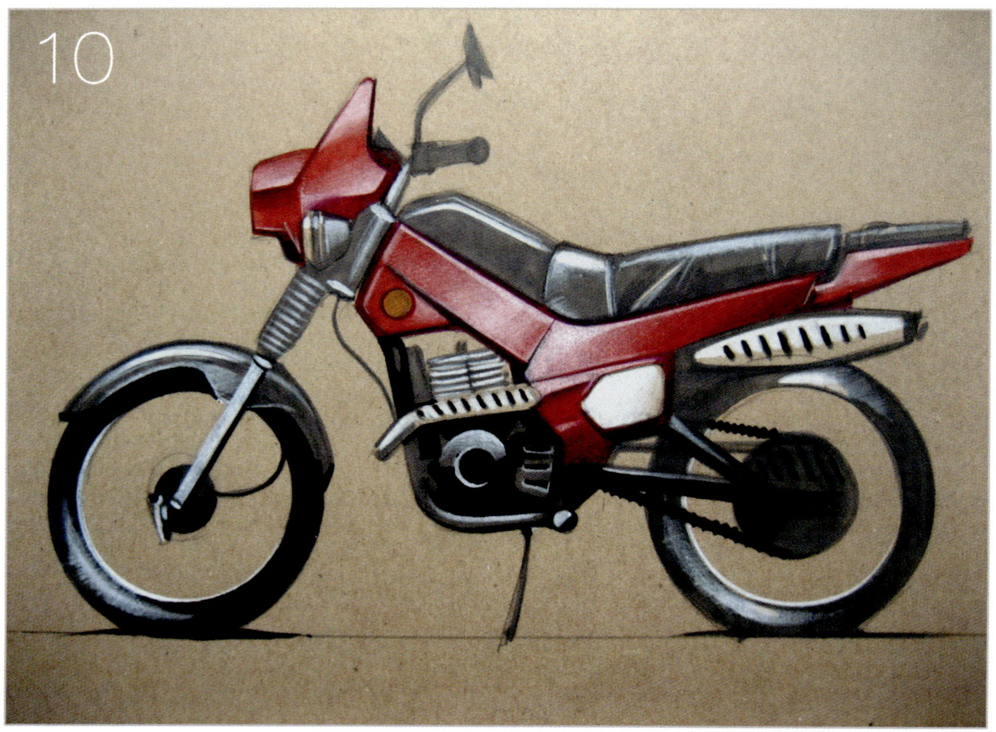

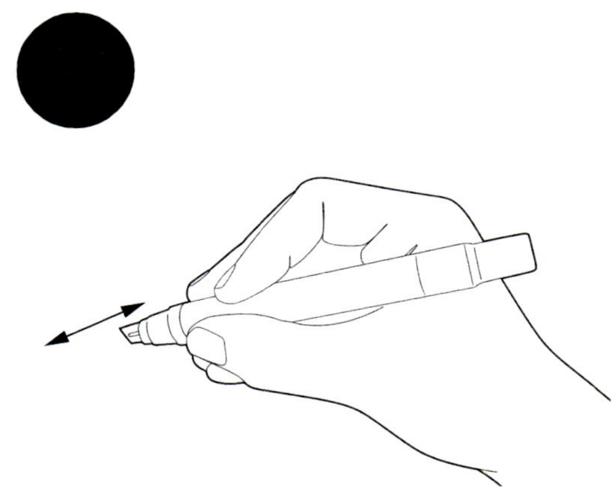

Continuing with the black marker, block in the darkest parts of all the mechanical elements. At this point, you can start to see the critical role that the grey marker plays in the rendering.

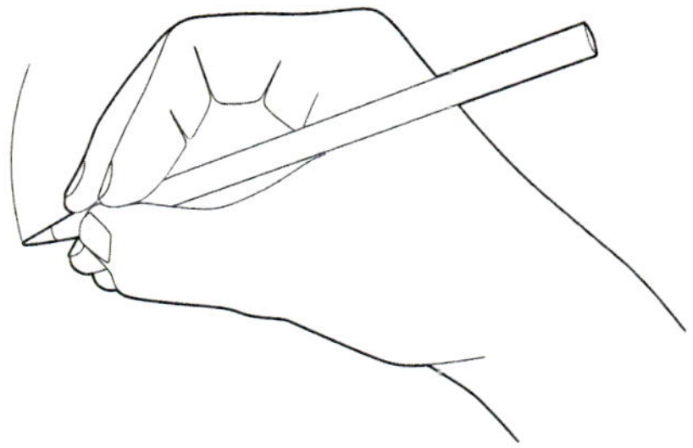

With a black felt-tipped pen, add some detail to the bike's front springs.

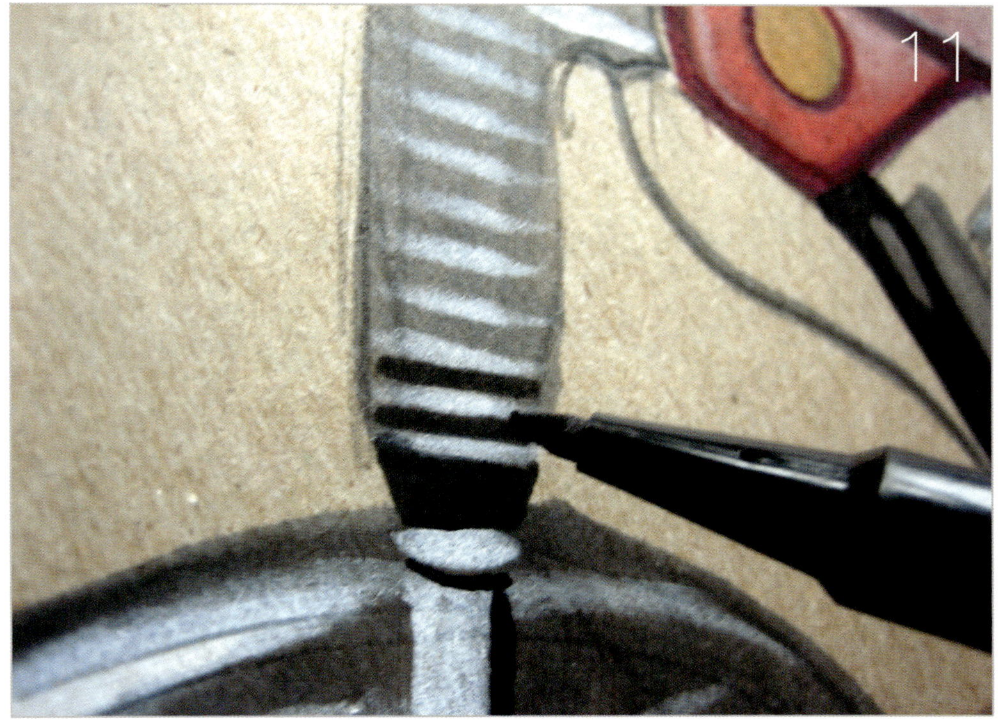

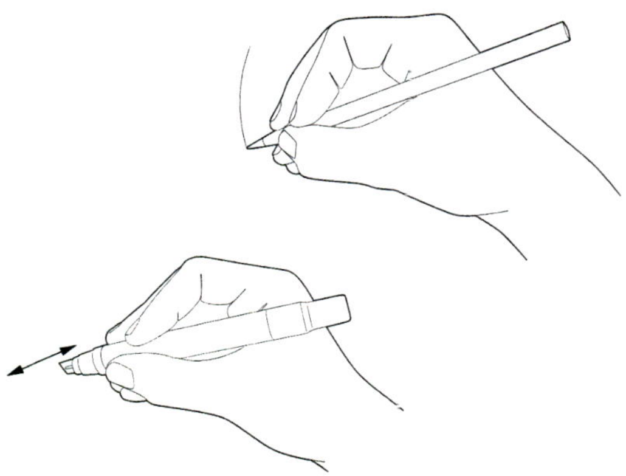

Use the same pen to accent the deep folds in the black leather seat. Add some Cerulean Blue marker on the top to indicate a bit of sky tone.

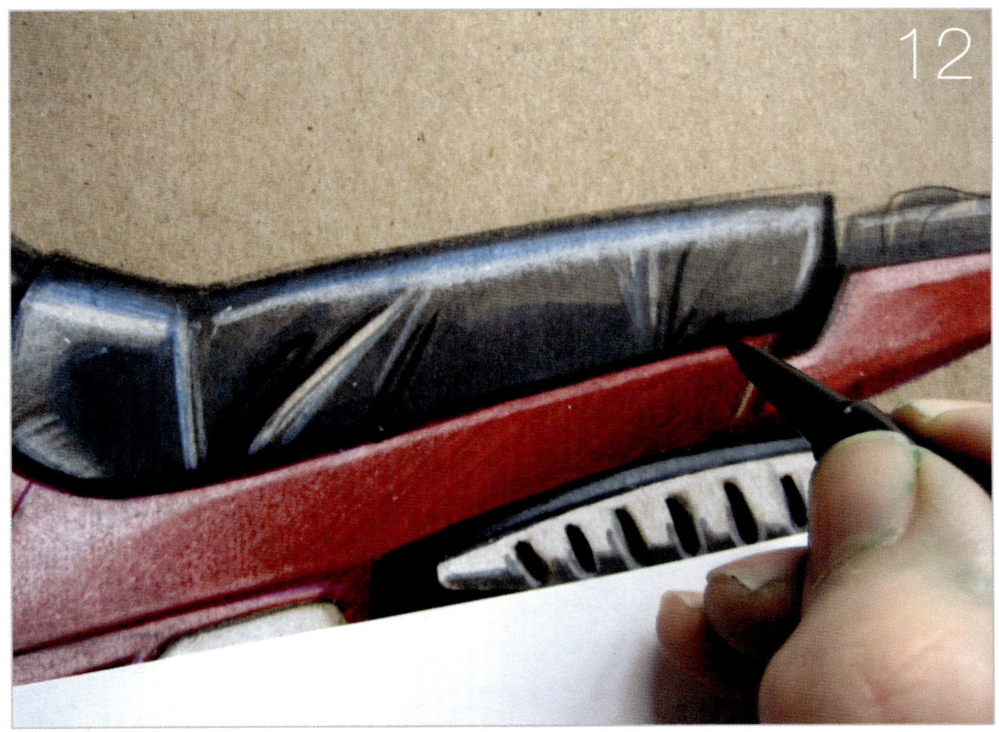

13

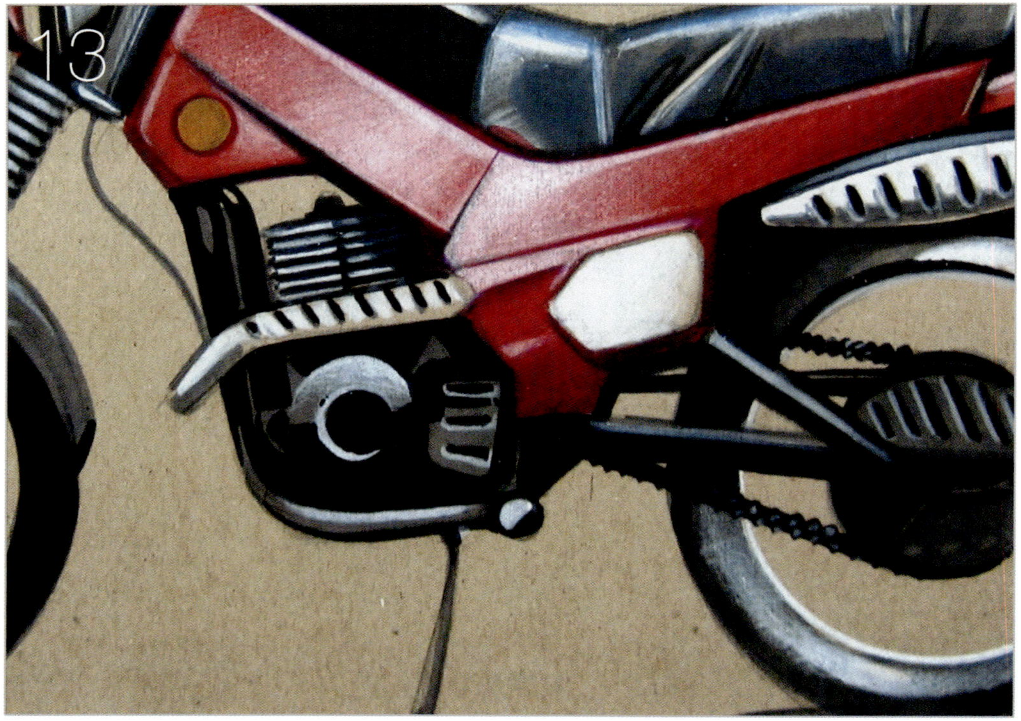

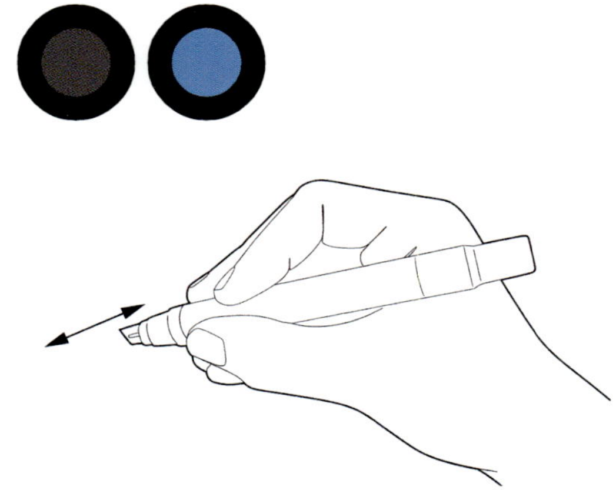

As you continue detailing the mechanical parts, remember that some items, like the engine, will be shadowed by the frame. To illustrate this, you can overlay white pencil highlights with a combination of grey and blue markers.

14

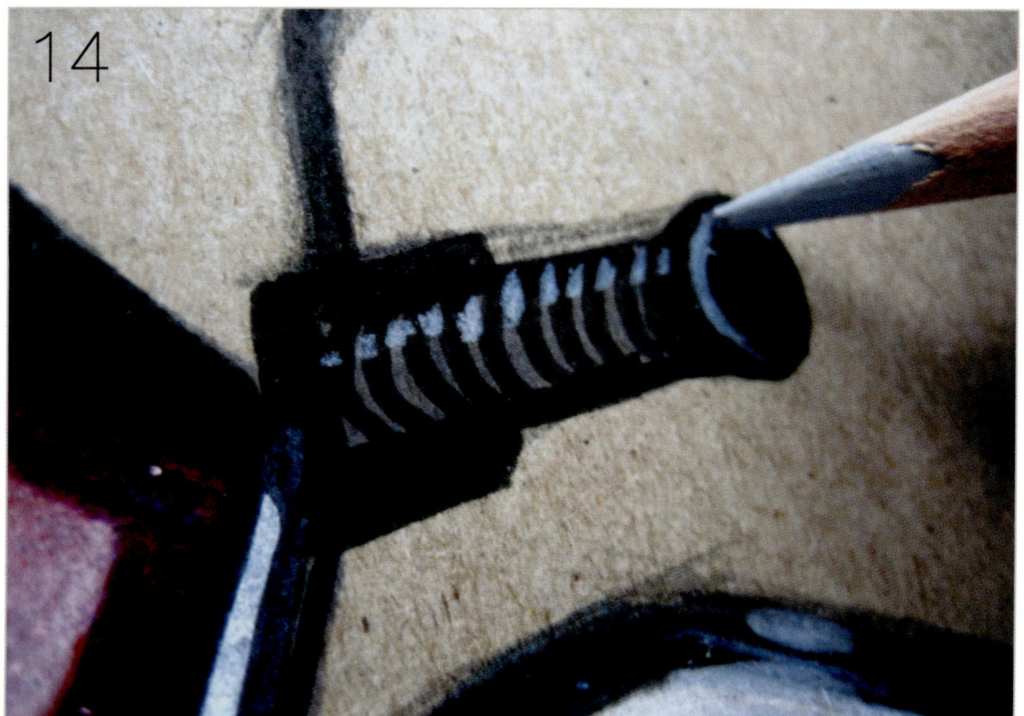

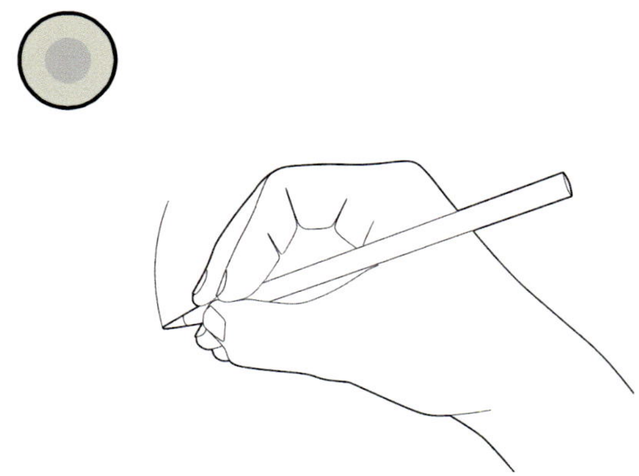

The hierarchy of details can sometimes come down to color choice. For the handlebar highlights, use a Cool Grey pencil instead of a white pencil. This will ensure that the handlebars won't compete with the chrome for attention.

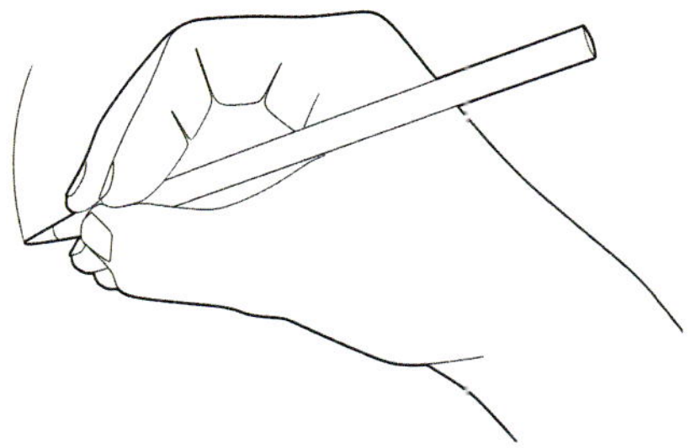

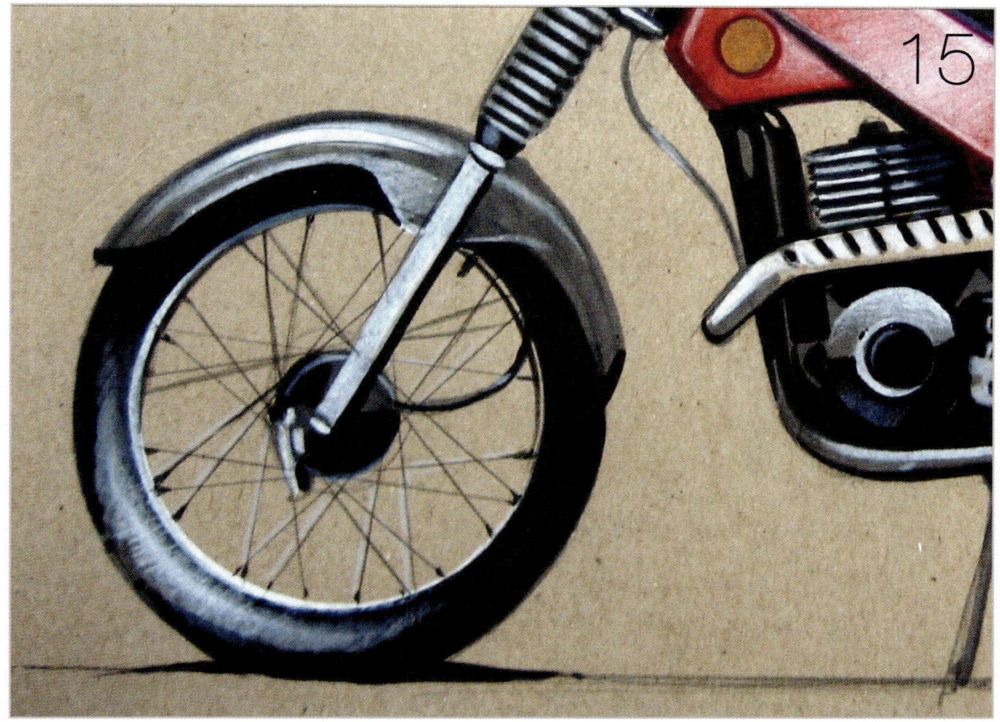

For the spokes, use a straight edge and strike light, black pencil lines. Then use a Cool Grey pencil to highlight the edges of some of the spokes.

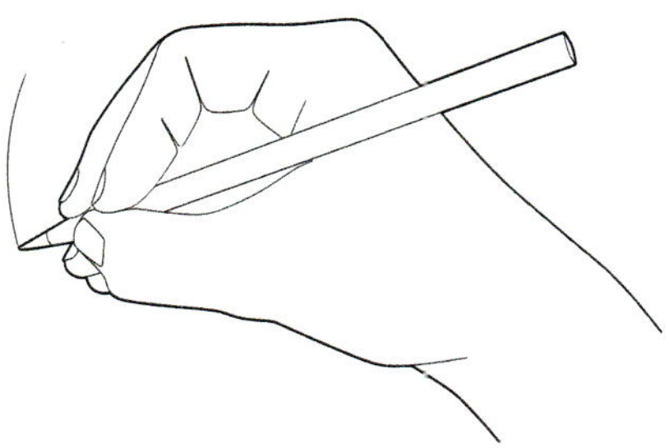

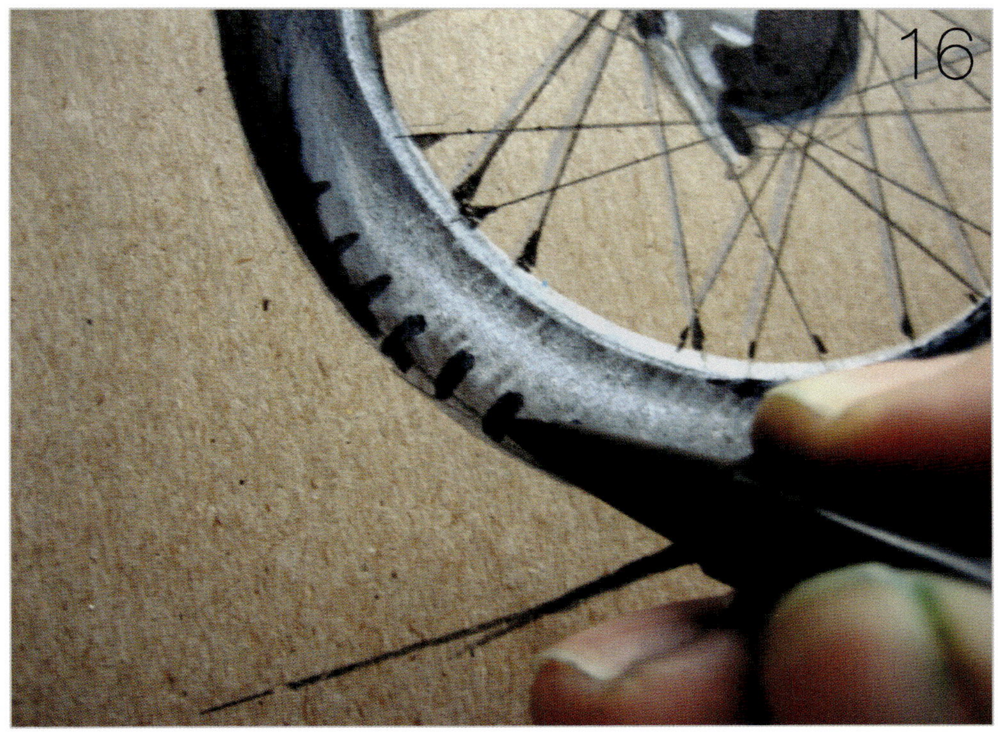

Use the black felt-tipped pen to indicate the tire tread.

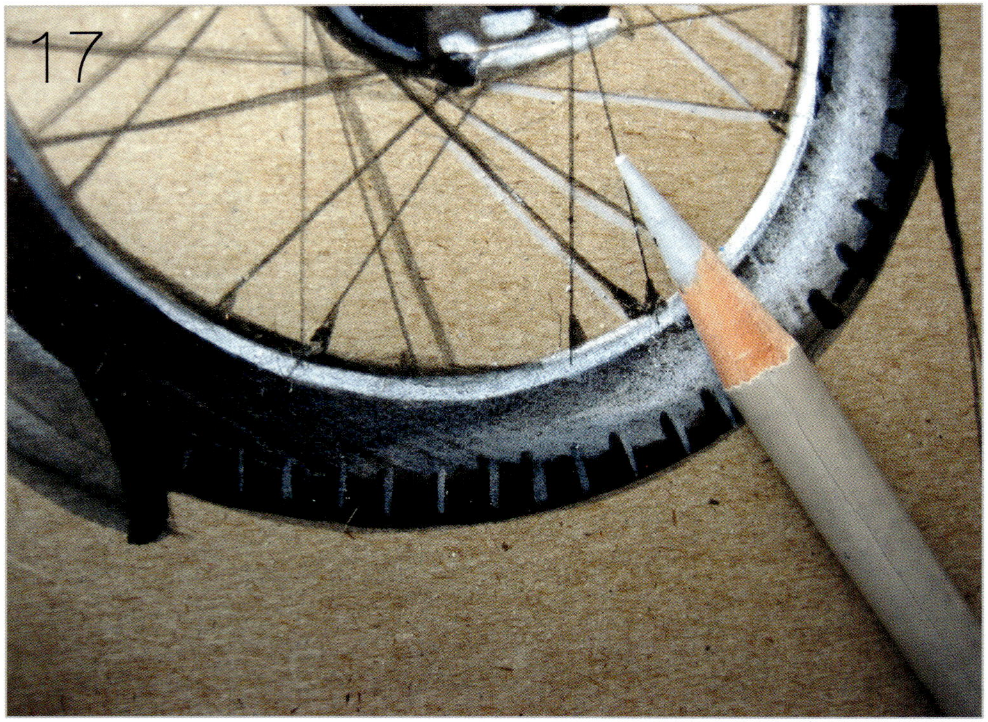

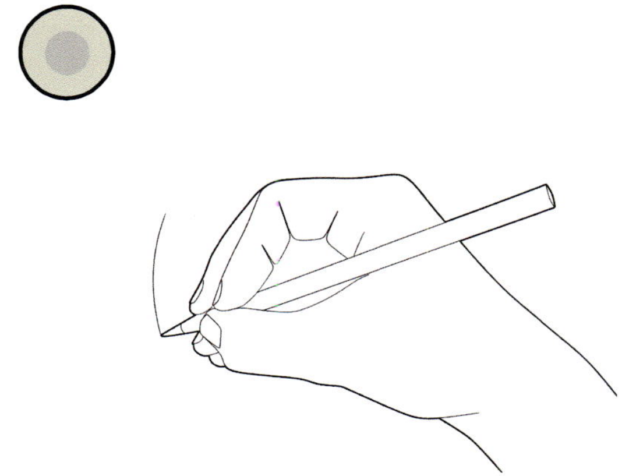

Similar to the spokes, use a Cool Grey pencil to indicate the lit edges of the tire tread.

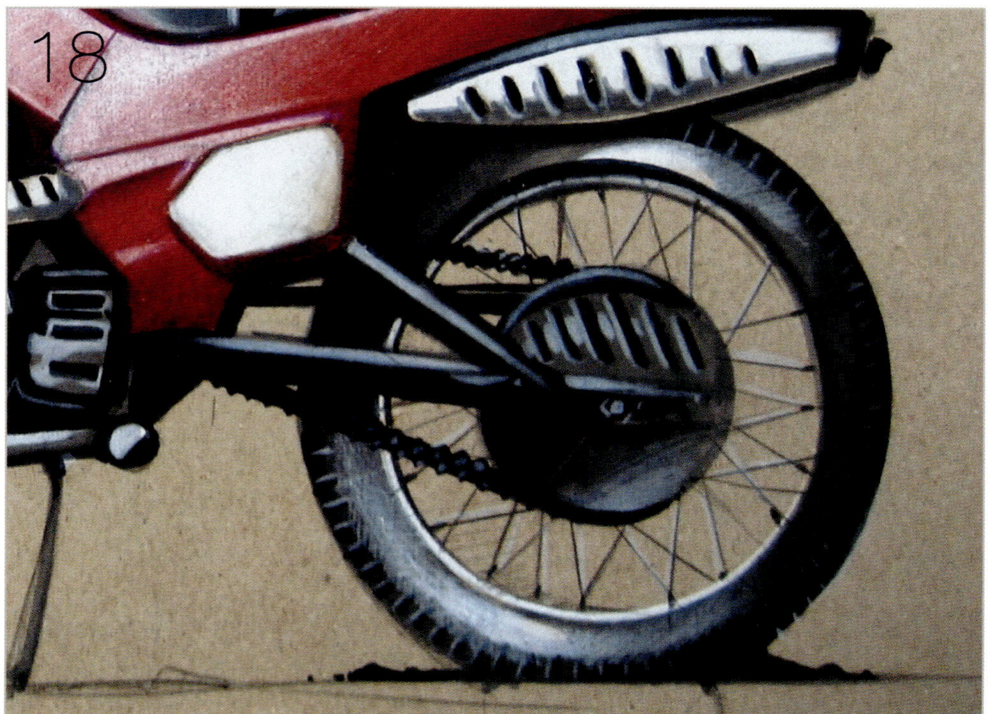

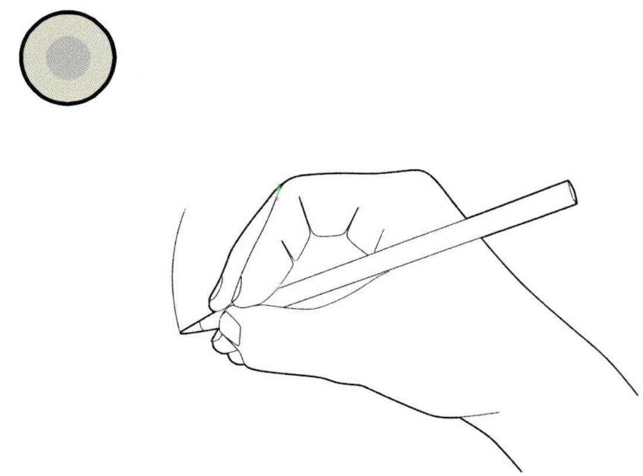

Survey the mechanicals and add some last-minute details: The lit edges of the exhaust vents, various bolts, chain drive highlights, etc.

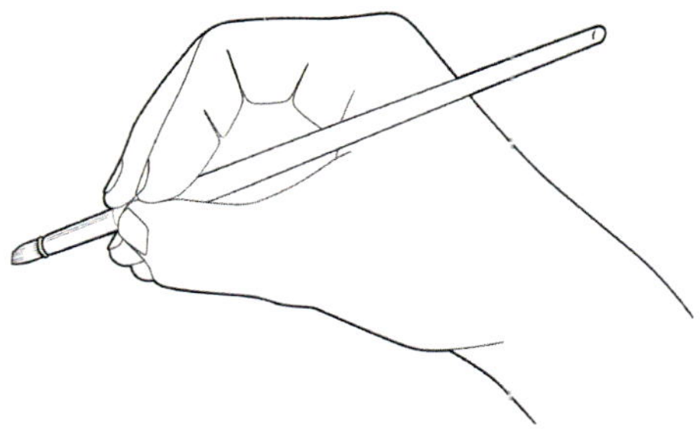

Concentrate the highlights in the chrome parts to help them stand out.

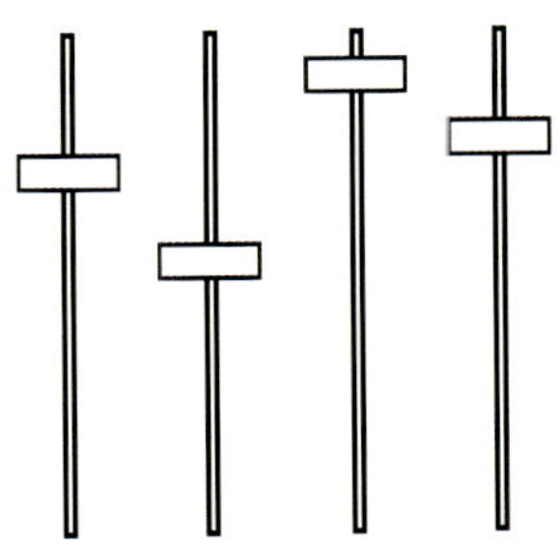

Scan in the rendering and adjust the levels to your liking in Photoshop.

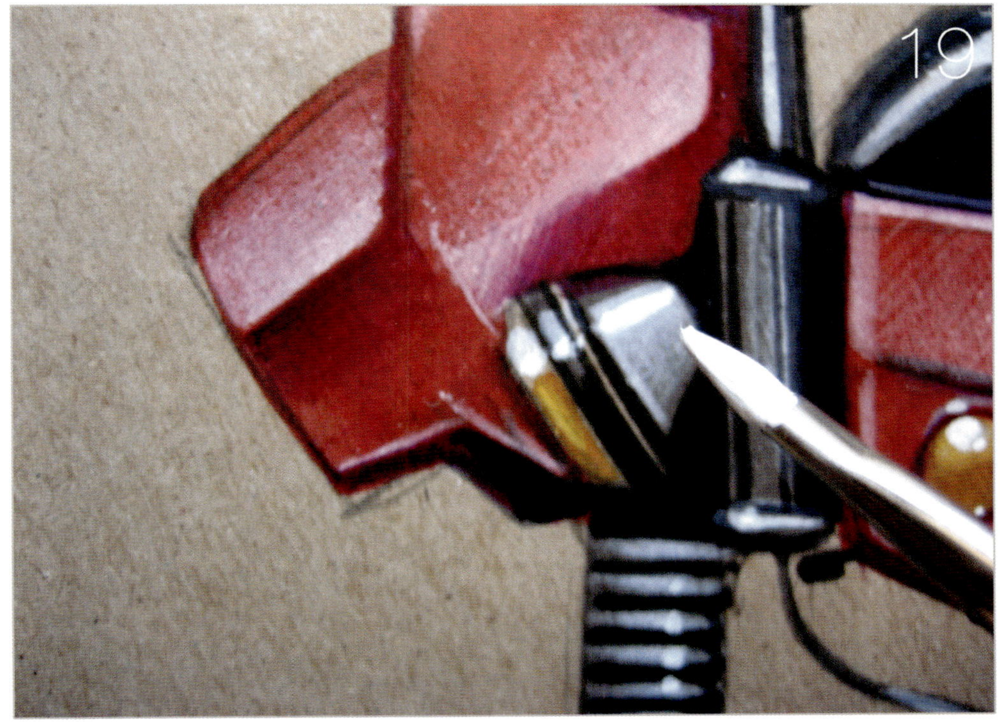

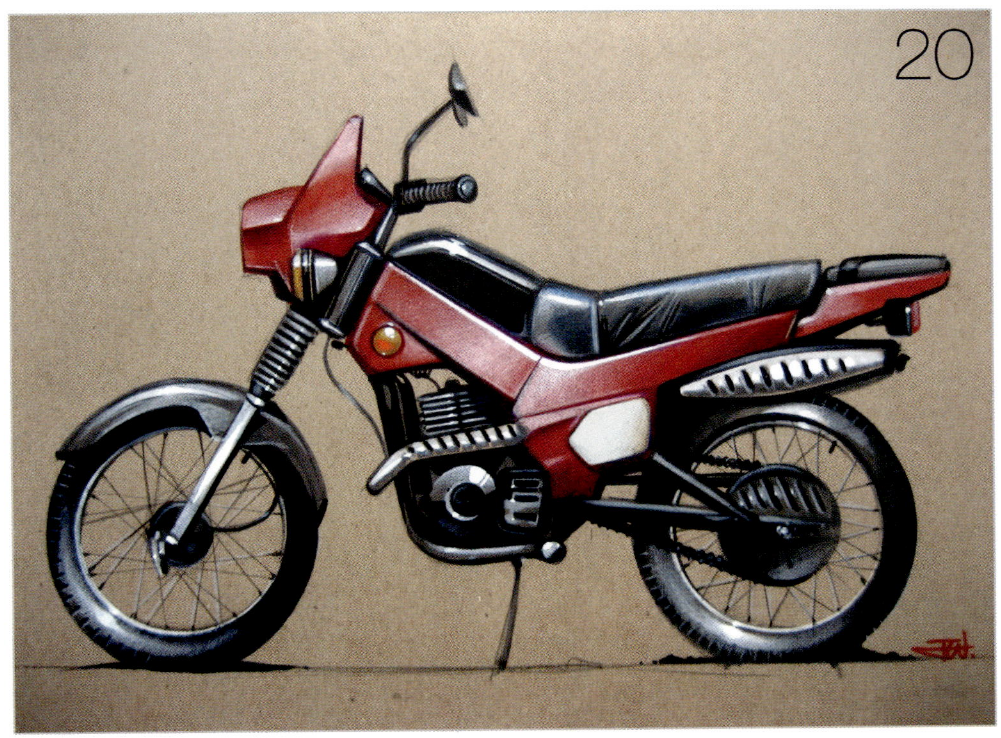

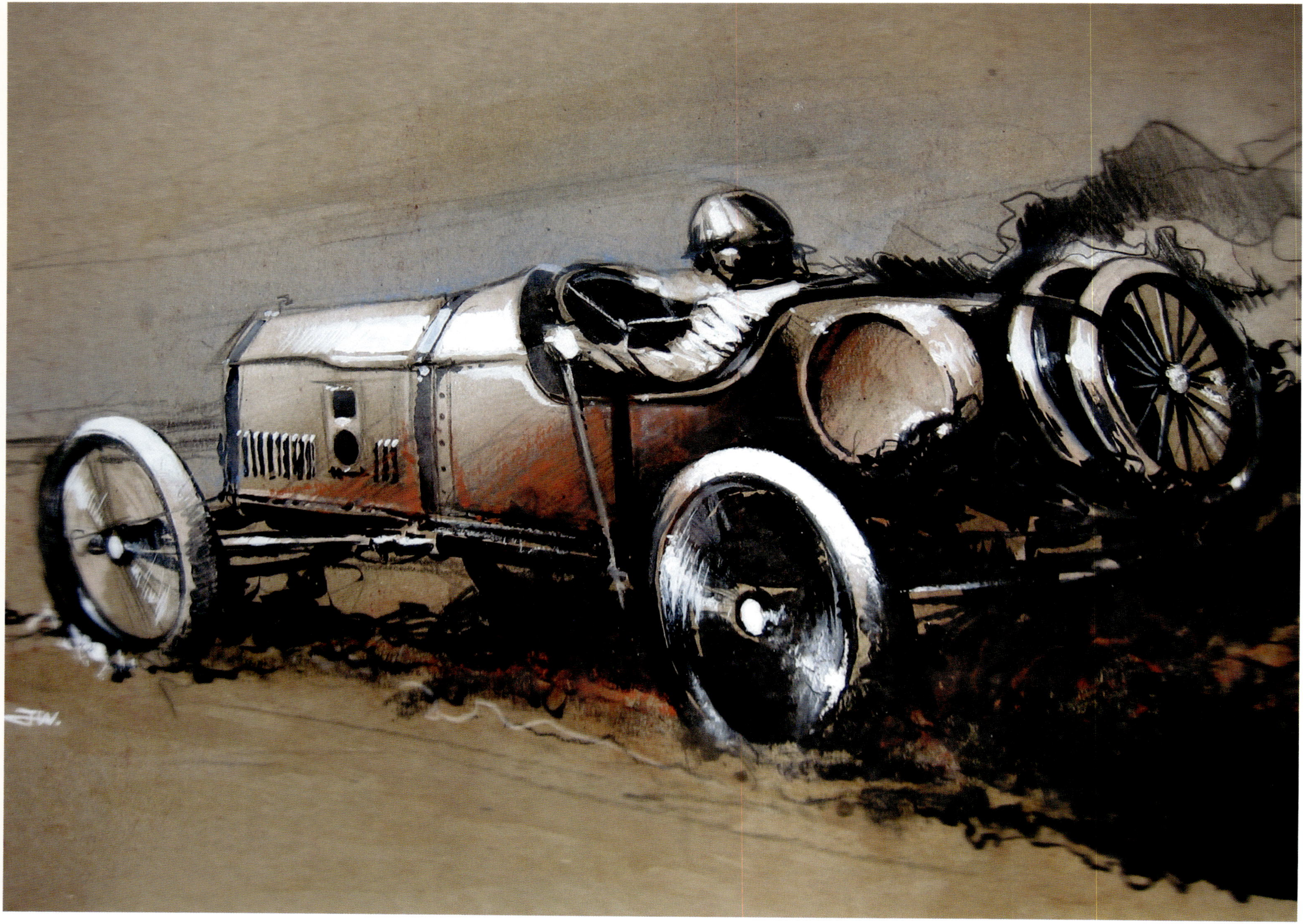

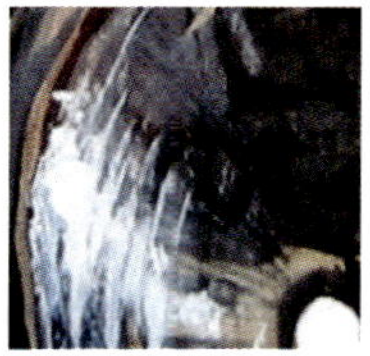

18. Experimentation

I *hate* undos. You'd think they would encourage designers to take a few risks in a rendering. But paradoxically, they don't. Without the danger of making a mistake, something is lost. Some kind of raw nerve. Passion. In real life, there are no undos. This is a tutorial rooted in the unexpected and the unforgiving; not unlike the subject matter, when you think about it.

The Tools

Prismacolors: White, Deco Pink, Black.

Nupastels: Orange, White, Light Blue, Black.

Verithins: Black.

Morning Coffee.

Gouache: White.

Prismacolor Markers: Black.

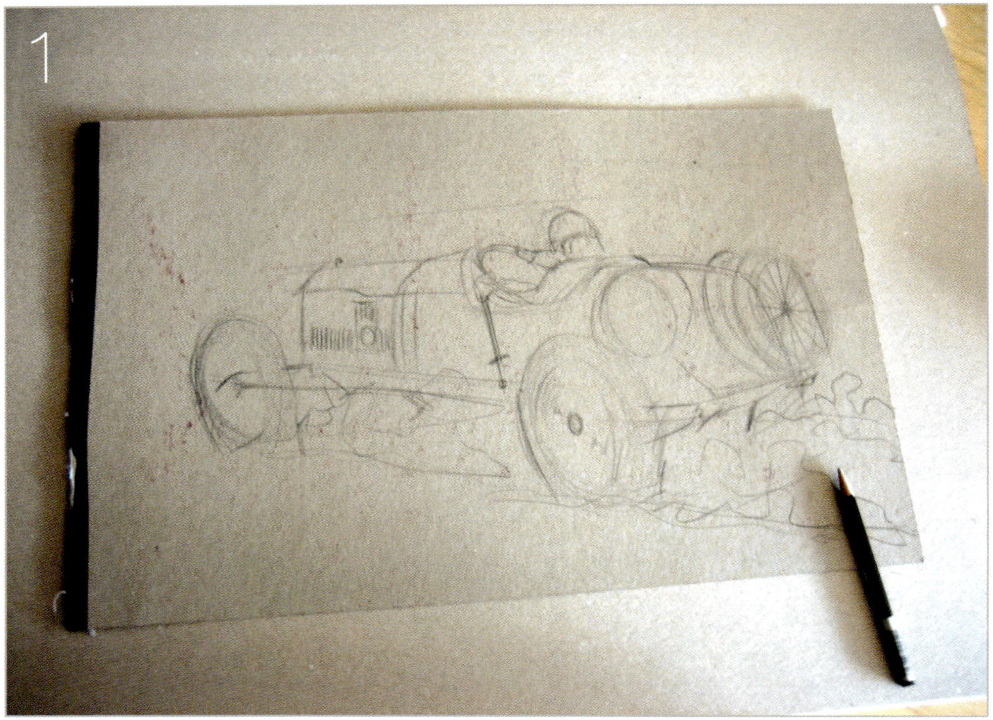

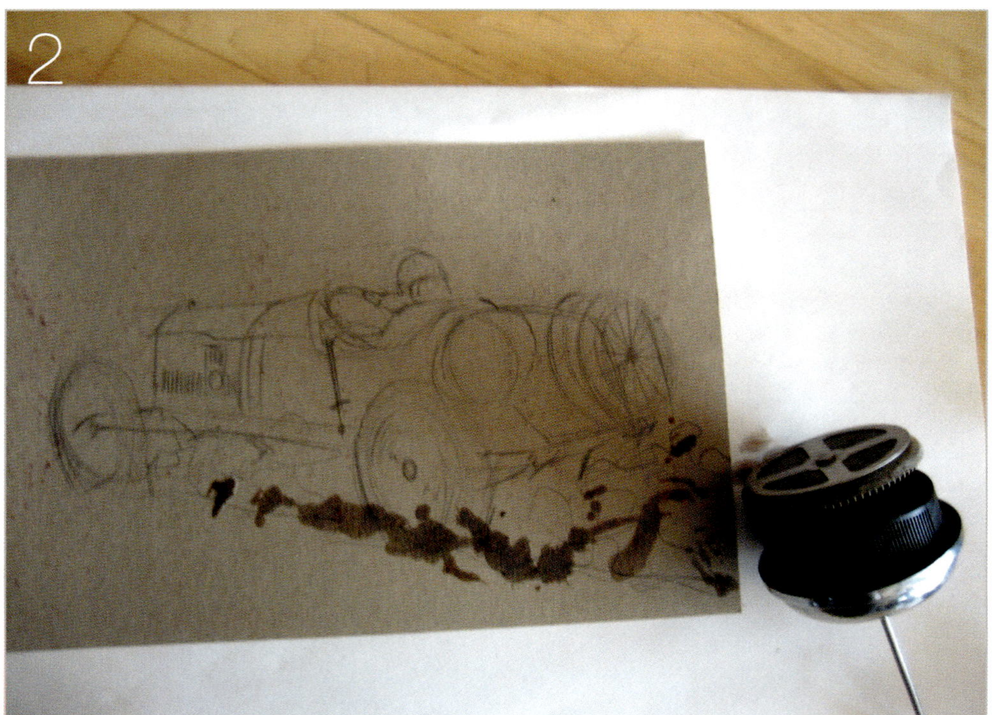

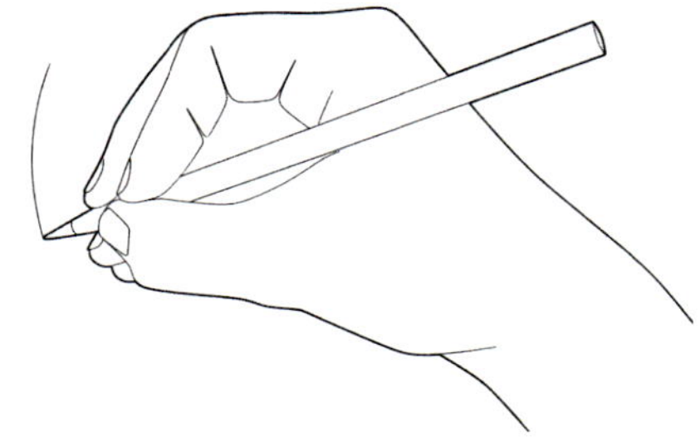

Start by laying out the basic lines of the car with a black Verithin pencil on a clean piece of chipboard.

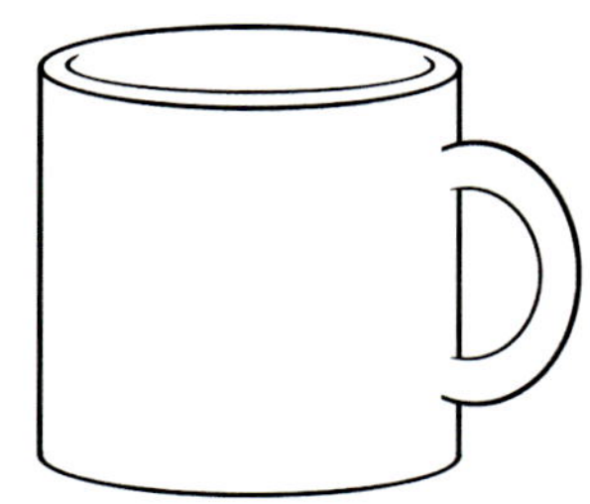

I wanted this rendering to have a painterly feel; almost impressionistic in nature. I had half a pot of leftover coffee that was going to be thrown out anyway, so in it went. Brown paint and water probably would have worked too, but this was more handy at the time...

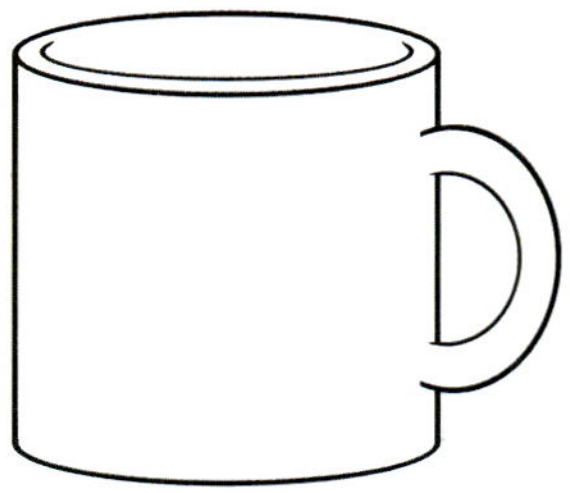

This vintage racing scene takes place on a dirt road, so play up the 'dirt' part of that equation. Allow yourself the freedom to make a sloppy mess and see where it leads.

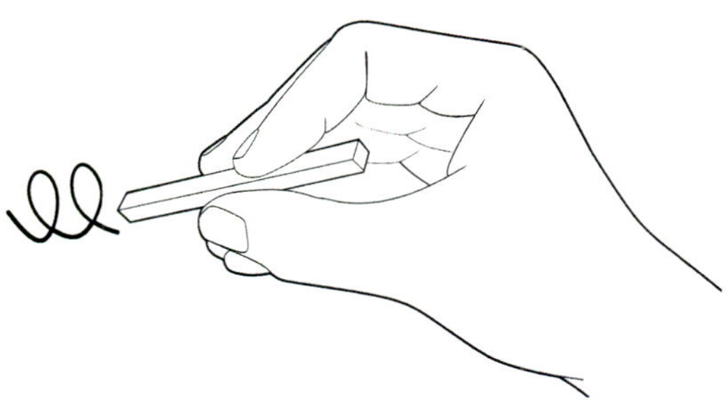

After the chipboard has dried, apply black pastel directly to the base in a sort of "roostertail" fashion. The board's billiard-table surface will have expanded thanks to step three, yielding an almost carpet-like texture to accept the pastel.

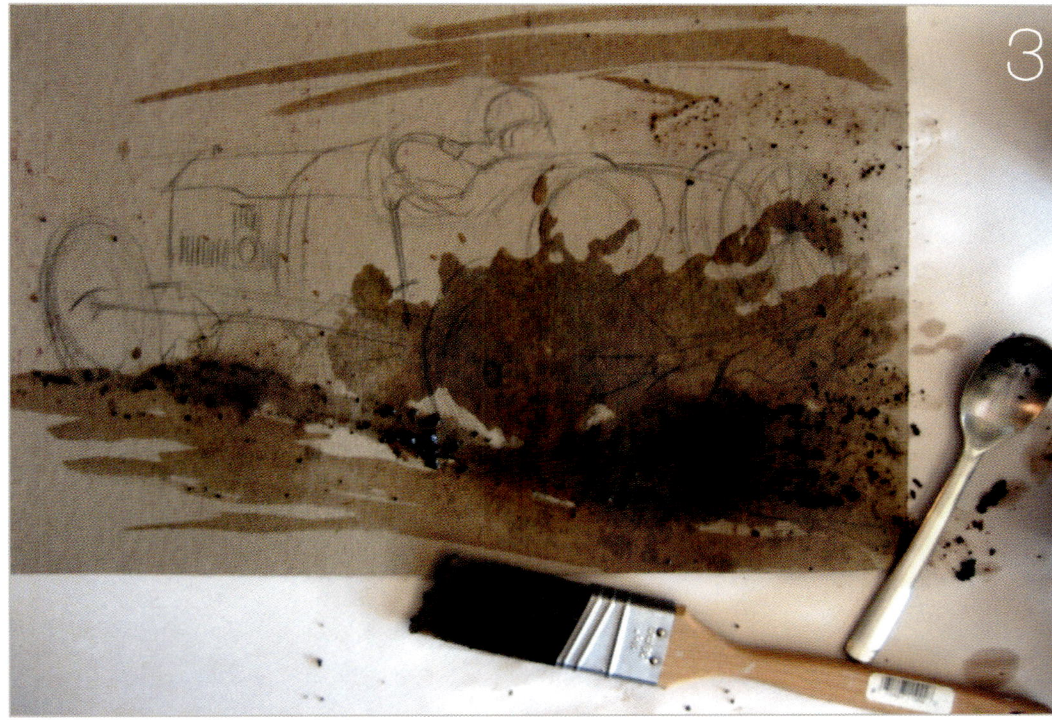

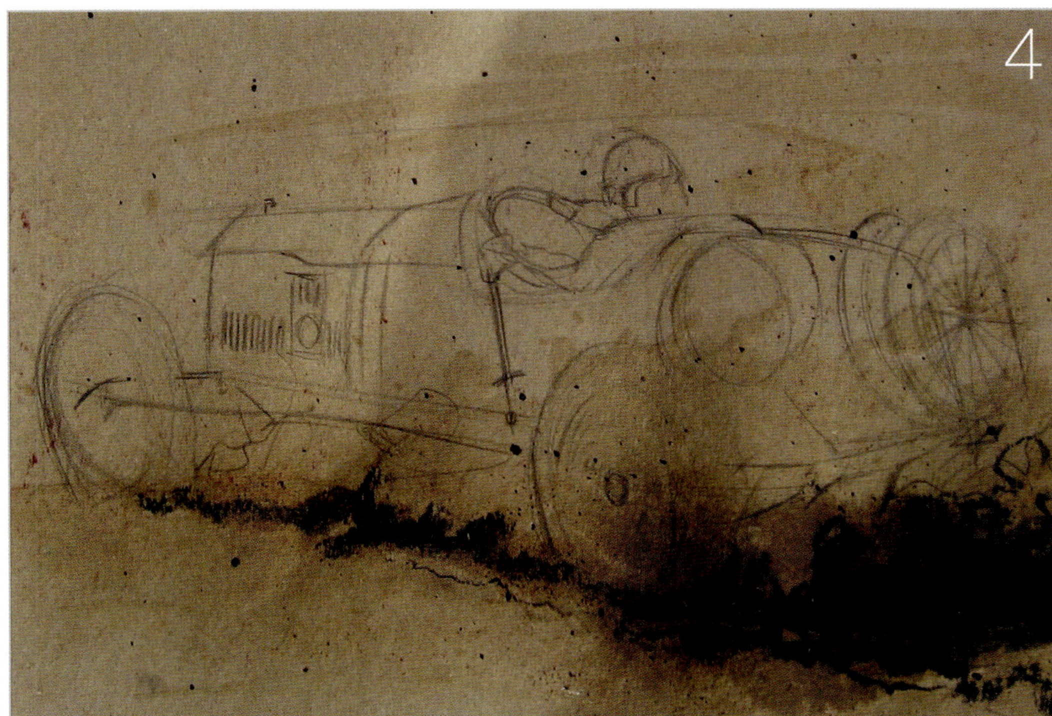

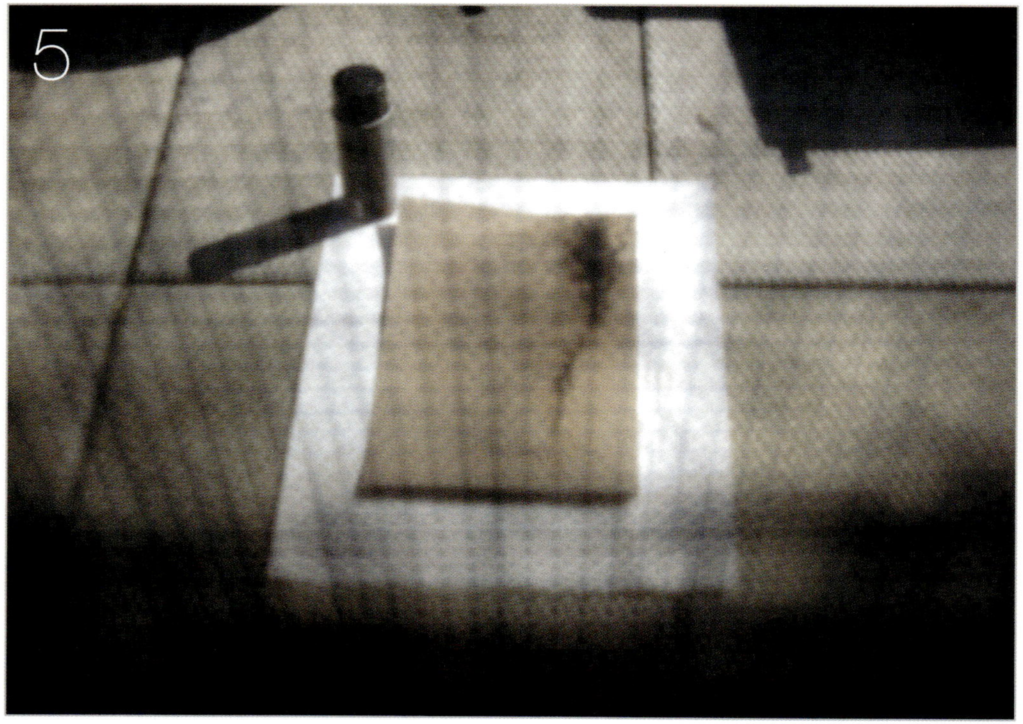

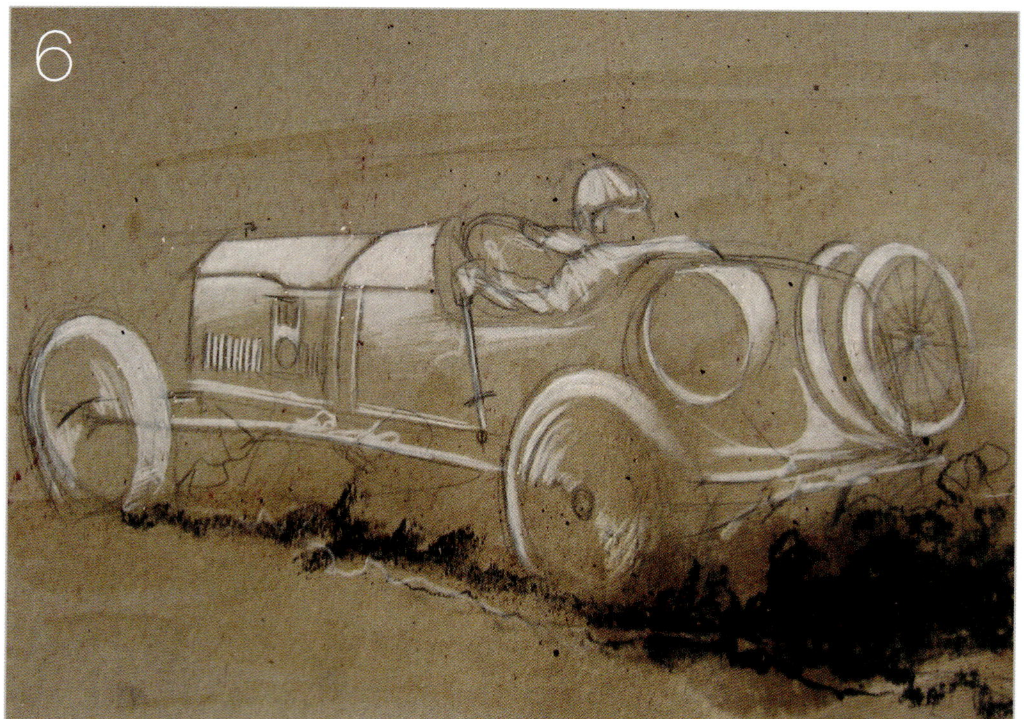

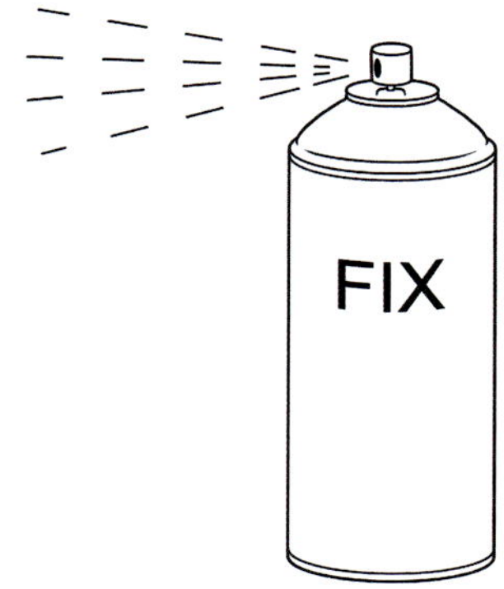

This would be a good time to hit the rendering with a bit of fixative – but be sure to read the warning labels first!

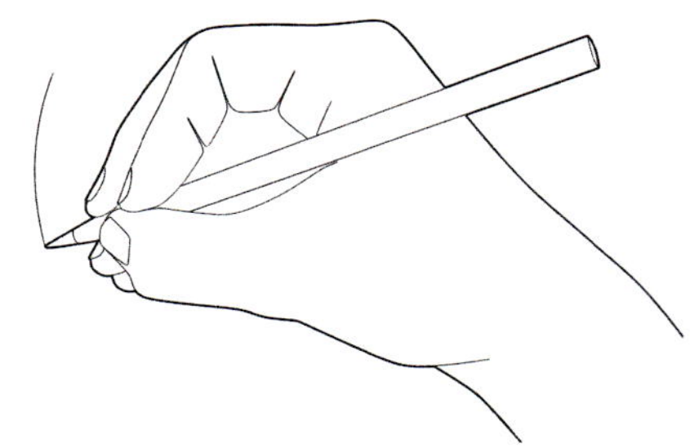

Begin laying in the lighter parts of the car with Deco Pink. This will work well with the warmth of the background and the period nature of the piece.

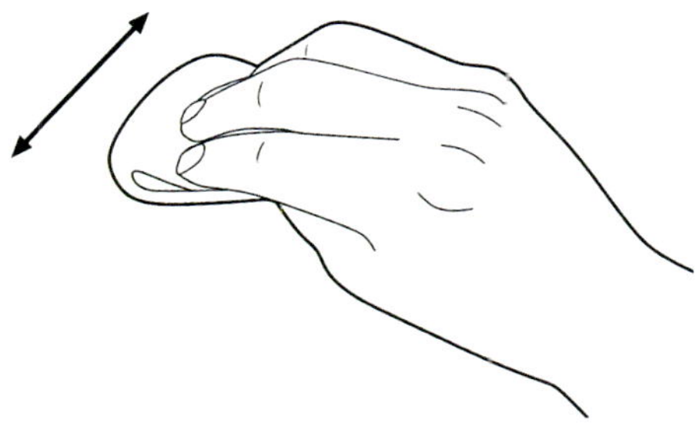

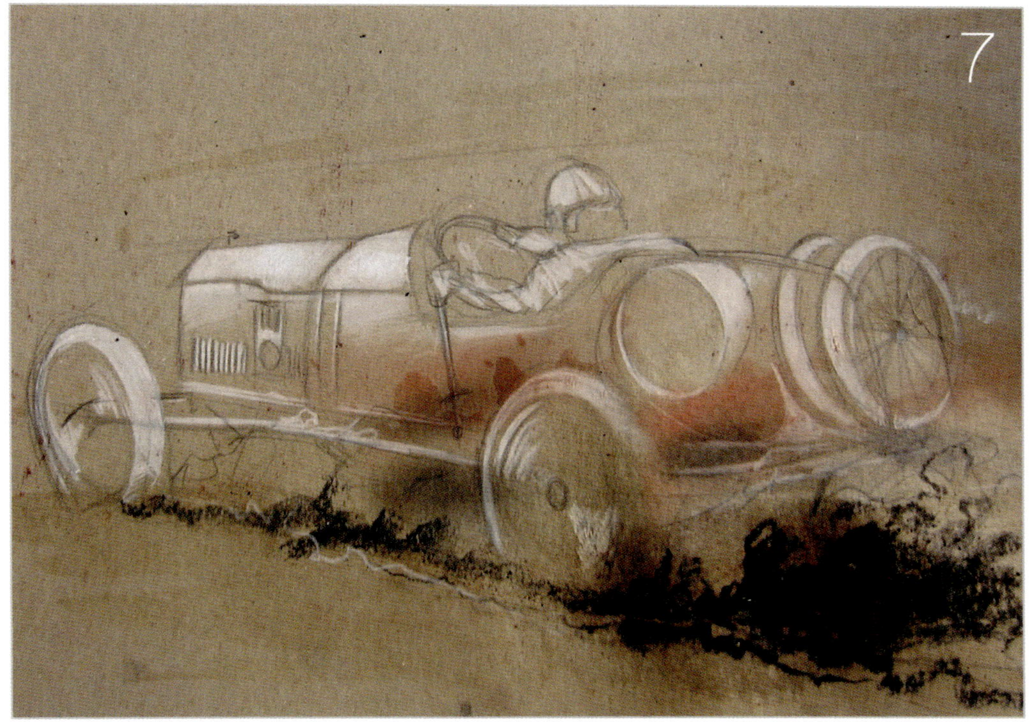

For the earth tone, ghost in a bit of orange pastel on the bodyside.

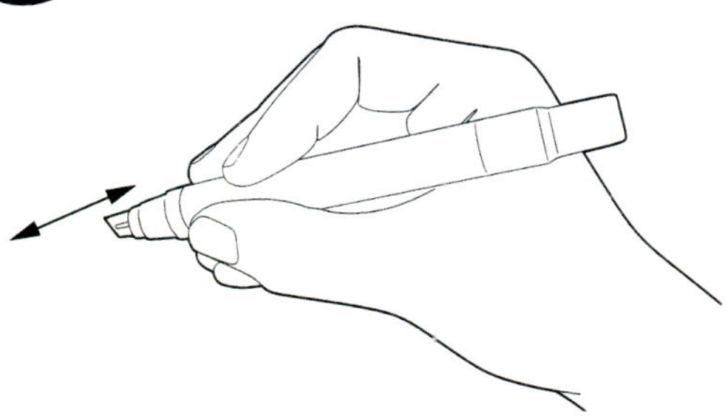

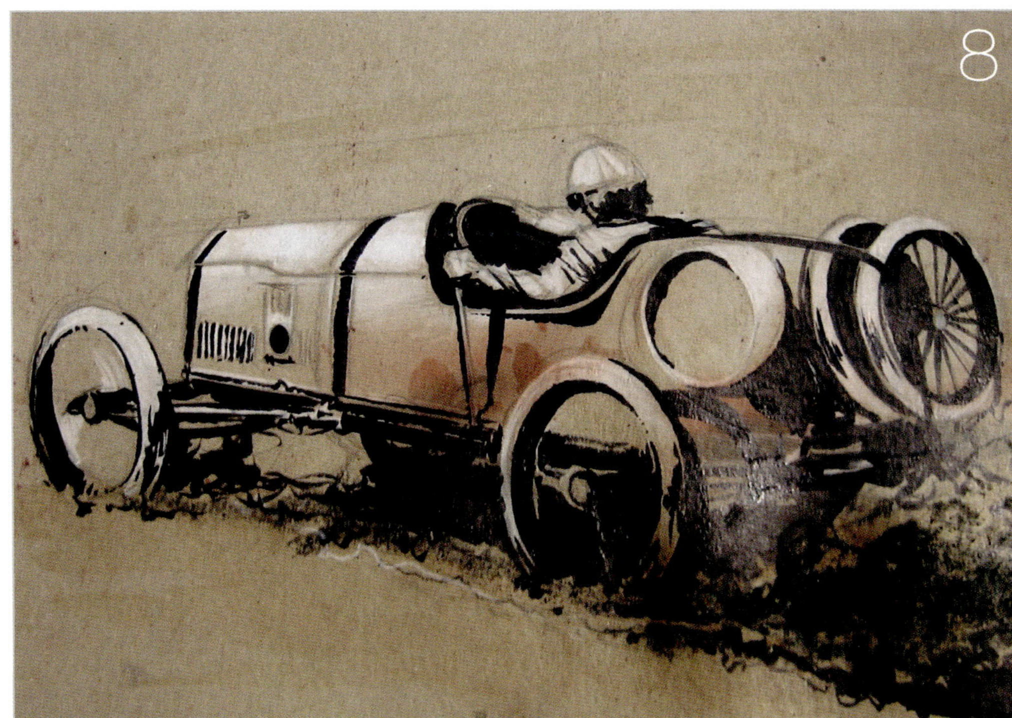

With a loose hand, block in the darkest parts of the rendering with a black marker. Try to emulate the look and feel of the black pastels.

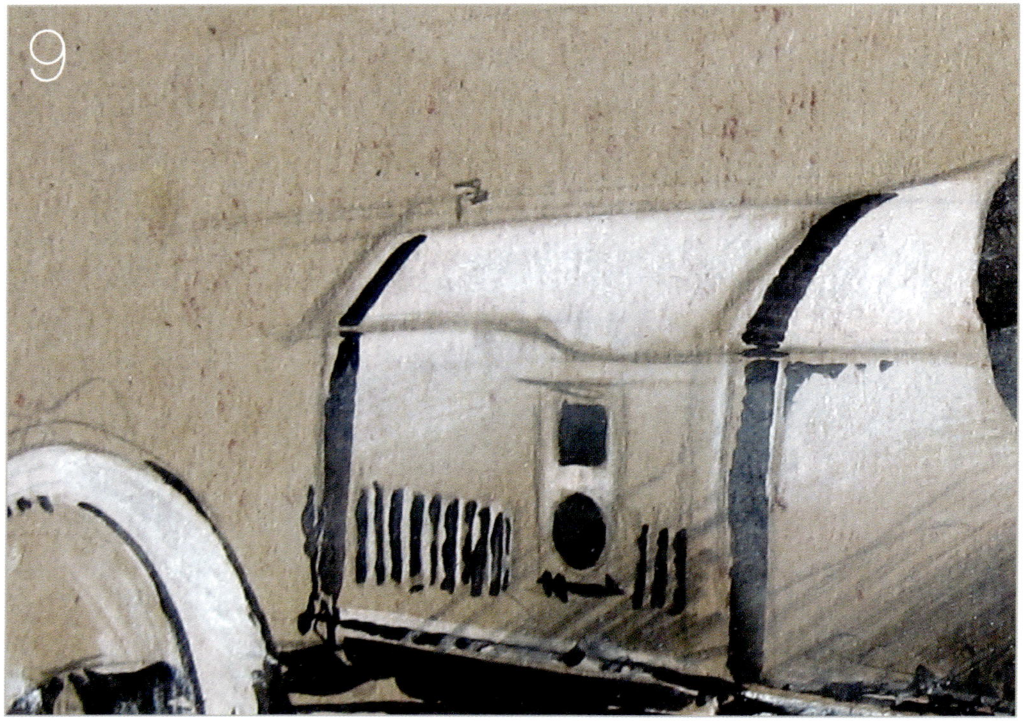

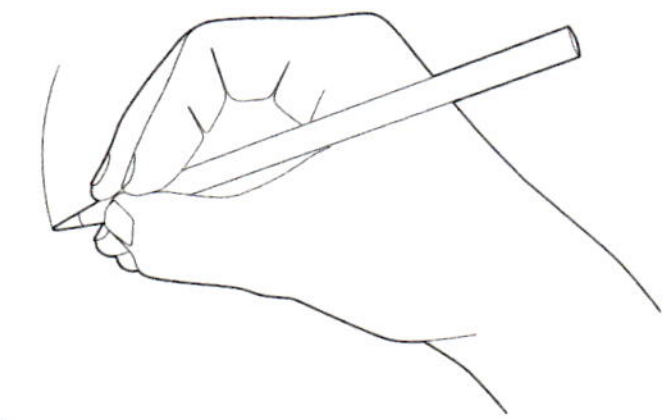

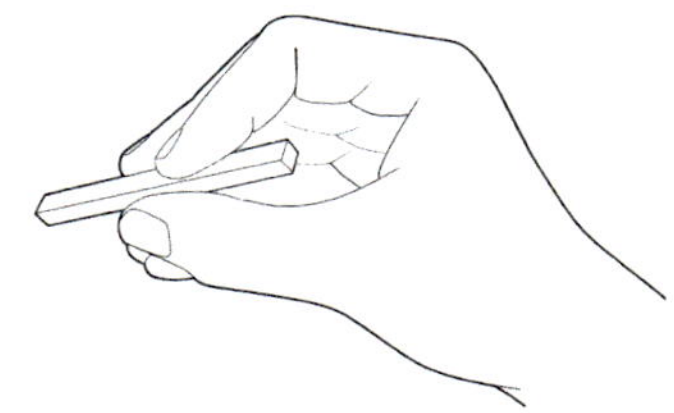

Add more texture to the body surface with black Verithin pencil and a little bit of white pastel.

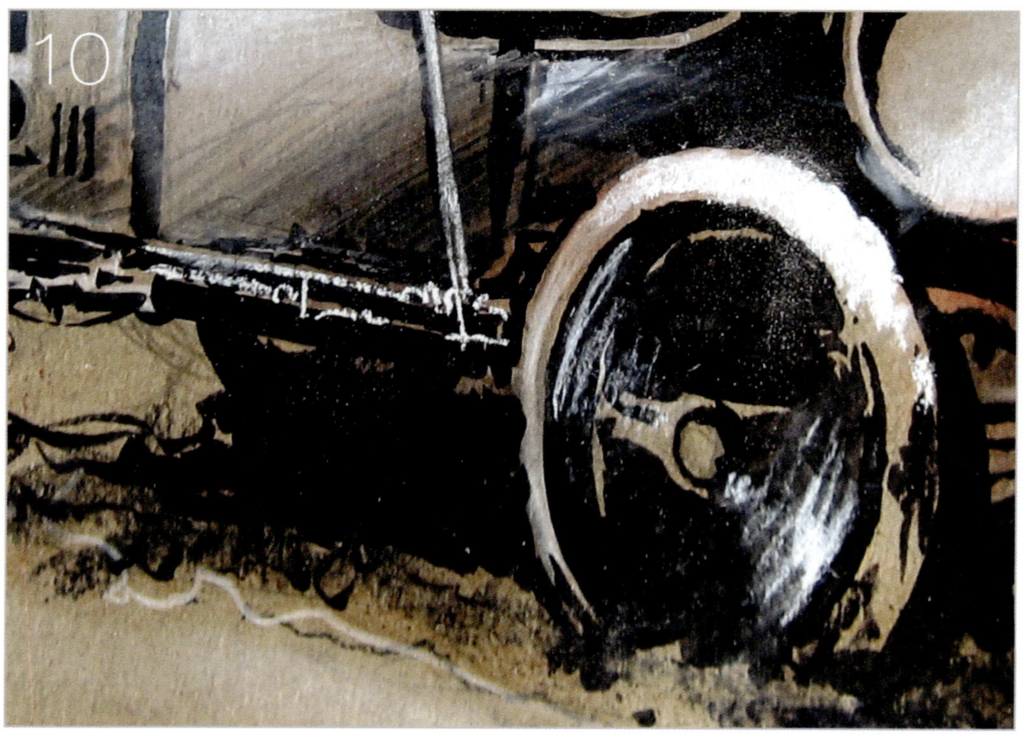

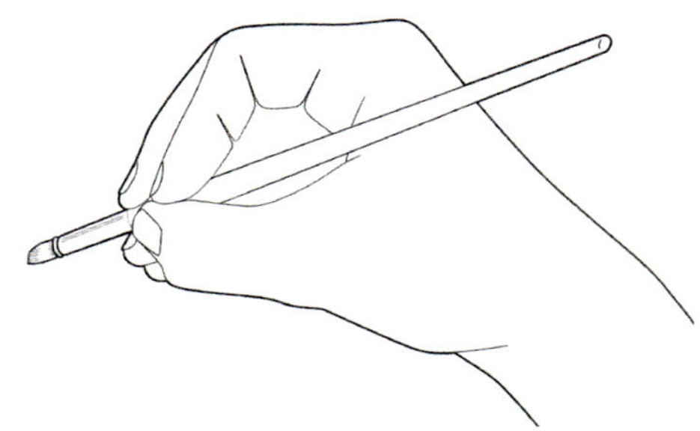

Try applying white gouache as you would a Verithin pencil, with sketch-like strokes. These loose highlights will help describe the motion of the spokes.

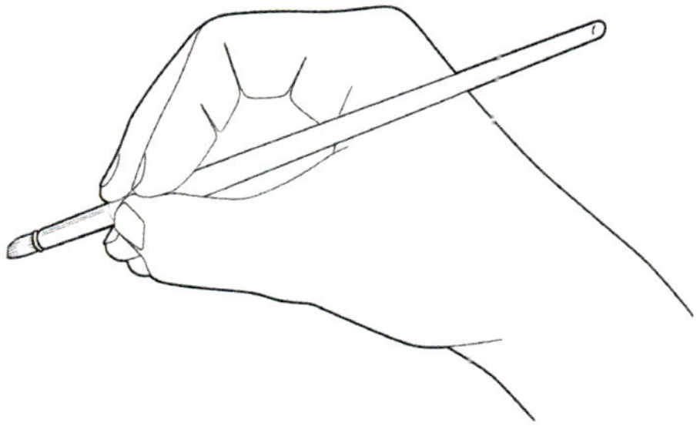

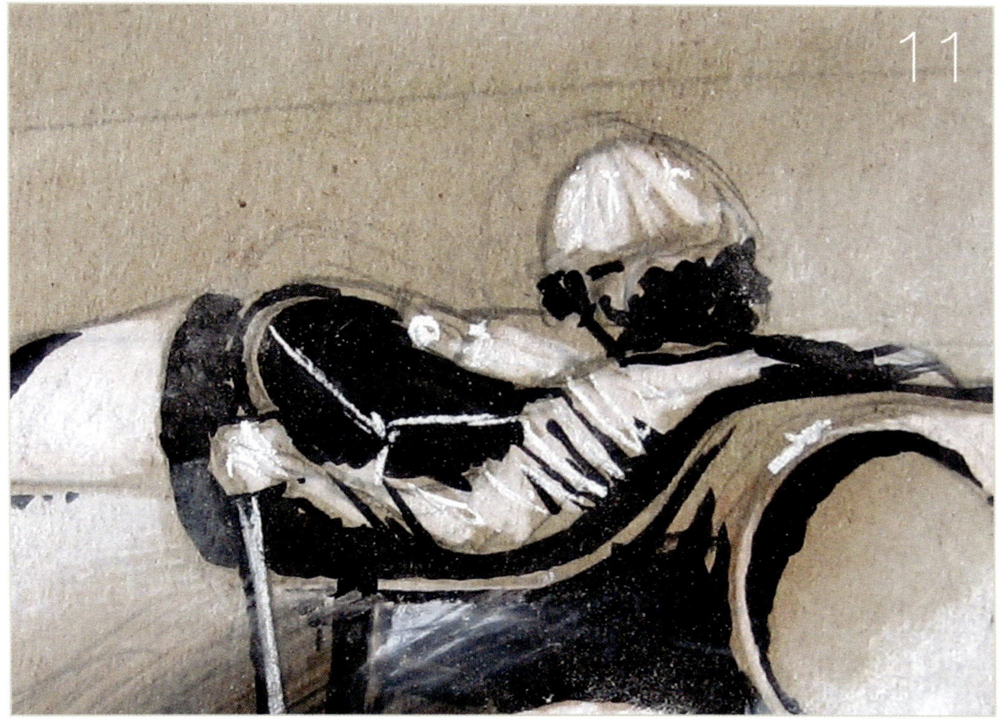

White gouache is also a great tool for describing the deep folds in the driver's sleeve.

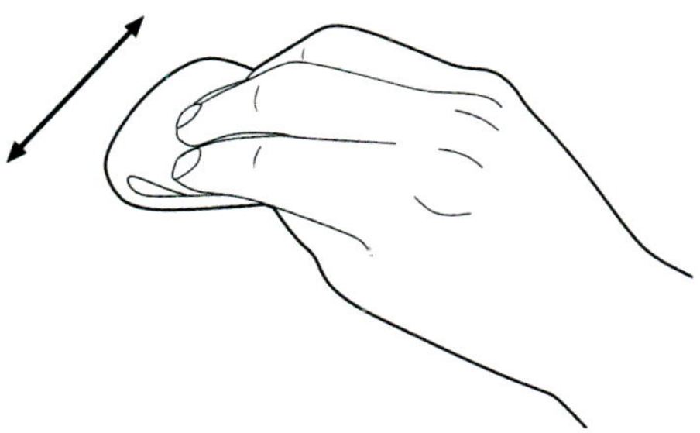

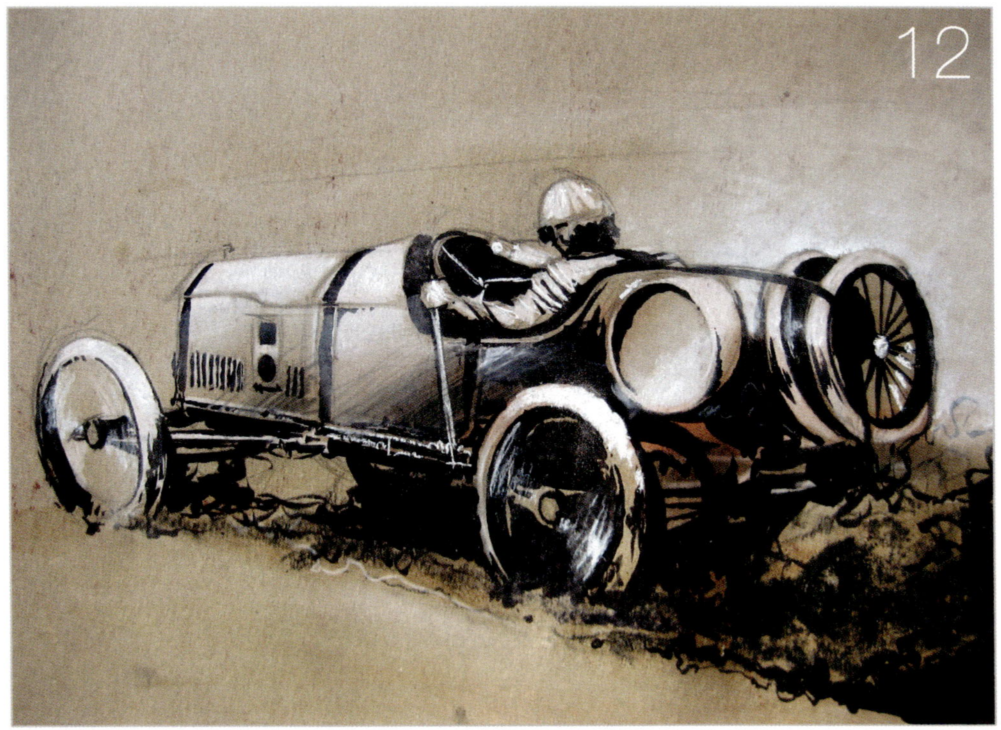

To give the suggestion of road dust, add a bit of white pastel directly to the background and blend it out with a clean Webril pad.

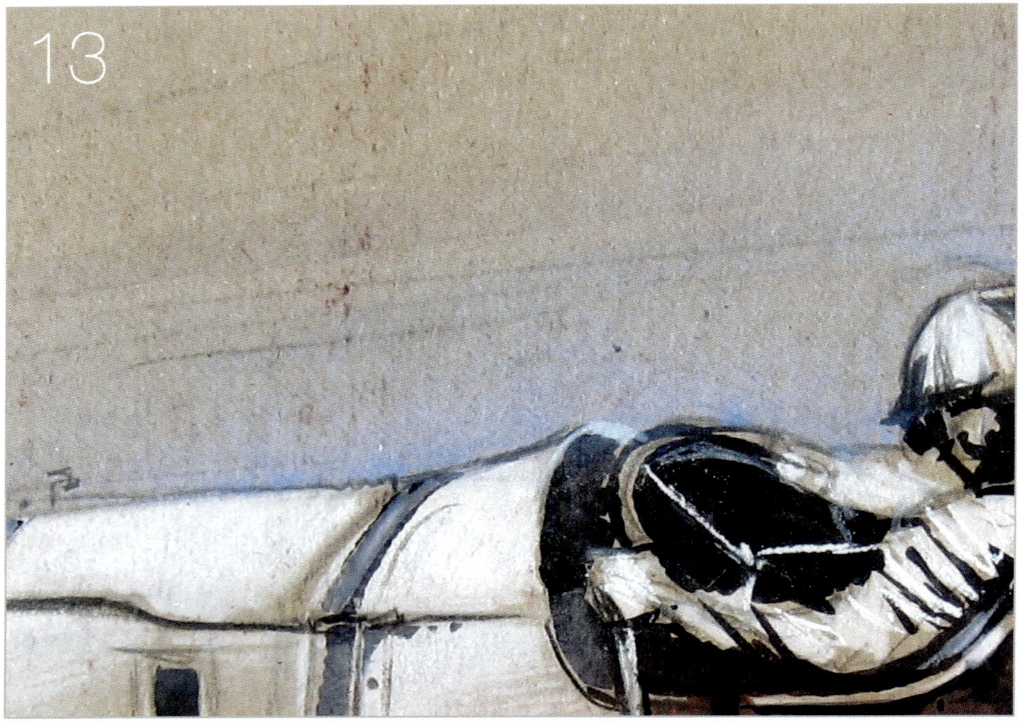

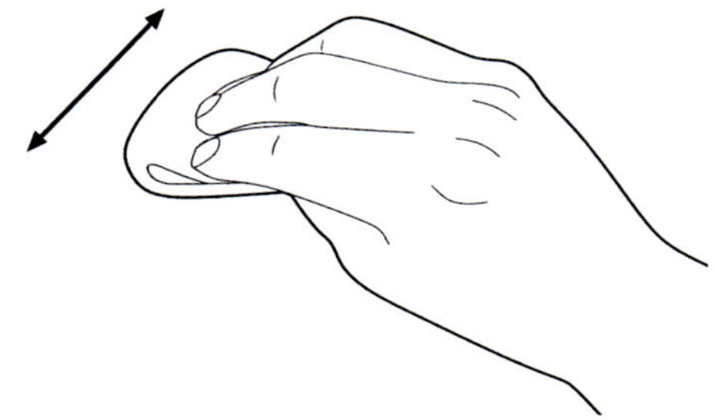

For a bit more color contrast, apply some light blue pastel along the back edge of the car, then lift it upward with a clean Webril pad.

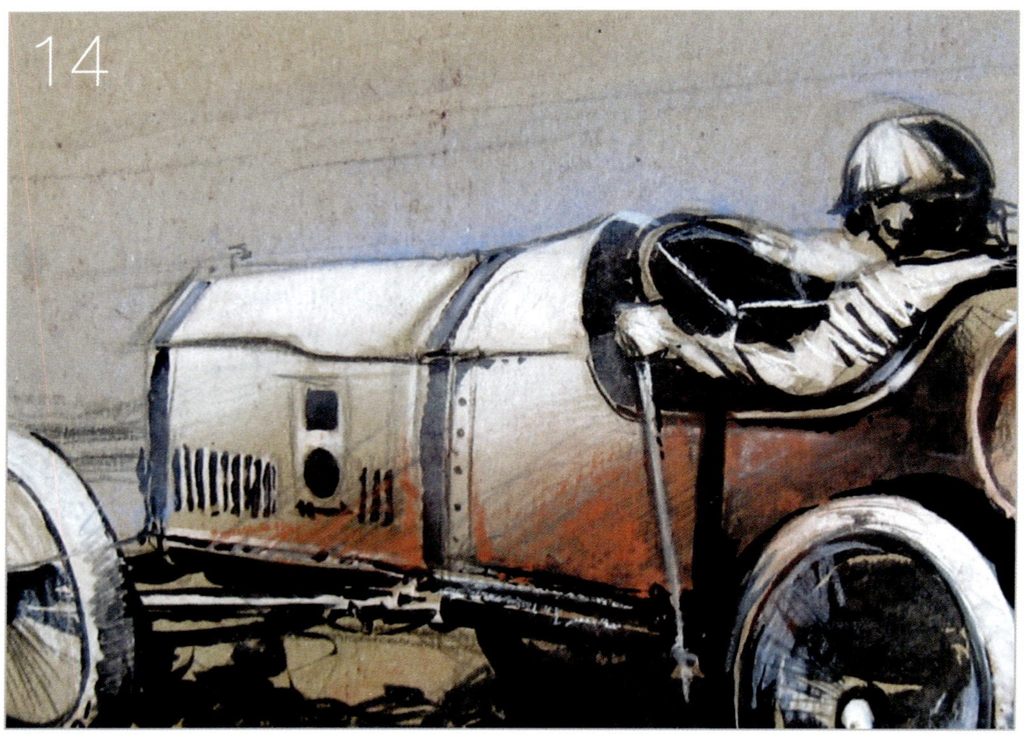

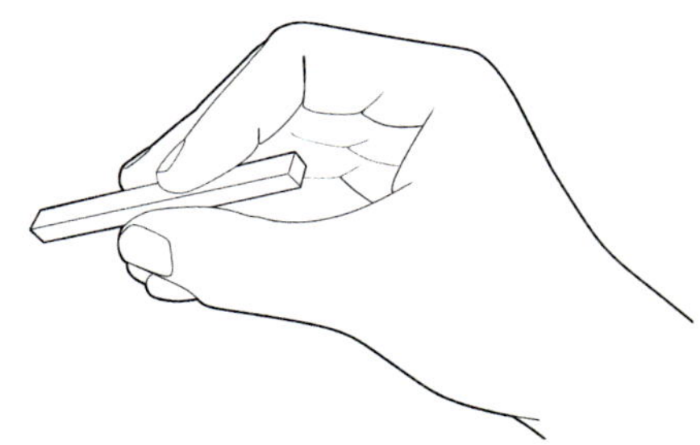

If the orange has become a bit subdued over the last few steps, punch it up by applying the pastel directly.

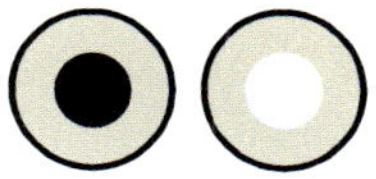

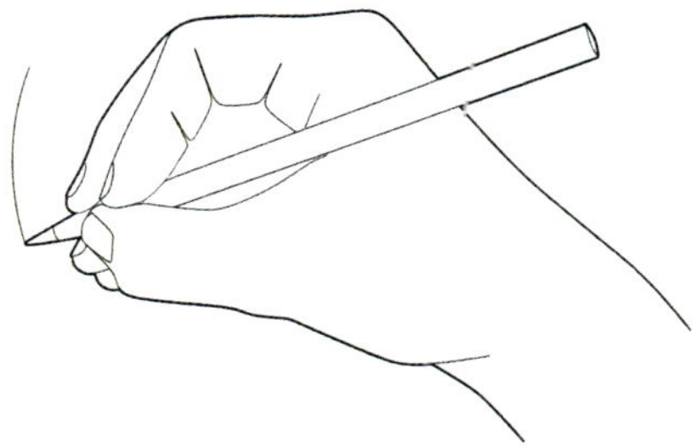

Use a bit of black and white Prismacolor to add another dimension to the spare tire, road dust and debris.

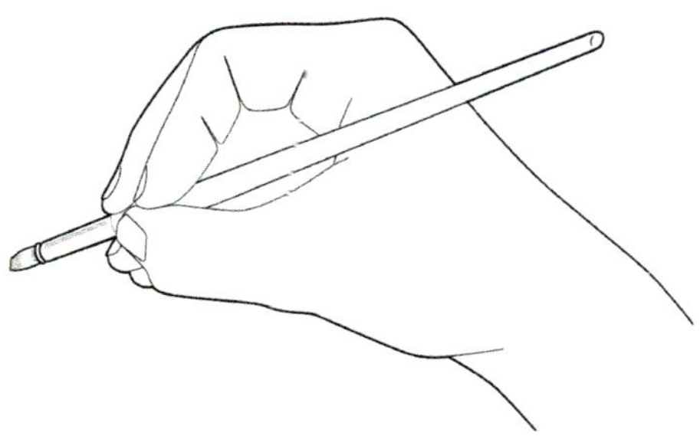

Return to the white gouache, this time with thicker, almost Van Gogh-like strokes. Here is an occasion where you can afford to be a bit more liberal with your highlights.

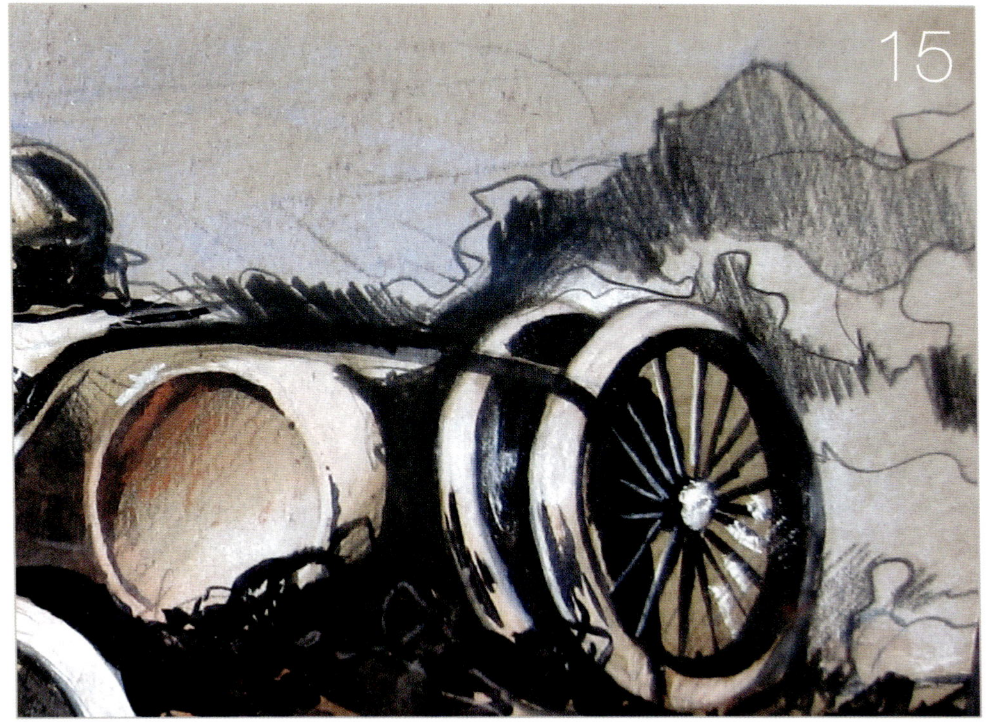

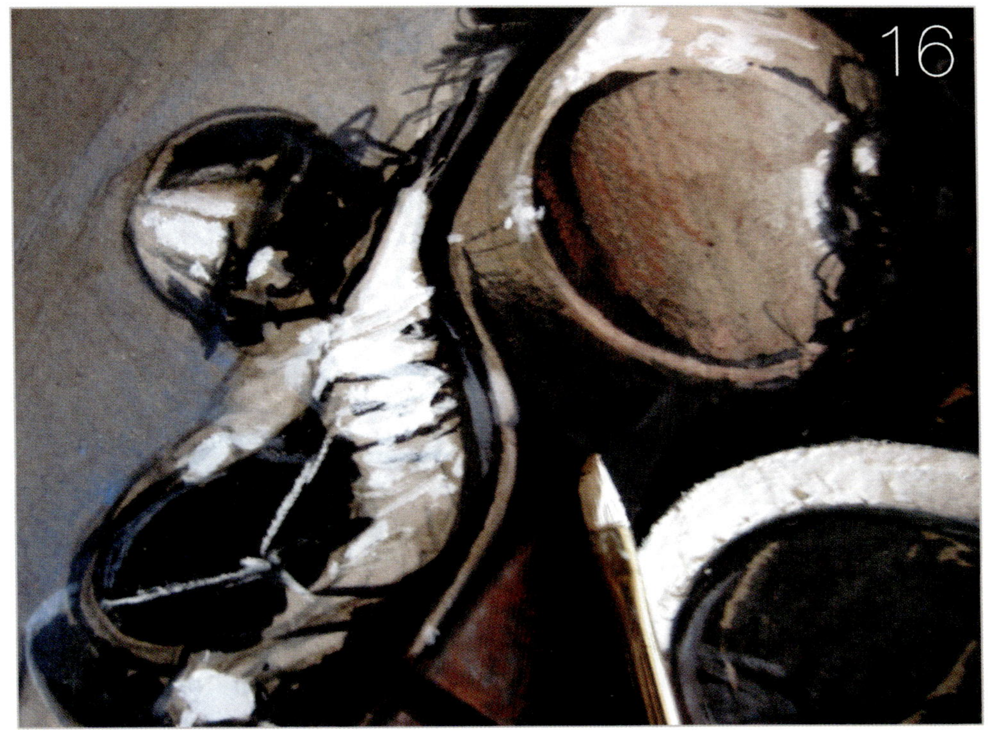

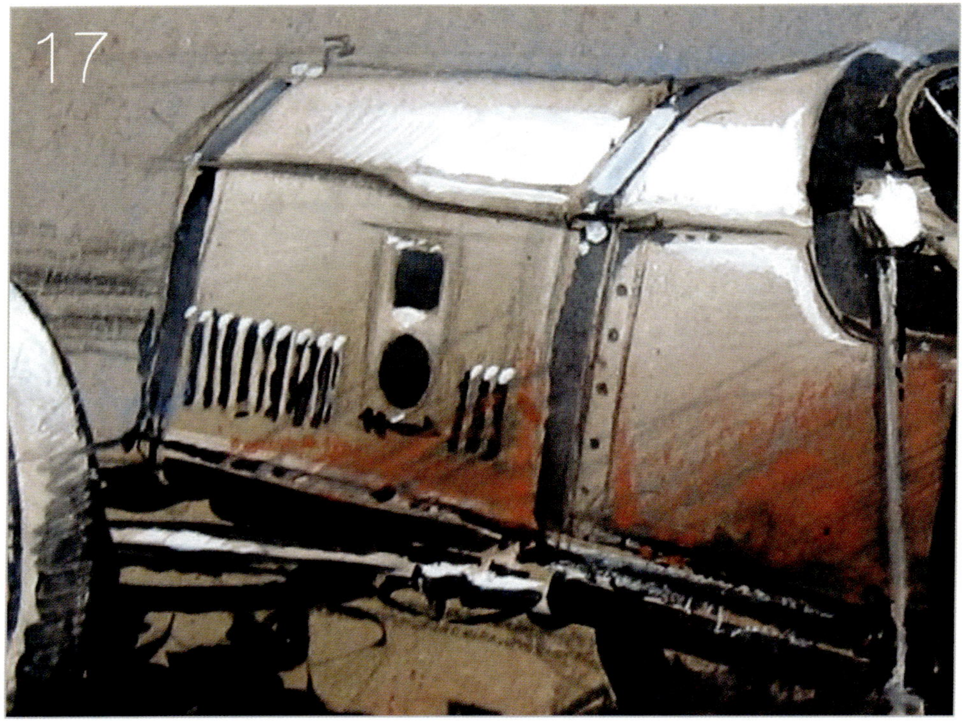

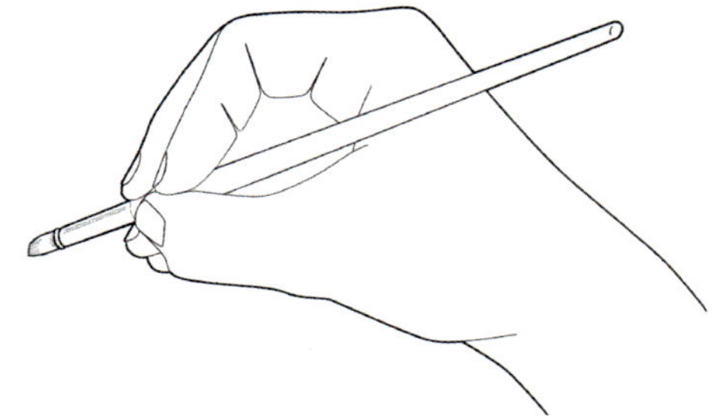

Indicate the flat upper surface of the engine cover with a large, geometric highlight. Echo this with sharp, angled blips of light at the tips of the vent slots.

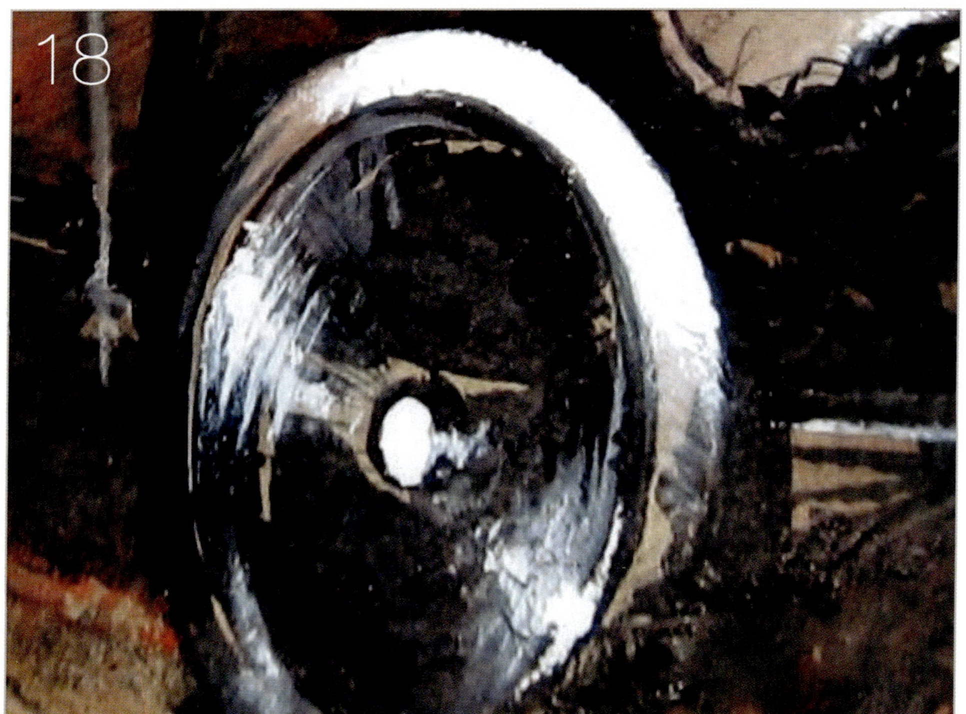

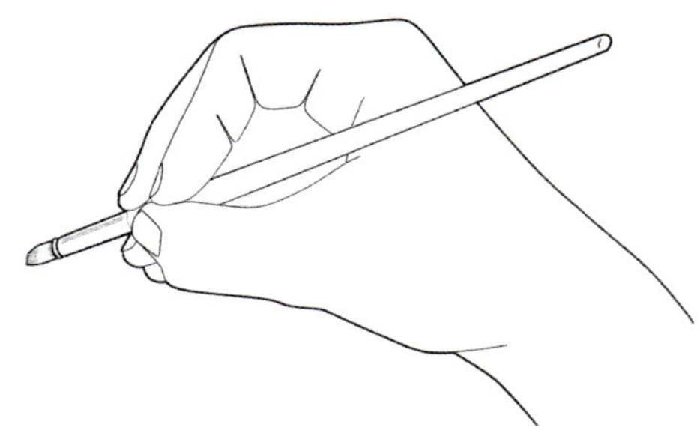

Revisit the spoke highlights, this time with tighter, more focused strokes.

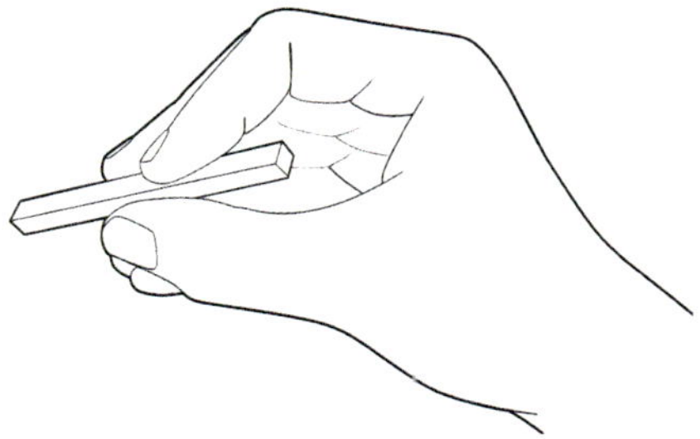

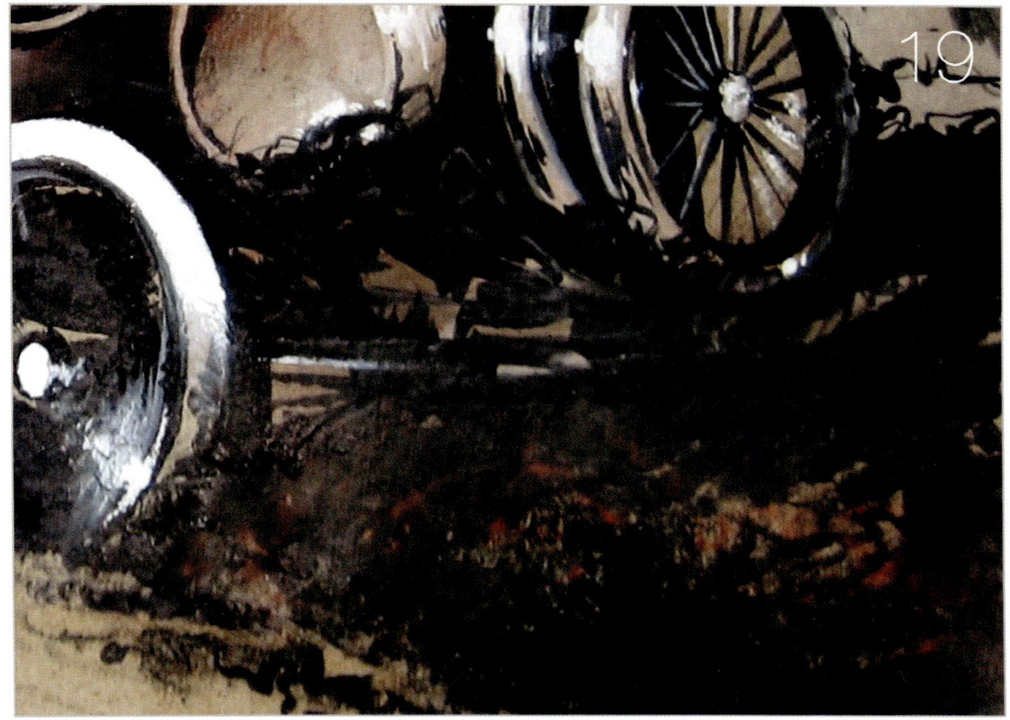

Add one more layer of detail by applying orange pastel directly to the dirt road. Try to mimic the feel of the marker.

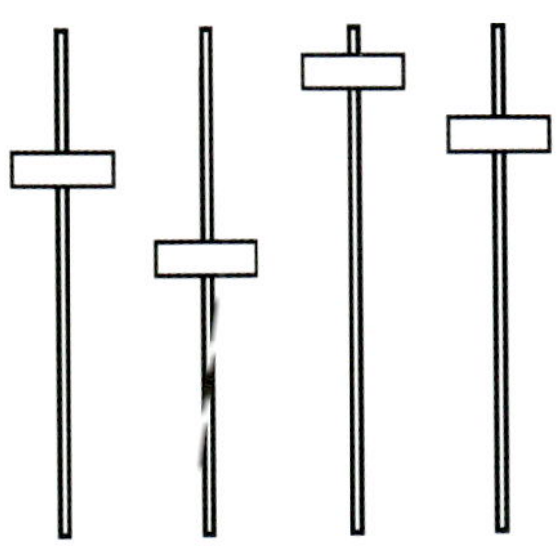

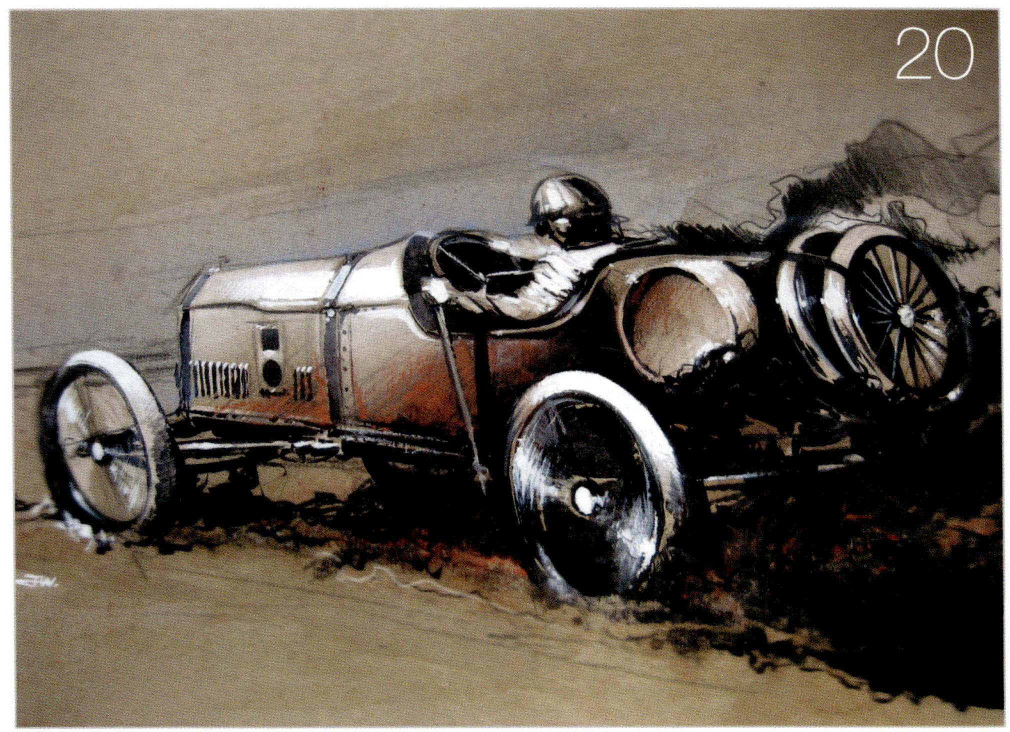

Scan in the rendering and adjust the levels to your liking in Photoshop. Use the blur effect on the front edge to enhance the sensation of speed.

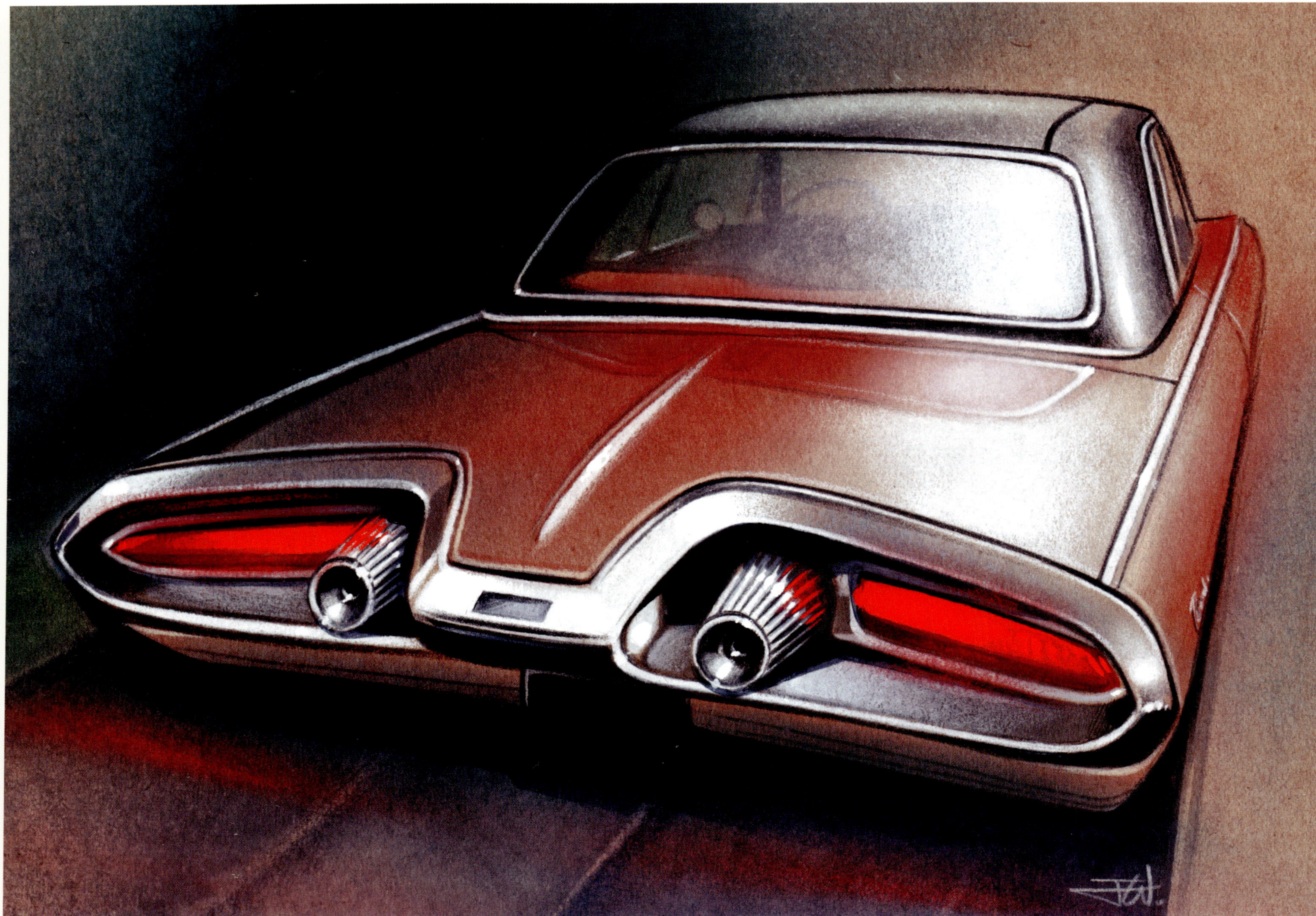

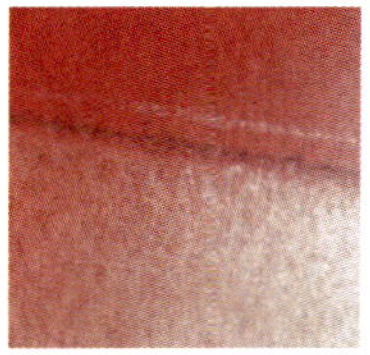

19. Finding the Middle Ground

Sometimes, all a subject needs is the right kind of middle ground to achieve its full visual potential. The beauty of craft paper is that it can provide a middle ground for a variety of colors, materials and finishes, not just one. In the following tutorial, you'll see just how versatile this material can be.

The Tools

Prismacolors: White, Black, Cool Grey 20%.

Nupastels: Orange, White, Indigo Blue, Black, Red.

Verithins: Black.

Gouache: Black, Red, White.

Prismacolor Markers: Indigo Blue, Salmon Pink, Cool Grey 70%, Crimson Red, Dark Green.

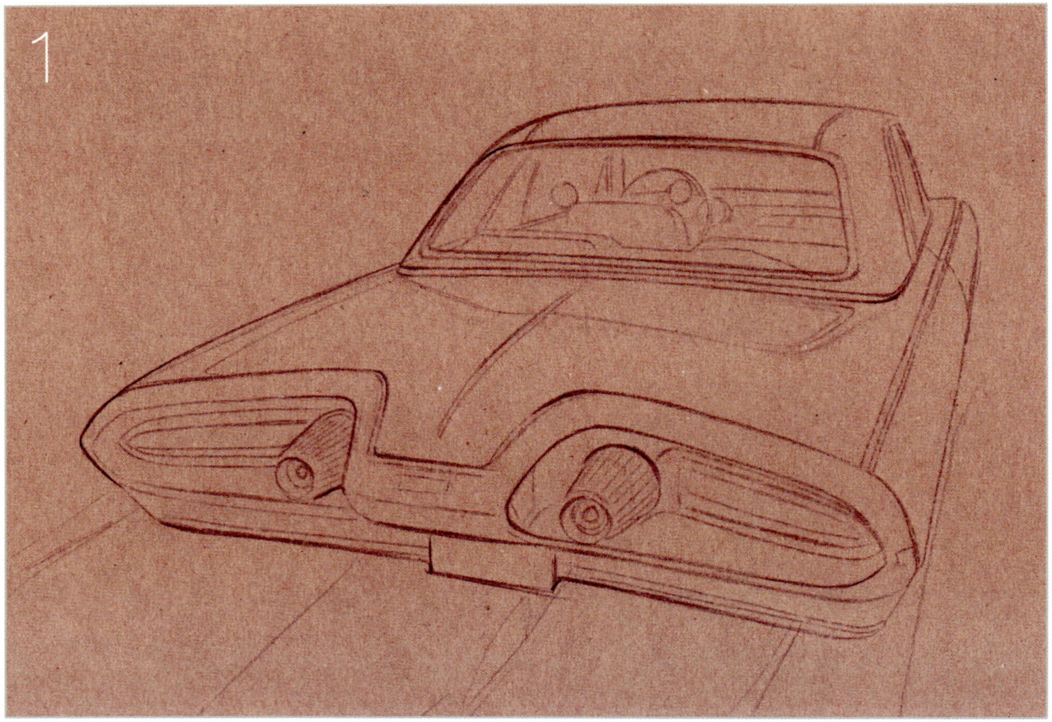

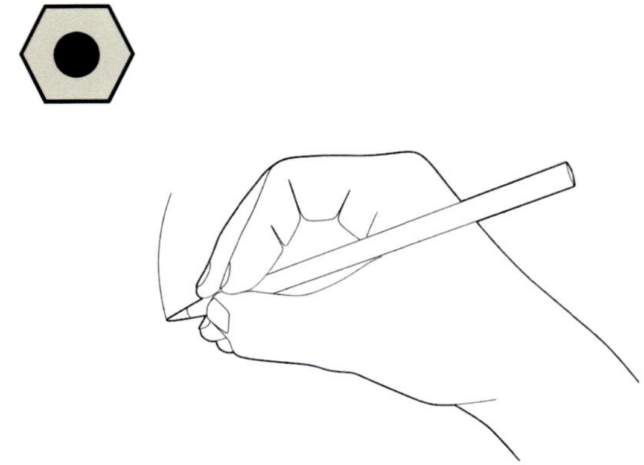

Start by laying out the basic lines of the car with a black Verithin pencil on a clean piece of craft paper.

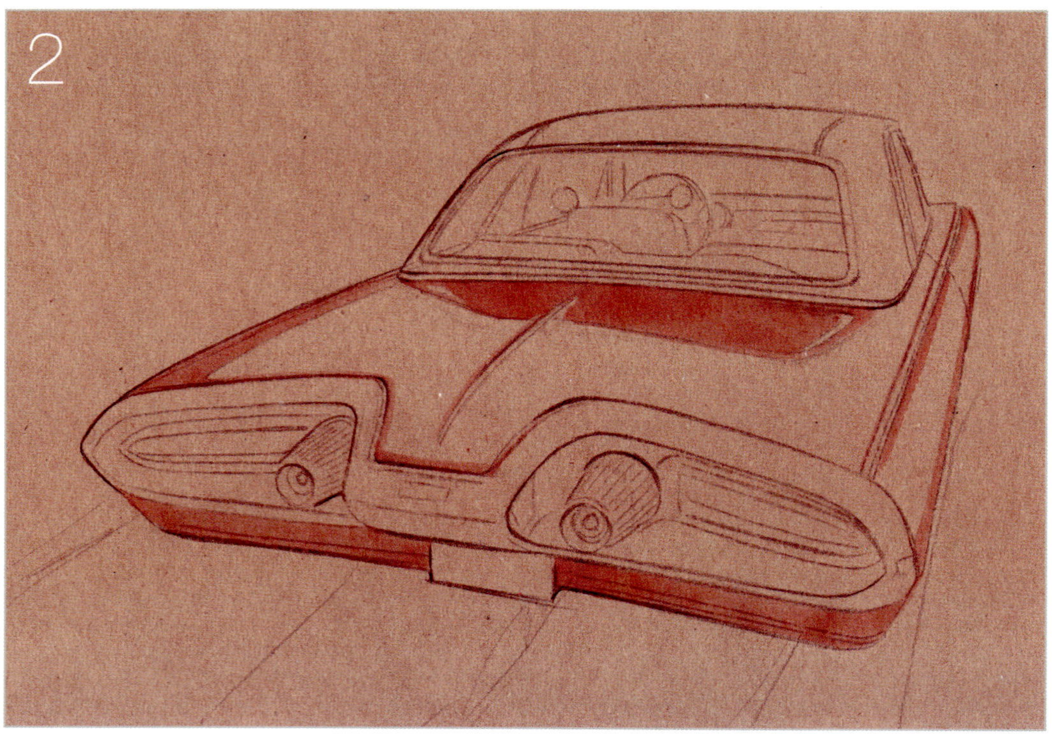

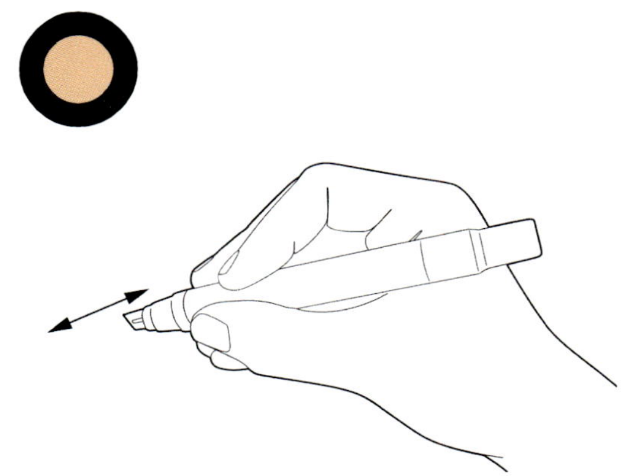

Block in the body reflections with a Salmon marker. Since this is almost a direct rear view, there won't be many to speak of – but don't forget about the rear window reflection in the deck lid.

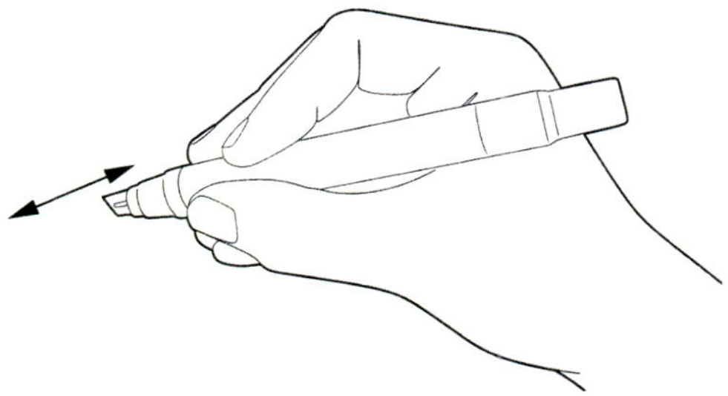

Now block in the darkest parts of the chrome and the vinyl roof with Cool Grey 70%. Be sure to leave a bit of the paper showing through in the right-rear pillar; this will help with highlights later.

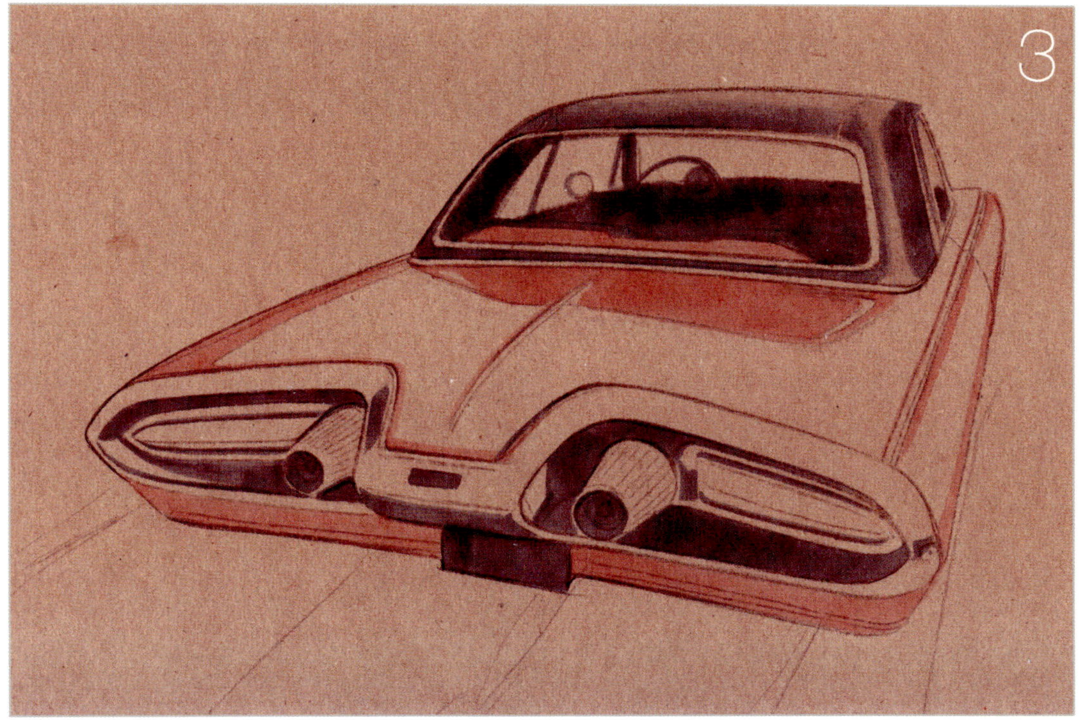

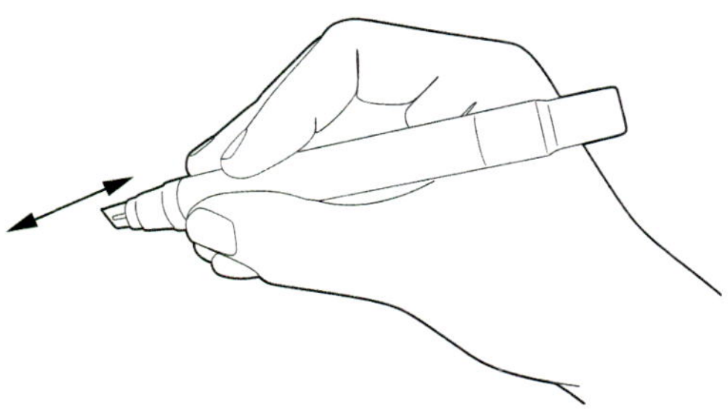

Staying with the Cool Grey 70%, lay in a bold shadow streaking away from the rear end. Allow the grey to fade out towards the left corner; this ensures that the shadow will not feel too heavy.

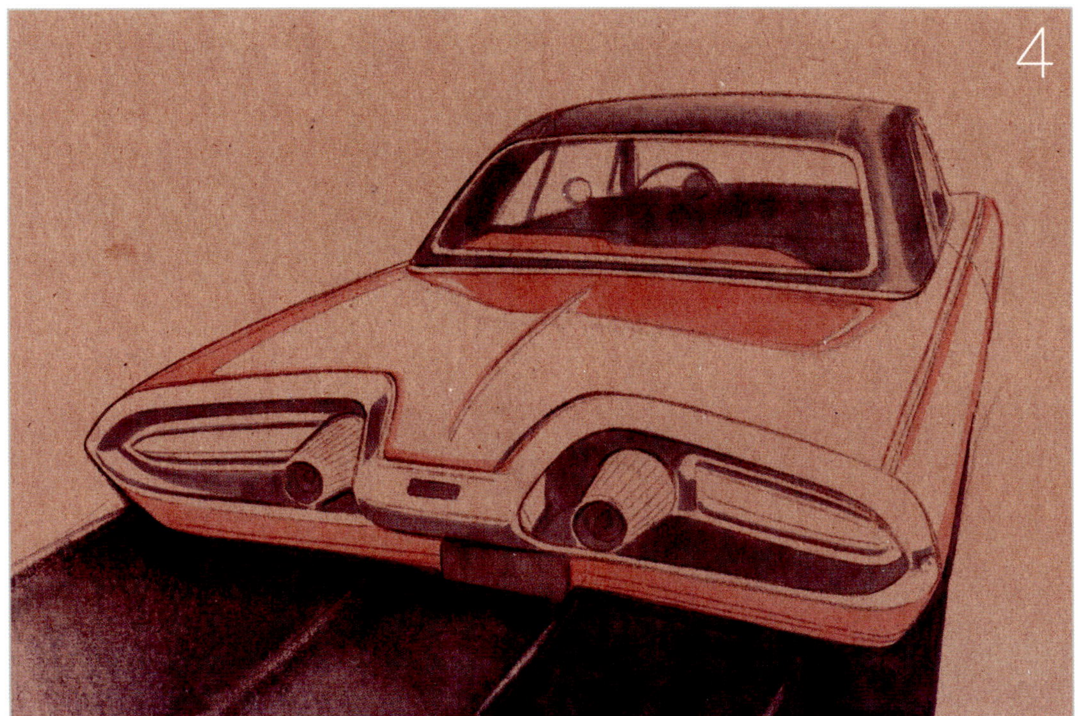

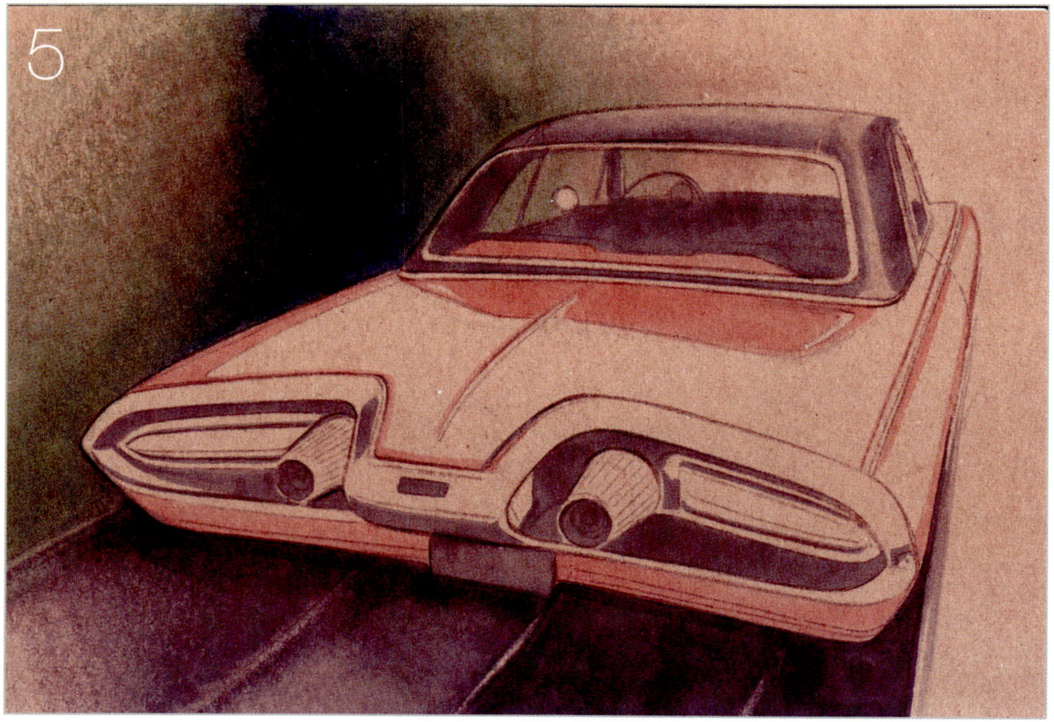

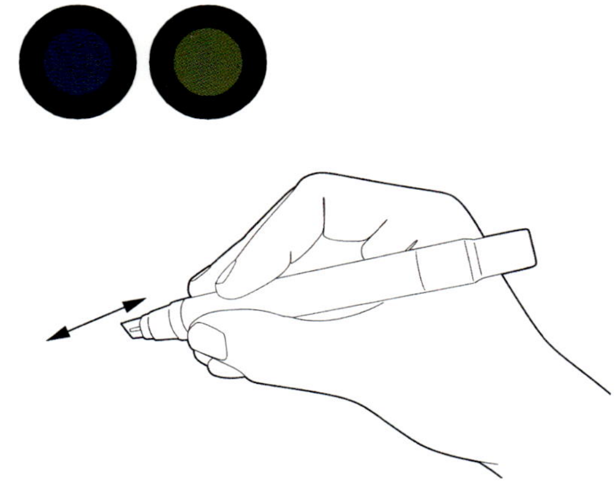

With a combination of green and blue markers, lay in a vertical vignette behind the left side of the car.

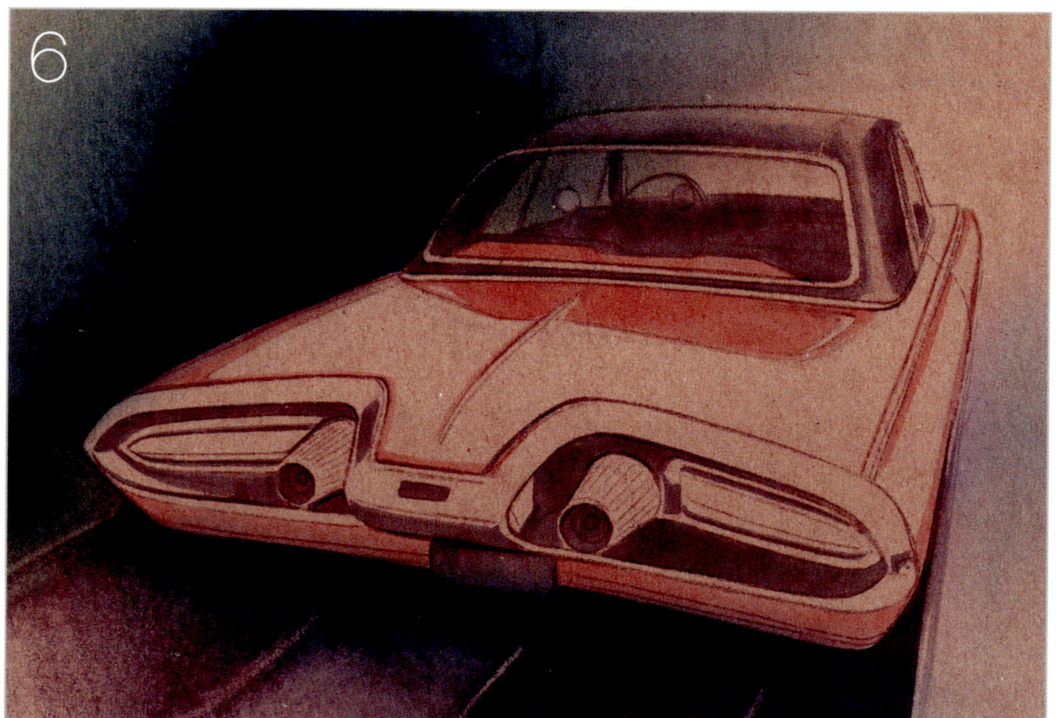

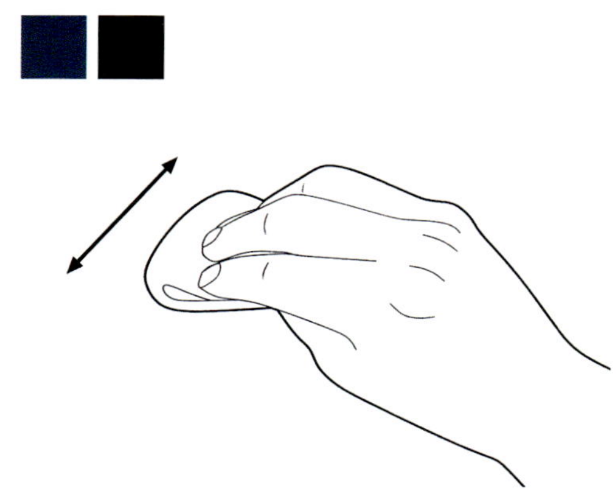

After the marker portion, soften the edges of the vignette with some blue pastel powder. You may wish to mix in some black pastel to add more punch.

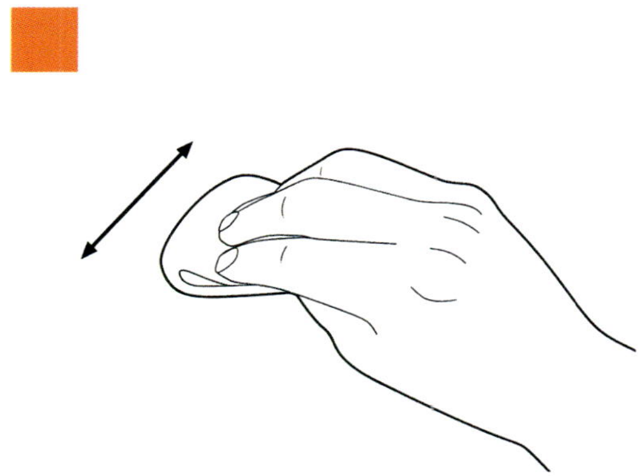

For the car's bronze paint, brush in some orange pastel. Combined with the light brown tone of the paper, this will yield a close approximation of the color.

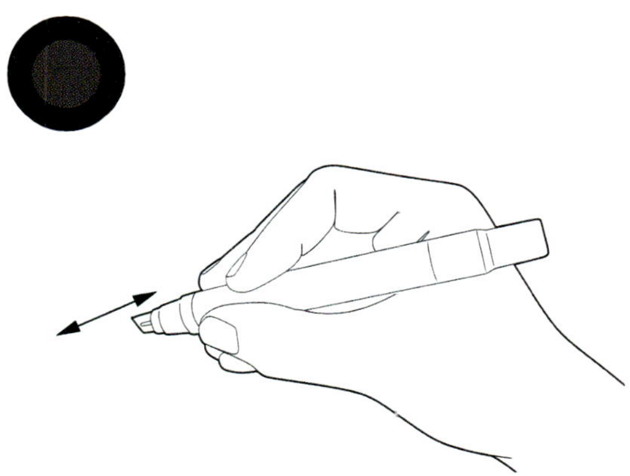

Return to Cool Grey 70% for the immediate bumper shadow and the more cavernous parts of the car.

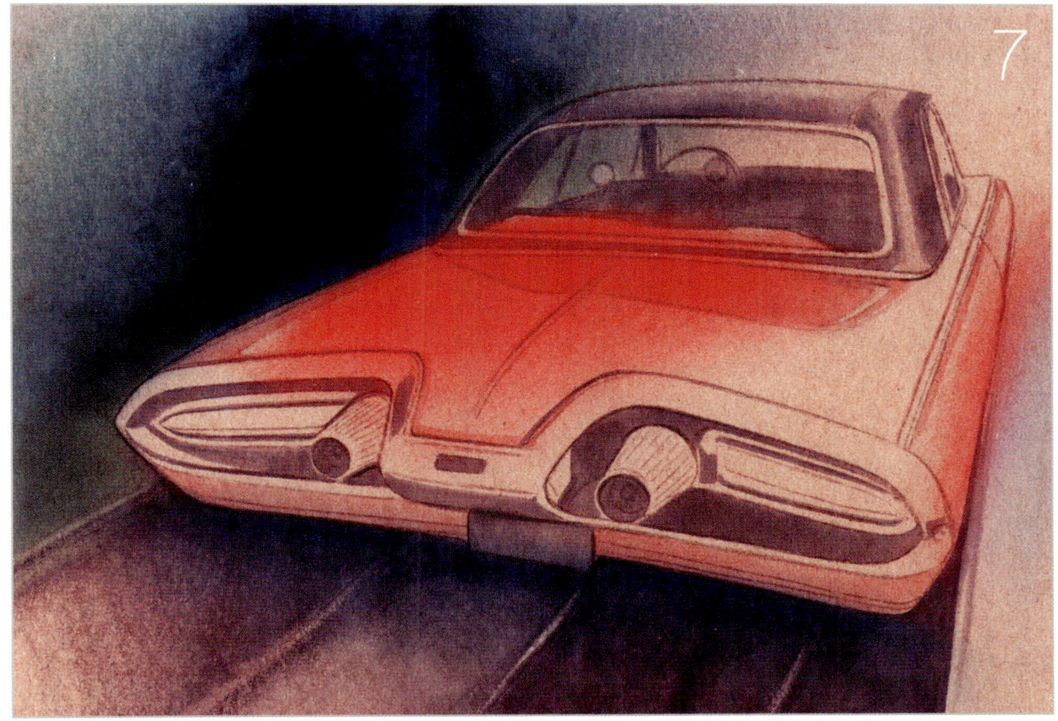

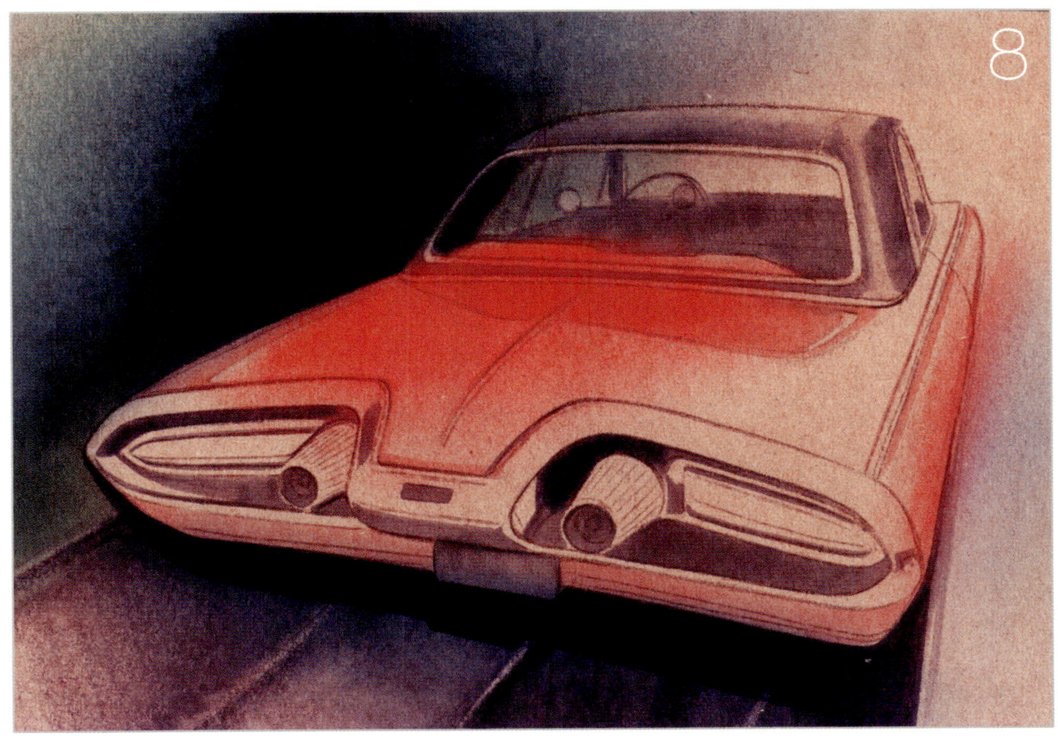

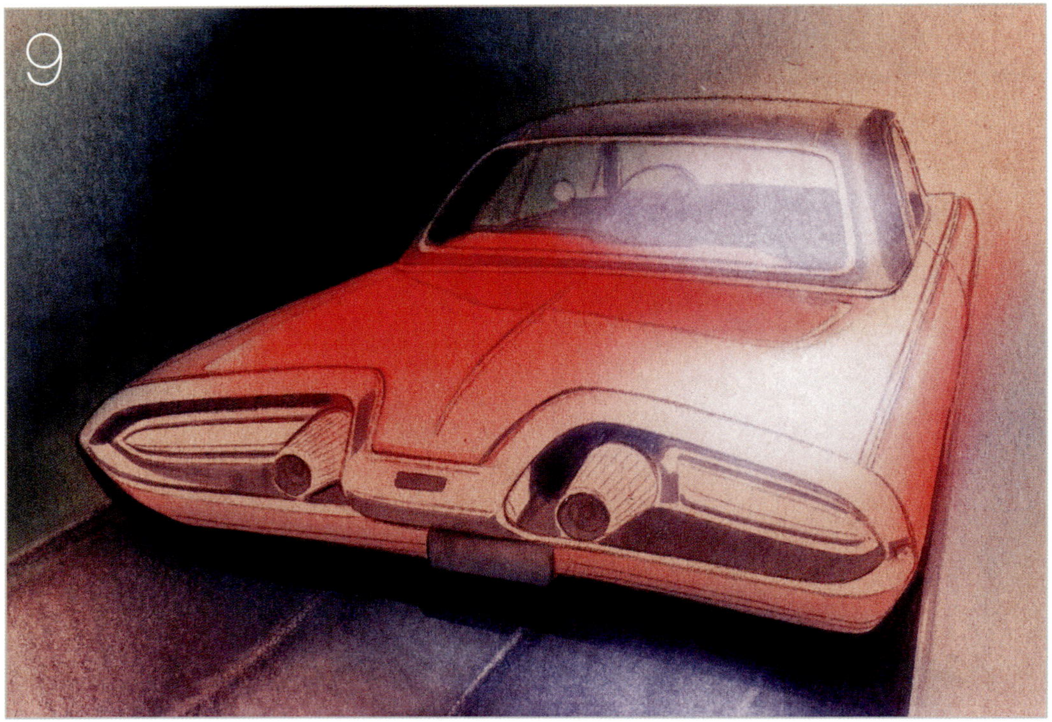

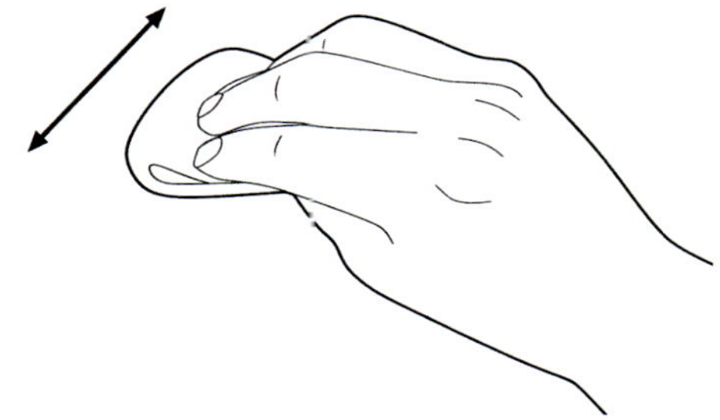

Now that the darkest tones have been accounted for, it's time to focus on the light tones. Lay in some white pastel on the deck lid and rear window of the car. You may chose to apply the pastel directly to the page at first, then blend it out with a Webril pad. This will yield a bit more punch.

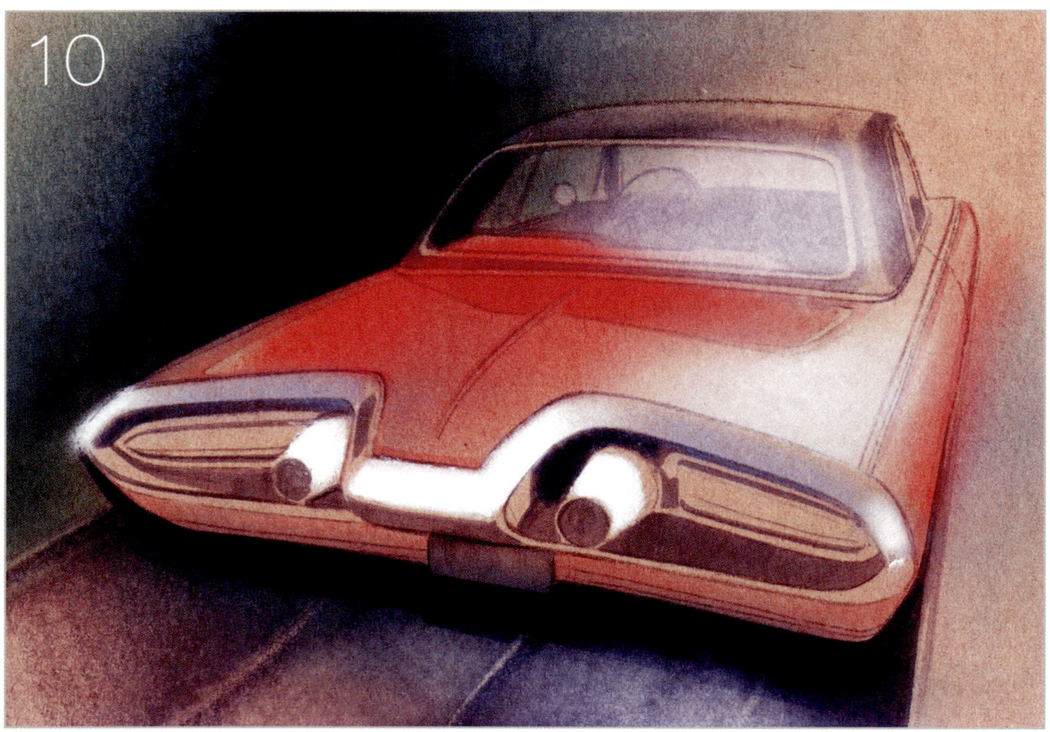

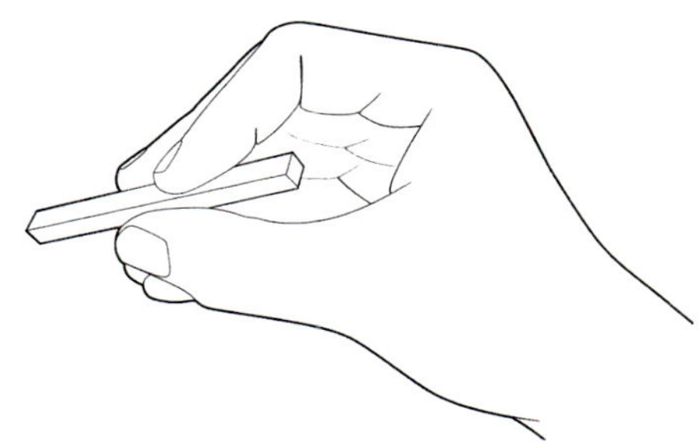

For the brightest parts of the chrome details, apply the white pastel directly to the page.

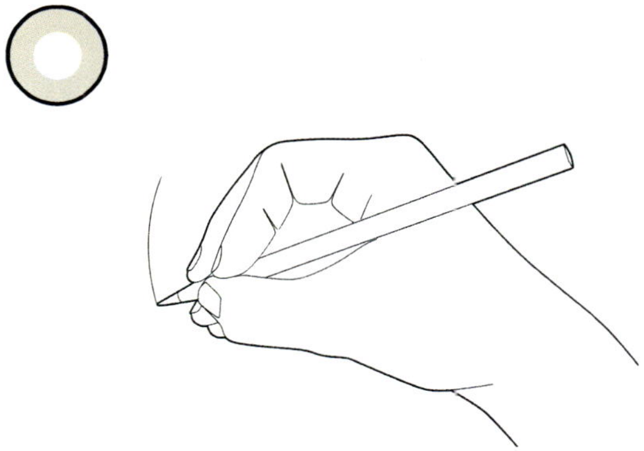

Now switch to a sharp, white Prismacolor pencil for the more delicate highlights. These include the edges of the chrome parts as well as the window frames. Use a lighter touch to reflect the rear window frame into the deck lid.

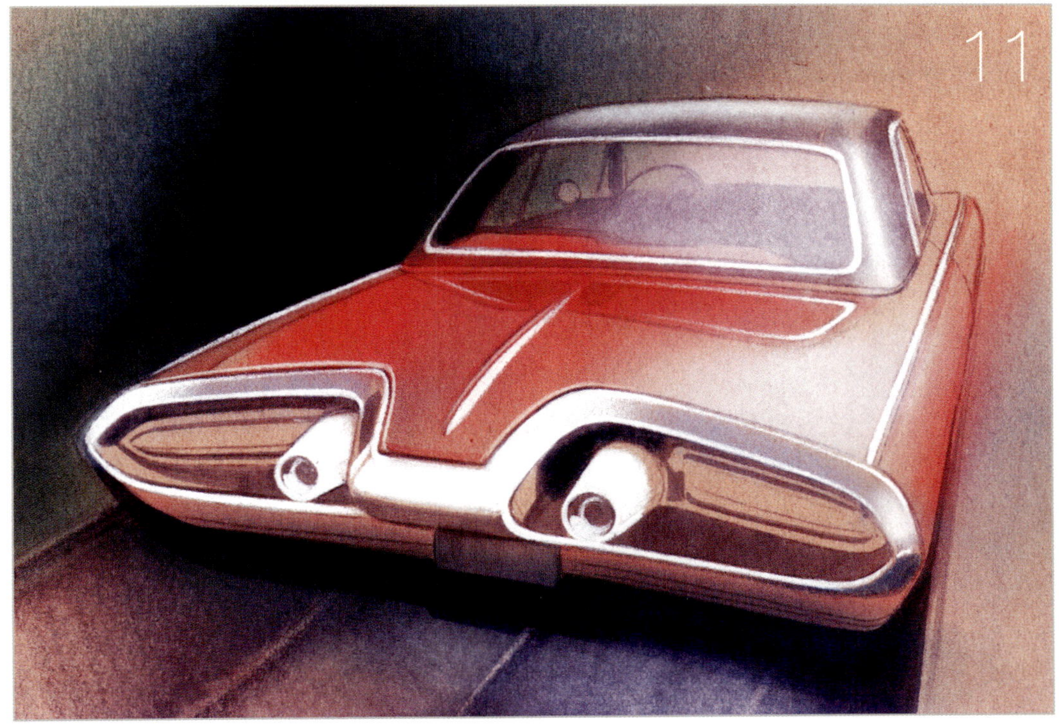

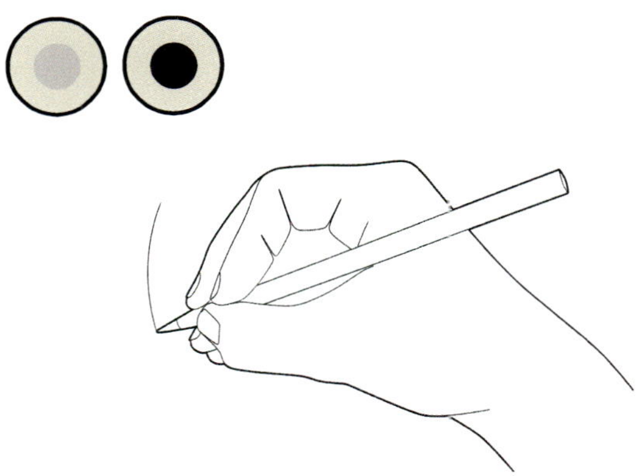

For the turbine details in the lamp bezels, try using a sharp Cool Grey pencil. Notice how the resulting lines appear darker than the highlights, but lighter than the shadow.

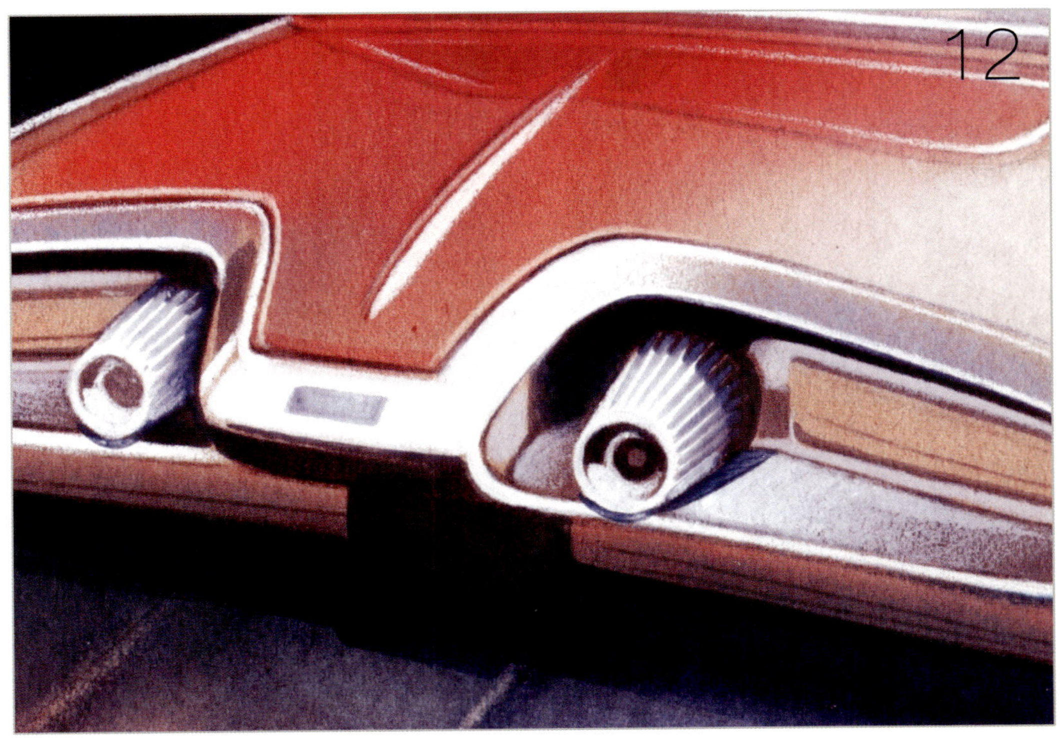

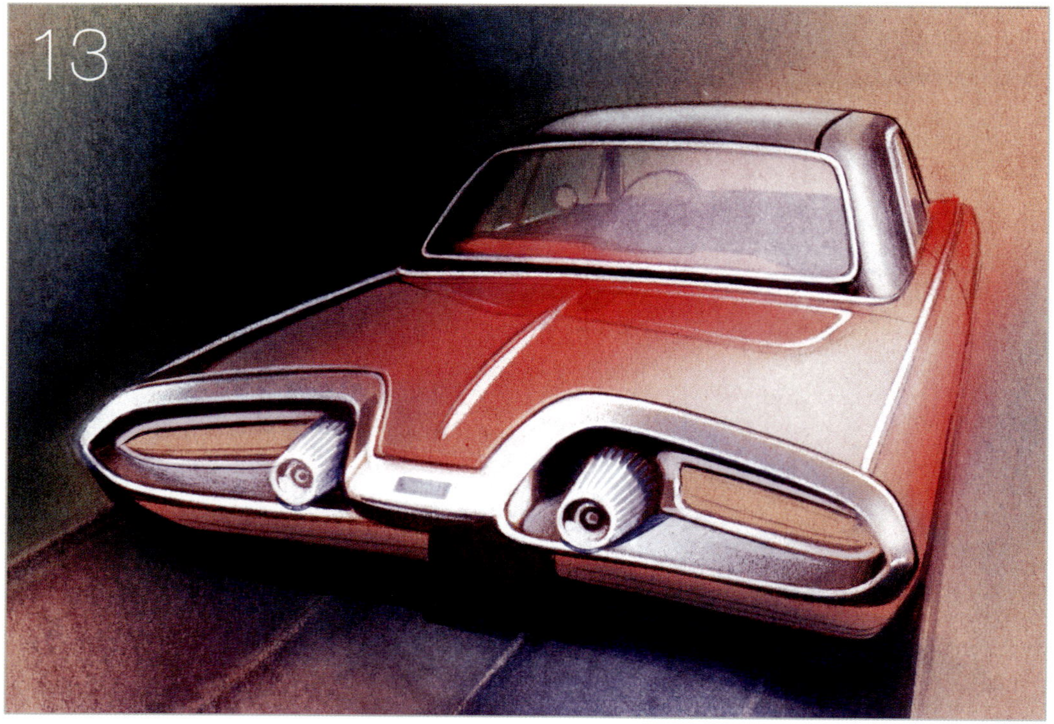

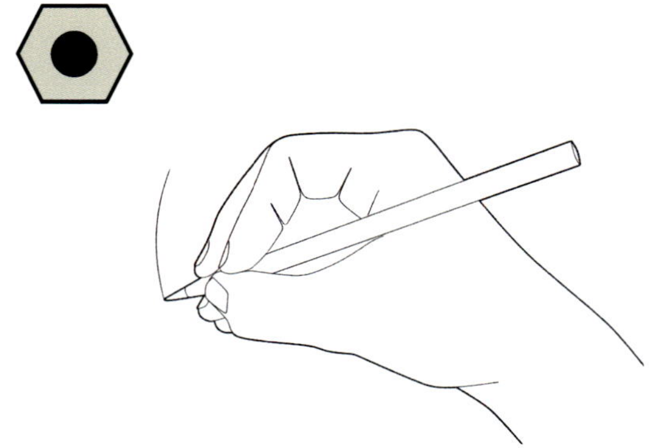

Now's a good time to sharpen up some of the car's shut lines. For this, use a sharp, black Verithin pencil.

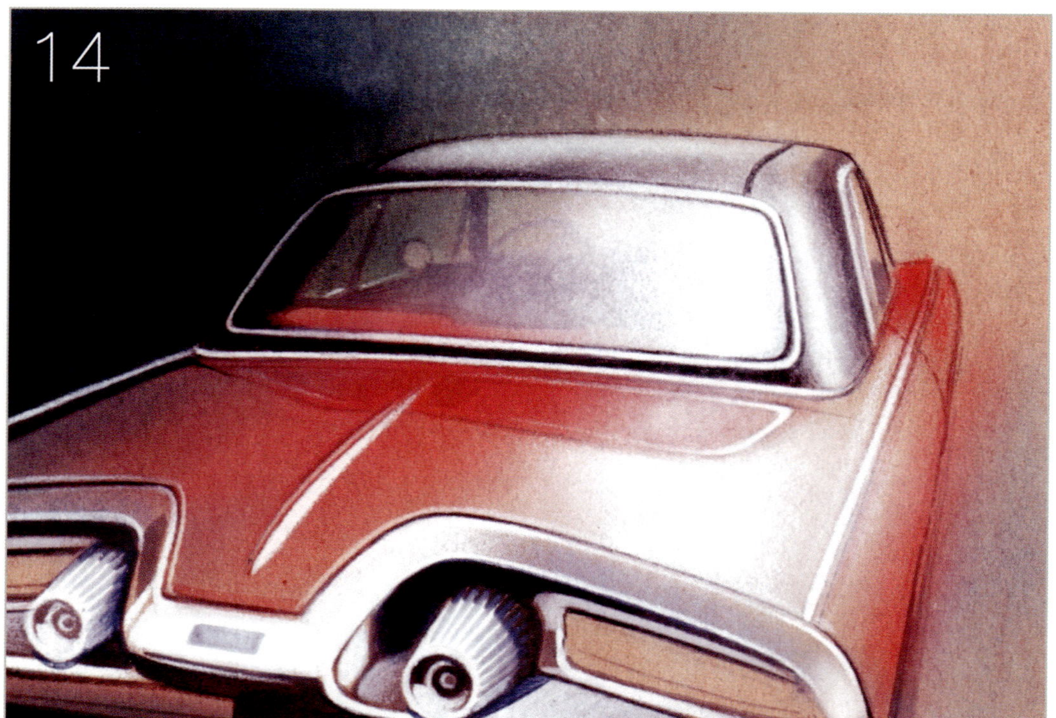

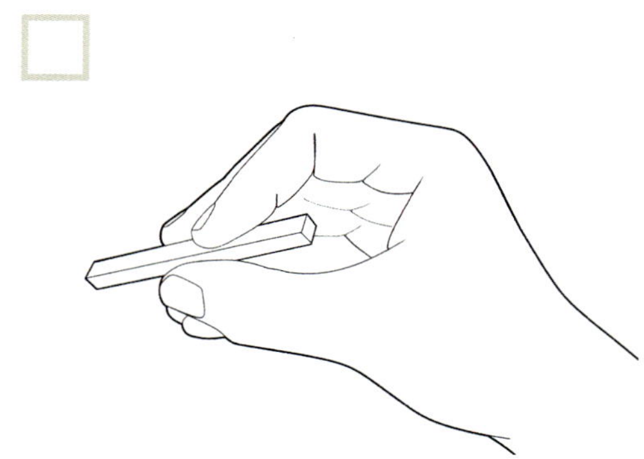

If the body highlights seem out of step with the contrast of the chrome, bump them up with more white pastel.

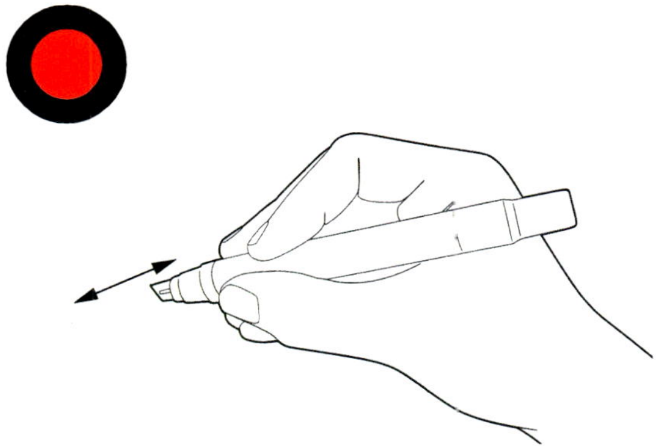

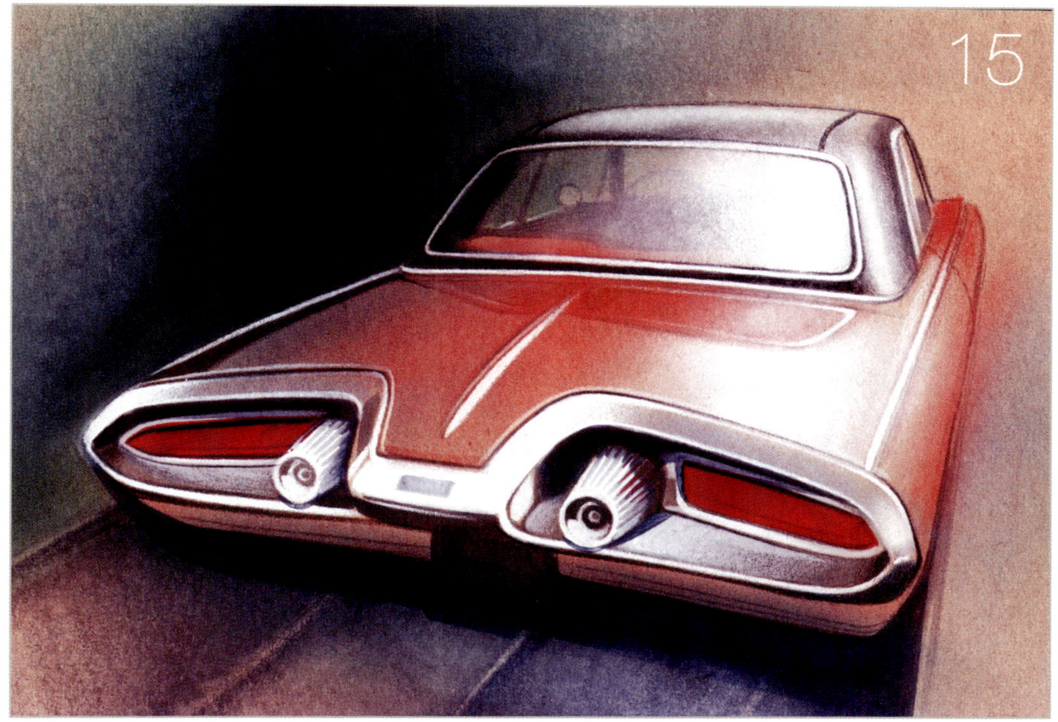

Block in the taillamp lenses with a red marker. Avoid picking up any black pencil edges with the marker.

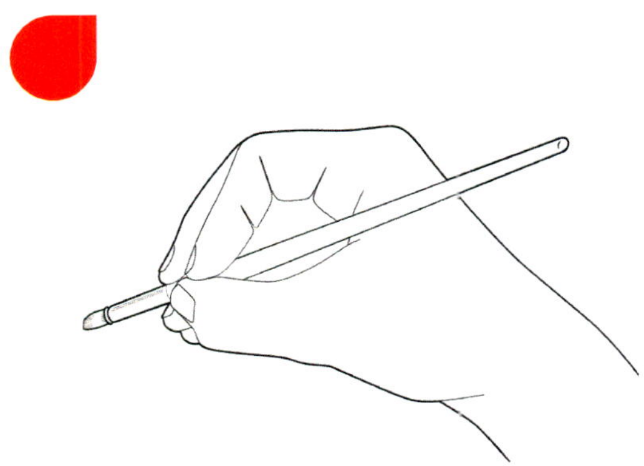

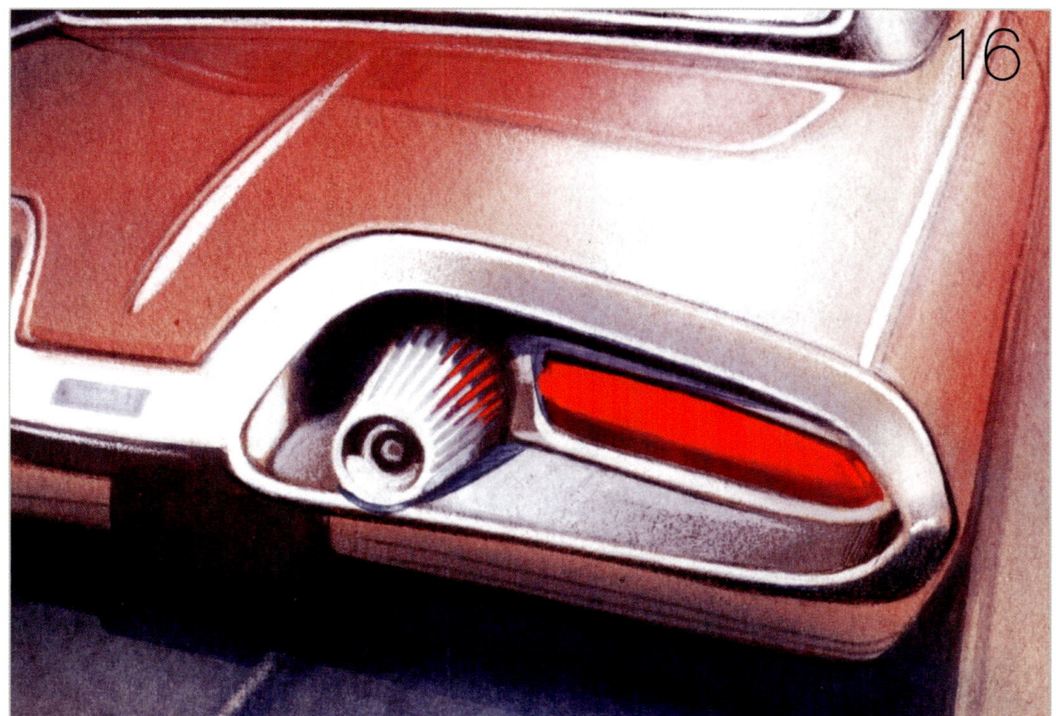

For the lit portion of the taillamps, carefully apply a swath of red gouache. Work the brush into a chisel shape and apply short, sharp strokes on the nearby bezel. Details like these separate good renderings from great renderings.

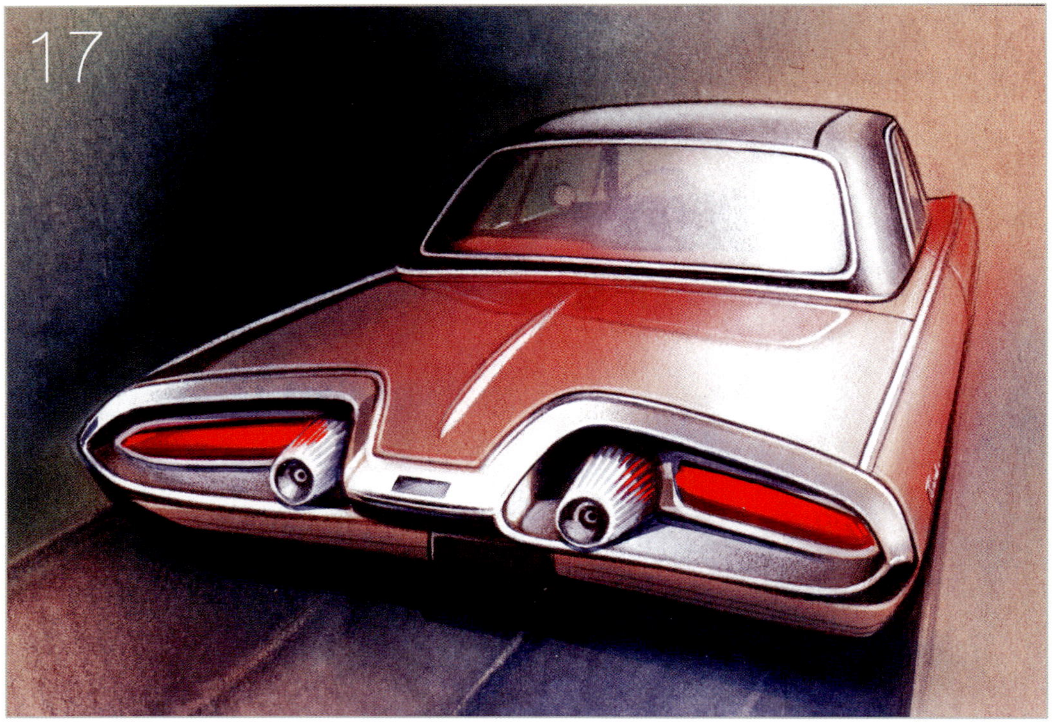

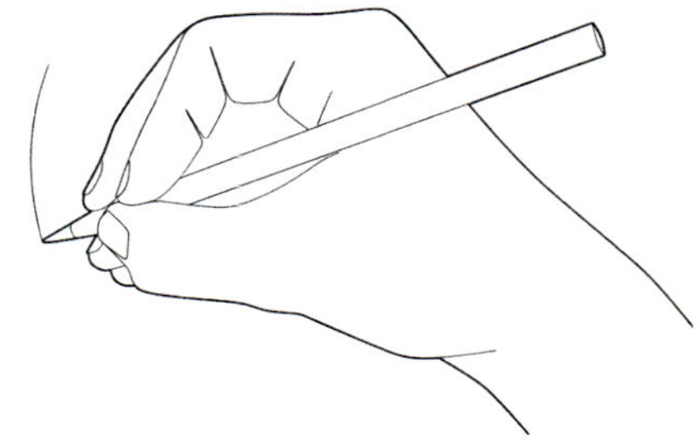

Return again to the parting lines and make one more pass with the black Verithin pencil. Sometimes taking a break from a task to concentrate on another area can be helpful; you may come back later and see things that you didn't see before.

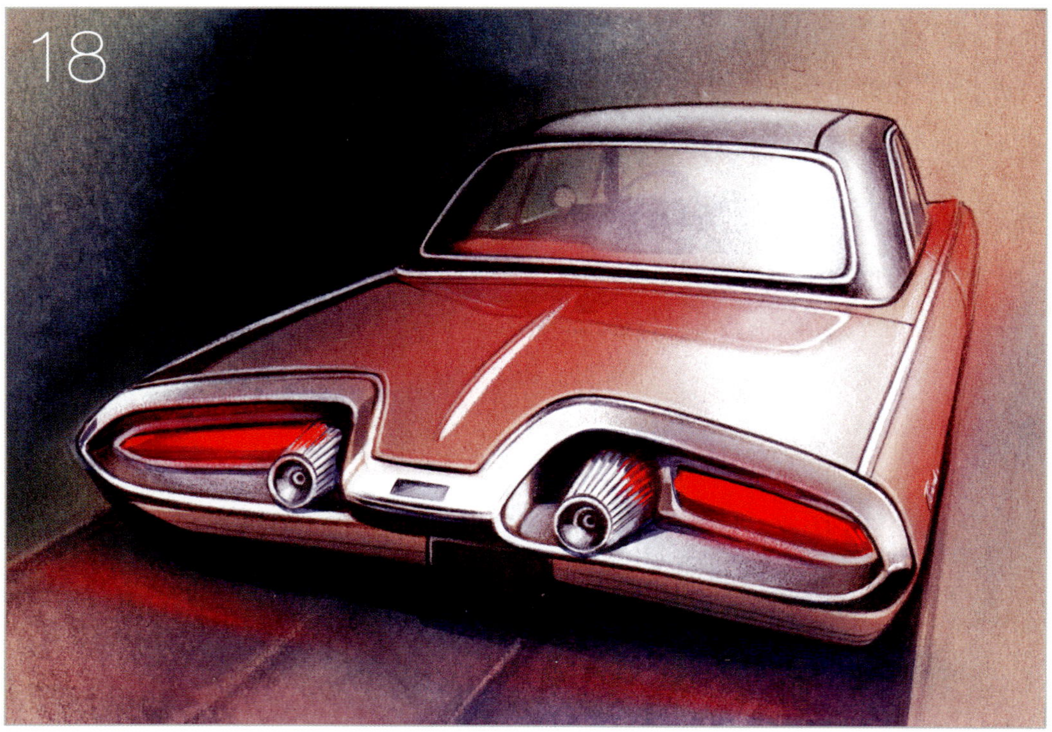

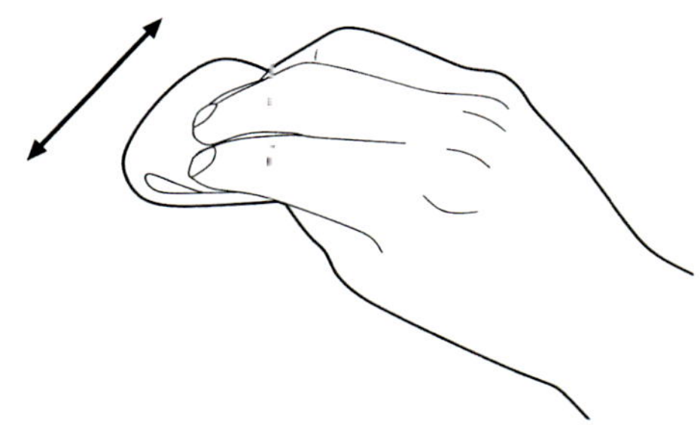

Apply a bit of red pastel around the taillamps themselves and on the pavement directly below.

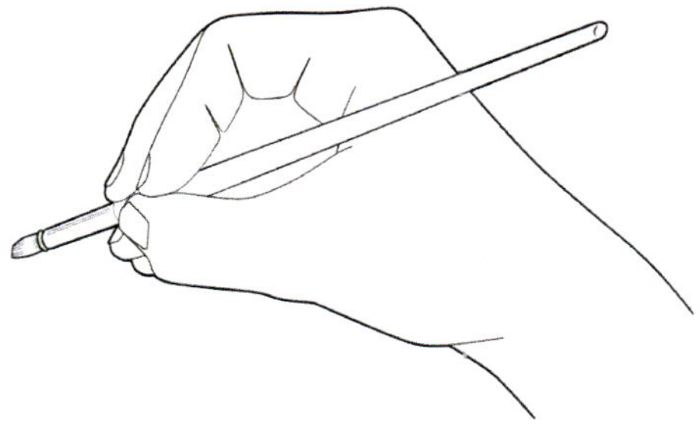

Concentrate the white gouache in the chrome parts, as these will be the most shiny elements in the rendering. Use a small amount of black gouache in the darkest areas.

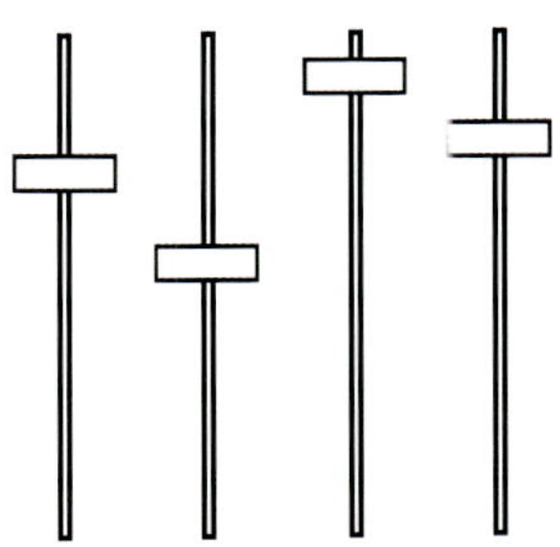

Scan in the rendering and adjust the levels to your liking in Photoshop. Use a white airbrush tool to soften the white highlights and add a bit of glow.

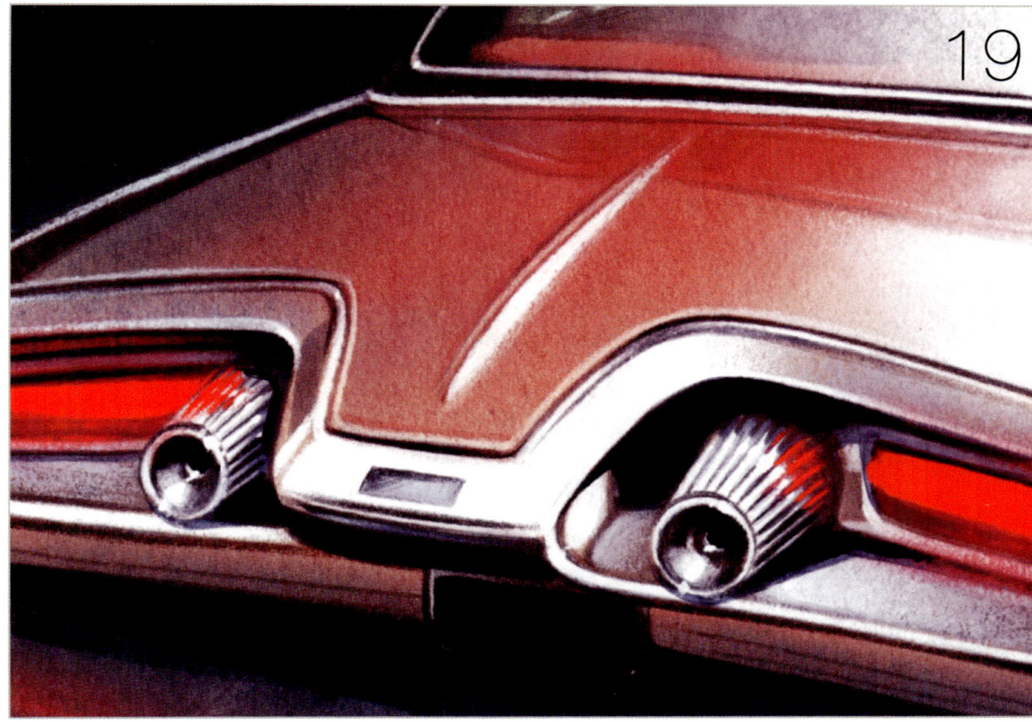

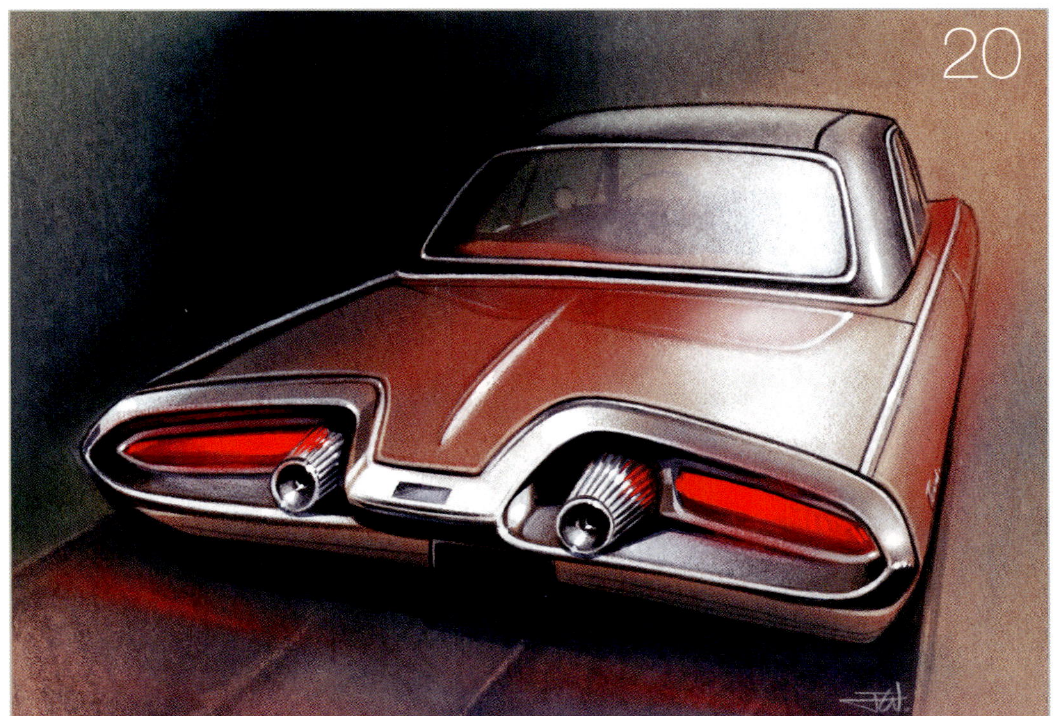

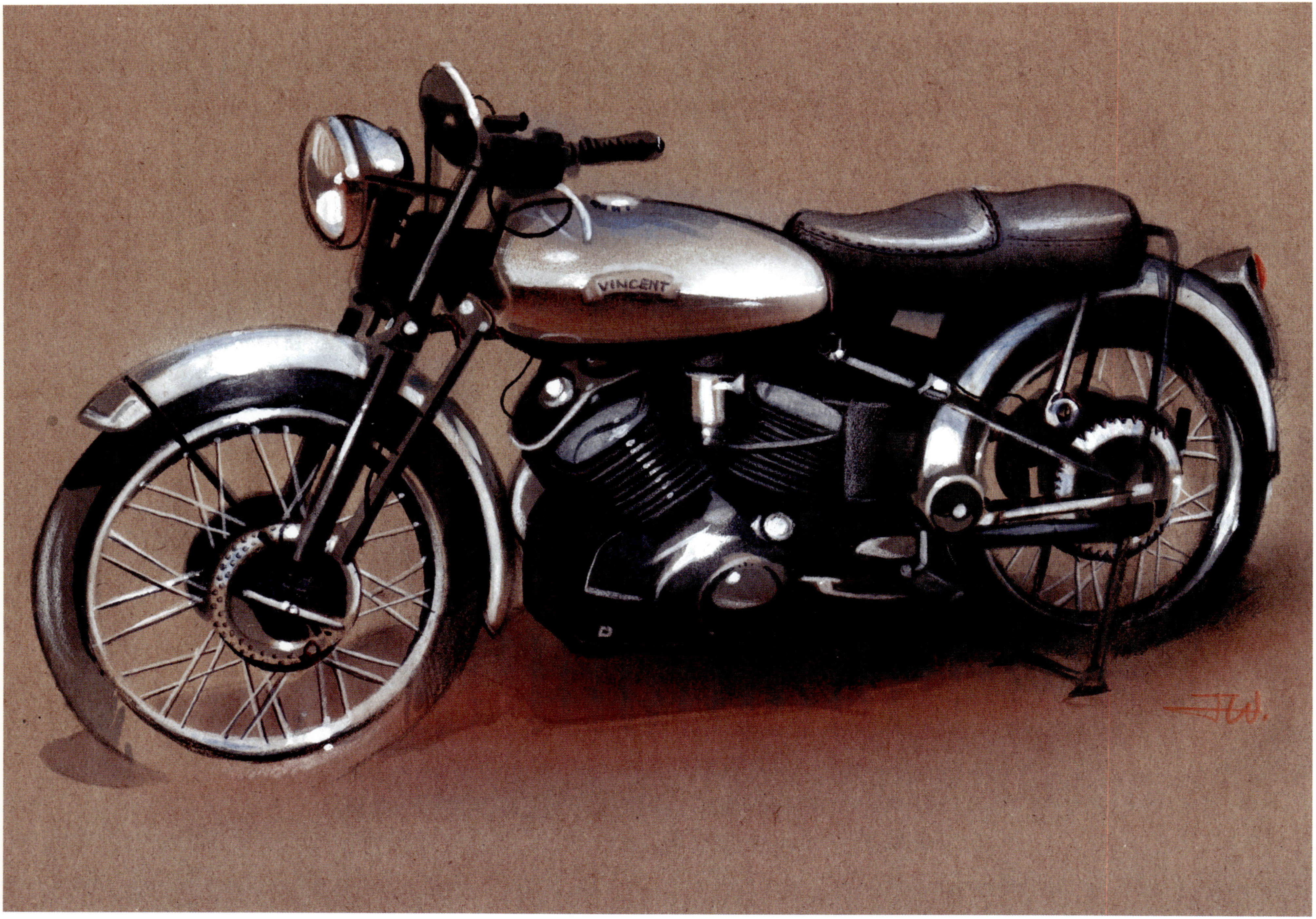
VINCENT

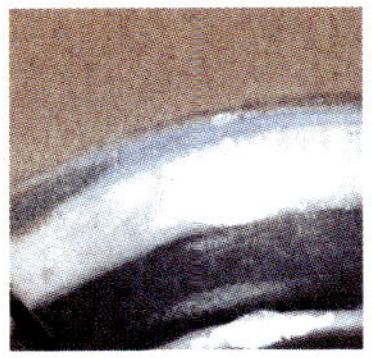

20. Vintage Chrome

Nothing attracts the eye like chrome. In the case of vintage chrome, there can be so much of it that it's hard to decide where to place emphasis. In this tutorial you'll learn how to give hierarchy to your chrome details so the eye is dazzled, not confused.

The Tools

Prismacolors:
White,
Cool Grey 20%,
Crimson Red.

Nupastels:
Orange, White,
Light Blue,
Black.

Verithins:
Black.

Gouache: White.

Prismacolor Markers: Warm Grey 30%, Cerulean Blue, Cool Grey 40%, Cool Grey 70%, Black.

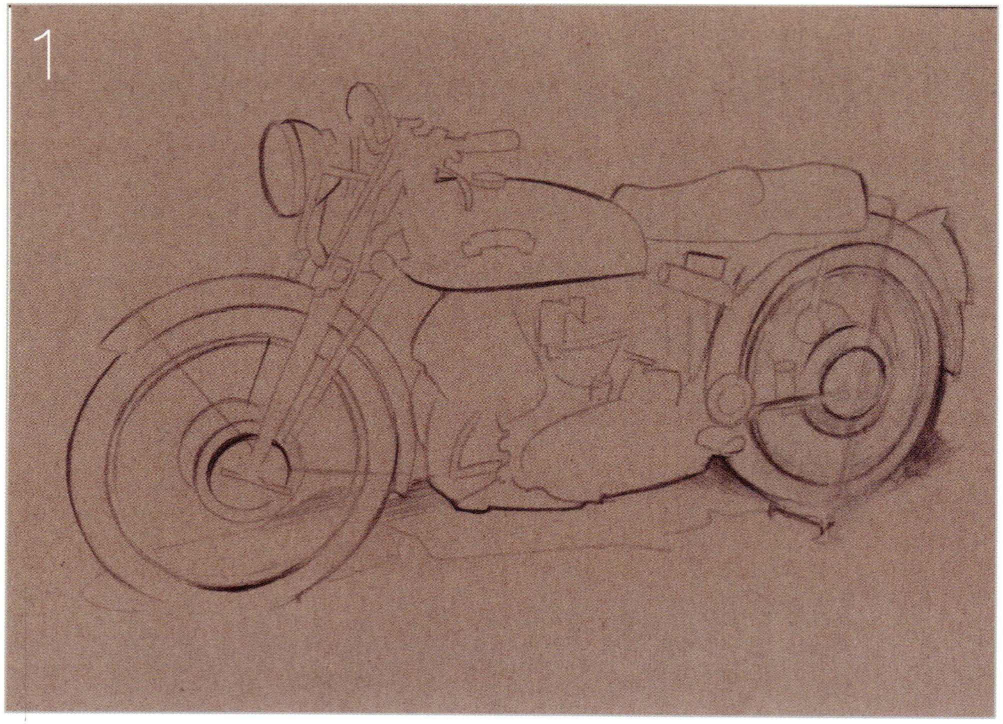

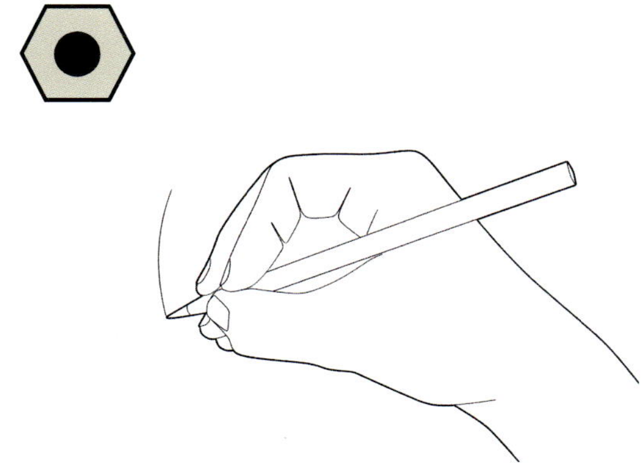

Start by laying out the basic lines of the bike with a black Verithin pencil on a clean piece of craft paper.

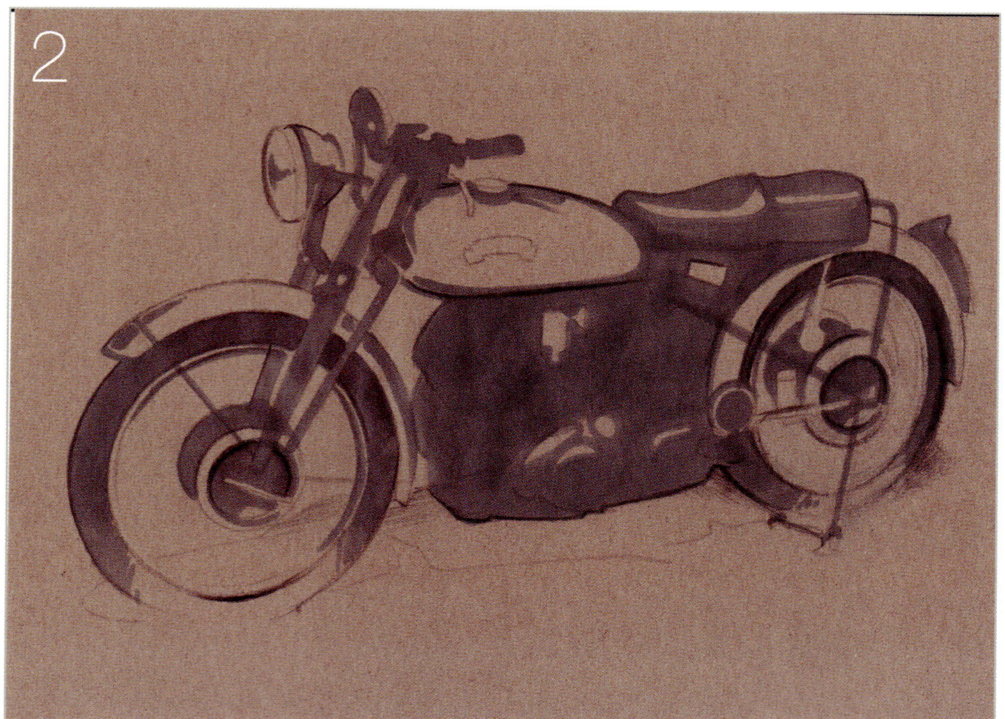

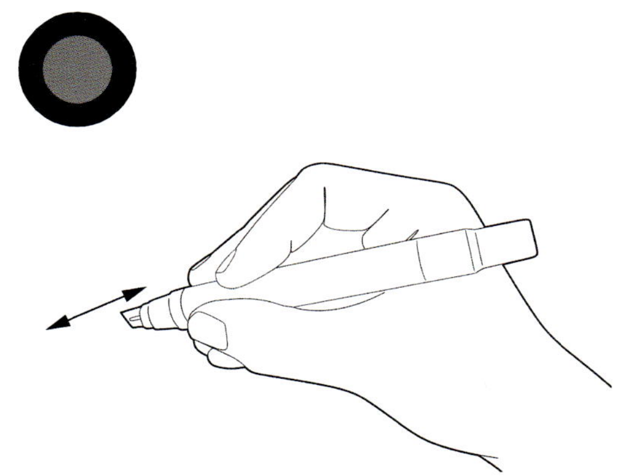

Block in the darkest parts of the bike with a Cool Grey marker. These areas include the tires, engine, various mechanical parts and the leather seat.

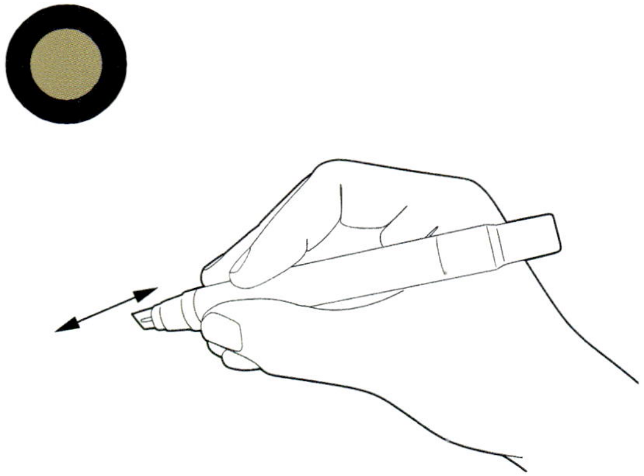

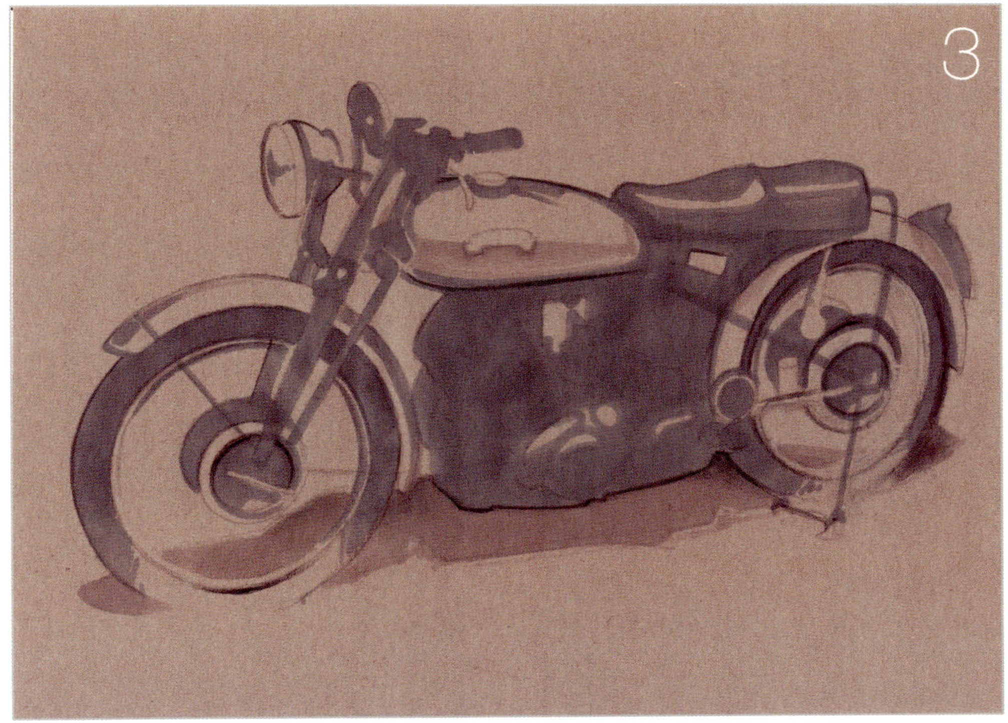

For the ground shadows and the chrome horizon lines, switch to a Warm Grey marker. This temperature shift will help differentiate the mechanical parts from the environment.

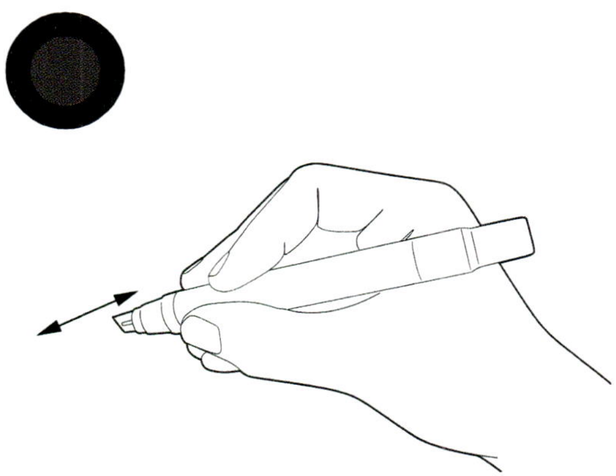

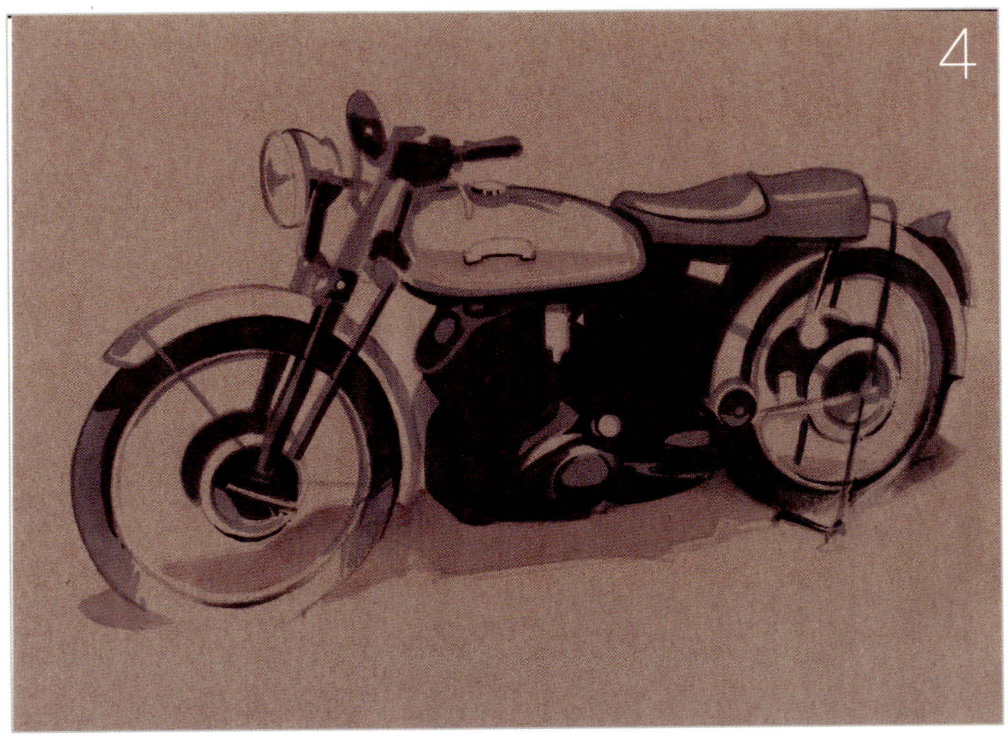

Return to the mechanicals and the tires with Cool Grey 70%; be sure to leave some of the lighter grey showing through as a mid-tone.

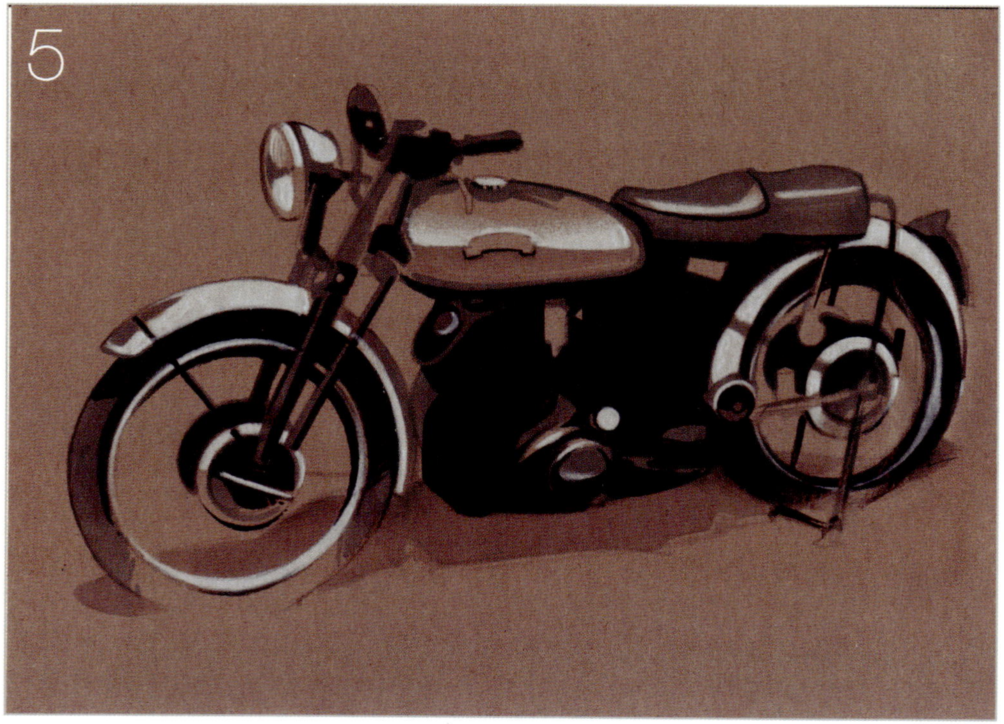

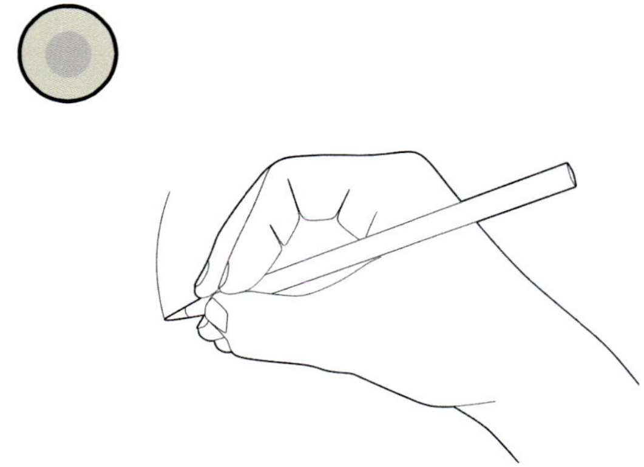

Before you break out the white pencils, describe the lighter parts of the chrome with a Cool Grey Prismacolor.

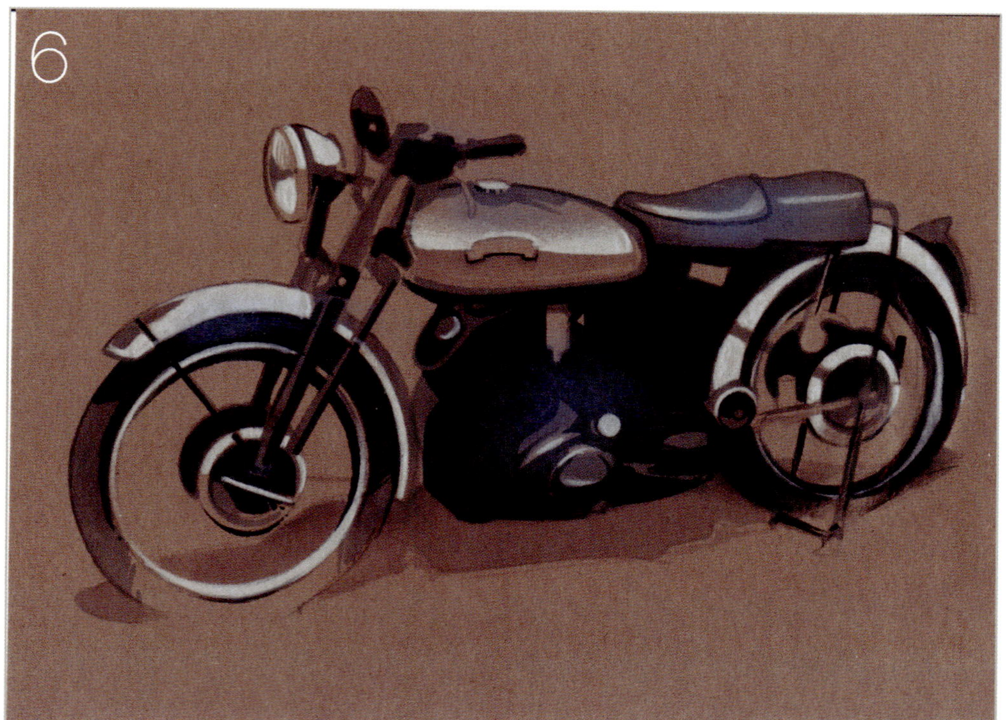

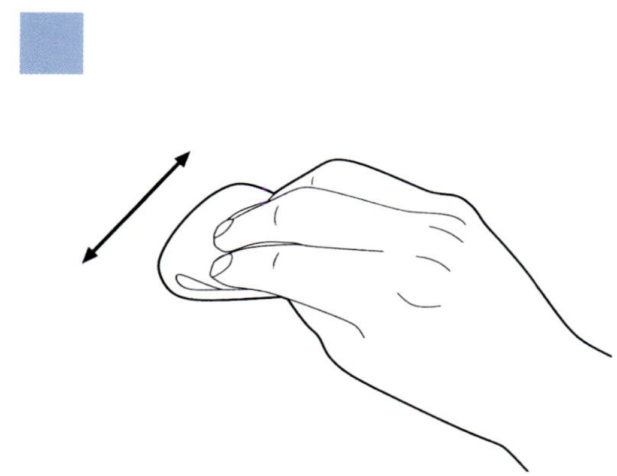

Before you start detailing the engine, ghost in some light blue pastel with a Webril pad. You can also apply this color in the fenders, the gas tank and seat as an indication of sky tone.

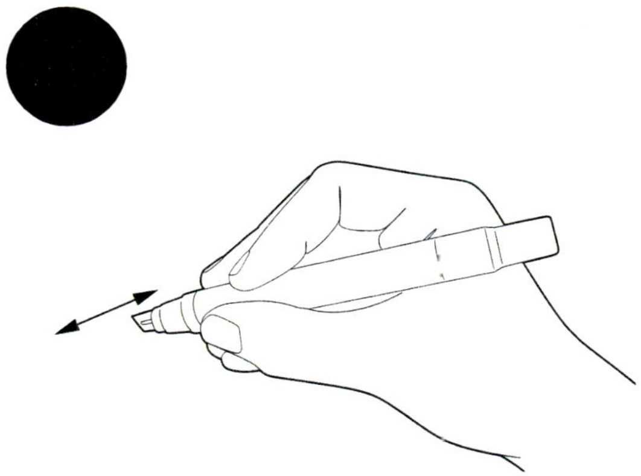

Now, using the fine point of a black marker, begin detailing the engine. Notice how the blue pastel provides a soft, cool mid-tone for the engine block.

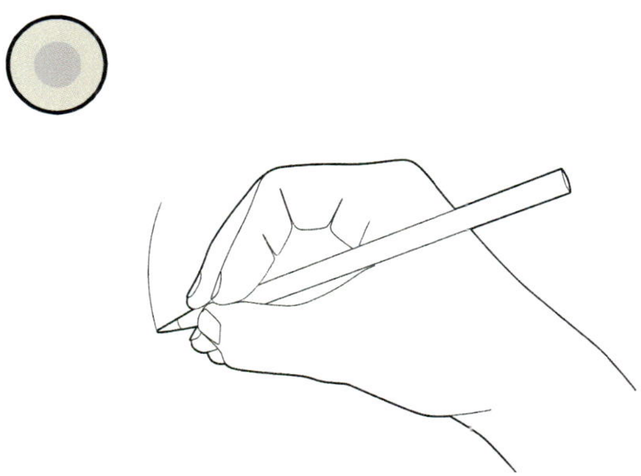

Use a Cool Grey Prismacolor pencil to describe the engine highlights. Thanks to some careful planning, you now have a light, mid-tone and dark for the engine in three simple steps.

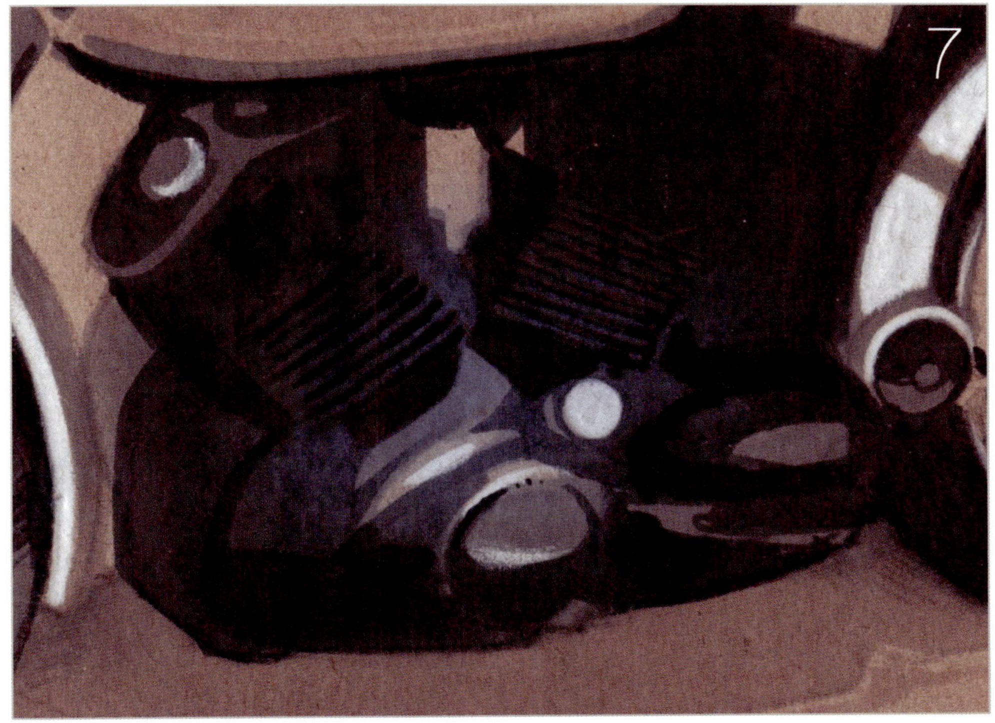

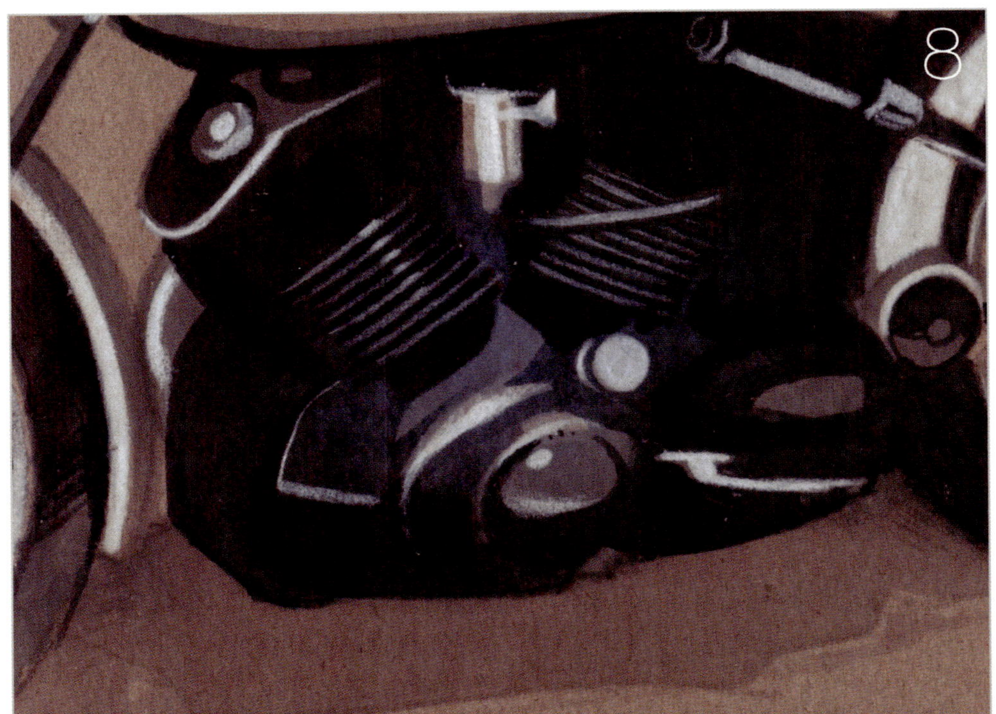

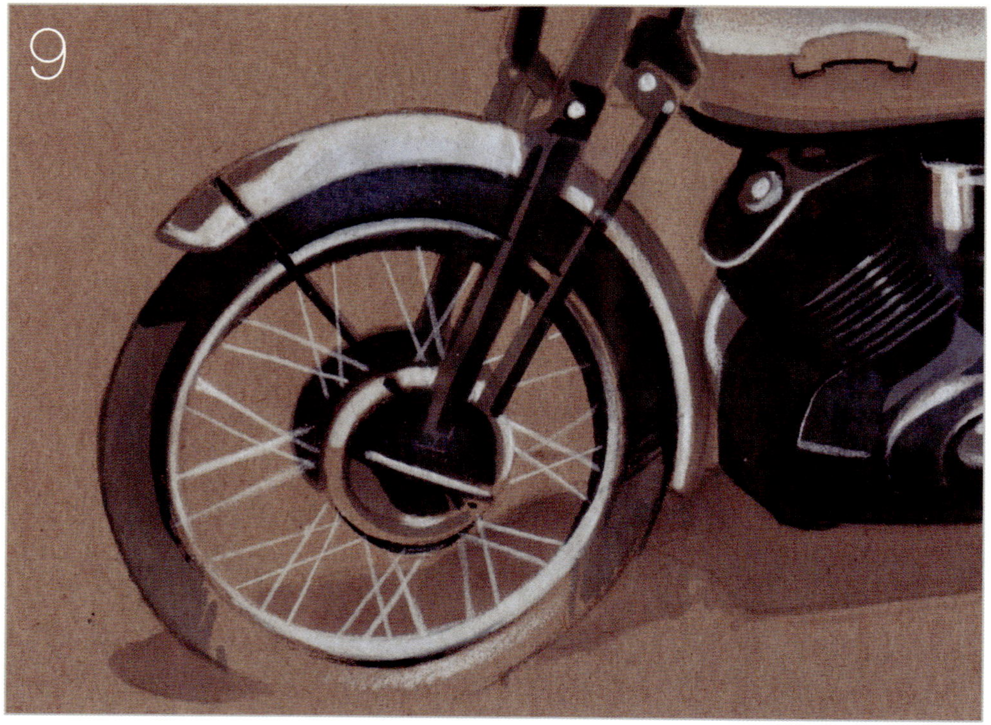

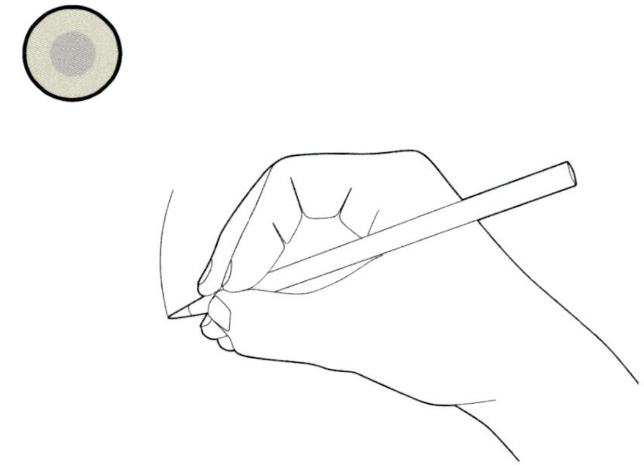

For the wheel spokes, use a sharp, Cool Grey Prismacolor pencil and a clean straight edge.

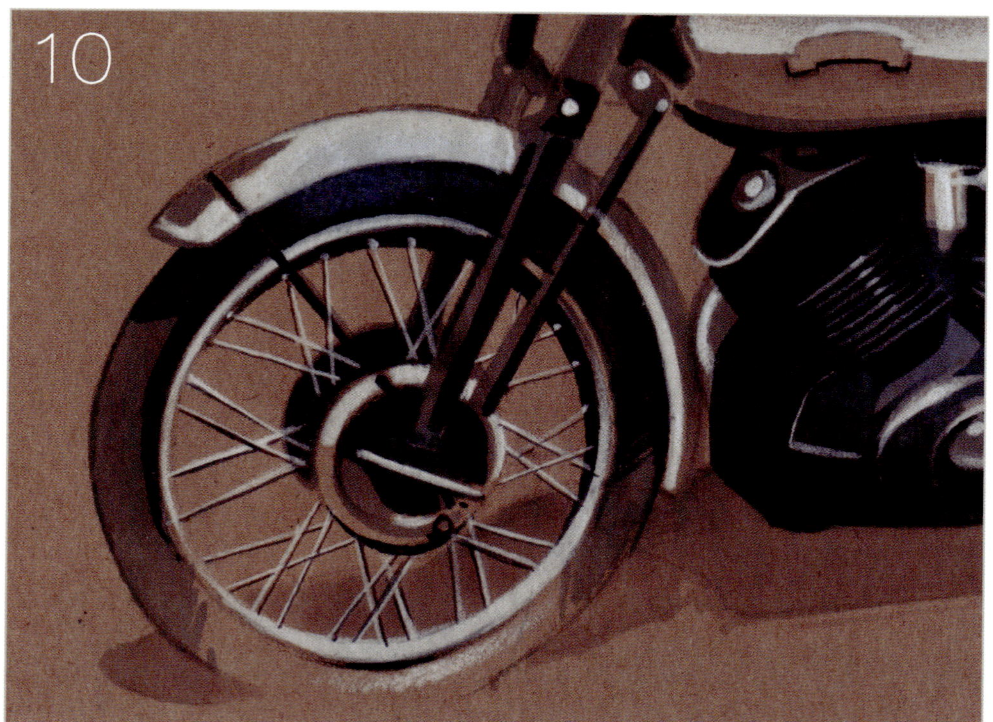

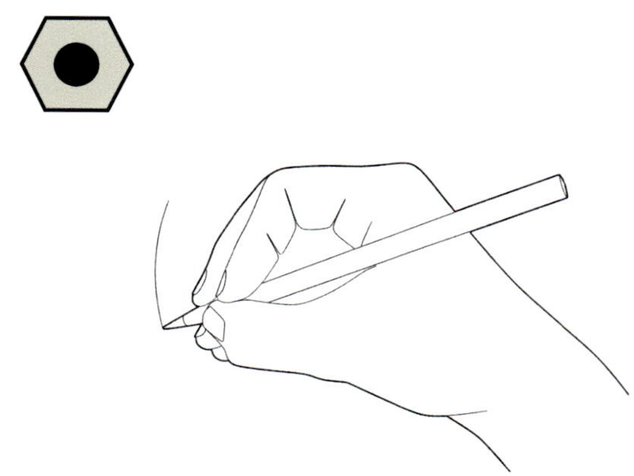

Now switch to a sharp, black Verithin pencil for the dark side of the spokes. Try to keep the light side of the spokes facing upward.

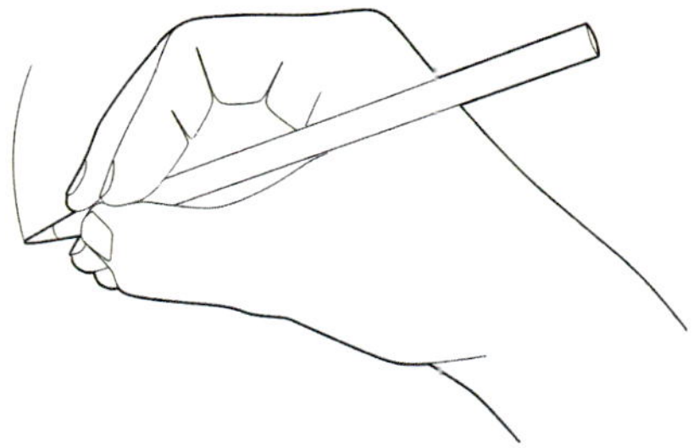

As you complete the spokes on the back wheel, make sure you give attention to the surrounding elements like the stand and the curved underside of the leather seat.

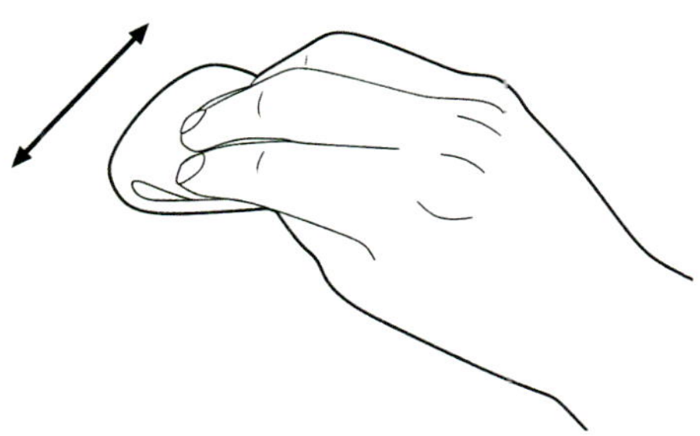

Add some depth and softness to the shadow with a bit of black pastel.

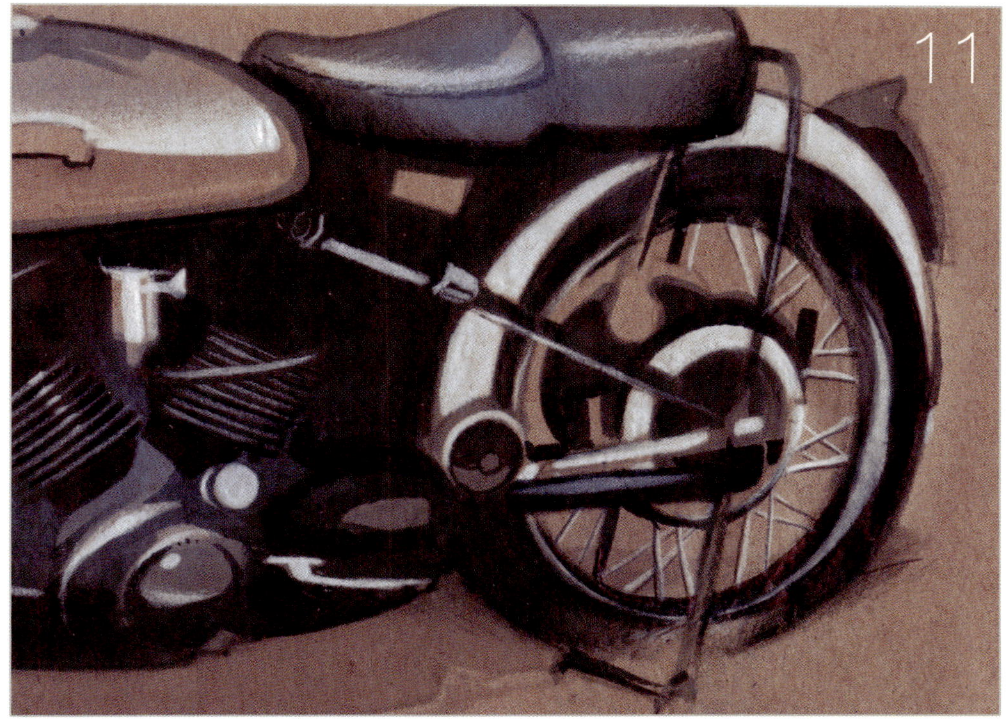

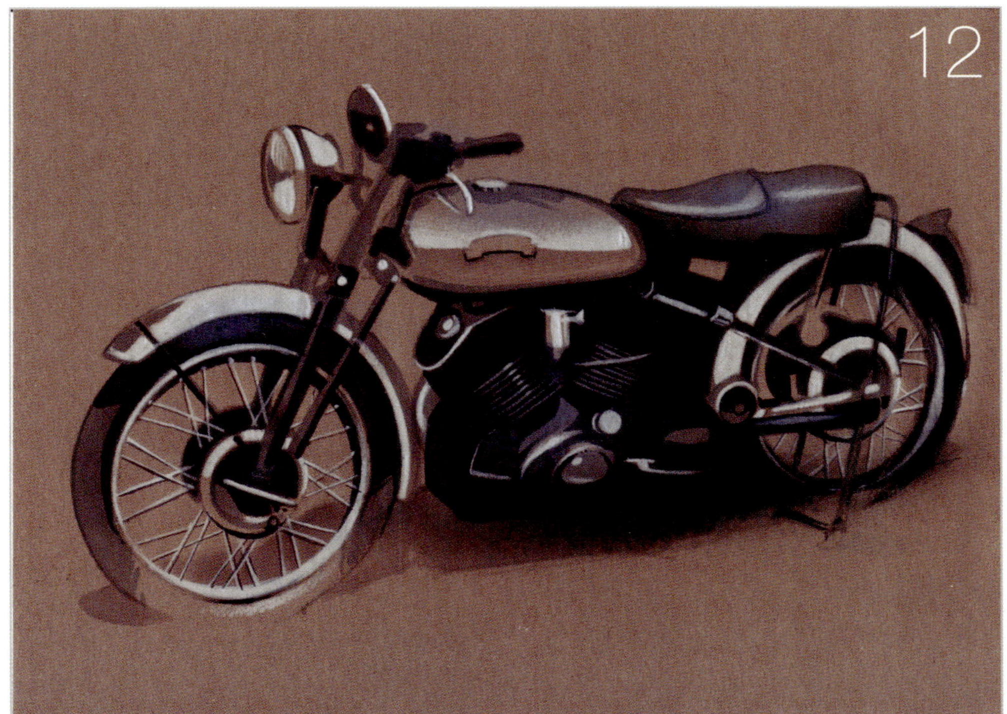

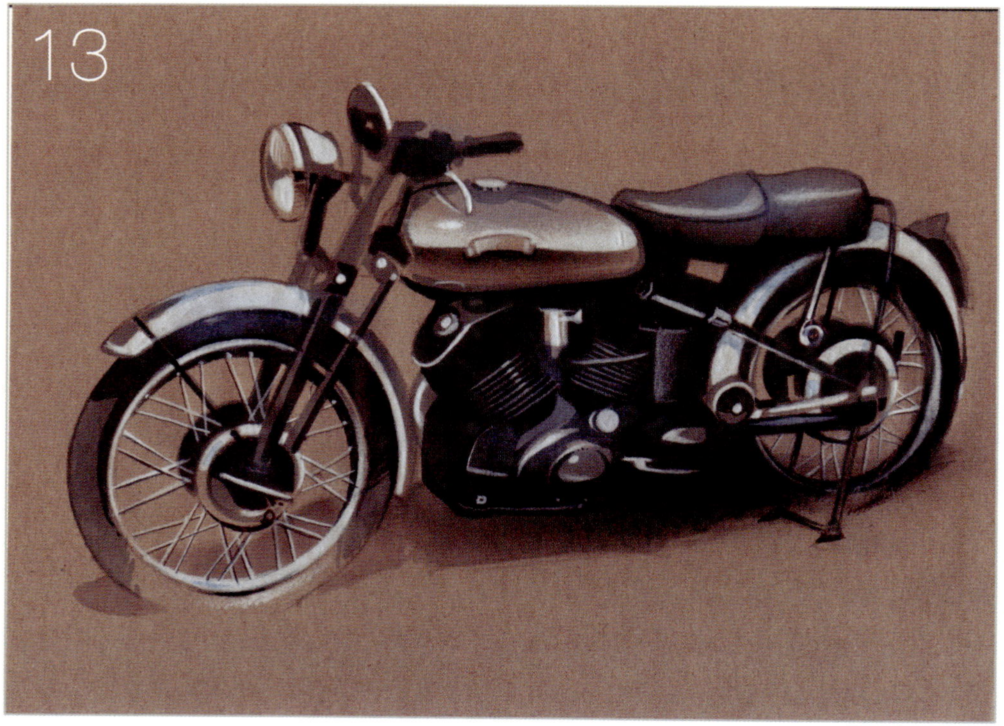

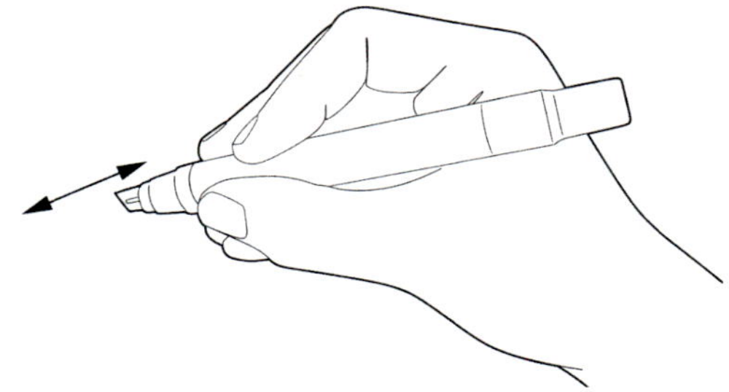

Add a touch of Cerulean Blue to the highlighted parts of the chrome; this will add some punch to the sky tone.

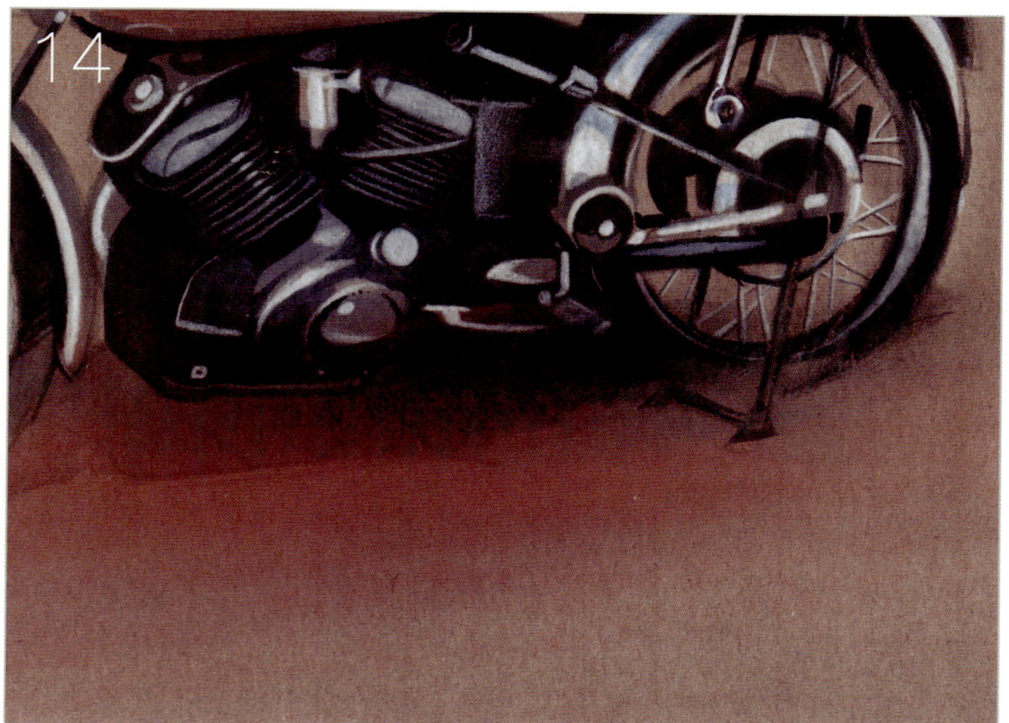

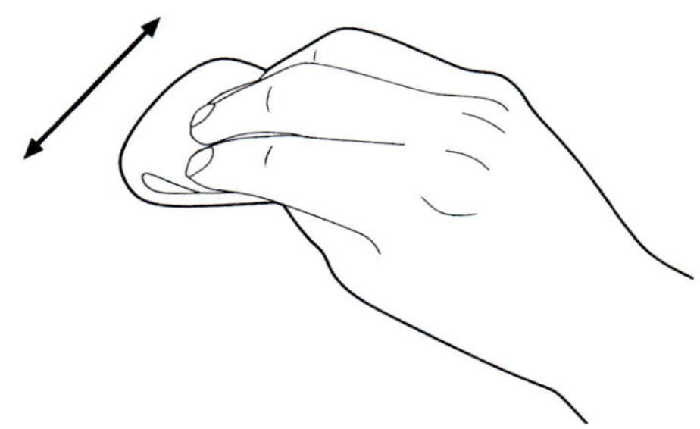

For the warm earth tone, lay in some orange pastel under the shadow. You can also add a hint of orange beneath the gas tank's horizon line.

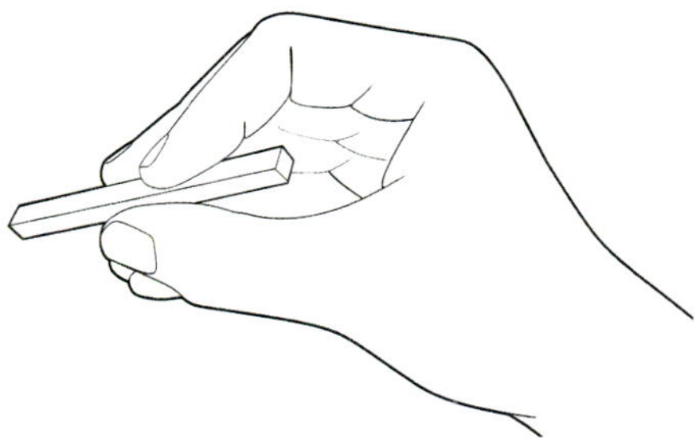

Punch up the middle of the front chrome fender and the gas tank horizon with white pastel. To ensure maximum contrast, apply it directly to the page and blend it out with your fingertip.

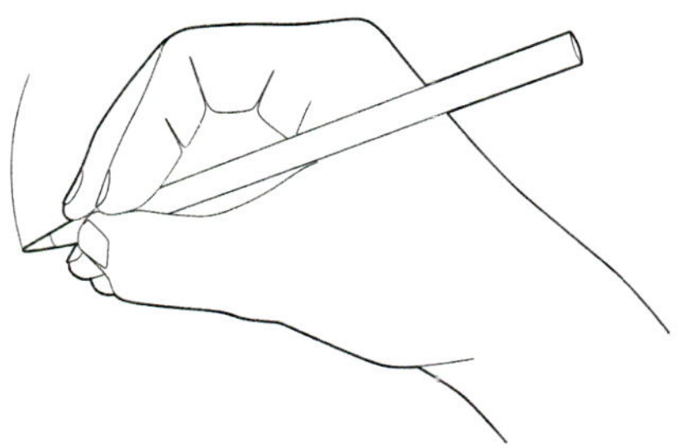

Add a touch of red pencil to indicate the convex taillamp lens.

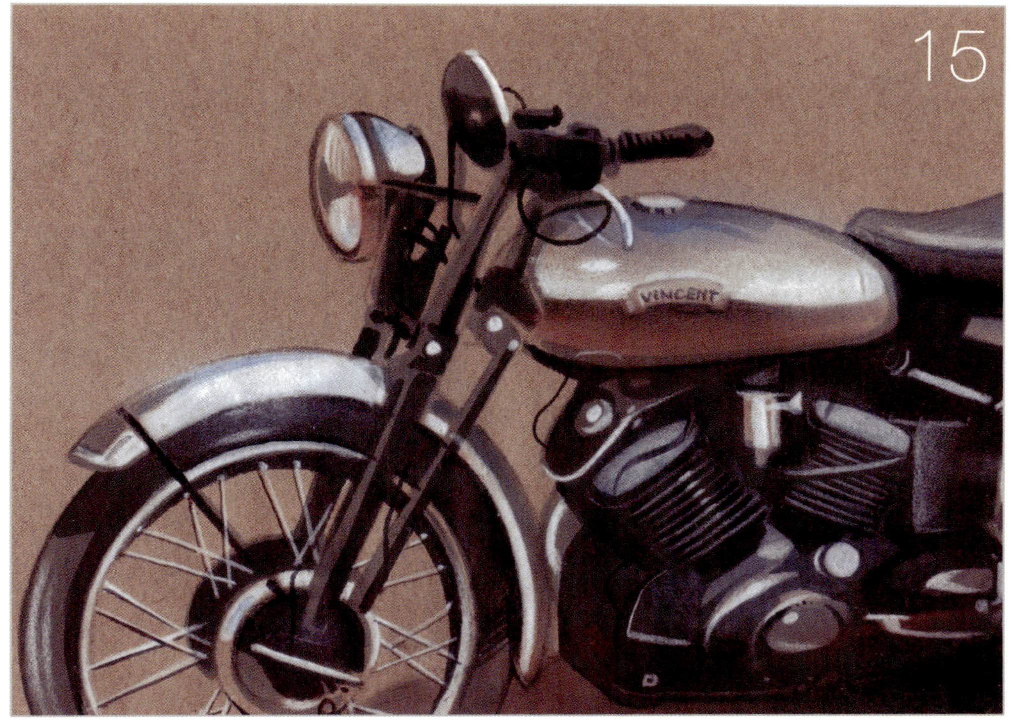

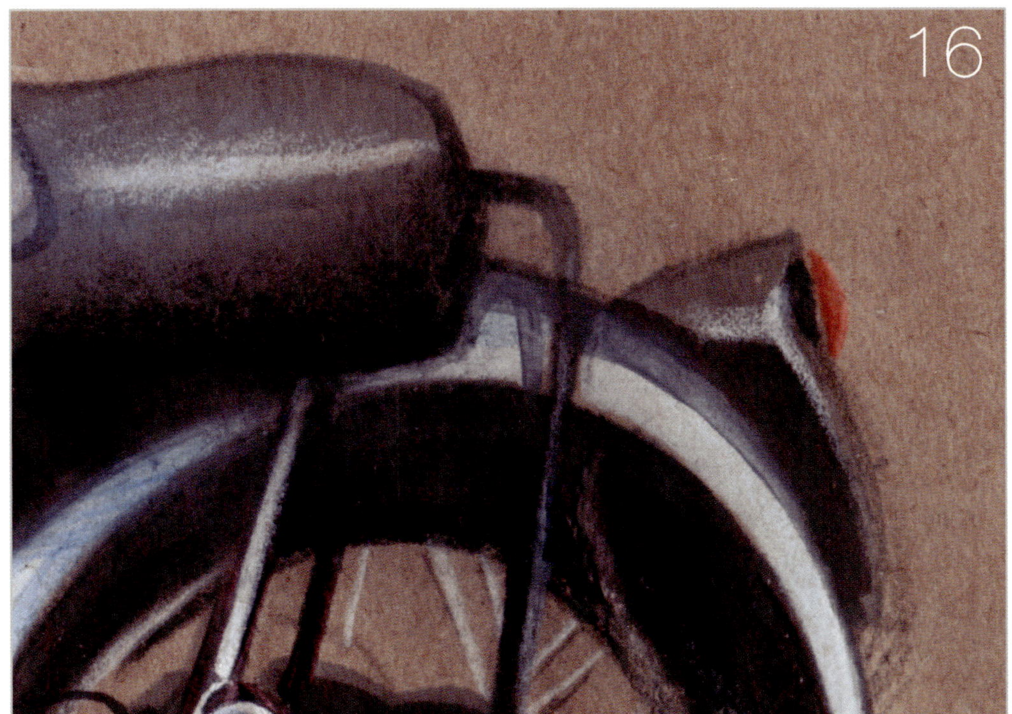

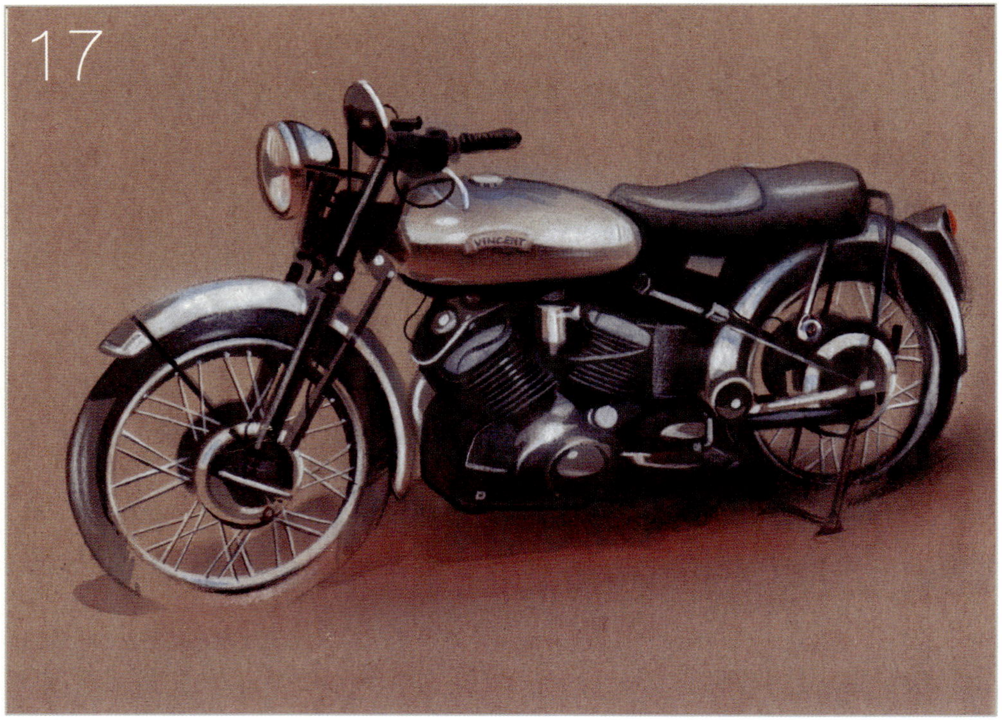

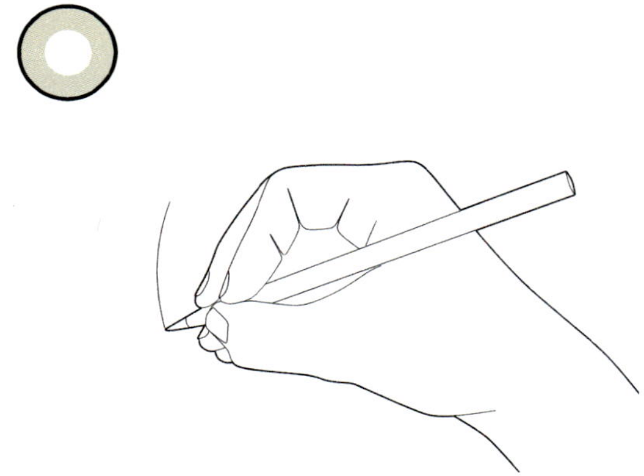

Now go around the bike with a white Prismacolor pencil and punch up the select highlights that you want the viewer to perceive as the brightest – those on the front fender and gas tank, for example. By starting out with Cool Grey, it is now far easier to create a hierarchy of highlights.

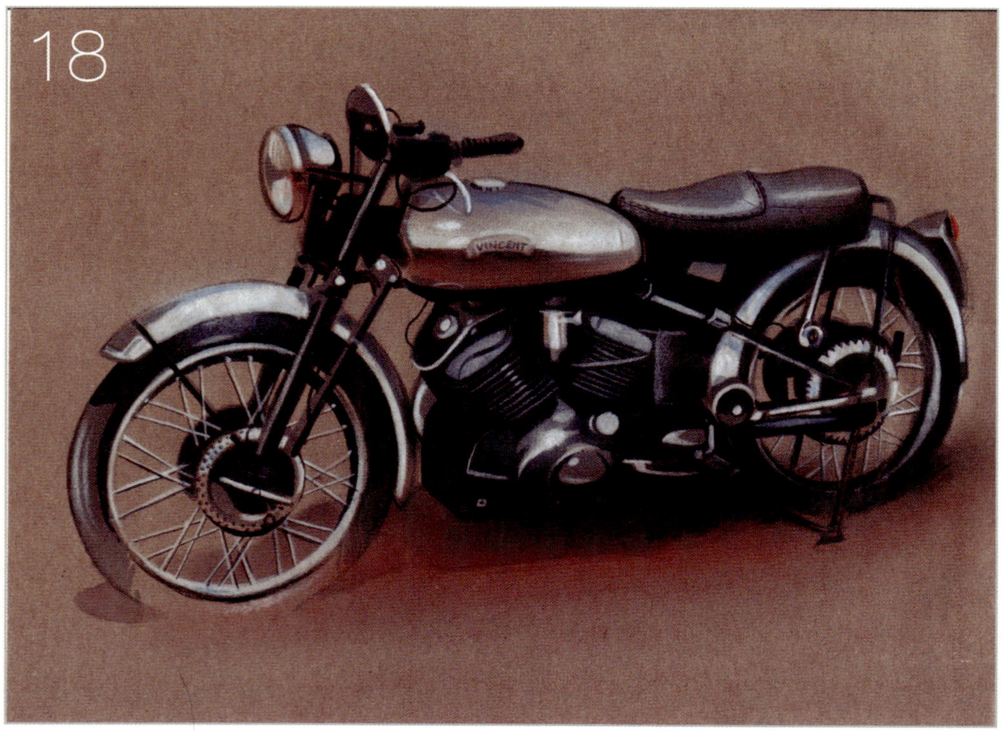

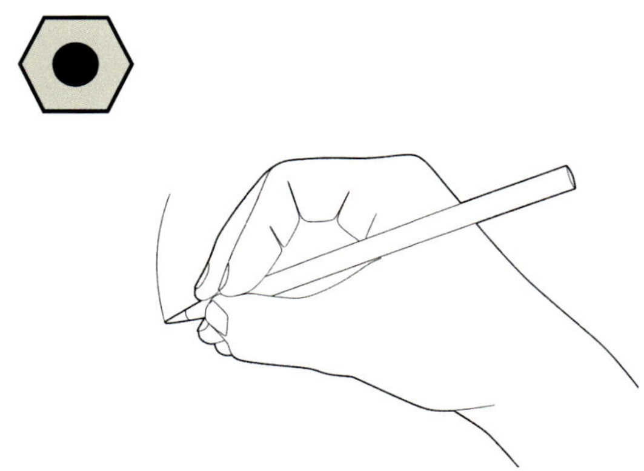

Switching to the harder and less waxy Verithin pencil, add a few last-minute details like the brake disc holes, the sprocket teeth and various bolts.

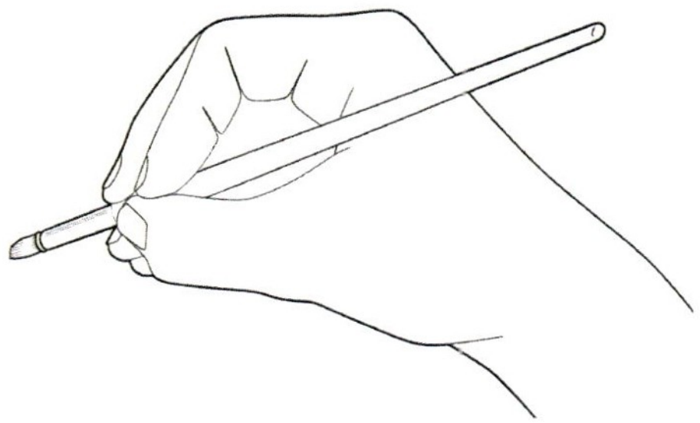

Now complete the chrome effect with white gouache. Remember to keep the fender and wheel highlights in line with the hub. As tempting as it may be, resist the urge to 'go overboard.'

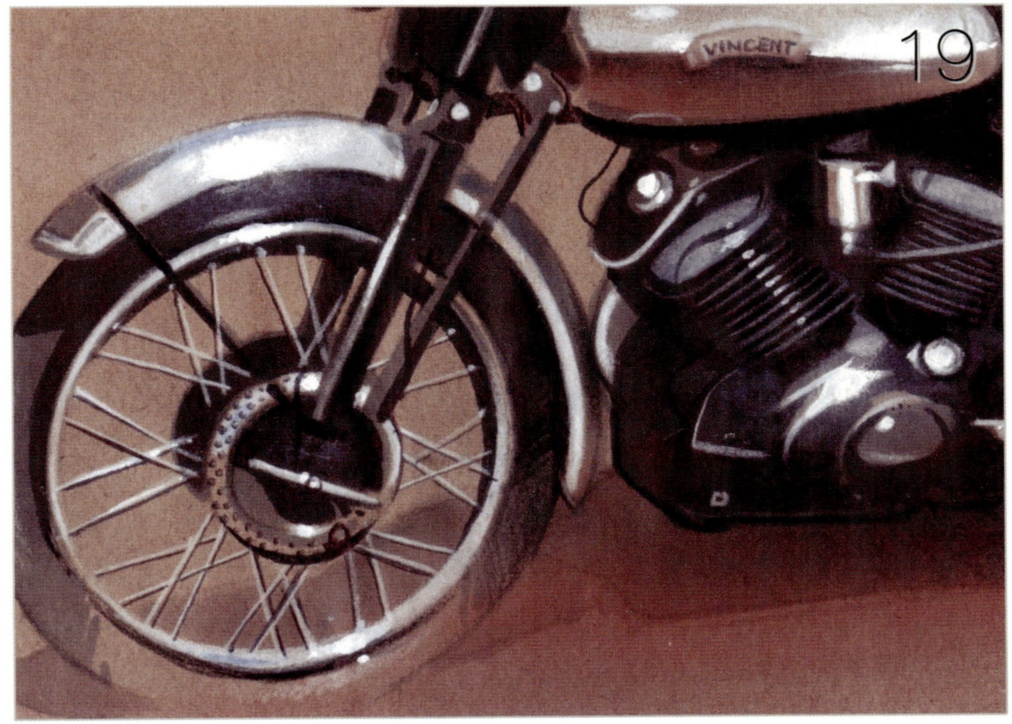

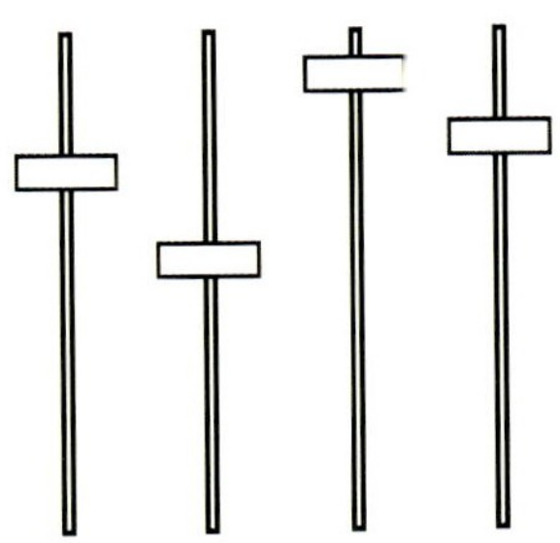

Scan in the rendering and adjust the levels to your liking in Photoshop. Use a white airbrush tool to soften the white highlights and add a bit of glow.

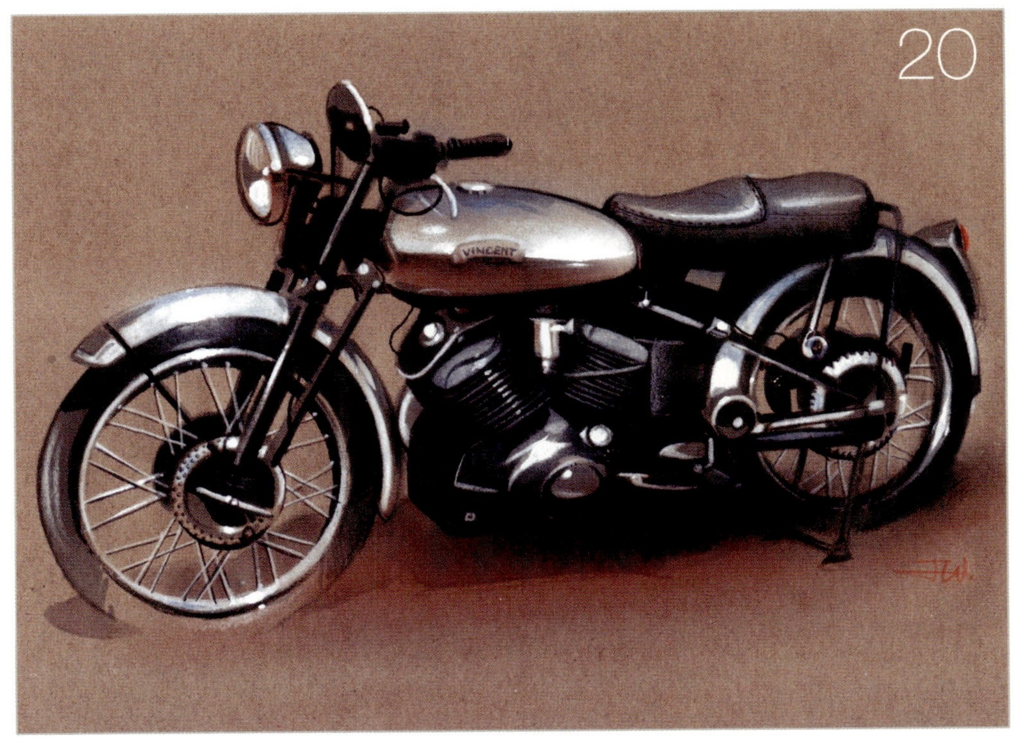

Acknowledgments

The author would like to acknowledge the following people for their inspiration, their influence and their excellent professional example:

Wayne Kady, Charles and Ray Eames, Elwood Engel, Pablo Ferro, Giorgetto Giugiaro, Matt Jeffries, Erik Klimisch, Ib Kofod Larsen, Raymond Loewy, David McIntosh, Syd Mead, Carl Olsen, Sung Paik, Dwayne Patterson, Bill Robinson and Alex Tremulis.

A special 'thank you' to the following people, whose encouragement and expertise helped make this book possible:

Jean Jennings, Marc Bardeen, John, Delores and Holly White.

About the Author

Jason White graduated from Detroit's College for Creative Studies in 1999 with a Bachelor of Fine Arts in Transportation Design. While at CCS, he was awarded internships at Chrysler Corporation and William M. Schmidt Associates. He worked as a designer at Ford Motor Company in Dearborn, Michigan from 1999 to 2006. His contributions there include the 2008 Ford Escape/Mercury Mariner interior and the 2007 Ford Super Duty instrument panel. After a three year stint at Hyundai in Ann Arbor, Michigan, Jason returned to Ford as a contract designer. He has taught freshman transportation design courses at his alma mater since 2007. He is also a voice over artist for AM 1610 The Station in Hamtramck, Michigan, and is a contributing writer for the website CarDesignFetish.com.